Future Trends in Advanced Materials and Processes

Future Trends in Advanced Materials and Processes

Editor

Petrica Vizureanu

MDPI • Basel • Beijing • Wuhan • Barcelona • Belgrade • Manchester • Tokyo • Cluj • Tianjin

Editor
Petrica Vizureanu
Gheorghe Asachi Technical
University of Iasi
Romania

Editorial Office
MDPI
St. Alban-Anlage 66
4052 Basel, Switzerland

This is a reprint of articles from the Special Issue published online in the open access journal *Materials* (ISSN 1996-1944) (available at: https://www.mdpi.com/journal/materials/special_issues/Future_Trends_Adv_Mater_Process).

For citation purposes, cite each article independently as indicated on the article page online and as indicated below:

LastName, A.A.; LastName, B.B.; LastName, C.C. Article Title. *Journal Name* **Year**, *Volume Number*, Page Range.

ISBN 978-3-0365-6470-8 (Hbk)
ISBN 978-3-0365-6471-5 (PDF)

Contents

About the Editor

Petrica Vizureanu

Prof. Dr. Petrica Vizureanu is Head of the Department of Technology and Equipment for Materials Processing, Faculty of Materials Science and Engineering, "Gheorghe Asachi" Technical University of Iasi.

Technical Sciences Academy of Romania, Bulevardul Dacia 26, 030167 București, Romania

peviz2002@yahoo.com, http://afir.org.ro/peviz/.

Vizureanu is a professor and researcher at "Gheorghe Asachi" Technical University of Iasi, with more than 30 years of experience. He earned his Ph.D. degreein Materials science and engineering in 1999, and has been a Ph.D. Supervisor in the Materials Engineering domain from 2010 to present. He has over 200 publications, more than 150 of these indexed in Web of Science. He has extensive experience in the fields of composite materials; ceramic materials; insulating materials; geopolymers; biomaterials; optimization of materials characteristics; expert systems; and energy. His H-index is 20.

Preface to "Future Trends in Advanced Materials and Processes"

This Special Issue contains original high-quality research papers covering the most recent advances in materials' properties, as well as comprehensive reviews addressing the relevant state-of-the-art topics in the area of materials processing, with examples of relevant practical applications. The Special Issue, titled "Future Trends in Advanced Materials and Processes", covers the characterization of various materials, focusing on relevant or innovative applications such as radiological hazard evaluation of non-metallic materials, composite materials' characterization, geopolymers, metallic biomaterials, etc. We would like to gratefully acknowledge the authors and reviewers who participated in the elaboration of this Special Issue.

Petrica Vizureanu
Editor

materials

MDPI

Editorial

Future Trends in Advanced Materials and Processes

Petrica Vizureanu [1,2]

[1] Faculty of Material Science and Engineering, Gheorghe Asachi Technical University of Iasi, 41 D. Mangeron St., 700050 Iasi, Romania; peviz@tuiasi.ro

[2] Materials Science and Engineering Department, Technical Sciences Academy of Romania, Bulevardul Dacia 26, 030167 București, Romania

Citation: Vizureanu, P. Future Trends in Advanced Materials and Processes. *Materials* **2022**, *15*, 6554. https://doi.org/10.3390/ma15196554

Received: 15 September 2022
Accepted: 17 September 2022
Published: 21 September 2022

The main objective of this Special Issue was to publish original high-quality research papers covering the most recent advances in materials properties, as well as comprehensive reviews addressing the relevant state-of-the-art topics in the area of materials processing, with relevant practical applications.

Nowadays, advanced materials and processes are available next to advanced characterization techniques. The special issue managed to gather several outstanding articles in a broad field, from non-metallic, concretes and porous materials to metallic biomaterials or molded products.

An important number of studies have shown that there are many differences in the expansibility of different dolostones, and the key factors determining the expansibility of alkali carbonate rocks. Some rocks were selected from five different geological ages: Jixianian, Cambrian, Ordovician, Devonian, and Triassic ages, and after determining the morphology of dolomites, it was concluded that there is a good positive correlation between ordering degree and the molar fraction of $MgCO_3$ of dolomites [1]. The assessment of the radiological hazards associated with applying the investigated granite in the building materials for infrastructures application. Ref. [2] After classifying the granites, investigations were made related to the environmental parameters such as absorbed dose rate, annual effective dose, radium equivalent activity, and external and internal hazard indices. Main findings are that the investigated types of granites with high radioactivity concentration cannot be applied in the different applications of building materials and ornamental stone. Another radiological hazard evaluation of some egyptian magmatic rocks used as ornamental stone was made [3] and a number of nineteen samples were prepared from seven rock types for assessment in order to be used as ornamental stones. Using equipment as gamma-ray spectrometer or radiological hazard indices for natural radioactivity, the main conclusion was that the investigated rocks are suitable for use as ornamental stone in the construction buildings.

A novel process to recover gypsum from phosphogypsum [4] was achieved taking into consideration harmful elements, such as silicon, phosphorus, and fluorine. Gypsum was recovered using a direct flotation method and limited amounts of other minerals (muscovite, zoisite) were found in the gypsum concentrate during the flotation process because of mechanical displacement.

The effects of Niobium and Molybdenum on the microstructures and corrosion properties of high-entropy $CrFeCoNiNb_xMo_x$ and $CrFeCoNiNb_xMo_{1-x}$ alloys were investigated [5] by preparing in arc melting equipment under an argon atmosphere. Together with growing Nb and Mo percentages, it will increase the hardness of the alloys because of formation of the hexagonal close packing phase of the solid solution [6]. An interesting approach was to re-melt using a TIG welding process and to determine the behavior and wear resistance of Vanadium carbide precipitating $Cr_{27.5}Co_{14}Fe_{22}Mo_{22}Ni_{11.65}V_{2.85}$ High Entropy Alloy (HEA). The wear resistance was found to be similar to Stellite 6, which is a cobalt base alloy consisting of a matrix containing dispersed carbides.

Experimental results for thermal shock resistance and thermal insulation properties were obtained for lanthanum magnesium hexaluminate ($LaMgAl_{11}O_{19}$)/yttria-stabilised zirconia (YSZ) thermal barrier coating (FG-TBC) [7]. In the as-sprayed and laser-glazed conditions, microstructural changes were observed as a formation of cracks on coated surfaces. This procedure can significantly enhance the thermal performance of the coatings and can be used for different applications that require effective heat management (gas turbines). Moreover, transmission, reflection and dissipation of microwaves in magnetic composites were investigated for two compositions [8]. It was calculated the microwave losses and obtained the values for the dielectric permittivity and magnetic permeability. Another property investigated was electrical conductivity of a woven structure by describing a mixing model [9]. The woven structure was treated as a complex multiphase mixture being composed by two conducting phases and one non-conducting phase, with low connectivity in the conductive phases, as the results have shown. The main findings are regarding the strips' contact phases that play an important role in the structure of the composite, also being very important is how the phases were arranged in the whole composite.

Another approach was developed by a comprehensive review, regarding the nonmetallic materials, which are introducing alkali activated based steel waste in the steelmaking industry [10]. The review focuses on the last ten years and is showing the most important providers of steel waste (blast, electric arc and basic oxygen furnaces), together with the possibilities of using the steel waste to improve the Ordinary Portland Cements properties through an alkali activation system. Another review on the same topic raises the issue of the innovative geopolymer composites designed for water and wastewater treatment [11]. The aim of the article is to make an assessment and to explain to the scientific world the main concerns in geo-synthesis, regarding the properties and applications of geopolymer composites that can be used for the elimination of hazardous contaminants. Moreover, the development of the self-compacting concrete mixtures based on some modifiers at a nanoscale was determined [12] by laboratory investigations and mathematical modelling applied to technological process and raw materials for concrete mixtures. In addition, mechanical properties of porous materials become important for various industries, such as energy storage [13]. The equivalent property of a material with pores was determined using a software and a comparison with measurement results were completed.

In the field of concretes and porous materials, an experimental development and mathematical dependencies of the gas release process in the production of non-autoclaved aerated were completed [14], trying to fill theoretical gaps in this phenomenon during the formation of the structure of aerated concrete. In this way, a new method to produce aerated concrete was published by early hydrating of the aerated concrete mixture, improving the gas-holding capacity. It was obtained via an eco-friendly material (aerated concrete) with better properties.

Titanium alloys were another focus of this special issue [15,16]; the main achievement was electrochemical impedance spectroscopy (EIS) characterization, since corrosion resistance is very important for this type of alloys. A new titanium alloy with addition of Ta in various percentages (5–30%) was investigated using scanning tunneling microscopy (STM) and energy dispersive X-ray spectroscopy (EDS). These binary alloys [15] can be considered for medical applications due to increasing of a passive layer resistance with the formation of the passive layer of TiO_2 and Ta_2O_5. The obtaining of a new and promising titanium alloy [16] with 20% Molibdenum and various percentages of Sillicon (0.5 to 1.0) was followed by a characterization of the microsctructure, and it was investigated via their electrochemical responses in Ringer´s solution by linear polarization, cyclic potential dynamic polarization and EIS.

A different approach was discussed [17] in order to evaluate rapid tooling (RT) in rapid heat cycle molding (RHCM) by reviewing the implementation of this method in the molding industry. There are still limited studies available on molds fabricated using RT in RHCM to understand the mechanical properties and aesthetic properties (surface appearance) of the molded part, but RHCM represents an alternative to increase the quality of those products.

All this published researches will offer a new approach for future studies, in order to create important progresses in materials science and engineering.

Funding: This research received no external funding.

Acknowledgments: The Guest Editors of this Special Issue would like to thank all the Authors from all over the world (Romania, Egypt, Russia, Germania, Korea, India, China, Malaysia, Cyprus, Saudi Arabia, Spain, Taiwan, Poland, UK, Algeria), who contributed with their valuable works to the accomplishment of the Special Issue. Special thanks are due to the Reviewers for their constructive comments and thoughtful suggestions. Finally, the editor is grateful to the Materials Editorial Office for their kind assistance.

Conflicts of Interest: The authors declare no conflict of interest.

References

1. Fan, Z.; Mao, Z.; Liu, X.; Yi, L.; Zhang, T.; Huang, X.; Deng, M. Microstructure of Dolostones of Different Geological Ages and Dedolomitization Reaction. *Materials* **2022**, *15*, 4109. [CrossRef] [PubMed]
2. El Dabe, M.M.; Ismail, A.M.; Metwaly, M.; Taalab, S.A.; Hanfi, M.Y.; Ene, A. Hazards of Radioactive Mineralization Associated with Pegmatites Used as Decorative and Building Material. *Materials* **2022**, *15*, 1224. [CrossRef] [PubMed]
3. Lasheen, E.S.R.; Rashwan, M.A.; Osman, H.; Alamri, S.; Khandaker, M.U.; Hanfi, M.Y. Radiological Hazard Evaluation of Some Egyptian Magmatic Rocks Used as Ornamental Stone: Petrography and Natural Radioactivity. *Materials* **2021**, *14*, 7290. [CrossRef] [PubMed]
4. Xiao, J.; Lu, T.; Zhuang, Y.; Jin, H. A Novel Process to Recover Gypsum from Phosphogypsum. *Materials* **2022**, *15*, 1944. [CrossRef] [PubMed]
5. Tsau, C.-H.; Chen, Y.-H.; Tsai, M.-C. The Effects of Niobium and Molybdenum on the Microstructures and Corrosion Properties of CrFeCoNiNbxMoy Alloys. *Materials* **2022**, *15*, 2262. [CrossRef] [PubMed]
6. Treutler, K.; Lorenz, S.; Wesling, V. Re-Melting Behaviour and Wear Resistance of Vanadium Carbide Precipitating Cr27.5Co14Fe22Mo22Ni11.65V2.85 High Entropy Alloy. *Materials* **2021**, *14*, 1871. [CrossRef] [PubMed]
7. Anaz Khan, M.; Vivek Anand, A.; Duraiselvam, M.; Srinivas Rao, K.; Arvind Singh, R.; Jayalakshmi, S. Thermal Shock Resistance and Thermal Insulation Capability of Laser-Glazed Functionally Graded Lanthanum Magnesium Hexaluminate/Yttria-Stabilised Zirconia Thermal Barrier Coating. *Materials* **2021**, *14*, 3865. [CrossRef]
8. Rinkevich, A.B.; Perov, D.V.; Ryabkov, Y.I. Transmission, Reflection and Dissipation of Microwaves in Magnetic Composites with Nanocrystalline Finemet-Type Flakes. *Materials* **2021**, *14*, 3499. [CrossRef] [PubMed]
9. Tokarska, M. A Mixing Model for Describing Electrical Conductivity of a Woven Structure. *Materials* **2022**, *15*, 2512. [CrossRef] [PubMed]
10. Aziz, I.H.; Abdullah, M.M.A.B.; Salleh, M.A.A.M.; Ming, L.Y.; Li, L.Y.; Sandu, A.V.; Vizureanu, P.; Nemes, O.; Mahdi, S.N. Recent Developments in Steelmaking Industry and Potential Alkali Activated Based Steel Waste: A Comprehensive Review. *Materials* **2022**, *15*, 1948. [CrossRef]
11. Luhar, I.; Luhar, S.; Abdullah, M.M.A.B.; Razak, R.A.; Vizureanu, P.; Sandu, A.V.; Matasaru, P.-D. A State-of-the-Art Review on Innovative Geopolymer Composites Designed for Water and Wastewater Treatment. *Materials* **2021**, *14*, 7456. [CrossRef]
12. Stel'makh, S.A.; Shcherban', E.M.; Beskopylny, A.; Mailyan, L.R.; Meskhi, B.; Beskopylny, N.; Zherebtsov, Y. Development of High-Tech Self-Compacting Concrete Mixtures Based on Nano-Modifiers of Various Types. *Materials* **2022**, *15*, 2739. [CrossRef]
13. Pyo, C.; Kim, Y.; Kim, J.; Kang, S. A Study to Derive Equivalent Mechanical Properties of Porous Materials with Orthotropic Elasticity. *Materials* **2021**, *14*, 5132. [CrossRef] [PubMed]
14. Shcherban', E.M.; Stel'makh, S.A.; Beskopylny, A.; Mailyan, L.R.; Meskhi, B.; Shuyskiy, A.; Beskopylny, N.; Dotsenko, N. Mathematical Modeling and Experimental Substantiation of the Gas Release Process in the Production of Non-Autoclaved Aerated Concrete. *Materials* **2022**, *15*, 2642. [CrossRef] [PubMed]
15. Socorro-Perdomo, P.P.; Florido-Suárez, N.R.; Mirza-Rosca, J.C.; Saceleanu, M.V. EIS Characterization of Ti Alloys in Relation to Alloying Additions of Ta. *Materials* **2022**, *15*, 476. [CrossRef] [PubMed]
16. Baltatu, M.S.; Vizureanu, P.; Sandu, A.V.; Florido-Suarez, N.; Saceleanu, M.V.; Mirza-Rosca, J.C. New Titanium Alloys, Promising Materials for Medical Devices. *Materials* **2021**, *14*, 5934. [CrossRef] [PubMed]
17. Huzaim, N.H.M.; Rahim, S.Z.A.; Musa, L.; Abdellah, A.E.-h.; Abdullah, M.M.A.B.; Rennie, A.; Rahman, R.; Garus, S.; Błoch, K.; Sandu, A.V.; et al. Potential of Rapid Tooling in Rapid Heat Cycle Molding: A Review. *Materials* **2022**, *15*, 3725. [CrossRef] [PubMed]

materials

MDPI

Article

Microstructure of Dolostones of Different Geological Ages and Dedolomitization Reaction

Zhiyuan Fan [1], Zhongyang Mao [1,2], Xiang Liu [1,2], Lei Yi [1,2], Tao Zhang [1,2], Xiaojun Huang [1,2] and Min Deng [1,2,*]

1 College of Materials Science and Engineering, Nanjing Tech University, Nanjing 211800, China; 201961203172@njtech.edu.cn (Z.F.); mzy@njtech.edu.cn (Z.M.); 201961203150@njtech.edu.cn (X.L.); 201961203130@njtech.edu.cn (L.Y.); 201961203162@njtech.edu.cn (T.Z.); 5967@njtech.edu.cn (X.H.)

2 State Key Laboratory of Materials-Oriented Chemical Engineering, Nanjing 211800, China

* Correspondence: dengmin@njtech.edu.cn; Tel.: +86-136-0518-4865

Abstract: Dolostone is widely distributed and commonly used as concrete aggregates. A large number of studies have shown that there are significant differences in the expansibility of different dolostones, and the key factors determining the expansibility of alkali carbonate rocks have not been clarified. In this paper, rocks were selected from five different geological ages: Jixianian, Cambrian, Ordovician, Devonian, and Triassic ages. The ordering degree and the content of $MgCO_3$ of dolomites in rocks of different geological ages were determined by X-ray diffraction (XRD). The degree of dedolomitization reaction in rocks cured in 80 °C, 1 mol/L NaOH solution was determined by quantitative X-ray diffraction (QXRD). The morphology of dolomites in rocks was determined by a polarizing microscope. The products of the dedolomitization reaction were determined by field emission electron microscopy (FESEM-EDS). According to the test results, the following conclusions are drawn. There is a good positive correlation between ordering degree and the molar fraction of $MgCO_3$ of dolomites. When the $MgCO_3$ mole fraction of dolomites varies from 47.17% to 49.60%, the higher the $MgCO_3$ mole fraction, the greater the ordering degree of dolomite. By analyzing the degree of the dedolomitization reaction of different dolostone powders cured at 80 °C in 1 mol/L NaOH solution, it is found that the older the geological age of dolostone, the slower the dedolomitization reaction rate and the lower the degree of dedolomitization reaction. The lower the ordering degree of dolomite crystal in the same geological age, the faster the rate of dedolomitization reaction and the higher the degree of dedolomitization reaction.

Keywords: geological ages; dolostone; ordering degree; dedolomitization reaction

Citation: Fan, Z.; Mao, Z.; Liu, X.; Yi, L.; Zhang, T.; Huang, X.; Deng, M. Microstructure of Dolostones of Different Geological Ages and Dedolomitization Reaction. *Materials* **2022**, *15*, 4109. https://doi.org/10.3390/ma15124109

Academic Editor: Petrica Vizureanu

Received: 9 May 2022
Accepted: 7 June 2022
Published: 9 June 2022

1. Introduction

Carbonate rocks are widely distributed throughout China, accounting for about one-fifth of the total surface sedimentary rock distribution [1]. It is often used as aggregate in engineering construction, but due to its alkali dolomite reaction (ADR) activity, it has caused serious damage to engineering construction all over the world, resulting in serious hidden dangers to safety and huge repair costs [2,3]. Therefore, it is of great significance to further study the structure of carbonate rocks. ADR is also known as a dedolomitization reaction. Dedolomitization refers to the calcification of dolomite, a common process by which dolomite is converted to calcite by alkali, which may cause rocks to swell [4]. There are two main types of carbonate rocks: dolostone and limestone [5]. The main mineral of dolostone is dolomite, in addition to a small amount of calcite, quartz, feldspar, and clay minerals. The main mineral of limestone is calcite. The structural characteristics of dolostone mainly include mineral content, grain size, and crystallinity. The molar ratio of Ca^{2+} and Mg^{2+} in ideal dolomite crystals is the same, but Mg^{2+} is often replaced by Ca, Fe, and Mn plasma in nature, which does not conform to the structure of ideal dolomite crystals [6]. The ordering degree is a common method to characterize the crystallinity of dolomite crystals. So far, the intensity ratio of I (015)/I (110) obtained by XRD analysis is the most commonly used way to express the ordering degree of dolomite [7–10]. Previous studies on the crystallographic characteristics of dolomite have been carried out in detail, but they tend to be pure crystallographic studies, and rarely relate the crystallographic characteristics of different dolomites with their dedolomitization reaction [5,9,11]. Gao [12] analyzed a large amount of data on the ordering degree of dolomite and concluded that the ordering degree of penecontemporaneous dolomite is the worst, and that of deep burial dolomite is the best. Zhong [13] analyzed the Triassic dolomite through XRD, analyzed various factors affecting the ordering degree of dolomite and pointed out that the ordering degree of dolomite formed in the burial environment is the best, while the ordering degree of penecontemporaneous dolomite is the worst. Hadley [14] used XRD to study the relationship between ADR activity of aggregates and crystallinity of dolomite crystals, and the results showed that aggregates with and without ADR activity had poor crystallinity of dolomite, so he pointed out that crystallinity would not affect ADR activity of aggregates. In addition, Deng [15] analyzed the ordering degree of dolomite crystals in different rocks and found that the ordering degree will have a certain impact on ADR activity. Gillott [16,17] and Swenson [17,18] studied dolomite structure, and they found that dolomite grain size had a certain influence on ADR reaction rate. Niu [19] studied dolomite with different grain sizes and found that the larger the grain size of dolomite in dolostone, the higher the degree of dedolomitization reaction. Feng [20] analyzed the ordering degree of dolomite at different burial depths, and the results showed that the ordering degree of dolomite crystals increased with burial depth, while the dedolomitization gradually decreased.

In this study, XRD and polarizing microscope are used to determine the microstructure of dolostones of different geological ages, and the influence of factors such as the ordering degree of dolomite crystal on the degree of dedolomitization reaction was discussed, which provided data support for exploring the ACR expansion law of dolostones. Identifying the types of rocks that are likely to cause damage can lead to the more efficient and rational use of resources.

2. Materials and Methods

2.1. Raw Materials

Rocks

By consulting the 1:50,000 scale geological map of China and referring to the regional geological literature, the dolostones of different geological ages are determined. Twelve rock samples from five geological ages were selected for this experiment, as shown in Figure 1. Among them, the rocks WMS-6 and WMS-8 from Tianjin belong to Jixianian age, the rocks BFL-7 and BFL-12 from Linqu county in Shandong Province belong to Cambrian age, the rocks ZC from Taiyuan belong to Ordovician age, the rocks JF, DH-1 and DH-2 from Baoding belong to Ordovician age, the rocks SFP-1, SFP-2 and SFP-3 from Dushan County in Guizhou Province belong to Devonian age and the rocks DJY from Yibin City in Sichuan Province belong to Triassic age.

The chemical analysis method of rock is carried out according to the standard GB/T 176-2017(CIS, 2017). Table 1 shows the chemical composition of rocks. It can be seen from Table 1 that the MgO content in rocks WMS-6, WMS-8, DH-1, DH-2, SFP-1, SFP-2, SFP-3, and JF is relatively similar, about 20%, the MgO content in rocks DJY and BFL-12 is about 15%, and the MgO content in rocks ZC and BFL-7 is only about 5%. The contents of SiO_2 are below 10.0% except for rock JF. The content of Fe_2O_3 and Al_2O_3 in rock DJY is slightly more than 1%.

Figure 1. The appearance of dolostones.

Table 1. Chemical composition of rocks.

Geological Age	Rocks	Chemical Composition/%								
		SiO_2	CaO	MgO	Al_2O_3	Fe_2O_3	K_2O	Na_2O	Loss	Total
Jixianian	WMS-6	0.18	29.99	22.61	0.12	0.32	0.03	0.07	46.52	99.84
	WMS-8	6.16	26.56	19.77	0.41	1.22	0.03	0.07	45.48	99.70
Cambrian	BFL-7	2.68	44.04	4.81	0.50	0.22	0.08	0.12	47.11	99.56
	BFL-12	3.38	34.74	15.86	0.84	0.52	0.06	0.13	43.26	98.79
Ordovician	ZC	2.38	47.03	5.13	0.24	0.33	0.13	0.18	43.56	98.58
	JF	10.51	26.50	19.10	0.24	0.33	0.03	0.08	41.2	97.99
	DH-1	0.63	29.13	21.85	0.24	0.33	0.04	0.04	46.11	98.37
	DH-2	4.86	25.34	18.06	0.25	0.32	0.12	0.09	47.81	96.85
Devonian	SFP-1	3.83	28.38	20.53	0.25	0.50	0.09	0.11	44.28	97.97
	SFP-2	1.31	29.23	20.67	0.25	0.50	0.19	1.17	42.81	96.13
	SFP-3	0.18	30.58	20.64	0.13	0.25	0.41	0.15	45.62	97.96
Triassic	DJY	6.78	29.83	16.47	1.04	1.49	0.73	0.23	42.31	98.88

Figure 2 shows the minerals of the studied dolostones determined by XRD (the scanning speed is 10°/min, the scanning angle is 5~80°). From Figure 2, it can be seen that the main minerals in the dolostones WMS-6, WMS-3, BFL-12, DH-1, DH-2, JF, SFP-1, SFP-2, and DJY are dolomite and a small amount of quartz, while the main minerals in the dolomitic limestone ZC and BFL-7 are dolomite, calcite, and quartz.

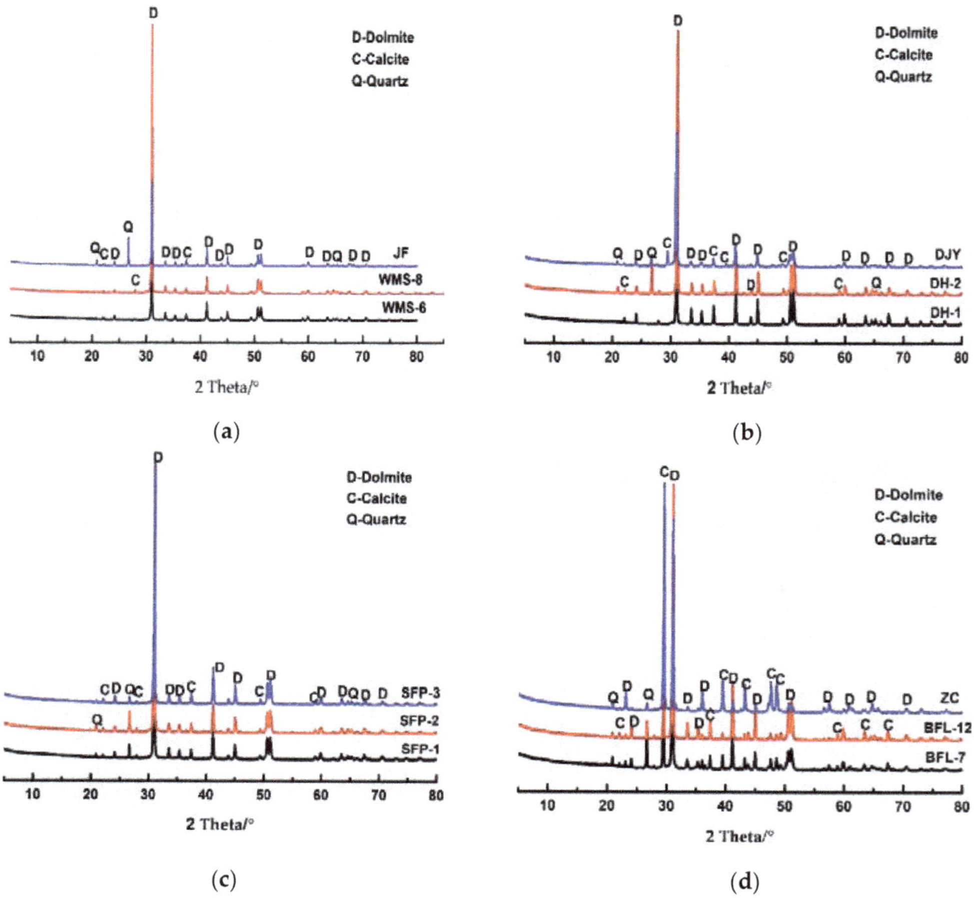

Figure 2. XRD patterns of rocks: (**a**–**c**) dolomite; (**d**) dolomitic limestone.

2.2. Methods

2.2.1. Ordering Degree of Dolomite

The rocks are ground to about 60 μm to avoid damage to crystallinity caused by excessive grinding, dried and then XRD (the scanning angle is 25~40°, the scanning speed is 2°/min) was used to measure the ordering degree of dolomites. XRD data of rocks were analyzed through the Jade software. The integral intensity value I (015) and I (110) of the diffraction peak of the crystal plane of dolomite (015) and (110), respectively, were read [21,22], and ordering degree of dolomites was calculated according to Formula (1):

$$\delta = \frac{I(015)}{I(110)} \quad (1)$$

The d (104) value of the diffraction peak of dolomite (104) crystal plane on the measurement spectrum was read, and the mole fraction of $MgCO_3$ and $CaCO_3$, respectively, was calculated by using Formulas (2) and (3) [23,24]:

$$\text{Mole fraction of } MgCO_3\ (\%) = -333.33\ d\ (104) + 1011.99 \quad (2)$$

$$\text{Mole fraction of } CaCO_3\ (\%) = 100\% - \text{Mole fraction of } MgCO_3\ (\%) \quad (3)$$

2.2.2. Thin Section Petrography

The dolomite grain sizes in rocks of different geological ages were examined by thin slices of different rocks for optical microscope observation. About a hundred pictures were taken of each rock. A polarizing optical microscope (Optiphot-II Pol reflecting light apparatus, 25–400×, Nikon, Tokyo, Japan) with transmitted light was used. The preparation of thin sections was performed according to the section "Thin section specimen preparation" [25].

2.2.3. Dedolomitization Reaction

The dolomite powder was cured in the condition of 80 °C, 1 mol/L NaOH solution, and taken out regularly. After drying, ZnO was used as the internal standard for quantitative analysis by XRD, and then Jade was used to fitting and analyze the XRD pattern to obtain the peak area ratio of ZnO and dolomite, and the content of brucite was calculated by internal standard method. The degree of dedolomitization reaction in rocks is calculated by the ratio of the reduced amount of dolomite to the content of original dolomite, and the scanning speed is 1°/min, the scanning angle is 30~33°.

2.2.4. Microstructure of Dedolomitization Reaction Products

Firstly, 5~10 mm rock particles were taken out after curing in 80 °C, 1 mol/L NaOH solution for 7 days. Then, rock particles were solidified in resin and polished by vibration polishing machine. After rough grinding, fine grinding, polishing, and drying for 24 h, the samples were observed by FE-SEM.

3. Results

3.1. Analysis of Morphology and Distribution Characteristics of Dolomites

Figure 3 shows the distribution of dolomite grains of Jixianian rocks WMS-6 and WMS-8. It can be seen that the dolomite grains are mosaic distribution, which is mainly caused by the existence of a large number of dolomite grains with different grain sizes in the rock. The dolomite crystals are dominated by anhedral crystals.

(a)

(b)

Figure 3. Distribution of dolomites in Jixianian rocks (**a**) WMS-6 and (**b**) WMS-8.

Figure 4 shows the distribution of dolomite grains of Cambrian rocks BFL-7 and BFL-12, and the dolomite grains of rock BFL-7 are dispersed in the calcite matrix. The dolomite grains in BFL-12 are mainly distributed in a mosaic state, and a few dolomite grains are dispersed in the calcite matrix. The dolomite grains in BFL-7 and BFL-12 are mainly anhedral with a little subhedral.

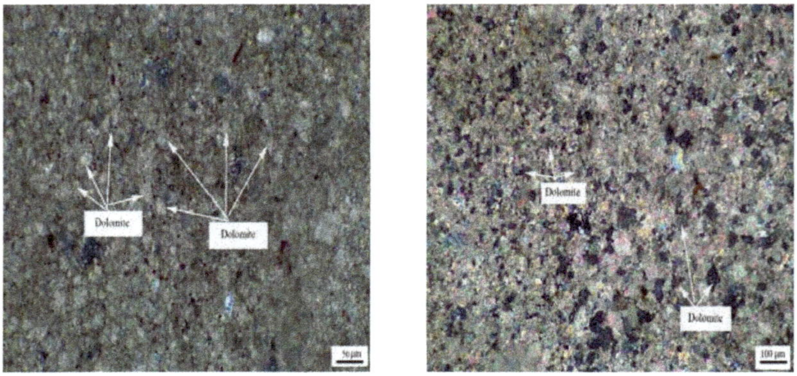

(**a**) (**b**)

Figure 4. Distribution of dolomites in Cambrian rocks (**a**) BFL-7 and (**b**) BFL-12.

Figure 5 shows the distribution of dolomite grains of Ordovician rocks ZC, JF, DH-1, and DH-2. It can be seen from Figure 5a that dolomite grains of rock ZC are dispersed in the calcite matrix. Dolomite grains have a high degree of idiomorphic crystal, and dolomite crystals are mainly euhedral crystals and subhedral crystals. The crystals support each other in a network shape. It can be seen from Figure 5b that the dolomite crystals of rock JF are dominated by anhedral crystals. It can be seen from Figure 5c,d that dolomite crystals of rocks DH-1 and DH-2 are mosaic distribution. The dolomite crystals in rock DH-1 are mainly euhedral crystals and subhedral crystals, while the dolomite crystals in DH-2 are anhedral crystals.

(**a**) (**b**)

Figure 5. *Cont.*

(**c**) (**d**)

Figure 5. Distribution of dolomites in Ordovician rocks (**a**) ZC, (**b**) JF, (**c**) DH-1, and (**d**) DH-2.

Figure 6 shows the distribution of dolomite grains of Devonian rocks SFP-1, SFP-2, and SFP-3. It can be seen from Figure 6a–c that the dolomite grains in rocks SFP-1, SFP-2, and SFP-3 are mosaic distribution. The dolomite crystals in rocks SFP-1, SFP-2 and SFP-3 are mainly euhedral crystals with a small number of subhedral crystals.

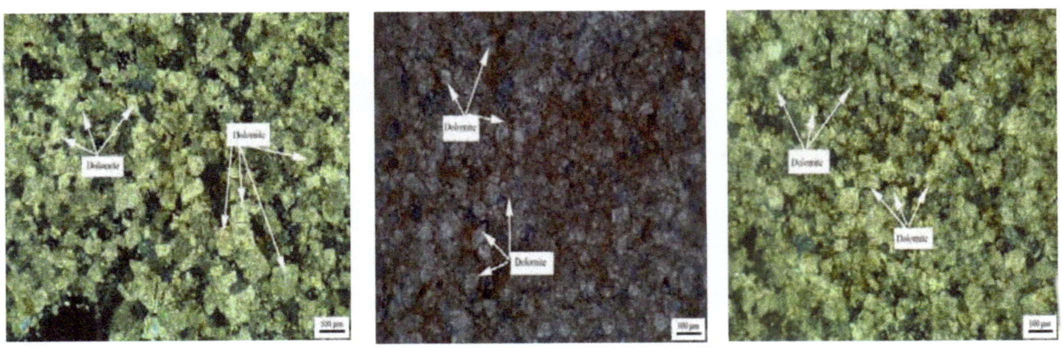

(**a**) (**b**) (**c**)

Figure 6. Distribution of dolomites in Devonian rocks (**a**) SFP-1, (**b**) SFP-2, and (**c**) SFP-3.

Figure 7 shows the distribution of dolomite grains of Triassic rock DJY. It can be seen from Figure 7 that the dolomite grains in DJY are mainly mosaic distribution, and a few dolomite grains are dispersed in the calcite matrix. The dolomite crystals are mainly anhedral crystals.

Figure 7. Distribution of dolomites in Triassic rock DJY.

3.2. The Ordering Degree of Dolomite

The ordering degree and $MgCO_3$ mole fraction of dolomites were analyzed by XRD to characterize the crystallinity of dolomite crystals. Table 2 shows the analysis of rocks of different geological ages by XRD, and the calculation of $MgCO_3$ mole fraction and ordering degree of corresponding dolomites. It can be seen from Table 3 that the ordering degree and the content of $MgCO_3$ of dolomites are significantly different in rocks of different geological ages.

Table 2. The $MgCO_3$ mole fraction and ordering degree of dolomite crystals in rocks.

Rocks	Geological Age	d (104) /Å	Ordering Degree I (015)/I (110)	Mole Fraction/ mol/%	
				$CaCO_3$	$MgCO_3$
WMS-6	Jixianian	2.8915 ± 0.0532	0.7352 ± 0.0883	51.83	48.17
WMS-8	Jixianian	2.8893 ± 0.0611	0.7821 ± 0.0774	51.10	48.90
BFL-7	Cambrian	2.8910 ± 0.0722	0.7710 ± 0.1103	51.67	48.33
BFL-12	Cambrian	2.8884 ± 0.0631	0.7528 ± 0.0757	50.79	49.21
ZC	Ordovician	2.8888 ± 0.1050	0.7694 ± 0.0934	50.93	49.07
JF	Ordovician	2.8910 ± 0.0611	0.7182 ± 0.1032	51.77	48.33
DH-1	Ordovician	2.8894	0.7789	51.13	48.87
DH-2	Ordovician	2.8887 ± 0.0933	0.7394 ± 0.0952	50.90	49.10
SFP-1	Devonian	2.8876	0.8067	50.53	49.47
SFP-2	Devonian	2.8862 ± 0.0612	0.8284 ± 0.0544	50.07	49.93
SFP-3	Devonian	2.8872 ± 0.0403	0.8279 ± 0.0448	50.40	49.60
DJY	Triassic	2.8945 ± 0.0972	0.6059 ± 0.1176	52.83	47.17

Table 3. The ordering degree of dolomite and degree of dedolomitization reaction.

Rocks	The Ordering Degree of Dolomite	The Degree of Dedolomitization Reaction/%			
		1 d	4 d	7 d	14 d
WMS-6	0.74	9.92	27.86	37.66	40.66
WMS-8	0.78	13.45	35.83	36.76	38.16
BFL-7	0.77	18.34	34.25	36.67	48.88
BFL-12	0.75	17.54	32.11	41.24	54.23
ZC	0.77	35.79	59.10	81.44	100.00
JF	0.72	31.94	78.00	100.00	100.00
DH-1	0.78	7.44	23.92	40.39	100.00
DH-2	0.74	39.34	81.15	91.92	100.00
SFP-1	0.81	22.63	51.62	85.24	100.00
SFP-2	0.83	11.21	31.61	56.28	100.00
SFP-3	0.83	23.75	38.20	53.27	100.00
DJY	0.61	82.10	97.30	100.00	100.00

Figure 8 shows the relationship between the ordering degree of dolomites in rocks of different geological ages and the $MgCO_3$ mole fraction of dolomites. As can be seen from Figure 7, the ordering degree of dolomite crystals in Devonian rocks SFP-1, SFP-2, and SFP-3 are the highest, up to 0.83, while the ordering degree of dolomite crystals of Triassic rocks DJY is the lowest, only 0.61. It can also be seen that when the mole fraction of $MgCO_3$ varies from 47.17% to 49.60%, there is an approximately linear relationship between the $MgCO_3$ mole fraction and the ordering degree of dolomite crystals. The ordering degree of dolomites increases with the increase in the $MgCO_3$ mole fraction. When the molar fraction of $MgCO_3$ of dolomite crystals is close to 50%, the order degree is higher. When the content of Mg^{2+} and Ca^{2+} in dolomite crystals is close to each other, Mg^{2+} and Ca^{2+} are more likely to be regularly distributed. Dolomite is formed in a low-salinity environment with a low ratio of Mg^{2+} to Ca^{2+}, so the crystallization rate of dolomite is relatively slow, so the ordering degree of dolomite is high.

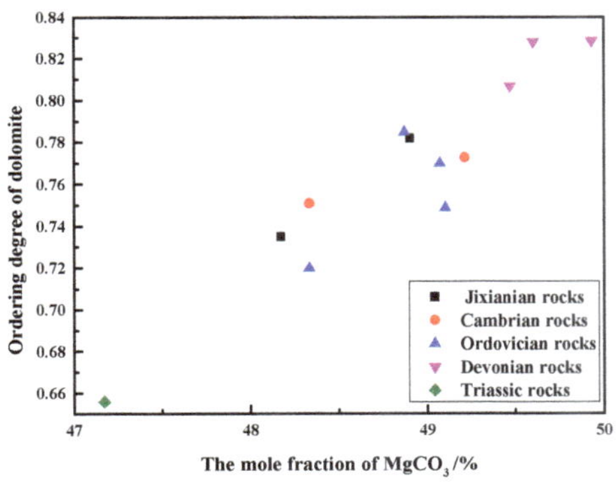

Figure 8. Relationship between ordering degree and the mole fraction of $MgCO_3$ of dolomites.

3.3. The Degree of Dedolomitization Reaction

Figure 9 shows the degree of dedolomitization reaction in rocks of different geological ages at different reaction times. It can be seen that there are certain differences in the reaction rates of dolomite in dolostones. The dedolomitization reaction rate of Jixianian rocks WMS-6 and WMS-8 is relatively slow in the early stage. After curing for 4 days, the degree of dedolomitization reaction almost does not change at about 30%. The degree of dedolomitization of the Cambrian rocks BFL-7 and BFL-12 is relatively slow and tends to be stable at about 40%. The dolomite of Ordovician rocks ZC, DH-1, and DH-2 react completely after curing for about 10 days, while the dolomite of Ordovician rocks JF reacted completely at 4 days. The dolomite in Devonian rocks SFP-1, SFP-2, and SFP-3 reacted completely at about 10 days. The dedolomitization reaction rate of Triassic rocks is the highest in the early stage and can be fully reacted at 7 days. The dedolomitization reaction rate of the rocks in the Jixianian and Cambrian ages are the slowest. The dedolomitization reaction rate of Ordovician rocks is slightly faster than that of Devonian rocks and can be fully reacted with the increase in curing time. Therefore, geological age has a certain influence on the dedolomitization reaction in rocks, and the older the geological age of rocks, the lower the degree of dedolomitization reaction and the slower the reaction rate.

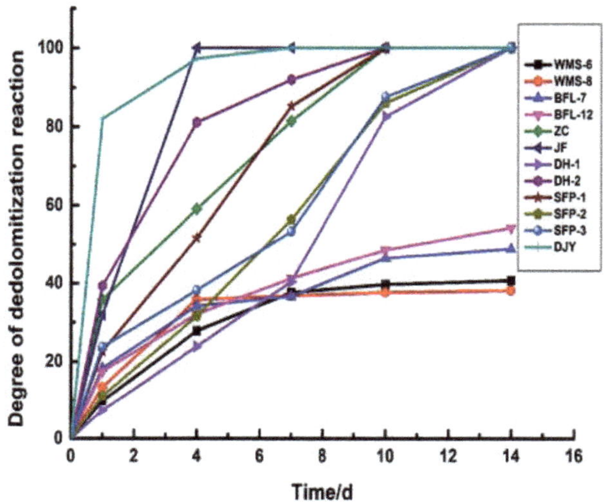

Figure 9. The degree of dedolomitization reaction in rocks.

3.4. Influence of Ordering Degree of Dolomite on Degree of Dedolomitization Reaction

The degree of dedolomitization reaction of different dolostones is different, and the reason for the difference is not clear. By analyzing the ordering degree of dolomites, the reason for the difference in the degree of dedolomitization reaction is explained.

Table 3 shows the ordering degree of dolomites and the degree of dedolomitization reaction in the dolostones. The ordering degree of dolomite of Jixianian rocks WMS-6 and WMS-8 is 0.74 and 0.78, respectively, and the degree of dedolomitization reaction in WMS-6 is greater than that in WMS-8 at 14 d. The ordering degree of dolomite of Cambrian rocks BFL-7 and BFL-12 is 0.77 and 0.75, respectively. The degree of dedolomitization reaction of dolostone in BFL-7 and BFL-12 is close. The ordering degrees of dolomites of Ordovician rocks ZC, JF, DH-1, and DH-2 are 0.77, 0.72, 0.78, and 0.74, respectively. The dedolomitization rate of dolomites of JF is the fastest, and that of dolomites of DH-1 is the slowest. The ordering degree of dolomites of Devonian rocks SFP-1, SFP-2, and SFP-3 are 0.81, 0.83, and 0.83, respectively. The dedolomitization reaction rate of dolomites in SFP-1 is the fastest, and the dedolomitization reaction rate of dolomites of SFP-2 and SFP-3 are close to each other. The ordering degree of dolomite of Triassic rock DJY is only 0.61, and the dedolomitization reaction rate of rock DJY is the highest. It can be seen that the lower the ordering degree of dolomites in the same geological age, the faster the dedolomitization reaction rate of dolomite, and the higher the degree of dedolomitization reaction.

3.5. Microscopic Analysis of Products after Dedolomitization

Figure 10 shows the SEM-EDS diagram of the products generated after the dedolomitization reaction. It can be seen from Figure 10a,b that a large number of flaky and granular substances are formed around dolomite after the dedolomitization reaction. The results of Figure 10c,d EDS show that flaky substance (point 1) is mainly composed of Mg and O, which can be judged as brucite after dedolomitization reaction, while granular substance (point 2) is mainly composed of Ca, C and O, which is another product calcite after de dolomitization reaction. The shape of calcite is mostly granular and columnar, and the shape of brucite is mainly flake. These reaction products are distributed and accumulated around dolomite crystals, and it can be seen that there are a large number of pores between brucite and calcite.

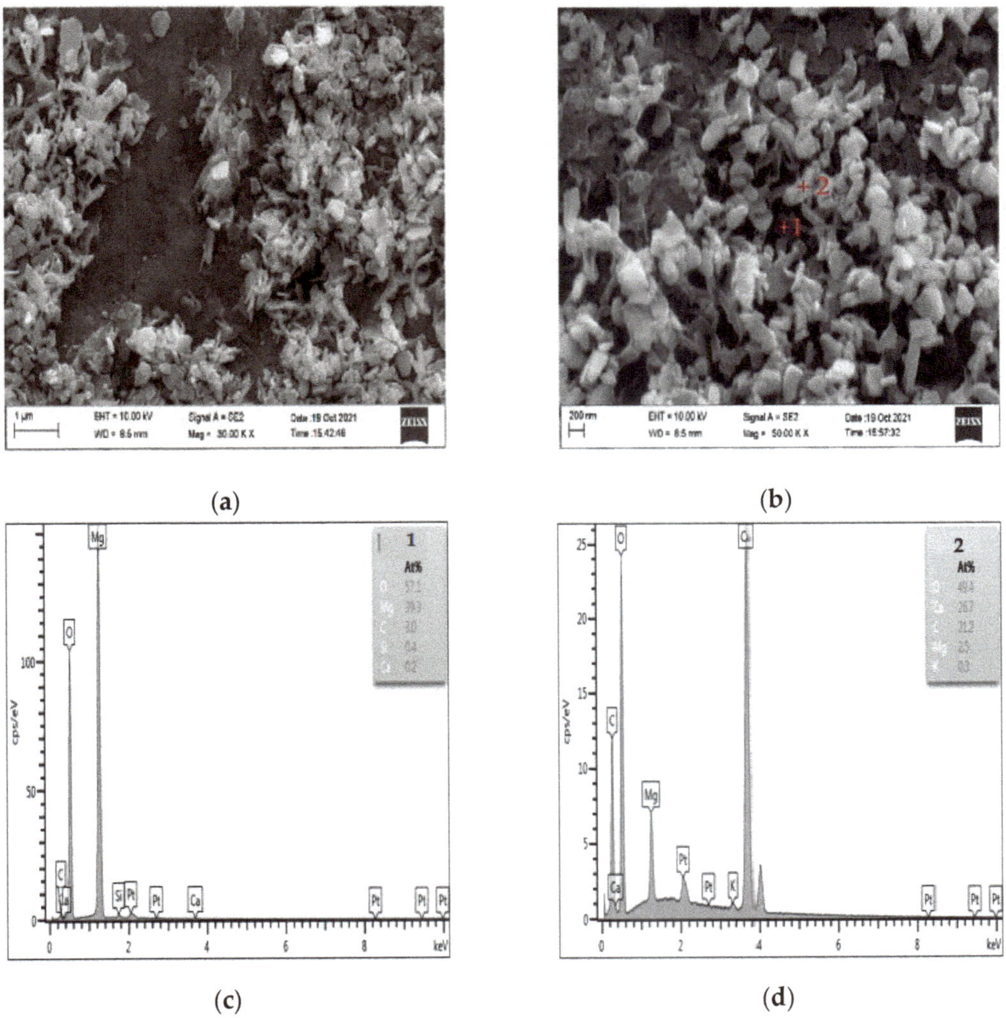

Figure 10. SEM-EDS images of dedolomitization reaction products. (**a**,**b**) The secondary electron image of dedolomitization reaction product (**c**) The elements composition of point 1 (**d**) The elements composition of point 2.

4. Discussion

The influence of the morphology of dolomite on the dedolomitization reaction was discussed. The dolomites of rocks WMS-6, WMS-8, Bfl-7, Bfl-12, JF, DH-2, and DJY are mainly anhedral crystals. Among them, the degree of dedolomitization reaction of dolostones WMS-6, WMS-8, BFL-7, and BFL-12 is relatively low, and the degree of dedolomitization reaction is only about 40% at 14 d. The degree of dedolomitization reaction of dolostones

DH-2 and DJY is close to 100% at 7 days. The degree of dedolomitization reaction of dolostone ZC, SFP-1, SFP-2, and SFP-3 with a good euhedral degree is close to 100% at about 14 d, and there is a great difference in the degree of dedolomitization reaction at the initial stage of dedolomitization reaction, among which, the degree of dedolomitization reaction of rock ZC reaches 59.1% at 4 d. However, the degree of dedolomitization reaction of rock SFP-2 is only 31.6% in 4 d. There is no correlation between the crystal morphology of dolomites and the degree of dedolomitization reaction, so the crystal morphology of dolomite in dolostones is not the key factor to determine the degree of dedolomitization reaction.

5. Conclusions

Through the study of rocks' microstructure and dedolomitization reaction in dolostones of different geological ages, the conclusions are as follows.

1. The ordering degree of dolomite crystals in Devonian rocks is the highest, which is 0.83. The ordering degree of dolomite crystals in Jixianian, Cambrian, and Ordovician rocks is about 0.75. The ordering degree of dolomite crystals in Triassic rocks is the lowest, which is 0.61. The $MgCO_3$ mole fraction of dolomite crystals in the test rocks ranges from 0.4717 to 0.4960. With the increase in the $MgCO_3$ mole fraction, the ordering degree of dolomite increases, and the relationship between the $MgCO_3$ mole fraction and the ordering degree is approximately linear.
2. After curing in a 1 mol/L NaOH solution at 80 °C, the dedolomitization reaction rate of dolomite in 0.045–0.080 mm Triassic rock powders is the fastest, and the dolomite completely reacts at 7 days. The dedolomitization rate in the Jixianian rock powder sample is the slowest, and the dedolomitization reaction degree reaches about 40% at 7 days, and then hardly changes. The dedolomitization reaction degree of the dolomite in the Cambrian rock powder sample reaches about 50% at 10 d and then tends to be stable. The dedolomitization reaction rates of powder samples in Ordovician and Devonian rocks are similar, and the dolomites in the rocks can fully react after 10 days and 14 days, respectively. Therefore, it can be inferred that the older the geological age of rocks, the slower the dedolomitization reaction rate, and the lower the degree of dedolomitization reaction.
3. The effects of the ordering degree of dolomites on the degree of dedolomitization reaction in different dolostones are analyzed. The lower the ordering degree of dolomite crystals in rocks of the same geological age, the faster the rate of dedolomitization reaction and the higher the degree of dedolomitization reaction.
4. The products of the dedolomitization reaction of dolostones were determined by SEM-EDS, and the calcite and brucite were distributed around the dolomite crystals, and there were many tiny pores between the calcite and brucite.

Author Contributions: Conceptualization, M.D.; data curation, Z.M., Z.F., X.L., L.Y., T.Z. and X.H.; writing—original draft preparation, Z.F.; writing—review and editing, M.D. and Z.M. All authors have read and agreed to the published version of the manuscript.

Funding: This work was supported by the Open Fund Project of the National Laboratory for High-Performance Civil Engineering Materials (2021CEM012) and the Priority Academic Program Development of Jiangsu Higher Education Institutions (PAPD).

Institutional Review Board Statement: Not applicable.

Informed Consent Statement: Not applicable.

Data Availability Statement: The data presented in this study are available on request from the corresponding author.

Acknowledgments: The authors gratefully acknowledge the assistance from Zhongyang Mao, Tao Zhang, Xiang Liu, and Lei Yi from NJTECH, and the staff from the State Key Laboratory of Materials-Oriented Chemical Engineering.

Conflicts of Interest: The authors declare no conflict of interest.

References

1. He, J.; Kang, Y.; Liu, D. Experimental research on drilling alkali sensitivity of carbonate gas reservoirs in Sichuan—Chongqing region. *Nat. Gas Ind.* **2005**, *25*, 60–61.
2. Alexander, M.G. Alkali–aggregate reaction. In *Developments in the Formulation and Reinforcement of Concrete*, 2nd ed.; Elsevier: Amsterdam, The Netherlands, 2019; pp. 87–113.
3. Hird, K. Petrography and Geochemistry of Some Carboniferous and Precambrian Dolomites. Ph.D. Thesis, University of Durham, Durham, UK, 1985.
4. Donatelli, J.L. Dedolomitization and Other Diagenesis in the Backreef Setting of the Permian Reef Complex in Dark Canyon, New Mexico. Master's Thesis, New Mexico Institute of Mining and Technology, Socorro, NM, USA, 2016.
5. Banerjee, A. Estimation of dolomite formation: Dolomite precipitation and dolomitization. *J. Geol. Soc. India* **2016**, *87*, 561–572. [CrossRef]
6. Zhang, J.; Shou, J.; Zhang, T.; Pan, L.; Zhou, J. New Approach on the Study of Dolomite Origin: The crystal structure analysis of dolomite. *Acta Sedimentol. Sin.* **2014**, *32*, 550–559.
7. Zucchini, A.; Comodi, P.; Katerinopoulou, A.; Balic-Zunic, T.; McCammon, C.; Frondini, F. Order–disorder–reorder process in thermally treated dolomite samples: A combined powder and single-crystal X-ray diffraction study. *Phys. Chem. Miner.* **2012**, *39*, 319–328. [CrossRef]
8. McKenzie, J.A. Holocene Dolomitization of Calcium Carbonate Sediments from the Coastal Sabkhas of Abu Dhabi, U.A.E.: A Stable Isotope Study. *J. Geol.* **1981**, *89*, 185–198. [CrossRef]
9. Alonso-Zarza, A.M.; Martin-Perez, A. Dolomite in caves: Recent dolomite formation in oxic, non-sulfate environments. Castanar Cave, Spain. *Sediment. Geol.* **2008**, *205*, 160–164. [CrossRef]
10. Pina, C.M.; Pimentel, C.; Crespo, Á. Dolomite cation order in the geological record. *Chem. Geol.* **2020**, *547*, 119667. [CrossRef]
11. Geske, A.; Goldstein, R.H.; Mavromatis, V.; Richter, D.K.; Buhl, D.; Kluge, T.; John, C.M.; Immenhauser, A. The magnesium isotope (δ26 Mg) signature of dolomites. *Geochim. Et Cosmochim. Acta* **2015**, *149*, 131–151. [CrossRef]
12. Gao, Z.; Zhu, S. Controlling factors of dolomite ordering degree. *Sci. Technol. Innov. Her.* **2018**, *15*, 7.
13. Zhong, Q.; Huang, S.J.; Zou, M.L. Controlling factors of order degree of dolomite in carbonate rocks: A case study from Lower Paleozoic in Tahe Oilfield and Triassic in northeastern Sichuan Basin. *Lithol. Reserv.* **2009**, *21*, 50–55.
14. Hadley, D.W. Alkali-reactivity of carbonate rocks-expansion and dedolomitization. *Highw Res. Board* **1961**, *40*, 462–474.
15. Deng, M.; Qian, G. Ordered index and dedolomitization of dolomite crystals. *J. Nanjing Univ. Chem. Technol.* **2001**, *23*, 1–5.
16. Gillott, J.E.; Duncan, M.A.G.; Swenson, E.G. Alkali-aggregate reaction in nova Scotia IV. Character of the reaction. *Cem. Concr. Res.* **1973**, *3*, 521–535. [CrossRef]
17. Swenson, E.G.; Gillott, J.E. Alkali-carbonate rock reaction. *Highw. Res. Rec.* **1964**, *45*, 21–40.
18. Swenson, E.G. A reactive aggregate undetected by ASTM tests. *ASTM Bull.* **1957**, *226*, 48–51.
19. Niu, L.L.; Deng, M.; Yang, Y.F. Influence of Grain Size of Dolomites on Alkali-Dolomite Reaction and Expandability of Dolostone. *Bull. Chin. Ceram. Soc.* **2015**, *34*, 2757–2763.
20. Feng, S.H.; Hong, L.I.; Jiang, J.J.; Lei, Y.; Niu, Y.Z.; Yang, R.; Liu, Y.J. The Multiple Dolomitizations in Ordovician Majiagou Carbonate Rocks in Liujiang Basin, Qinhuangdao Area, North China. *Acta Sedimentol. Sin.* **2017**, *35*, 664–680.
21. Kaczmarek, S.E.; Sibley, D.F. A Comparison of Nanometer-Scale Growth and Dissolution Features on Natural and Synthetic Dolomite Crystals: Implications for the Origin of Dolomite. *J. Sediment. Res.* **2007**, *77*, 424–432. [CrossRef]
22. Deffeyes, K.S.; Lucia, F.J.; Weylt, P.K. Dolomitization: Observations on the Island of Bonaire, Netherlands Antittes. *Science* **1964**, *143*, 678–679. [CrossRef]
23. Huang, S.J. Experimental research methods of carbonate rocks. *Mineral. Petrol.* **1990**, *10*, 114–117.
24. Zeng, L.; Wan, M.X.; Peng, Y. Dolomite Sequentiality and Its Application to Petroleum Geology. *Nat. Gas Explor. M Dev.* **2004**, *27*, 64–66.
25. French, W.J. Concrete petrography: A review. *Q. J. Eng. Geol.* **1991**, *24*, 17–48. [CrossRef]

materials MDPI

Article

Hazards of Radioactive Mineralization Associated with Pegmatites Used as Decorative and Building Material

Mohamed M. El Dabe [1], A. M. Ismail [1], Mohamed Metwaly [1], Sherif A. Taalab [2], Mohamed Y. Hanfi [1,3,*] and Antoaneta Ene [4,*]

1 Nuclear Materials Authority, Maadi, Cairo 11381, Egypt; Mohamedeldabe@yahoo.com (M.M.E.D.); a.m.ismail302@gmail.com (A.M.I.); dr_mohamedmetwaly@yahoo.com (M.M.)
2 Faculty of Science, Al-Azhar University, Cairo 11884, Egypt; sheriftaalab@azhar.edu.eg
3 Institute of Physics and Technology, Ural Federal University, St. Mira, 19, 620002 Yekaterinburg, Russia
4 INPOLDE Research Center, Department of Chemistry, Physics and Environment, Faculty of Sciences and Environment, Dunarea de Jos University of Galati, 47 Domneasca Street, 800008 Galati, Romania
* Correspondence: mokhamed.khanfi@urfu.ru (M.Y.H.); Antoaneta.Ene@ugal.ro (A.E.); Tel.: +7-(982)-646-63-57 (M.Y.H.)

Abstract: The present study aimed to assess the radiological hazards associated with applying the investigated granite in the building materials and the infrastructures applications. The investigated granites are classified into four categories: El-Urf, barren, colourful and opaque. El Urf monzogranite intrudes metagabbro diorite complex with sharp contacts. Based on the activity concentrations, the environmental parameters such as absorbed dose rate (D_{air}), annual effective dose (AED), radium equivalent activity (Ra_{eq}), external (H_{ex}) and internal (H_{in}) hazard indices were measured. The mineralized pegmatite is located in the southwestern foothill of the Gabal El Urf younger granite. It displays well-defined zonation of three zones: outer, middle and inner zones represented by potash feldspar, quartz and mica, respectively. The isorad map showed that El Urf monzogranite is barren (Up to 100 cps) surrounding an excavation of the studied pegmatite that exhibits moderate colorful mineralization (phase-I = 500–1500 cps) and anomalous opaque mineralization (phase-II = 1500–3500 cps) pegmatites. The obtained results of radionuclides activity concentrations illustrated that the Opaque granites have the highest values of ^{238}U (561 ± 127 Bq kg^{-1}), ^{232}Th (4289 ± 891 Bq kg^{-1}) and ^{40}K (3002 ± 446 Bq kg^{-1}) in the granites, which are higher than the recommended worldwide average. Many of the radiological hazard parameters were lesser than the international limits in the younger granites and barren pegmatites. All of these parameters were higher in the colorful and opaque mineralized pegmatites. The high activity and the elevated radiological hazard parameters in the mineralized pegmatites are revised to the presence of radioactive and radioelements bearing minerals, such as thorite, meta-autunite, kasolite, phurcalite, columbite, fergusonite, Xenotime and fluorapatite. Other instances of mineralization were also recorded as cassiterite, atacamite, galena, pyrite and iron oxide minerals. Thus, the granites with high radioactivity concentration cannot be applied in the different applications of building materials and ornamental stones.

Keywords: Gabal El Urf; pegmatitic rocks; zoned pegmatite; poly-phased mineralization uranothorite; radiological hazard parameters; radioactive minerals

Citation: El Dabe, M.M.; Ismail, A.M.; Metwaly, M.; Taalab, S.A.; Hanfi, M.Y.; Ene, A. Hazards of Radioactive Mineralization Associated with Pegmatites Used as Decorative and Building Material. *Materials* **2022**, *15*, 1224. https://doi.org/10.3390/ma15031224

Academic Editor: Petrica Vizureanu

Received: 5 December 2021
Accepted: 2 February 2022
Published: 6 February 2022

1. Introduction

Granites are igneous rocks generally made up of quartz, K-feldspar and mica, and are used for internal and exterior decorative uses, including building and ornamental materials. Because of their nature, these rocks contain radionuclides. Exposure to the radioactive series ^{238}U and ^{232}Th, as well as ^{40}K, produces external irradiation. Internal doses from radon inhalation and the aforementioned radioactive chains' short-lived products are concentrated in respiratory tract tissues [1–3]. Uranium (U) and thorium (Th) series of natural radionuclides can be found in various levels in all terrestrial materials, depending on the geological and geographical conditions of the study area [4,5]. They can be found in almost every environment and can even be identified in the human body [3]. The terrestrial radionuclides and their daughters and cosmic radiation contribute to background radiation in the environment. Mineralogical, geochemical and physicochemical factors all play a role in its presence in the environment [6,7]. In recent years, there has been a lot of discussion about the radiological risk posed by building materials [6].

Moreover, the radiological impact of the general public is a major topic of research in radioecology, where the data will provide importantly and required information in monitoring environmental contamination, allowing the public to access more appropriate and effective protection advice [8,9]. The production of gamma radiation from natural radionuclides must be closely monitored in order to safeguard humans against gamma radiation, which can be caused by various diseases [10,11]. According to the ATSDR (Agency for Toxic Substances and Disease Registry), long-term radioactive exposure causes significant ailments that include oral necrosis, chronic lung disease, leukopenia and anaemia [12,13]. Several studies have been carried out to estimate the radiation risk and yearly dose supply of natural radioactivity in building materials [14,15]. Implementing a radiological impact assessment for construction materials in order to analyze and control radioactive consequences on humans and the environment is a critical and complex task that must be carried out in order to meet the criteria for sustainable development. Radiation effects should be assessed using quantifiable values that can be utilized as input parameters for designing environmental distribution and estimating radiation dose [16,17]. The present paper concerns the geological and mineralogical composition of the studied mineralized pegmatite and its environmental impacts on humans and the environment. Some of the radiological risks such as radium equivalent activity (Ra_{eq}), absorbed dose rate (D_{air}), annual effective dose (AED), external (H_{ex}) and internal (H_{in}) hazard indices and gamma index ($I\gamma$) are computed.

2. Materials and Methods

2.1. Geological Setting

The studied mineralized pegmatite located in the eastern part of Gabal El Urf granite, Central Eastern Desert of Egypt, and bounded by latitudes 26°37′58″–26°38′11″ N and longitudes 33°26′51″–33°28′09″ E (Figure 1).

Gabal El Urf younger granite has an elongate shape, nearly striking NE–SW. It is monzogranite with medium to coarse-grains, and has calc-alkaline to alkaline nature affinity, with a nearly estimated Sr-Nd age of 600 ± 11 Ma [18,19]. Many pegmatite bodies and masses had intruded the metagabbro diorite complex, representing El Urf monzogranite's country rocks [20]. Many pegmatite bodies and masses had intruded the metagabbro diorite complex, representing El Urf monzogranite's country rocks [21]. They display zoned pegmatites constituting a source for the mineralization of radioactive and rare metals (Y, Th, Nb, and Zr) [22]. The pegmatites derived from metaluminous to peralkaline magma fall within the plate granite type and are enriched with cheralite (Ca-rich monazite) and zircon [23]. Hydrothermal processes are enriched with rare metals mineralization and radioactive minerals [21]. In general, all pegmatite rocks in Gabal El Urf younger granite have been recorded in their country rocks. The latter revealed that a huge zoned mineralized pegmatite body had intruded the El Urf younger granite, with an average of nearly 14×7m in size. It is characterized by potash feldspar, quartz and mica minerals, and outer, middle and inner zones. They (Optic.) recoded earlier colorful mineralization (phase-I) and latter opaque stages (phase-II). It can be documented that the main difference between the two mineralized phases is attributable to time gapping, not the spatial issue. Both colourful and opaque mineralization stages can be found in the same location in the pegmatite zones. However, obviously, the opaque minerals phase-II (latter) cut the earlier colourful phase-I, indicated by both field investigations as well petrographic studies. El Urf monzogranite intrudes metagabbro diorite complex with sharp contacts [18,21]. The studied mineralized pegmatite is located in the southwestern foothill of the Gabal El Urf younger granite. It displays well definite zonation and consists of three zones: outer, middle and inner zones represented by potash feldspar, quartz and mica, respectively. Generally, the huge studied pegmatite had been noticed by the diggers who look after the potash feldspar masses. They excavate all the masses they can find, which are used in the ceramic industry. After excavating and removing quartz pockets, some unexposed potash masses had appeared, in which some radioactive minerals were contained (Figure 2).

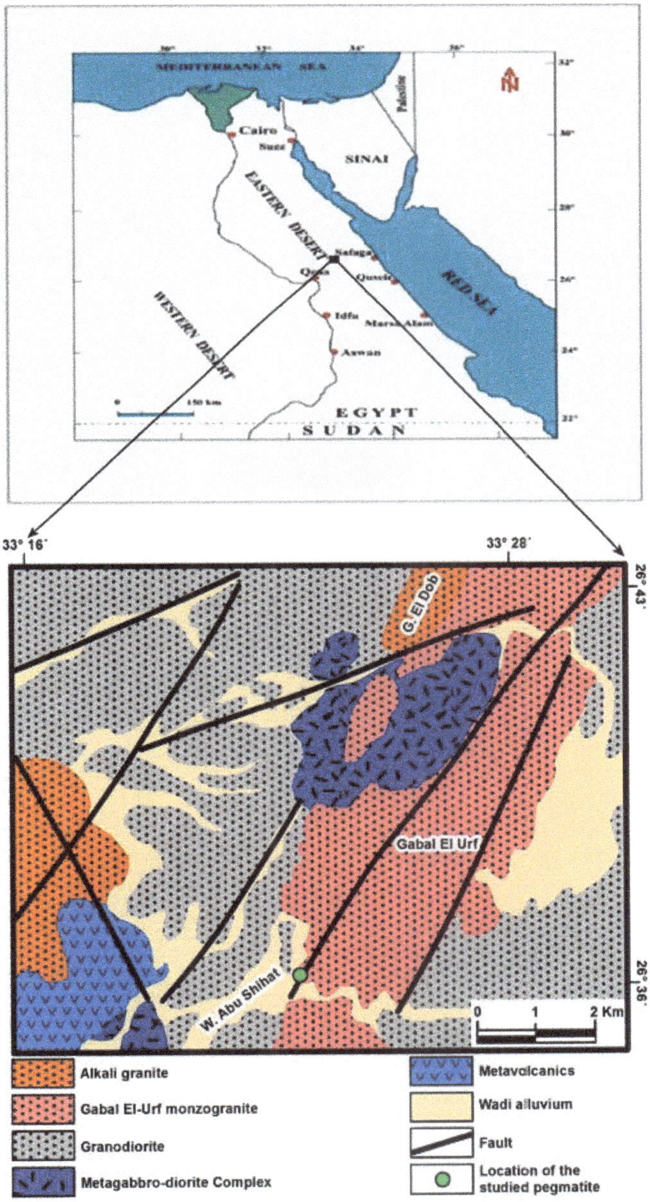

Figure 1. Location map of Gabal El Urf younger granite, Central Eastern Desert, Egypt. [18].

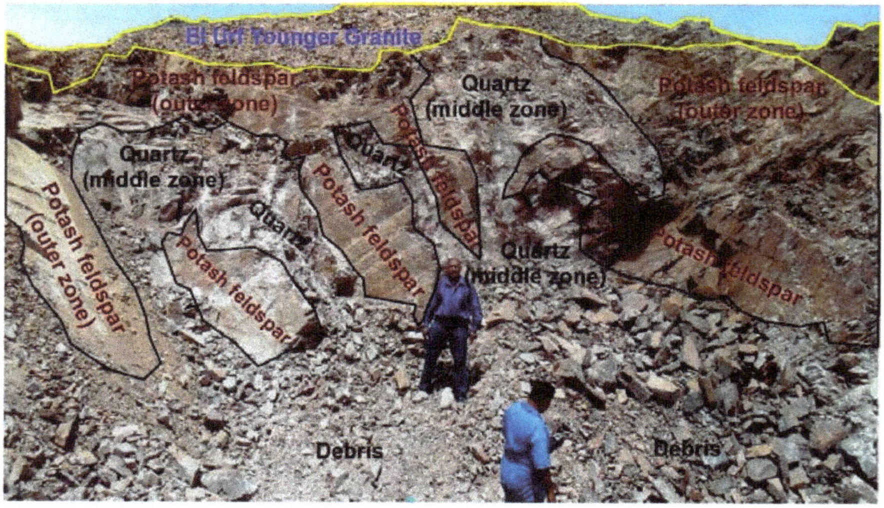

Figure 2. A close view of the western part of the huge studied mineralized pegmatite shows newly potash exposure masses containing opaque mineralization after excavation in the mineralized pegmatite of Gabal El Urf area, looking NE.

The studied mineralization includes both colorful (phase-I) and opaque mineralization (phase-II); rarely they occur consistently in the same place (Figure 3). Generally, the colorful phase-I occurs as clots of disseminated minute crystals with bright colors ranging from yellow to green in quartz and potash feldspar (Figure 4). Opaque mineralization phase-I displays as a network of fracture-filling iron associated with a vast array of accessory minerals (Figure 5). Opaque mineralization (phase-II) displays a coarser grain size of minerals than (phase-I). It includes iron oxides and mega crystals of colorless and purple fluorite associated with black radioactive minerals (Figure 6); purple fluorite is an indication to the radioactive influence.

Figure 3. Both the colorful phase-I and opaque phase-II mineralization between the outer and middle zones, in the mineralized pegmatite of Gabal El Urf area, looking NE.

Figure 4. A close view of the colorful mineralization (phase-I) disseminated in potash feldspar associated with a quartz dyke-like body, looking N.

Figure 5. A hand specimen showing parallel bands of iron oxides with high radioactivity minerals (phase-II), associated with quartz veinlets.

According to the petrographic and mineralogical studies, Phase-I encloses accessory minerals such as thorite, fluorite, zircon and xenotime, whereas the latter (phase-II) has another array of the accessory minerals, for example, fluorapatite, cassiterite, atacamite, Nb-minerals and sulfide minerals, besides the Th-minerals (thorite, uranothorite,), U-minerals (meta-autunite and uranophane) and REE-bearing minerals (pyrochlore and bastnasite), associated mainly with fractures filled by iron oxides.

Figure 6. A hand specimen showing iron oxides associated with megacrystals of colorless and purple fluorite with high radioactivity minerals (phase-II).

Gabal El Urf monzogranite is bounded from the south by the elliptical pegmatitic body that distinguished the moderate radioactive pegmatite phase-I surrounded by the anomalous radioactive pegmatite phase-II (Figure 7). In addition, structurally, both the colorful mineralization phase-I and opaque phase-II are mainly located near or along with definite fractures, leading to easy migration or removal of uranium ions, especially at the oxidizing regime.

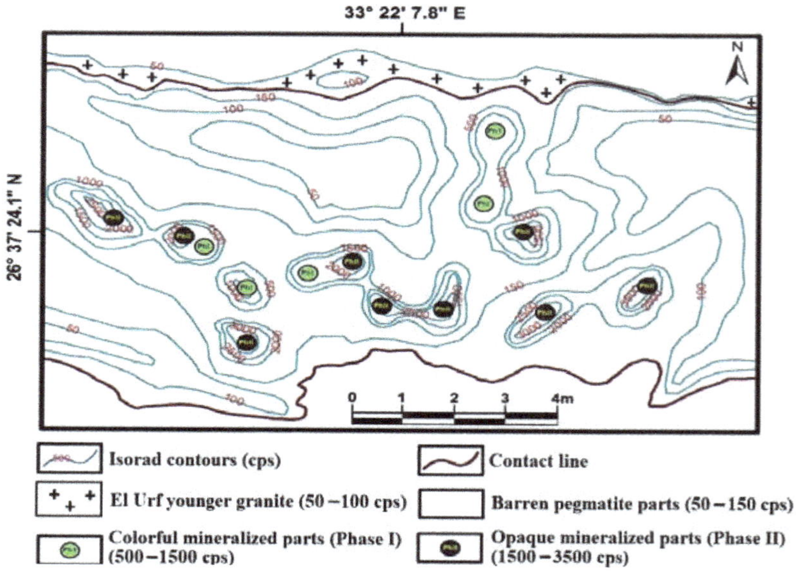

Figure 7. Radiometric map showing the distribution of radioactive measuring values of the two phases of mineralization.

2.2. Radiometric and Mineral Analysis

The radiometric field survey of the El Urf younger granite and its related pegmatite was carried out using the portable scintillometer (UG-130), measuring in terms of count per second (Cps), and also determined as equivalent uranium (eU), thorium (eTh) and potassium (K). Before the measurements were carried out in the field, the portable scintillator was calibrated using the calibration pads which are certified by IAEA. The calibration experiment was designed by Matolin's (1990) [24]. The obtained data by UG-130 were in agreement with the NaI (Tl) detector. A Nickon polarized microscope (Olympus-BZ70) mainly examined the petrographic studies to recognize the radioactive minerals and radioelement-bearing minerals of the studied mineralized pegmatite. The X-ray diffraction technique (XRD), using a Philips PW 3710/31 diffractometer, scintillation counter, Cr & Cu target tube and Ni filter at 40 kV and 30 mA. This instrument is connected to a computer system using the APD program and PDF-2 database for mineral identification. An scanning electron microscope (SEM model Philips XL30) supported by an energy dispersive spectrometer (EDX) unit was used at 25-30 kV accelerating voltage, 1–2 mm beam diameter and 60–120 s counting time. All the analyses were carried out at the labs of the Nuclear Materials Authority (NMA), Cairo, Egypt. Table 1 summarizes how to calculate radiological risk factors using activity concentrations of ^{238}U, ^{232}Th and ^{40}K, and the mathematical equations.

Table 1. Important radiological parameters and indices.

Parameter	Symbol	Definition	Formula
Radium equivalent activity	Ra_{eq}	Radium equivalent activity is a weighted sum of the ^{226}Ra, ^{232}Th and ^{40}K activities according to the hypothesis that 370 Bq kg^{-1} of ^{226}Ra, 259 Bq/kg of ^{232}Th and 4810 Bq/kg of ^{40}K attain the same dose rates of gamma rays	Ra_{eq} (Bq kg^{-1}) = A_{Ra} + 1.43 A_{Th} + 0.077 A_K
External hazard index	H_{ex}	The external hazard index is the radiological parameters applied to assess the hazard of γ-radiation [25,26]	$H_{ex} = \frac{A_U}{370} + \frac{A_{Th}}{259} + + \frac{A_K}{4810}$
Internal hazard index	H_{in}	The internal hazard index is applied to the internal exposure from radon and its decay products [27,28]	$H_{in} = \frac{A_U}{185} + \frac{A_{Th}}{259} + + \frac{A_K}{4810}$
Radiation level index	$I\gamma$	The other index used to estimate the level of γ-radiation hazard associated with the natural radionuclides in the samples was suggested by a group of experts due to the different combinations of specific natural activities in the sample [29,30]	$I_\gamma = \frac{A_{Ra}}{150} + \frac{A_{Th}}{100} + + \frac{A_K}{1500}$
Absorbed dose rate	D (nGy/h)	The absorbed dose rate is the radioactive factor that was applied to detect the effect of gamma radiation at 1 m from the radiation sources in the air due to the concentrations of ^{238}U, ^{232}Th and ^{40}K	D_{air} (nGy h^{-1}) = 0.430 A_U + 0.666 A_{Th} + 0.042 A_K
Outdoor annual effective dose	AED	The annual effective dose is a radioactive factor utilized to detect the exposure level for radiation during a stationary duration (1 year)	AED (mSv y^{-1}) = D_{air} (nGy/h) × 0.2 × 8760 (h/y) × 0.7 (Sv/Gy) × 10^{-6} (mSv/nGy)

3. Results and Discussion

3.1. Mineralogical Studies

The mineralogical studies of the pegmatite rocks of Gabal El Urf were carried out to determine the minerals that cause radioactive anomalies and identify the minerals that contain rare earth elements associated with uranium and thorium elements present in the two mineralized phase-I and phase-II. In phase-I sections, radioactivity refers to the mica minerals that include an array of radioelement-bearing minerals in addition to the presence of xenotime, zircon and fluorite. Phase-II mineralized pegmatite spots are characterized by an array of significant minerals comprising thorite, meta-autunite, kasolite and phurcalite. Nb-minerals include columbite and fergusonite, xenotime, fluorapatite, cassiterite, atacamite, sulfide minerals galena and pyrite, and iron oxide minerals.

3.1.1. Thorite ((Th, U) SiO_4)

Thorite mineral is presented in phase-I as minute grains included in the mica minerals and as a fracture filling. El Dabe (2022) illustrated the EDX analysis of thorite minerals containing thorium (35.64%) and uranium (10.51%), representing the main constituents with the silicate (10.94%) [31]. Yttrium is the sole trace element (5.34%) occupying a limited percentage of U-cote, according to the similarity of the ionic radii (Figure 8). Thorite exists in phase-II relatively more than in phase-I. It is presented in phase-II as small grains, disseminated clusters and microfracture filling. The XRD analysis shows that the d-spacing of Thorite characteristic peaks (3.56, 4.71, and 2.65 A^O) at 2theta (24.98, 18.79, and 33.78), respectively, matches PDF-2 Card No. (11–17) (Figure 9). Its composition is confirmed by EDX analysis, where thorium (61.14) and uranium (13.98) represent the main constituents, along with silica (13.74) and Yttrium (8.43) (Figure 8). It is clear that thorium and uranium in phase-II exceed the corresponding one in phase-I. This coincides with the entire measuring e(Th) and e(U) values belonging to both (phase-I) and (phase-II) pegmatite parts.

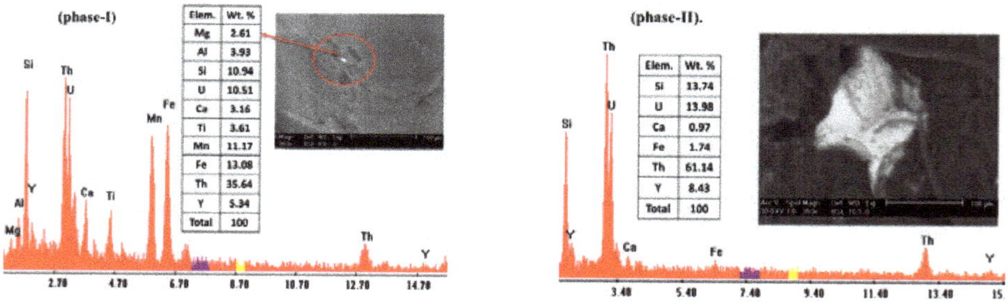

Figure 8. EDX-analysis of thorite, El-Urf mineralized pegmatite (phase-I), and (phase-II).

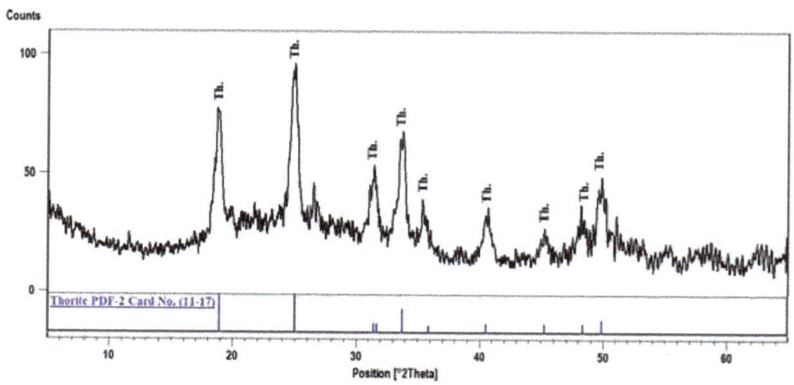

Figure 9. XRD pattern of thorite, El Urf mineralized pegmatite (phase-II).

3.1.2. Phurcalite $Ca_2(UO_2)_3(PO_4)_2(OH)_4 \cdot 4H_2O$

Phurcalite is a secondary calcium uranium phosphate mineral product of hydrothermal activity and is presented in (phase II) mineralized spots of pegmatites. The XRD pattern (Figure 10) shows the d-spacing of its characteristic peaks (8.00, 3.09, and 2.88 A^O) at 2theta (16.45, 43.45 and 46.79), respectively, and matches PDF-2 Card No. (30-284).

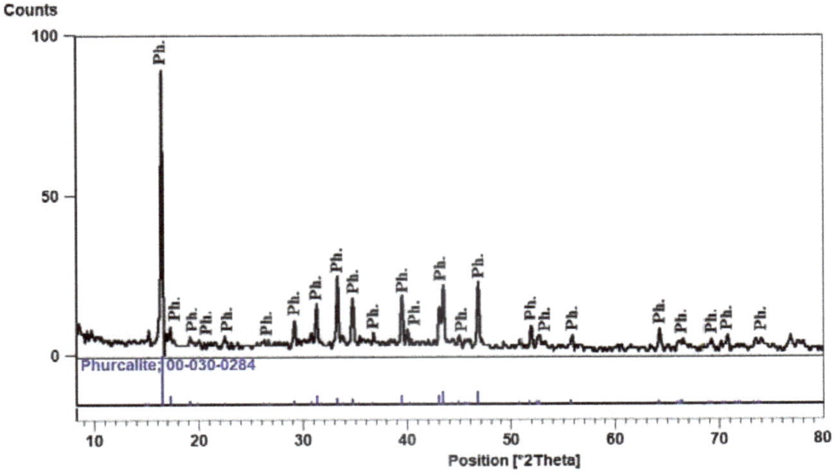

Figure 10. XRD pattern of Phurcalite, El Urf mineralized pegmatite (phase-II).

3.1.3. Meta-Autunite ($Ca(UO_2)_2(PO_4)_2 \cdot 6H_2O$)

Uranyl phosphate minerals are present in phase-II as Meta-autunite minerals formed by the dehydration of autunite that crystallized from the hydrous magma. An XRD investigation analysis was carried out. The XRD pattern shows that the d-spacing of its characteristic peaks (8.62, 3.48, and 3.66 A^O) at 2theta (15.27, 38.39, and 36.37), respectively, matches PDF-2 Card No. (28-247). (Figure 11).

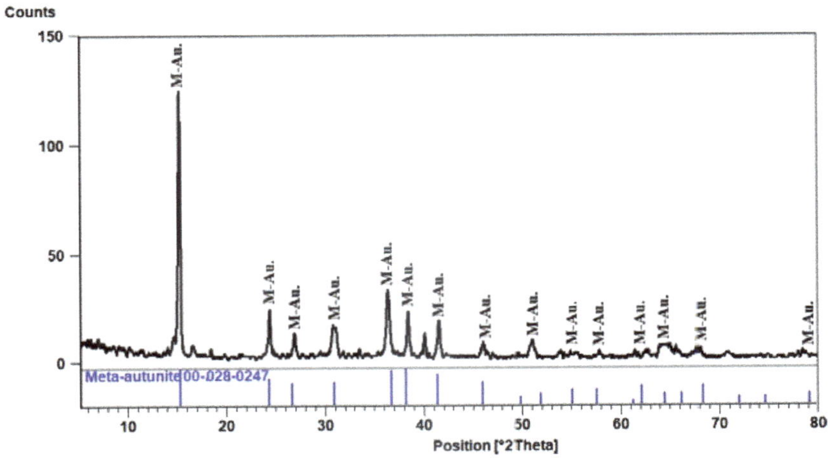

Figure 11. XRD pattern of Meta-autunite, El Urf mineralized pegmatite (phase-II).

3.1.4. Kasolite $Pb(UO_2)SiO_4 \cdot H_2O$

Kasolite is the uranyl silicate mineral riches by lead. It is presented in the (phase-II) mineralization of pegmatites. The XRD pattern of kasolite shows that the d-spacing of its characteristic peaks (2.92, 3.07, and 3.26 A^O) at 2theta (30.58, 29.06, and 27.30), respectively, matches PDF-2 Card No. (29-788) (Figure 12).

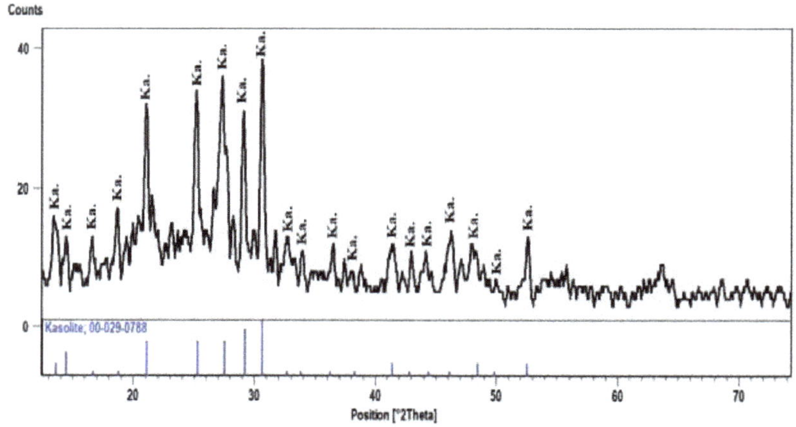

Figure 12. XRD pattern of Kasolite, El Urf mineralized pegmatite (phase-II).

3.1.5. Xenotime (YPO_4)

Xenotime is an yttrium phosphate presented in phase-I. The mineral is enriched in rare earth elements (REE) and is associated with thorium and uranium minerals. The EDX-analysis indicates the presence of appreciable contents of Th (8.84%) and U (2.47%). The XRD pattern of Xenotime shows that the d-spacing of its characteristic peaks (3.45, 2.56, and 1.76 A^O) matches PDF-2 Card No. (83-658) (Figure 13).

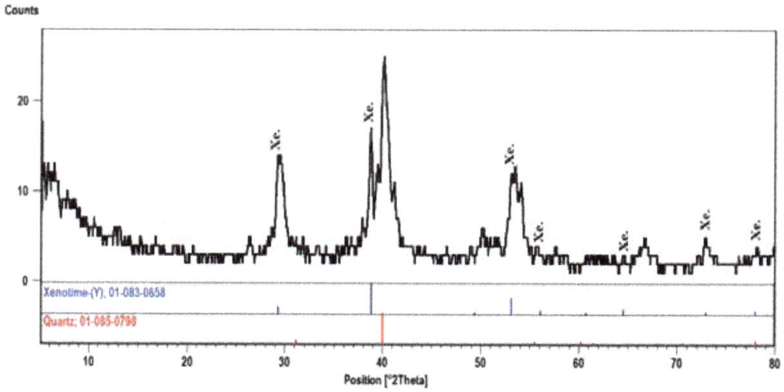

Figure 13. XRD pattern of xenotime, El-Urf mineralized pegmatite parts (phase-I).

3.2. Radioactivity and Radiological Hazards Assessment

Generally, the difference among uranium and thorium contents, as well as their ratio values of El Urf monzogranite, barren pegmatite and mineralized pegmatite phase-I and phase-II, suggest the removal or migration concept of uranium ions from the whole pegmatite parts by different manners. Two uranium migration styles were revalued. The colorful mineralization (phase-I) with the barren parts migrated its uranium content through the regular method, whereas the uranium ion charge was removed from opaque mineralization (phase-II) and the El Urf monzogranite due to its disturbance and irregular style. Table 2 displays the radiometric data to clarify the distribution of radioactivity and locate the three levels of radioactivity. The obtained results showed that El Urf monzogranite is barren (Up to 100 cps), surrounding an excavation of the studied pegmatite that exhibits moderate colorful mineralization (phase-I = 500–1500 cps) and anomalous opaque mineralization (phase-II = 1500–3500 cps) of pegmatites.

The eU and eTh contents values in ppm, as well as K, in %, were converted to activity concentration, Bq kg^{-1}, using the conversion factors (12.35 and 4.06 Bq kg^{-1}/ppm for ^{238}U and ^{232}Th, respectively, as well as 313 Bq kg^{-1}/ % for ^{40}K) [32], where A_U, A_{Th} and A_K are the average activity concentrations of ^{238}U, ^{232}Th and ^{40}K in Bq kg^{-1}, respectively. The ^{238}U activity concentration ranges between 16 and 46, averaging 34 Bq kg^{-1}, and ^{232}Th activity concentration varies between 16 and 45 Bq kg^{-1} with 30 Bq kg^{-1} as an average. ^{40}K ranges between 406.90 and 1201.92 Bq kg^{-1} with an average of 914 Bq kg^{-1}. The studied El Urf younger samples have slightly lower activity than the worldwide average values for ^{238}U and ^{232}Th but with higher amounts of ^{40}K. The worldwide average values are 33, 45 and 412 Bq kg^{-1} for ^{238}U, ^{232}Th and ^{40}K, respectively [3]. Activity concentrations of ^{238}U, ^{232}Th and ^{40}K in the barren pegmatites range from (64 to 123), (69 to 89) and (1083 to 2075) with averages (93, 78, 1568), respectively. All the barren pegmatite samples have activity concentrations higher than the permissible levels, (Table 1). The radioactive pegmatites are classified according to the opaque mineral contents into colorful mineralized pegmatite phase and opaque mineralized phase. Activity concentrations ranges of ^{238}U, ^{232}Th and ^{40}K in the colorful pegmatite phase are (208–499), (566–1854) and (1561–3449), with averages of 321, 991.66 and 2354, respectively, whereas for opaque mineralized pegmatite, phases are (364–763), (2799–5733) and (2212–3506), with averages of 561, 4289 and 3002, respectively. The activity concentrations averages of both colorful and opaque mineralized pegmatites are much higher than the international averages mentioned later (Table 1).

Radiological Hazards Parameters

The mean Ra_{eq} values for the granitic rocks of El Urf are 147, 324, 1919 and 6919 Bq kg^{-1} for the younger granite, barren pegmatite and colorful and opaque mineralized pegmatites, respectively. However, colorful and opaque mineralized pegmatites have much higher values than the criterion limit of 370 Bq kg^{-1}; however, the younger granite and barren pegmatite are lower. However, colorful and opaque mineralized pegmatites display much higher values than the criterion limit of 370 Bq kg^{-1}, whereas the younger granite and barren pegmatite are lower (Figure 14).

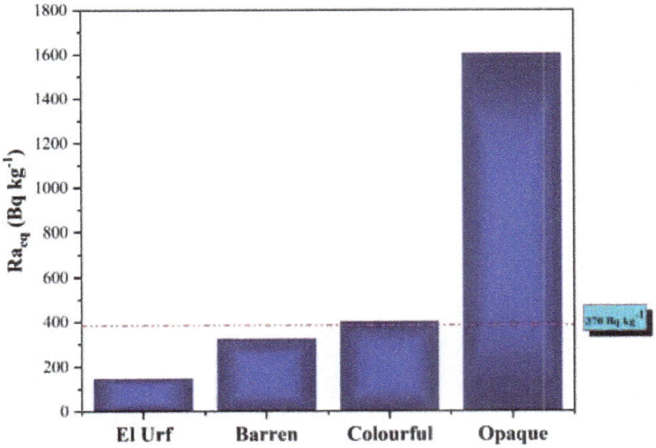

Figure 14. An illustrating histogram shows the average radium equivalent activity (Raeq) values for the El Urf younger granite and different mineralized pegmatite parts.

Table 2. Results of radionuclide activity concentrations, the dose rate (D_{air}), the annual effective dose (AED), radium equivalent activity (Ra_{eq}), external (H_{ex}), internal (H_{in}) hazard indices and gamma index ($I\gamma$) for younger granite and barren pegmatite samples.

Rock Type	Series of Observations No #.	A_U Bq kg^{-1}	A_{Th} Bq kg^{-1}	A_K Bq kg^{-1}	Dair (nGyh^{-1})	AED mSv	Ra_{eq} Bq kg^{-1}	H_{ex}	H_{in}	I_γ
El Urf younger granite	1	25	23	664	53.31	0.07	109	0.30	0.36	0.84
	2	16	16	407	33.85	0.04	70	0.19	0.23	0.54
	3	42	37	1099	87.65	0.11	180	0.49	0.6	1.38
	4	40	28	898	72.63	0.09	149	0.4	0.51	1.14
	5	46	45	1202	98.79	0.12	203	0.55	0.67	1.56
	Av	34	30	914	71.75	0.09	147	0.4	0.49	1.13
	SD	11	10	291	23	0.03	48	0.13	0.16	0.37
	min	16	16	407	33.85	0.04	70	0.19	0.23	0.54
	max	46	45	1202	98.79	0.12	203	0.55	0.67	1.56
Barren pegmatite parts	6	64	71	1083	117.75	0.14	249	0.67	0.85	1.86
	7	87	82	1474	151.26	0.19	318	0.86	1.09	2.38
	8	92	86	1552	158.85	0.19	333	0.9	1.15	2.5
	9	119	69	2009	180.23	0.22	372	1	1.33	2.82
	10	84	74	1424	142.87	0.18	299	0.81	1.04	2.25
	11	79	84	1340	142.97	0.18	302	0.82	1.03	2.26
	12	123	69	2075	184.98	0.23	381	1.03	1.36	2.89
	13	96	89	1621	165.46	0.2	347	0.94	1.2	2.61
	14	121	75	2050	186.85	0.23	386	1.04	1.37	2.93
	15	85	74	1434	143.95	0.18	301	0.81	1.04	2.27
	16	75	77	1277	134.29	0.16	283	0.76	0.97	2.12
	17	87	82	1474	150.95	0.19	317	0.86	1.09	2.38
	Av	93	78	1568	155.01	0.19	324	0.88	1.13	2.44
	SD	18	6	305	20	0.03	40	0.11	0.16	0.31
	min	64	69	1083	117.75	0.14	249	0.67	0.85	1.86
	max	123	89	2075	186.85	0.23	386	1.04	1.37	2.93
Colorful Mineralized parts (phase-I)	18	267	743	1969	132.22	0.16	305	0.82	0.88	2.16
	19	208	566	1562	101.01	0.12	233	0.63	0.67	1.65
	20	403	1327	2980	230.17	0.28	533	1.44	1.53	3.77
	21	282	791	2100	140.57	0.17	324	0.88	0.94	2.3
	22	500	1854	3449	316.7	0.39	734	1.99	2.09	5.19
	23	301	851	2225	151.01	0.19	348	0.94	1.01	2.47
	24	288	809	2197	143.66	0.18	331	0.9	0.96	2.35
	Av	321	992	2354	173.62	0.21	401	1.08	1.15	2.84
	SD	91	413	594	69	0.09	160	0.43	0.45	1.13
	min	208	566	1562	101.01	0.12	233	0.63	0.67	1.65
	max	500	1854	3449	316.7	0.39	734	1.99	2.09	5.19
Opaque Mineralized parts (phase-II)	25	365	2866	2194	457.27	0.56	1071	2.89	2.97	7.53
	26	700	5315	3409	849.92	1.04	1990	5.38	5.53	14
	27	508	3650	2936	585.77	0.72	1371	3.7	3.82	9.65
	28	444	3366	2570	538.35	0.66	1260	3.41	3.5	8.87
	29	365	2799	2132	447.29	0.55	1047	2.83	2.91	7.37
	30	565	4291	3271	686.2	0.84	1607	4.34	4.47	11.3
	31	683	5187	3446	829.35	1.02	1942	5.25	5.4	13.66
	32	544	4150	3155	663.33	0.81	1553	4.2	4.32	10.93
	33	764	5398	3506	867.37	1.06	2029	5.48	5.65	14.28
	34	612	4767	3205	760.95	0.93	1782	4.82	4.95	12.54
	35	365	2821	2113	450.51	0.55	1055	2.85	2.93	7.42
	36	569	4750	3296	755.14	0.93	1769	4.78	4.91	12.44
	37	754	5733	3440	916.67	1.12	2146	5.8	5.96	15.1
	38	631	4954	3393	790.43	0.97	1851	5	5.14	13.02
	Av	561	4289	3002	685.54	0.84	1605.05	4.34	4.46	11.29
	SD	127	891	446	143	0.18	335	0.90	0.93	2.36
	min	365	2799	2113	447.29	0.55	1047	2.83	2.91	7.37
	max	764	5733	3506	916.67	1.12	2146	5.8	5.96	15.1

These indices must be less than the average in order to keep the radiation hazard insignificant [33–35]. The radiation exposure due to radioactivity in construction materials must be limited to 1.5 mSv year^{-1} (Table 2). The values of the external hazard (H_{ex}) and internal hazard (H_{in}) for the studied granitic rocks are less than standard in El Urf younger granites, which agrees with the recommended values, whereas almost samples in the barren pegmatite and all samples in the colorful and opaque mineralized pegmatites are, to a greater extent, higher than international standards, suggesting that these samples cannot be used as a building or decorative material of dwelling (Table 2). The opaque mineralized pegmatites parts have the both the highest external hazard (H_{ex}) and internal hazard (H_{in}) values, reflecting the most dangerous used material among the studied rock types in the El Urf area (Figure 15). The safety value for this index is ≤ 1, whereas the obtained $I\gamma$ averages for the studied rocks are 1.13, 2.44, 13.63 and 48.63 for younger granites, barren pegmatite and colorful and opaque mineralized pegmatites, respectively. Most of the studied rocks have a value higher than the recommended safety value, to a great extent.

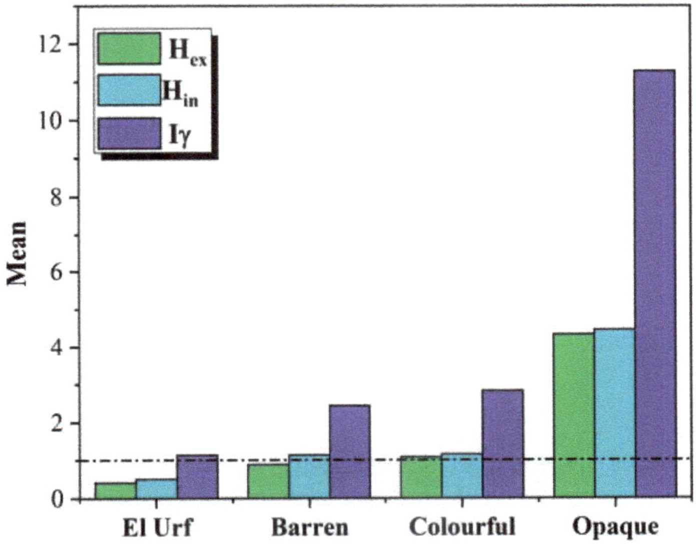

Figure 15. An illustrating histogram shows the average external hazard values (H_{ex}), internal hazard (H_{in}) and radiation level index ($I\gamma$) for the El Urf younger granite and different mineralized pegmatite parts.

Table clarifies the estimated gamma-absorbed dose rate values for the studied granitic rock samples. The D_{air} values for the younger granite samples range from 34 to 99 nGy h^{-1}, with a mean of 72 nGy h^{-1}. Barren pegmatite samples range from 118 to 187 nGy h^{-1}, with a mean of 99 nGy h^{-1}. D_{air} values for both colorful and opaque mineralized pegmatites are (502–1495) and (1948–3954), with averages of 845 and 2975, respectively. The mean D_{air} values for all the studied granitic rocks exceed the worldwide average value (59 nGy h^{-1}, UNSCEAR, 2000), (Table 2). This displays that the Gabal El Ur area is not appropriate for the stratification of various infrastructure applications, particularly building materials.

The mean values of the studied granitic rocks are 0.09, 0.19, 1.04 and 3.65, for the younger granite and barren pegmatite, colorful and opaque mineralized pegmatites, respectively, which are higher than the recommended worldwide average of the annual effective dose (0.07 $mSvy^{-1}$), as suggested by UNSCEAR (2000) [36], (Table 2). Heavy minerals found in granites, such as monazite, uraninite and thorianite, are responsible for the high doses. Furthermore, long-term exposure to high dosages might have negative health consequences such as cancer and cardiovascular disease, which are linked to tissue degradation and deoxyribonucleic acid (DNA) in genes [37].

The principal component analysis (PCA) employed Varimax rotations to identify the matrix connection between distinct components. The PC1 and PC2 components are shown in Figure 16.

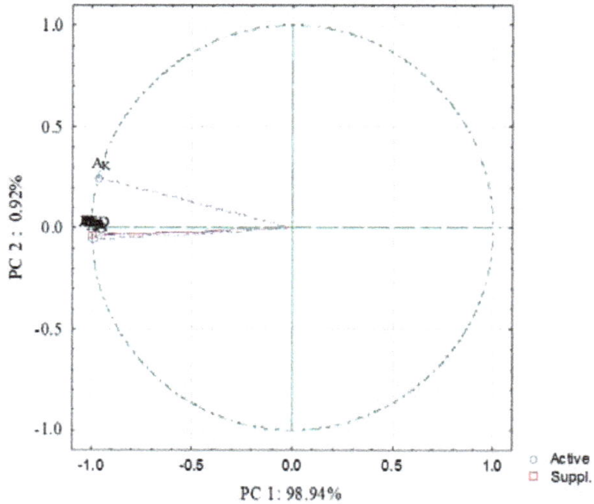

Figure 16. Principal component analysis (PC1 and PC2) for radiological data of opaque granite at the El Urf area.

In opaque granite samples, the activity concentrations of ^{238}U and ^{232}Th indicate a strong positive in PC1 loading, which is linked to all radiological factors and explains 98.94% of the variation. As a result, ^{238}U and ^{232}Th activity concentrations were the most common natural radioactive contributions in the opaque granite at the research location. PC2 accounts for 0.92 % of the variance [38,39].

The data of radiological variables are analyzed using a hierarchical clustering approach. Figure 17 depicts the relationship between all of the variables. The dendrogram of the examined data in the opaque granite at the El Urf area shows two clusters. Cluster I in the opaque granite at the analyzed location is made up of ^{238}U, ^{232}Th and radiological hazard factors. Although cluster II contains the ^{40}K, which are linked to cluster I, this analysis demonstrated that uranium and thorium minerals are responsible for the total radioactivity in the opaque granite. Finally, the cluster analysis results are consistent with PCA analysis.

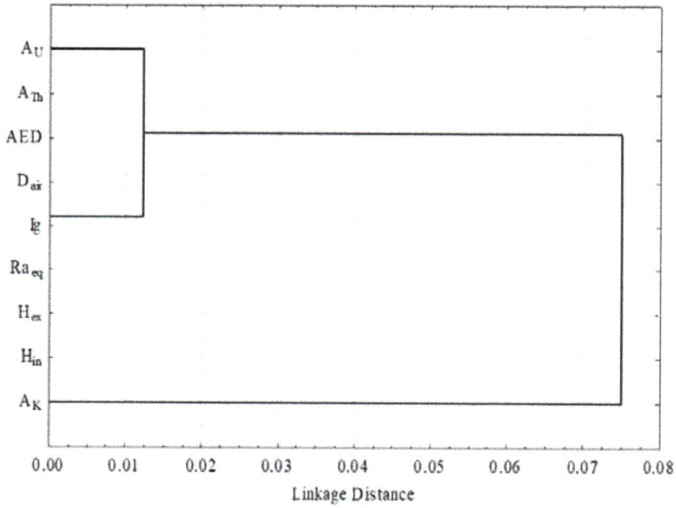

Figure 17. The clustering analysis of the radiological parameters of Opaque at EL Urf area.

4. Conclusions

The mineralized pegmatite is located in the southwestern foot hill of the Gabal El Urf younger granite and displays well-defined zonation of three zones: outer, middle and inner zones represented by potash feldspar, quartz and mica, respectively. The activity concentrations of ^{238}U, ^{232}Th and ^{40}K in the mineralized pegmatitites have higher values relative to the worldwide average. The highest values of ^{238}U (561±127 Bq kg^{-1}), ^{232}Th (4289±891 Bq kg^{-1}) and ^{40}K (3002 ± 446 Bq kg^{-1}) are found in the opaque mineralized pegmatites. Many of the radiological hazard parameters were lesser than the international limits in the younger granites and barren pegmatites. All these parameters were higher in the colorful and opaque mineralized pegmatites. This is attributed to the alteration of radioactive minerals such as radioactive earing minerals such as thorite, meta-autunite, kasolite, phurcalite, columbite, fergusonite, xenotime and fluorapatite. Other instances of mineralization were also recorded as cassiterite, atacamite, galena, pyrite and iron oxide minerals. The statistical analysis was conducted to illustrate the geological processes that lead to an increase in the radioactive concentration in the granite rocks. Thus, the granite rocks of the studied area are not safe, pose negative health risks and are not applied in the building materials and the application of various infrastructures.

Author Contributions: Conceptualization, M.M.E.D., A.M.I. and M.Y.H.; methodology, M.M.E.D., M.M. and S.A.T.; software, A.M.I. and M.Y.H.; validation, M.Y.H. and A.E.; formal analysis, M.M. and S.A.T.; investigation, M.M.E.D. and M.Y.H.; resources, A.M I. and S.A.T.; data curation, M.M.E.D. and M.Y.H.; writing—original draft preparation, A.M.I., S.A.T. and M.Y.H.; writing—review and editing, A.M.I., S.A.T., M.Y.H. and A.E.; visualization, M.M. and S.A.T.; supervision, M.M.E.D. and A.E.; project administration, M.M.E.D.; funding acquisition, A.E. All authors have read and agreed to the published version of the manuscript.

Funding: This research was funded by the Nuclear Materials Authority, Egypt. The APC was covered by "Dunarea de Jos" University of Galati, Romania, through grant no. RF3621/2021.

Institutional Review Board Statement: Not applicable.

Informed Consent Statement: Not applicable.

Data Availability Statement: The data presented in this study are available on request from the corresponding authors.

Acknowledgments: The authors are indebted to Mohamed G. El Feky, Professor of Geochemistry, Nuclear Materials Authority, Egypt, for supporting this work. The author A.E. acknowledges the support of "Dunarea de Jos" University of Galati, Romania.

Conflicts of Interest: The authors declare no conflict of interest.

References

1. Akpanowo, M.A.; Umaru, I.; Iyakwari, S.; Joshua, E.O.; Yusuf, S.; Ekong, G.B. Determination of natural radioactivity levels and radiological hazards in environmental samples from artisanal mining sites of Anka, North-West Nigeria. *Sci. Afr.* **2020**, *10*, e00561. [CrossRef]
2. Sivakumar, S.; Chandrasekaran, A.; Ravisankar, R.; Ravikumar, S.M.; Prince Prakash Jebakumar, J.; Vijayagopal, P.; Vijayalakshmi, I.; Jose, M.T. Measurement of natural radioactivity and evaluation of radiation hazards in coastal sediments of east coast of Tamilnadu using statistical approach. *J. Taibah Univ. Sci.* **2014**, *8*, 375–384. [CrossRef]
3. UNSCEAR. *Sources And Effects of Ionizing Radiation—Exposures of the Public And Workers From Various Sources of Radiation—UNSCEAR 2008 Report*; UNSCEAR: Vienna, Austria, 2010; Volume I, ISBN 9789211422740.
4. Pavlidou, S.; Koroneos, A.; Papastefanou, C. Natural radioactivity of granites used as building materials. *J. Environ. Radioact.* **2006**, *89*. [CrossRef] [PubMed]
5. Khandaker, U.M.; Asaduzzaman, K.; Bin Sulaiman, A.F.; Bradley, D.A.; Isinkaya, M.O. Elevated concentrations of naturally occurring radionuclides in heavy mineral-rich beach sands of Langkawi Island, Malaysia. *Mar. Pollut. Bull.* **2018**, *127*, 654–663. [CrossRef]
6. Boumala, D.; Mavon, C.; Belafrites, A.; Tedjani, A.; Groetz, J.E. Evaluation of radionuclide concentrations and external gamma radiation levels in phosphate ores and fertilizers commonly used in Algeria. *J. Radioanal. Nucl. Chem.* **2018**, *317*, 501–510. [CrossRef]

7. Hassan, N.M.; Mansour, N.A.; Fayez-Hassan, M.; Sedqy, E. Assessment of natural radioactivity in fertilizers and phosphate ores in Egypt. *J. Taibah Univ. Sci.* **2016**, *10*, 296–306. [CrossRef]
8. Calin, M.R.; Radulescu, I.; Calin, M.A. Measurement and evaluation of natural radioactivity in phosphogypsum in industrial areas from Romania. *J. Radioanal. Nucl. Chem.* **2015**, *304*, 1303–1312. [CrossRef]
9. Hanfi, M.Y.M. Radiological assessment of gamma and radon dose rates at former uranium mining tunnels in Egypt. *Environ. Earth Sci.* **2019**, *78*, 113. [CrossRef]
10. Ravisankar, R.; Chandrasekaran, A.; Vijayagopal, P.; Venkatraman, B.; Senthilkumar, G.; Eswaran, P.; Rajalakshmi, A. Natural radioactivity in soil samples of Yelagiri Hills, Tamil Nadu, India and the associated radiation hazards. *Radiat. Phys. Chem.* **2012**, *81*, 1789–1795. [CrossRef]
11. Abbady, A.; Ahmed, N.K.; El-arabi, A.M.; Michel, R.; El-kamel, A.H.; Abbady, A.G.E. Estimation of radiation hazard indices from natural radioactivity of some rocks. *Nucl. Sci. Tech.* **2006**, *17*, 118–122. [CrossRef]
12. ATSDR. *Toxicological Profile for Uranium*; Public Health Services Department of Health & Human Services: Washington, WA, USA, 1999; pp. 1–145.
13. ATSDR. *Draft Toxicological Profile for Radon: Agency for Toxic Substances and Disease Registry*; Public Health Services Department of Health & Human Services: Washington, WA, USA, 2012; 9–11, pp. 161–167.
14. La Verde, G.; Raulo, A.; D'Avino, V.; Roca, V.; Pugliese, M. Radioactivity content in natural stones used as building materials in Puglia region analysed by high resolution gamma-ray spectroscopy: Preliminary results. *Constr. Build. Mater.* **2020**, *239*, 117668. [CrossRef]
15. Sahoo, B.K.; Nathwani, D.; Eappen, K.P.; Ramachandran, T.V.; Gaware, J.J.; Mayya, Y.S. Estimation of radon emanation factor in Indian building materials. *Radiat. Meas.* **2007**, *42*, 1422–1425. [CrossRef]
16. Sabbarese, C.; Ambrosino, F.; Onofrio, A.D.; Roca, V. Radiological characterization of natural building materials from the Campania region (Southern Italy). *Constr. Build. Mater.* **2020**, 121087. [CrossRef]
17. Imani, M.; Adelikhah, M.; Shahrokhi, A.; Azimpour, G.; Yadollahi, A.; Kocsis, E.; Toth-Bodrogi, E.; Kovács, T. Natural radioactivity and radiological risks of common building materials used in Semnan Province dwellings, Iran. *Environ. Sci. Pollut. Res.* **2021**, *28*, 41492–41503. [CrossRef]
18. Moghazi, A.M. Magma source and evolution of the late Proterozoic granitiods in the Gabal El-Urf area, Eastern Desert, Egypt: Geochemical and Sr-Nd isotopic constraints. *Geol. Mag.* **1999**, *136*, 285–300. [CrossRef]
19. Abdel Ghani, I.M. Geology, Petrology, and Radioactivity of Gabal El-Urf Area, Central Eastern Desert. Ph.D. Thesis, South Valley University, Qena, Egypt, 2001.
20. Asran, A.; El-Mansi, M.M.; Ibrahim, M.E.; Abdelghani, I. Pegmatites of gabal el urf, central eastern desert. In Proceedings of the 7th International Conference on the Geology of Africa, Assiut, Egypt, 24–26 November 2013.
21. EL-Sherif, A.E. Mineralogical and Radioactive Characterization of Gabal El-Urf Pegmatites, Central Eastern Desert, Egypt. *Al-Azhar Bull. Sci.* **2015**, *26*, 85–96. [CrossRef]
22. Khalaf, E.A.; El-Azabi, M.; Mokhtar, H.; Bernard, K. Stratigraphy and facies architecture of the Neoproterozoic syn- and inter-eruptive succession: An example from Gabal El Urf, Northeastern Desert, Egypt. *Precambrian Res.* **2020**, *350*, 105905. [CrossRef]
23. Eliwa, H.; Breitkreuz, C.; Khalaf, I.; Gameel, K. El Depositional styles of Early Ediacaran terrestrial volcanosedimentary succession in Gebel El Urf area, North Eastern Desert, Egypt. *J. Afr. Earth Sci.* **2010**, *57*, 328–344. [CrossRef]
24. Matolin. *Construction and Use of Spectrometric Calibration Pads Laboratory γ-Ray Spectrometry*; NMA: New Cairo, Egypt, 1991.
25. ICRP. *60 Recommendations of the International Commission on Radiological Protection*; ICRP Publication 60; Pergamon Press Annals of the ICRP: Oxford, UK, 1990.
26. Al-Hamarneh, I.F.; Awadallah, M.I. Soil radioactivity levels and radiation hazard assessment in the highlands of northern Jordan. *Radiat. Meas.* **2009**, *44*, 102–110. [CrossRef]
27. Xinwei, L. Natural radioactivity in some building materials of Xi'an, China. *Radiat. Meas.* **2005**, *40*, 94–97. [CrossRef]
28. Beretka, J.; Mathew, P.J. Natural radioactivity of australian building materials, industrial wastes and by-products. *Health Phys.* **1985**, *48*, 87–95. [CrossRef] [PubMed]
29. Alam, M.N.; Chowdhury, M.I.; Kamal, M.; Ghose, S.; Islam, M.N.; Mustafa, M.N.; Miah, M.M.H.; Ansary, M.M. The 226Ra, 232Th and 40K activities in beach sand minerals and beach soils of Cox's Bazar, Bangladesh. *J. Environ. Radioact.* **1999**, *46*, 243–250. [CrossRef]
30. El Dabe, M.M.; Ismail, A.M.; Metwaly, M. Geology and Radioactivity of the Pegmatitic Rocks of Gabal El Urf, Northern Eastern Desert, Egypt. *J. Rad. Nuci. Appl.* **2022**, *1*, 1–13. [CrossRef]
31. NEA-OECD. *Exposure to Radiation from Natural Radioactivity in Building Materials*; Report by NEA Group of Experts of the Nuclear Energy Agency; OECD: Paris, France, 1979.
32. Abdel-Razek, Y.A.; Masoud, M.S.; Hanfi, M.Y.; El-Nagdy, M.S. Effective radiation doses from natural sources at Seila area South Eastern Desert, Egypt. *J. Taibah Univ. Sci.* **2016**, *10*, 271–280. [CrossRef]
33. Harpy, N.M.; El Dabour, S.E.; Sallam, A.M.; Nada, A.A.; El Aassy, A.E.; El Feky, M.G. Radiometric and environmental impacts of mill tailings at experimental plant processing unit, Allouga, Egypt. *Environ. Forensics* **2020**, *21*, 11–20. [CrossRef]
34. El Feky, M.G.; El Mowafy, A.A.; Abdel, A.W. Mineralogy, geochemistry, radioactivity and environmental impacts of Gabal Marwa granites, southeastern Sinai, Egypt. *Chin. J. Geochem.* **2011**, *30*, 175–186. [CrossRef]

35. Baykara, O.; Karatepe, Ş.; Doğru, M. Assessments of natural radioactivity and radiological hazards in construction materials used in Elazig, Turkey. *Radiat. Meas.* **2011**, *46*, 153–158. [CrossRef]
36. UNSCEAR. *Exposures from Natural Radiation Sources (Annex B). Sources and Effects of Ionizing Radiation*; UNSCEAR: New York, NY, USA, 2000; pp. 84–141.
37. USEPA EPA. *Radiogenic Cancer Risk Models and Projections for the U.S. Population*; EPA: Washington, DC, USA, 2011.
38. Ravisankar, R.; Chandramohan, J.; Chandrasekaran, A.; Prakash, J.P.; Vijayalakshmi, I.; Vijayagopal, P.; Venkatraman, B. Assessments of radioactivity concentration of natural radionuclides and radiological hazard indices in sediment samples from the East coast of Tamilnadu, India with statistical approach. *Mar. Pollut. Bull.* **2015**. [CrossRef]
39. Gaafar, I.; Elbarbary, M.; Sayyed, M.I.; Sulieman, A.; Tamam, N.; Khandaker, M.U.; Bradley, D.A.; Hanfi, M.Y. Assessment of Radioactive Materials in Albite Granites from Abu Rusheid and Um Naggat, Central Eastern Desert, Egypt. *Minerals* **2022**, *12*, 120. [CrossRef]

materials

MDPI

Article

Radiological Hazard Evaluation of Some Egyptian Magmatic Rocks Used as Ornamental Stone: Petrography and Natural Radioactivity

El Saeed R. Lasheen [1], Mohammed A. Rashwan [2], Hamid Osman [3], Sultan Alamri [3], Mayeen U. Khandaker [4] and Mohamed Y. Hanfi [5,6,*]

1 Geology Department, Faculty of Science, Al-Azhar University, Cairo P.O. Box 11884, Egypt; elsaeedlasheen@azhar.edu.eg
2 Geological Sciences Department, National Research Centre, 33 El Bohooth St. (Former El Tahrir St.), Dokki, Giza P.O. Box 12622, Egypt; ma.attia@nrc.sci.eg
3 Department of Radiological Sciences, College of Applied Medical Sciences, Taif University, Taif 21944, Saudi Arabia; ha.osman@tu.edu.sa (H.O.); s.alamri@tu.edu.sa (S.A.)
4 Centre for Applied Physics and Radiation Technologies, School of Engineering and Technology, Sunway University, Bandar Sunway 47500, Malaysia; mayeenk@sunway.edu.my
5 Institute of Physics and Technology, Ural Federal University, St. Mira, 19, 620002 Yekaterinburg, Russia
6 Nuclear Materials Authority, Maadi, Cairo P.O. Box 530, Egypt
* Correspondence: mokhamed.khanfi@urfu.ru or m.nuc2012@gmail.com; Tel.: +7-(982)-646-53-57

Abstract: Magmatic rocks represent one of the most significant rocks due to their abundance, durability and appearance; they can be used as ornamental stones in the construction of dwellings. The current study is concerned with the detailed petrography and natural radioactivity of seven magmatic rocks. All are commercial granitic rocks and are identified as black Aswan, Nero Aswan, white Halayeb, Karnak, Verdi, red Hurghada and red Aswan. Their respective mineralogical compositions are classified as porpheritic granodiorite, granodiorite, tonalite, monzogranite, syenogranite, monzogranite and syenogranite. A total of nineteen samples were prepared from these seven rock types in order to assess their suitability as ornamental stones. Concentrations of ^{226}Ra, ^{232}Th and ^{40}K radionuclides were measured using NaI (Tl) scintillation gamma-ray spectrometry. Among the studied magmatic rocks, white Halayeb had the lowest average values of ^{226}Ra (15.7 Bq/kg), ^{232}Th (4.71 Bq/kg) and ^{40}K (~292 Bq/kg), all below the UNSCEAR reported average world values or recommended reference limits. In contrast, the other granitic rocks have higher values than the recommended limit. Except for the absorbed dose rate, other radiological hazard parameters including radium equivalent activity, annual effective dose equivalent, external, and internal hazard indices reflect that the White Halyeb rocks are favorable for use as ornamental stone in the construction of luxurious and high-demand residential buildings.

Keywords: ornamental stone; gamma-ray spectrometer; radiological hazard indices; natural radioactivity

Citation: Lasheen, E.S.R.; Rashwan, M.A.; Osman, H.; Alamri, S.; Khandaker, M.U.; Hanfi, M.Y. Radiological Hazard Evaluation of Some Egyptian Magmatic Rocks Used as Ornamental Stone: Petrography and Natural Radioactivity. *Materials* **2021**, *14*, 7290. https://doi.org/10.3390/ma14237290

Academic Editor: Petrica Vizureanu

Received: 27 October 2021
Accepted: 25 November 2021
Published: 28 November 2021

1. Introduction

Ornamental stone represents one of the dominant industrial economies worldwide, and its demand shows geometric growth with the increasing construction of luxurious dwellings [1–3]. Egyptian basement rock constitutes the northwestern sector of the Arabian Nubian Shield (ANS), which crops out in the Eastern Desert, south Sinai and southwestern side of the Western Desert [4]. In the Eastern Desert, these rocks extend from the north, near Cairo, to the south along the Sudanese border. Granitic rocks represent the main magmatic rocks, making up ~60% of the total Egyptian basement rock of the Nubian Shield [4]. Granitic rocks represent one of the most important ornamental stones, along with marble [4]. The former is characterized by its durability and prestigious shape, and can be used as decorative stone in floors, stairs, walls, bridges, and sculptures [5,6].

Egyptian quarries represent one of the main producers globally as one of the top eight raw material-producing countries, producing about 3.2 million tons of quarried stone under twenty-five different brands. Moreover, Egypt represents the seventh-largest ornamental stone exporting country globally, amounting to 1.5 million tons/year [7].

The distribution of uranium in the Earth's crust is associated with magmatic activity through the creation of the Earth. Depending on the amount of natural radionuclides, the radioactivity concentration in soil is relatively higher in volcanic, phosphatic, granitic and salty rocks. These rocks break into pieces over time under environmental conditions and spread across the environment. Uranium, thorium and potassium concentrations are mostly higher in the granitic rocks, granitic pegmatites and syenites, and are closely linked to mineralogical composition and petrographic features [8–11]. In general, felsic igneous rocks have greater levels than sedimentary rocks. Primary uranium minerals such as uraninite, pitchblende and coffinite are formed during rock formation, while secondary ones (e.g., uranophane) are formed later due to hydrothermal solution [12].

Enrichment of some elements such as Rb, Nb, Ta, U, Th, Zr, K and REEs is related to protracted fractional crystallization of the magma toward felsic rocks. Therefore, the previous elements' abundance increases with increasing alkalinity ($K_2O + Na_2O$) [9]. In addition, felsic magmatic rocks require a high SiO_2 content; thus, felsic rocks can be viewed as crust-derived rocks. In addition, zircon, allanite, titanite, xenotime, apatite, monazite and thorite minerals are the main accessory minerals in felsic rocks, including granitic rocks, as U and Th are more easily able to enter the lattice structures of these minerals; they thus represent a U and Th reservoir [11]. On the other hand, enrichment of U and Th in felsic rocks may be ascribed to magmatic segregation and hydrothermal activities, which can lead to different alteration processes [9,13].

Radioactivity can be found in rocks, soil, beach sand, sediment, river bed, rivers and oceans, and even building materials and dwellings. Naturally occurring radioactive materials usually have a terrestrial origin (primordial radionuclides), left behind since the earth's creation [14]. They are typically long-lived, with half-lives ranging from hundreds of millions of years to billions of years. Natural sources of gamma radiation (background radiation) are predominantly attributable to primordial radionuclides, primarily the ^{232}Th and ^{238}U series and their decay products, as well as ^{40}K, all of which exist at trace amounts in the earth's crust. The natural radiation of soil and rocks depends on their mineralogical composition [15,16]. Measuring the concentration of radionuclides helps in monitoring environmental radioactivity. Radionuclide distributions vary from one rock to another depending on rock type. Radiological impacts on human health can be inferred by detecting the distributions and concentrations of natural radionuclides in those rocks/soil [17,18] generally used as raw materials in the construction of buildings and infrastructure.

The examined rocks were prepared as cubes in order to assess their suitability as ornamental stones through both physical (water absorption, bulk specific gravity, actual density) and mechanical (compressive strength, abrasion resistance) tests (to be reported elsewhere). Here, the current study aims to identify some magmatic rocks' mineralogical constituents and determine the concentrations of natural radionuclides (^{238}U, ^{226}Ra, ^{232}Th and ^{40}K). The measured data may help to assess the concomitant radiological hazards to human health resulting from exposure to emitted gamma radiation from those studied granitic rocks that show the greatest application for the construction of dwellings and infrastructure.

2. Materials and Methods

2.1. Radiation Shielding Capacity

The Egyptian Neoproteozoic rocks used in the present study are widely distributed along the Red Sea coast (Eastern Desert) as well as the Sinai. Particularly, granitoid rocks vary from older/syn-tectonic (OG, grey granites) to younger/late to post-tectonic granites (YG, pink granites) [19,20]. The Arabian Nubian Shield represents the northern zone of East Africa orogen, formed during the closing of the Mozambique Ocean and the collision of East and West Gondwana [9,21,22]. Although granitic rocks occur in the Oweinat area, the relative abundance of younger granites to older granites increases from 1:4 in the south to 1:1 in the north of the Eastern Desert and 12:1 in the Sinai [23] (Figure 1). Different commercial granitic rocks were collected from various distributed localities (seven quarries; Figure 1).

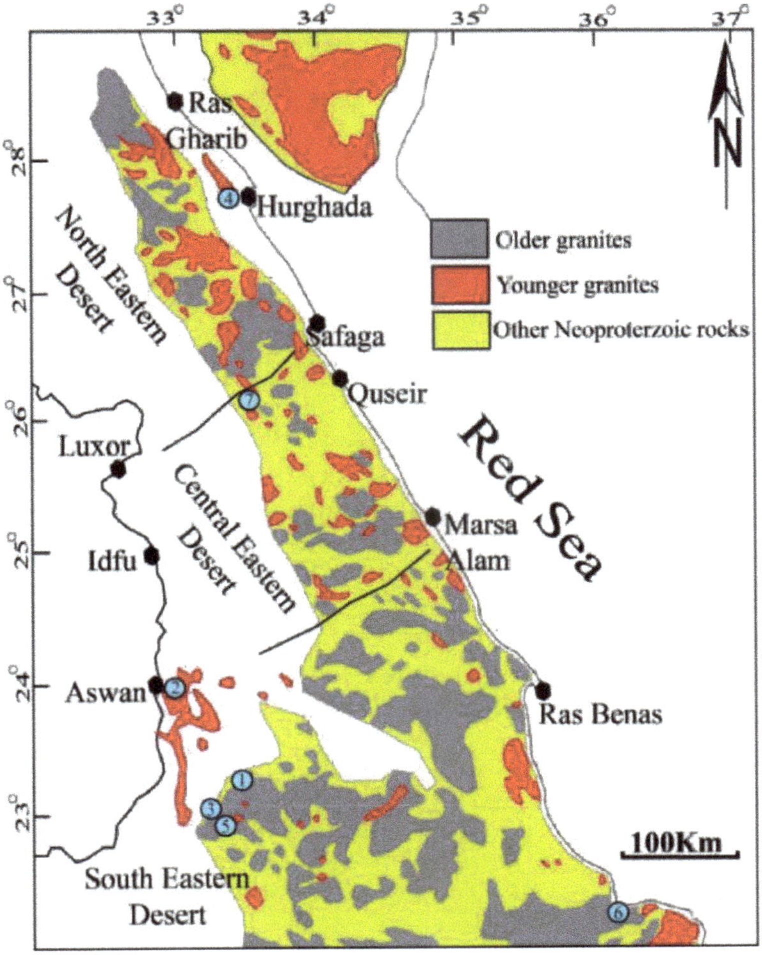

Figure 1. Distribution map of Neoproterozoic rocks in the Eastern Desert and Sinai after [10], including sample locations over a wide area: (**1**) black Aswan, (**2**) red Aswan, (**3**) Nero Aswan, (**4**) red Hurghada, (**5**) yellow Verdi, (**6**) white Halayeb, (**7**) Karnak.

The examined popular classes Nero Aswan, black Aswan, red Aswan, red Hurghada, yellow Verdi, white Halayeb and Karnak were collected from quarries (Figure 2). All samples were polished and prepared as equidimensional cubes to assess their suitability as ornamental stone, then girded to measure their natural radioactive concentrations.

Figure 2. Photographs of the examined granitic rocks.

2.2. Petrographic Investigation

The detailed petrographic investigation was carried out using point-counting techniques by polarizing microscope (Olympus bx53) for popular granitic rocks that were already used as an ornamental stone in order to detect their mineralogical composition textural relationships. Based on modal analysis of mineralogical composition, the examined magmatic rocks, including black Aswan, Nero Aswan, white Halayeb, Karnak, Verdi, red Hurghada and red Aswan, were classified as porpheritic granodiorite, granodiorite tonalite, monzogranite, syenogranite, monzogranite and syenogranite, respectively.

Black Aswan (1) was classified as porphyritic granodiorite. It has a medium to coarse grain and is composed essentially of plagioclase, quartz, potash feldspars, hornblende, biotite and a subordinate amount of augite. Plagioclase (58 vol.%) occurs as tabular, subhedral crystals that exhibit slight to extensive alteration (Figure 3a). Potash feldspars (14 vol.%) are represented by microcline and perthite. Microcline is common as fine-grained and subhedral crystals mostly corroded by quartz. Rapakivi textures are abundant. Quartz (22 vol.%) occurs as an anhedral crystal, fine- to medium-grained, that reveals undulose extinction. Hornblende occurs as a medium to coarse grained anhedral to subhedral crystal, partially altered to chlorite. Biotite occurs as fine to medium grained irregular flakes slightly altered to chlorite and muscovite, especially on cleavage planes and their periphery. Augite rarely occurs as anhedral crystals, high relief, and perfect perpendicular cleavage. Titanite, apatite and iron oxides as accessory minerals have also occurred (Figure 3b).

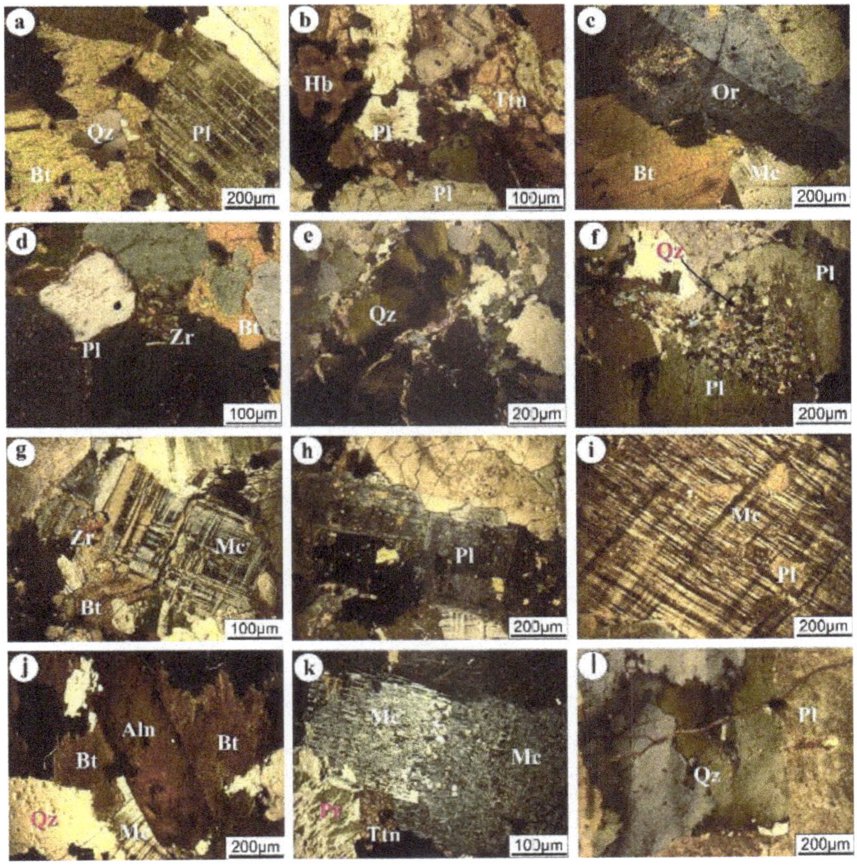

Figure 3. Photomicrographs of the studied magmatic rocks: black Aswan: (**a**) partially to completely chloritized biotite corroded by undulose quartz and plagioclase; (**b**) aggregation of titanite associated with plagioclase and titanite. Nero Aswan: (**c**) slightly sericitized orthoclase perthite corroded microcline and biotite; (**d**) well developed, euhedral zircon embedded in slightly saussuritized plagioclase. White Halayeb: (**e**) fine-to medium- grained undulose quartz fractured and filled by sericite; (**f**) zoned plagioclase engulfing fine-grained quartz. Karnak: (**g**) euhedral zircon enclosed in microcline, twisted and fractured biotite transformed to chlorite; (**h**) saussuritized and epidotized pericline plagioclase. Red Aswan: **i)** very coarse-grained, pristine, patchy microcline engulfing kaolinitized plagioclase; **j)** coarse-grained allanite surrounded by biotite crystals. Verdi: (**k**) titanite crystals enclosed by iron oxide, perthite and microcline perthite. Hurghada: (**l**) extended fracture filled by sericite intersecting extensive saussuritized plagioclase and undulose quartz.

Nero Aswan (2) was classified as granodiorite, which is composed mainly of plagioclase (55 vol.%), quartz (19 vol.%), and potash feldspars (15 vol.%), with subordinate hornblende and biotite based on their modal analysis. K-feldspar is represented by orthoclase perthite that is partially altered to sericite (Figure 3c). Titanite, allanite and zircon are present as accessory minerals. This is similar to black Aswan; however, the mineralogical constituents are slightly altered, certainly plagioclase and K-feldspars. Moreover, allanite and zircon are present accessory minerals (Figure 3d).

White Halayeb (6) was classified as tonalite. It is deformed, fine to medium grained and composed mainly of plagioclase, quartz, biotite, and a minor amount of muscovite. Epidote, chlorite and saussurite are the main secondary minerals. Quartz (16 vol.%) occurs as fine to coarse grains and reveals undulous extension, fractured and occasionally filled by sericite due to deformation processes (Figure 3e). Plagioclase (60.4 vol.%) is the most dominant mineral and is tabular with pericline, zoned, and lamellar twinning, altered to saussurite and epidote (Figure 3f). The main secondary mineral is epidote, which occurs as a well-developed subrounded mineral with high interference colour ranging from yellow to blue, scattered on the periphery of plagioclase as an alteration product.

Red Aswan (3) was classified as syenogranite, with megacrystals of K-feldspar, plagioclase and quartz. K-feldspar (36 vol.%) is represented by pristine, patchy and very coarse-grained microcline crystals (Figure 3i). Plagioclase (~20 vol.%) is subhedral and tabular, and partially or completely altered to saussurite, kaolinite and epidote. Normal, rarely-fractured quartz is present. Short zircon crystals occur as a euhedral, high-relief and embedded in biotite. Allanite occurs as a coarse-grained crystal and represents an alteration product of biotite (Figure 3j).

Hurghada (4) granitic rocks were classified as monzogranite. They are similar to Verdi in terms of their turbid, dusty surface, K-feldspar, and grain size. Coarse-grained and turbid surfaces represent potash Feldspars. Quartz occurs as a medium to coarse grained anhedral crystal and exhibits undulose extinction (Figure 3l). Occasionally, it is fractured and filled by secondary sericite. Most of the plagioclase crystals are kaolinitized and saussuritized and reveal lamellar and pericline twinning. In addition, biotite occurs as fine-grained flaky crystals, and is partially altered to chlorite and stained by iron oxides.

Verdi (5) was classified as syenogranite, with K-feldspar (~45 vol.%), plagioclase (~17 vol.%), and quartz (~33 vol.%) with normal extinction. K-feldspar occurs as coarse-grained, turbid (kaolinitized) microcline perthite (Figure 3k). Plagioclase mostly occurs as slightly kaolinitized tabular crystals. Titanite is the main accessory mineral, with a well-developed sphenoidal shape.

Karnak (7) granitic rocks were classified as monzogranite, consisting mainly of K-feldspars, quartz, plagioclase and biotite. K-feldspars (27 vol.%) are represented by pristine microcline (Figure 3g), medium to coarse grained and sometimes enclosing fine-grained saussuritized plagioclase. Quartz (31 vol.%) occurs as a medium to coarse grained subrounded crystal with normal extinction. Plagioclase (~35 vol.%) is represented by subhedral tabular crystals, and sometimes engulfs other constituents such as biotite and quartz (Figure 3h). Flaky biotite crystals are twisted and transformed into chlorite and iron oxides.

2.3. Analytical Techniques

Nineteen samples were collected from seven quarries: Nero Aswan (three samples), black Aswan (three samples), red Aswan (three samples), red Hurghada (three samples), Verdi (two samples), white Halayeb (three samples), and Karnak (two samples). Identification and nomenclature of the examined rocks were carried out using point-counting techniques by polarizing microscope (Olympus bx53), according to the relative amounts of potash feldspar, quartz and plagioclase. These samples were crushed and ground, and 350 gm from the ground samples were packed in a plastic container and sealed for about thirty days to attain secular equilibrium between parents and short-lived progeny. An NaI(Tl) scintillation gamma-ray spectrometer with a crystal size of 76 mm × 76 mm was used to determine the concentrations of radioelements (^{226}Ra, ^{232}Th and ^{40}K) in the granitic samples. A low background measurement environment was ensured by placing the detector in an arrangement that was enclosed in a cylindrical lead shield with a diameter of 15.7 cm, a length of 20.5 cm, and a thickness of 3.7 cm, with an attenuation factor of 0.16 (stopping at around 84% of input photons) for 2.6 MeV gamma rays. A spectroscopic amplifier and a multi-channel analyzer were part of the pulse processing and data analysis system, which was linked to an IBM-compatible computer. The corresponding gamma energies of ^{226}Ra, ^{232}Th and ^{40}K were 1764 keV (Iγ = 15.30%) from ^{214}Bi, 2614 keV (Iγ = 99.754%) from ^{228}Ac, and 1460 keV (Iγ = 10.66%), respectively [24,25]. Approved reference materials, such as RGU-1, RGTh-1, and RGK-1 were used, and their densities after pulverization were close to those of the building materials [26]. The design of the container was chosen based on the premise that the radioactivity in the measuring samples was evenly distributed. The samples were counted for 2000 s, with MDAs of 2, 4, and 12 Bq kg^{-1} for ^{226}Ra, ^{232}Th, and ^{40}K, respectively. The overall uncertainty of the radiation levels was calculated using the error propagation law of systematic and random measurement errors. Systematic errors of 0.5 to 2% existed in the efficiency calibration, and random errors of up to 5% existed in the radioactivity readings [27].

3. Results and Discussion

3.1. Radionuclides Concentrations

Radionuclides (^{226}Ra, ^{232}Th and ^{40}K) were measured by multichannel gamma-ray spectrometer, and are listed in Table 1. White Halayeb had the lowest average values of ^{226}Ra (15.7 Bq/kg), ^{232}Th (4.71 Bq/kg) and ^{40}K (~292 Bq/kg). In addition, it possessed radionuclide values lower than the recommended average worldwide values (32 Bq/kg for ^{226}Ra, 45 Bq/kg for ^{232}Th, and 412 Bq/kg for ^{40}K) according to [13] and [28]. Controversially, it is noticeable that the other studied samples had average radionuclide values higher than the average worldwide values. This may be ascribed to incorporation of radionuclides in the crystal structure of some accessory minerals, such as zircon, sphene and allanite [29]. On the other hand, the greater radium concentration is due to the modification of radioactive materials deposited within granite fractures. Furthermore, the leaching of uranium minerals (uranophane, uraninite, and betauranophane) through rainwater aids in their migration and precipitation along joints and faults [30].

Table 1. Activity concentrations of ^{226}Ra, ^{232}Th, ^{40}K (Bq/kg) and ^{232}Th/^{226}Ra ratio for the examined rocks.

Granite/Statistics Parameters	^{226}Ra	^{232}Th	^{40}K	^{232}Th/^{226}Ra
Black Aswan				
Mean	29.60	44.44	803.37	1.56
SD	6.41	4.04	160.62	0.40
Min	22.20	40.40	626.00	1.21
Max	33.30	48.48	959.00	2.00
Nero Aswan				
Mean	25.90	55.21	855.53	2.20
SD	6.41	6.17	95.62	0.52
Min	22.20	48.48	751.20	1.70
Max	33.30	60.60	959.00	2.73
White Halayb				
Mean	15.17	4.71	292.13	0.32
SD	5.25	1.17	41.68	0.04
Min	11.10	4.04	244.14	0.29
Max	21.09	6.06	319.26	0.36
Red Hurghada				
Mean	111.00	86.19	959.00	0.78
SD	11.10	6.17	93.90	0.08
Min	99.90	80.80	845.10	0.69
Max	122.10	92.92	1032.90	0.84
Red Aswan				
Mean	44.40	92.92	1042.29	1.92
SD	11.10	12.99	74.22	0.28
Min	33.30	68.68	898.31	1.60
Max	55.50	92.92	1042.29	2.09
Karnak				
Mean	55.50	46.46	616.61	0.85
SD	15.70	8.57	123.94	0.09
Min	44.40	40.40	528.97	0.79
Max	66.60	52.52	704.25	0.91
Verdi				
Mean	49.95	44.44	968.74	0.91
SD	7.85	5.71	104.02	0.26
Min	44.40	40.40	895.18	0.73
Max	55.50	48.48	1042.29	1.09
Worldwide (UNSCEAR, 2010)	32.00	45.00	412.00	

Assuming equilibrium between uranium and radium, the mean of the ^{232}Th/^{226}Ra ratio in the examined samples of magmatic rocks was less than the mean worldwide value of crustal rocks, 3.94 [15]. The lowest results contributed to the high Ra content relative to Th in the studied samples, reflecting Ra mobility relative to Th, especially in the secondary environment. This finding is linked to the slightly elevated uranium activity in the study area; radium content is derived from uranium-bearing granitic rocks [20].

Activity concentrations of ^{226}Ra, ^{232}Th, and ^{40}K in the examined granites were compared with others from Egypt and different countries (Table 2). The obtained results of radionuclide activity concentrations are observed to be higher than the values in Saudi Arabia and Oman, and lower than the other regions. The comparison shows how radioactivity varies from one country to the next, which is linked to differences in the geological structures of constituent rocks in each country.

Table 2. Comparison of ^{226}Ra, ^{232}Th and ^{40}K activity concentration in different areas.

Country	^{226}Ra	^{232}Th	^{40}K	References
Egypt	103–2047	12.4–101.2	831.6–1394.6	[31]
Egypt	165–27,851	71–274	1048–1230	[32]
Egypt	12.4–534.4	56.6–169.8	398–1113	[33]
Egypt	137	82	1082	[34]
Nigeria	63.29	226.67	832.59	[35]
Saudi Arabia	28.82	34.83	665.08	[36]
Saudi Arabia	11	22	641	[37]
Palestine	71	82	780	[38]
Oman	17	18	379	[39]
Iran	77.4	44.5	1017.2	[40]
Jordan	41.52	58.42	897	[41]

3.2. Radiation Hazard Assessment

The radiological hazard of the examined samples can be estimated using the obtained values of radionuclide (^{226}Ra, ^{232}Th and ^{40}K) concentration in order to determine their safe utilization as a decorative stone. The assessment indices include absorbed dose rate (D), annual effective dose (AED), radium equivalent activity (Ra_{eq}), and external ((H_{ex}) and internal (H_{in}) hazards.

3.2.1. Absorbed Dose Rate (D)

The rate of absorbed gamma dose can be calculated for the distribution of the unique radionuclide in the air at 1 m above the ground surface [42,43]:

$$D\ (\text{nGy/h}) = 0.462A_{Ra} + 0.604A_{Th} + 0.0417A_{K} \quad (1)$$

where A_{Ra}, A_{Th} and A_{K} are the specific activity concentrations of ^{226}Ra, ^{232}Th and ^{40}K in Bq/kg, respectively.

Among the calculated absorbed gamma dose rate, only the white Halayeb sample recorded the lowest value of 22.04 nGy/h (Figure 4), which lies within the safety limit (59 nGy/h) [42,44]. This shows that all granitic rocks except white Halayeb are unsuitable for infrastructure, especially building materials.

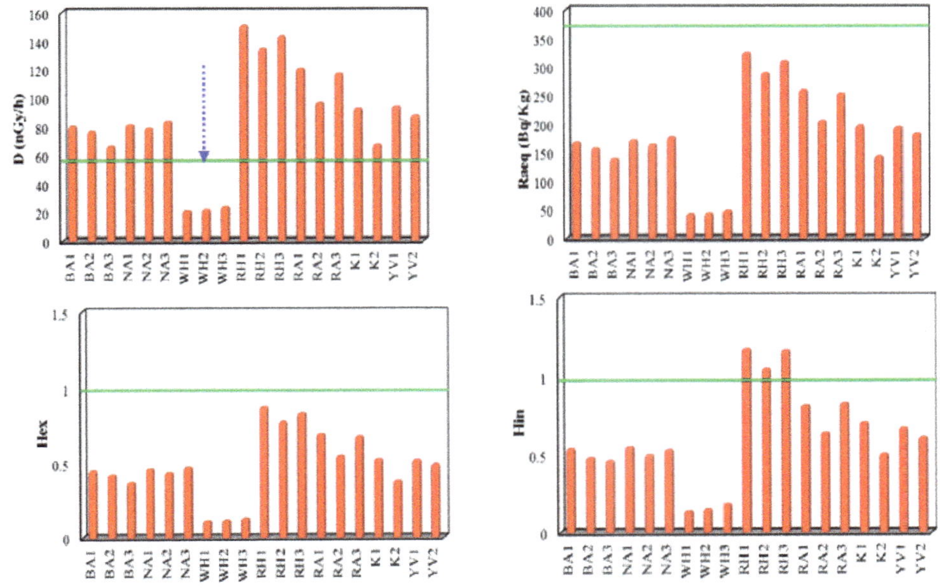

Figure 4. Absorbed dose rate (D), radium equivalent activity (Ra_{eq}), external (H_{ex}) and internal (H_{in}) hazards of the examined samples.

3.2.2. Annual Effective Dose (AED)

The annual effective dose (AED) was calculated from the absorbed dose by applying the dose conversion factor of 0.7 Sv/Gy and the outdoor occupancy factor of 0.2 [40]:

$$\text{AED (mSv/y)} = \text{D(nGy/h)} \times 8760\text{(h)} \times 0.2 \times 0.7\ \text{(Sv/Gy)} \times 10^{-6} \quad (2)$$

The average value of AED obtained from the white Halayeb samples was lower than the recommended limit (0.07 mSv y^{-1}, ref [39] (Figure 4), while the average of AED for the rest of granitic rocks exceeded the recommended limit. Heavy minerals found in granites, such as monazite, uraninite, and thorianite, can be responsible for high radiation exposure. Furthermore, long-term exposure to gamma radiation may have such negative health consequences as tissue degeneration, deoxyribonucleic acid (DNA) in genes, cancer, and cardiovascular disease. Refs [45–47].

3.2.3. Radium Equivalent Activity (Ra_{eq})

Activity levels of ^{226}Ra, ^{232}Th and ^{40}K in the examined samples can be determined using the radium equivalent activity (Ra_{eq}) index. This index can be calculated using Equation (3) [40]:

$$Ra_{eq}\ \text{(Bq/kg)} = A_{Ra} + 1.43A_{Th} + 0.077A_K \quad (3)$$

The average of the obtained values ranges from 36.10 Bq/kg in white Halayeb to 418 Bq/kg in red Hurghada. Furthermore, the constraints for all the calculated values are less than the maximum permissible value of 370 Bq/kg; [27], suggesting a negligible radium equivalent impact for the rocks (Figure 4). On the other hand, the highest values detected in the red Hurghada samples mean that it is not safe to utilize as an ornamental stone and building material.

3.2.4. External Hazard Index (H_{ex})

In order to assess the rate of radiation dose emitted from natural radionuclides in the examined samples, an external hazard index (H_{ex}) was applied using the following Equation [46]:

$$H_{ex} = A_{Ra}/370 + A_{Th}/259 + A_K/4810 \quad (4)$$

The mean calculated external hazard index ranged from 0.12 in white Halayeb to 0.83 in red Hurghada, which is less than unity and within the safety limit, reflecting a negligible radiation hazard (Figure 4).

3.2.5. Internal Hazard Index (H_{in})

Exposure to radon (^{222}Rn) and its radioactive daughters is detected by the internal hazard index (H_{in}), which can be used to measure impact on the respiratory organs and lungs [46]:

$$H_{in} = A_{Ra}/185 + A_{Th}/259 + A_K/4810 \quad (5)$$

All of the examined rocks had average internal hazard values lower than the limiting value of 1, except red Hurghada which had a value of 1.1. This reflects that the internal hazard values lie within the safety limit (Figure 4). Radiation hazard indices H_{ex} and H_{in}, with higher values, suggest a significant risk to human health. When granite is used as a construction material, external gamma rays and radon gas inhalation do not cause any radioactive health risks [42].

The present study discusses the link between petrographic studies and natural radioactivity. It is noticeable that there are strong associations between petrography (identifying the mineralogical constituents and textural relationships) and natural radioactivity. For example, White Halyeb is classified as tonalite, a type of older granite, and represents the oldest unit in this study. It consists mainly of plagioclase, quartz, biotite, and a minor amount of muscovite. A lack of accessory minerals (e.g., Allanite, monazite, titanite, thorite, and zircon) yields a low content of activity concentrations of natural radioactivity. Therefore, White Halyeb is more suitable as a decorative ornamental stone.

4. Conclusions

Nineteen samples of different rocks representing seven magmatic rocks in named commercial granitic classes were examined for their mineralogical, petrographic and radioactive compositions in this study. Mean activity concentrations of ^{226}Ra, ^{232}Th, and ^{40}K in all of the examined rocks except the White Halyeb exceeded the limiting range suggested by the UNSCEAR. Most of the calculated radiological hazard parameters suggest that White Halyeb is the most suitable rock for use as an ornamental stone or building materials, as it possesses a low level of natural radioactivity, and hence poses little or no radiation risk to human health.

Author Contributions: E.S.R.L. and M.Y.H. contributed equally to this work. Conceptualization, E.S.R.L., M.A.R. and M.Y.H.; methodology, E.S.R.L., M.A.R. and M.Y.H.; software, E.S.R.L. and M.A.R. validation, M.U.K., H.O. and S.A.; formal analysis, E.S.R.L., M.A.R. and M.Y.H.; investigation, M.U.K. and S.A.; resources, H.O. and S.A.; data curation, E.S.R.L., M.A.R. and M.Y.H.; writing—original draft preparation, E.S.R.L. and M.Y.H.; writing—review and editing, E.S.R.L., M.A.R. and M.Y.H.; visualization, H.O. and S.A.; supervision, M.U.K.; project administration, H.O. and S.A.; funding acquisition, H.O. and S.A. All authors have read and agreed to the published version of the manuscript.

Funding: We deeply acknowledge Taif University for supporting the researchers through Taif University Researchers Supporting Project number (TURSP-2020/287), Taif University, Taif, Saudi Arabia.

Institutional Review Board Statement: Not applicable.

Informed Consent Statement: Not applicable.

Data Availability Statement: Data sharing is not applicable for this article.

Acknowledgments: All authors would like to thank Nilly Kawady (Nuclear Material Authority) for her help in measuring the radionuclides. Authors also deeply acknowledge Taif University for supporting the researchers through Taif University Researchers Supporting Project number (TURSP-2020/287), Taif University, Taif, Saudi Arabia.

Conflicts of Interest: The authors declare no conflict of interest.

References

1. Gaafar, I.M.; Abdrabboh, A.; Alshami, A.S. Gamma-ray spectrometric investigation of north El-Tor area, southwestern. *NRIAG J. Astron. Geophys.* **2018**, *7*, 390–398. [CrossRef]
2. Rashwan, M.A.; Lasheen, E.S.R.; Shalaby, B.N. Incorporation of metagabbro as cement replacement in cement-based materials: A role of mafic minerals on the physico-mechanical and durability properties. *Constr. Build Mater.* **2019**, *210*, 256–268. [CrossRef]
3. Hanfi, M.Y.; Emad, B.M.; Sayyed, M.I.; Khandaker, M.U.; Bradley, D.A. Natural radioactivity in the prospecting tunnel in Egypt: Dose rate and risk assessment. *Radiat. Phys. Chem.* **2021**, *187*, 109555. [CrossRef]
4. Stern, R.J.; Ali, K.A.; Liegeois, J.P.; Johnson, P.R.; Kozdroj, W.; Kattan, F.H. Distribution and significance of pre-Neoproterozoic zircons in juvenile Neoproterozoic igneous rocks of the Arabian-Nubian Shield. *Am. J. Sci.* **2010**, *310*, 791–811. [CrossRef]
5. Costa, F.P.d.; Fernandes, J.V.; Melo, L.R.L.d.; Rodrigues, A.M.; Menezes, R.R.; Neves, G.d.A. The Potential for Natural Stones from Northeastern Brazil to Be Used in Civil Construction. *Minerals* **2021**, *11*, 440. [CrossRef]
6. El-Gamal, H.; Sidique, E.; El-Haddad, M. Spatial Distributions and Risk Assessment of the Natural Radionuclides in the Granitic Rocks from the Eastern Desert, Egypt. *Minerals* **2019**, *9*, 386. [CrossRef]
7. Mashaly, A.O.; El-Kaliouby, B.A.; Shalaby, B.N.; El–Gohary, A.M.; Rashwan, M.A. Effects of marble sludge incorporation on the properties of cement composites and concrete paving blocks. *J. Clean. Prod.* **2016**, *112*, 731–741. [CrossRef]
8. El Mezayen, A.M.; Heikal, M.A.; Abu Zeid, I.K.; Omar, S.A.; El-Feky, M.G.; Lasheen, E.S.R. Petrography, geochemistry and radioactivity of El-Gidami granitic Rocks, Central Eastern Desert, Egypt. *Al-Azhar Bullet. Sci. Confer.* **2017**, *9*, 25–40.
9. El Mezayen, A.M.; Heikal, M.A.; El-Feky, M.G.; Shahin, H.A.; Zeid, I.A.; Lasheen, S.R. Petrology, geochemistry, radioactivity, and M–W type rare earth element tetrads of El Sela altered granites, south eastern desert, Egypt. *Acta Geochim.* **2019**, *38*, 95–119. [CrossRef]
10. Jaupart, C.; Mareschal, J.C.; Bouquerel, H.; Phaneuf, C. The building and stabilization of an Archean Craton in the Superior Province, Canada, from a heat flow perspective. *J. Geophys. Res. Solid Earth* **2014**, *119*, 9130–9155. [CrossRef]
11. Bea, F. Residence of REE, Y5 Th and U in Granites and Grustal Protoliths; Implications for the Chemistry of Crustal Melts. *J. Petrol.* **1996**, *37*, 521–552. [CrossRef]
12. Abdel-Razek, Y.A.; Masoud, M.S.; Hanafi, M.Y.; El-Nagdy, M.S. Study of the parameters affecting radon gas flux from the stream sediments at Seila area Southeastern desert, Egypt. *Environ. Earth Sci.* **2015**, *73*, 8035–8044. [CrossRef]
13. Maithani, P.B.; Srinivasan, S. Felsic volcanic rocks, a potential source of uranium—An Indian overview. *Energy Procedia* **2011**, *7*, 163–168. [CrossRef]
14. Mudd, G.M. Radon releases from Australian uranium mining and milling projects: Assessing the UNSCEAR approach. *J. Environ. Radioact* **2008**, *99*, 288–315. [CrossRef] [PubMed]
15. Hanfi, M.Y.M.; Masoud, M.S.; Sayyed, M.I.; Khandaker, M.U.; Faruque, M.R.I.; Bradley, D.A.; Mostafa, M.Y.A. The presence of radioactive heavy minerals in prospecting trenches and concomitant occupational exposure. *PLoS ONE* **2021**, *16*, e0249329. [CrossRef]
16. Hanfi, M.Y.; Yarmoshenko, V.; Seleznev, A.A.; Malinovsky, G.; Ilgasheva, E.; Zhukovsky, M.V. Beta radioactivity of urban surface-deposited sediment in three Russian cities. *Environ. Sci. Pollut. Res.* **2020**, *27*, 40309–40315. [CrossRef] [PubMed]
17. Cumberland, S.A.; Douglas, G.; Grice, K.; Moreau, J.W. Uranium mobility in organic matter-rich sediments: A review of geological and geochemical processes. *Earth Sci. Rev.* **2016**, *159*, 160–185. [CrossRef]
18. Xinwei, L.; Lingqing, W.; Xiaodan, J. Radiometric analysis of Chinese commercial granites. *J. Radional. Nucl. Chem.* **2006**, *267*, 669–673. [CrossRef]
19. Zoheir, B.; Goldfarb, R.; Holzheid, A.; Helmy, H.; El Sheikh, A. Geochemical and geochronological characteristics of the Um Rus granite intrusion and associated gold deposit, Eastern Desert, Egypt. *Geosci. Front.* **2020**, *11*, 325–345. [CrossRef]
20. Gaafar, I.; Hanfi, M.; El-Ahll, L.S.; Zeidan, I.Z. Assessment of radiation hazards from phosphate rocks, Sibaiya area, central eastern desert, Egypt. *Appl. Radiat. Isot.* **2021**, *173*, 109734. [CrossRef]
21. El Mezayen, A.M.; Heikal, M.A.; Omar, S.A.; El Feky, M.G.; Lasheen, E.S. Petrology, Geochemistry and fractional modelling of El Gidami Neoproterozoic granitic rocks, Central Eastern Desert, Egypt. *Nat. Sci.* **2015**, *13*, 102–114.
22. Lasheen, E.S.R.; Saleh, G.M.; Khaleal, F.M.; Alwetaishi, M. Petrogenesis of Neoproterozoic Ultramafic Rocks, Wadi Ibib—Wadi Shani, South Eastern Desert, Egypt: Constraints from Whole Rock and Mineral Chemistry. *Appl. Sci.* **2021**, *11*, 10524. [CrossRef]
23. Yassin, A.A.; Masoud, S.M.; Moamed, Y.H.; Mohamed, S.E. Effective radiation doses from natural sources at Seila area South Eastern Desert, Egypt. *J. Taibah Univ. Sci.* **2016**, *10*, 271–280. [CrossRef]
24. Arunima, S.; Lekshmi, R.; Jojo, P.J.; Khandaker, M.U. A study on leaching of primordial radionuclides 232Th and 40K to water bodies. *Radiat. Phys. Chem.* **2021**, *188*, 109658. [CrossRef]
25. Monica, S.; Jojo, P.J.; Khandaker, M.U. Radionuclide concentrations in medicinal florae and committed effective dose through Ayurvedic medicines. *Int. J. Radiat. Biol.* **2020**, *96*, 1028–1037. [CrossRef]

26. Iqbal, M.; Tufail, M.; Mirza, S.M. Measurement of natural radioactivity in marble found in Pakistan using a NaI(Tl) gamma-ray spectrometer. *J. Environ. Radioact.* **2000**, *51*, 255–265. [CrossRef]
27. Papadopoulos, A.; Christofides, G.; Koroneos, A.; Papadopoulou, L.; Papastefanou, C.; Stoulos, S. Natural radioactivity and radiation index of the major plutonic bodies in Greece. *J. Environ. Radioact.* **2013**, *124*, 227–238. [CrossRef]
28. UNSCEAR. *Sources and Effects of ionizing Radiation—Exposures of The Public and Workers from Various Sources of Radiation—UNSCEAR 2008 Report*; United Nations Publication: New York, NY, USA, 2010.
29. Pavlidou, S.; Koroneos, A.; Papastefanou, C. Natural radioactivity of granites used as building materials. *J. Environ. Radioact.* **2006**, *89*, 48–60. [CrossRef] [PubMed]
30. Nagar, M.S.; Bayoumi, B.M.; Morsy, W.M. Characteristics and Evaluation of Leaching Behavior of Uranium Mineralization in Qash Amir Granite, South Eastern Desert, Egypt. *Am. J. Appl. Ind. Chem.* **2021**, *5*, 7–16. [CrossRef]
31. Nagar, M.S.; Shahin, H.A.; Bahige, M. Column Percolation Leaching of Uranium from El-Sela Area, South Eastern Desert, Egypt. *Res. Rev. J. Chem.* **2016**, *5*, 32–41.
32. El-Shershaby, A. Study of radioactivity levels in granite of Gable Gattar II in the north eastern desert of Egypt. *Appl. Radiat. Isot.* **2002**, *57*, 131–135. [CrossRef]
33. Awad, H.A.; Zakaly, H.M.; Nastavkin, A.V.; El Tohamy, A.M.; El-Taher, A. Radioactive mineralizations on granitic rocks and silica veins on shear zone of El-Missikat area, Central Eastern Desert, Egypt. *Appl. Radiat. Isot.* **2021**, *168*, 109493. [CrossRef]
34. Amin, R.M. Gamma radiation measurements of naturally occurring radioactive samples from commercial Egyptian granites. *Environ. Earth Sci.* **2012**, *67*, 771–775. [CrossRef]
35. Akpanowo, M.A.; Umaru, I.; Iyakwari, S.; Joshua, E.O.; Yusuf, S.; Ekong, G.B. Determination of natural radioactivity levels and radiological hazards in environmental samples from artisanal mining sites of Anka, North-West Nigeria. *Sci. Afr.* **2020**, *10*, e00561. [CrossRef]
36. AlZahrani, J.H.; Alharbi, W.R.; Abbady, A.G.E. Radiological impacts of natural radioactivity and heat generation by radioactive decay of phosphorite deposits from Northwestern Saudi Arabia. *Aust. J. Basic Appl.* **2011**, *5*, 683–690.
37. Al-Trabulsy, H.A.; Khater, A.E.M.; Habbani, F.I. Radioactivity levels and radiological hazard indices at the Saudi coastline of the Gulf of Aqaba. *Radiat. Phys. Chem.* **2011**, *80*, 343–348. [CrossRef]
38. Thabayneh, K.M. Measurement of Natural Radioactivity and Radon Exhalation Rate in Granite Samples Used in Palestinian Buildings. *Arab. J. Sci. Eng.* **2013**, *38*, 201–207. [CrossRef]
39. Zare, M.R.; Mostajaboddavati, M.; Kamali, M.; Abdi, M.R.; Mortazavi, M.S. ^{235}U, ^{238}U, ^{232}Th, ^{40}K and ^{137}Cs activity concentrations in marine sediments along the northern coast of Oman Sea using high-resolution gamma-ray spectrometry. *Mar. Pollut. Bull.* **2012**, *64*, 1956–1961. [CrossRef]
40. Abbasi, A. Calculation of gamma radiation dose rate and radon concentration due to granites used as building materials in Iran. *Radiat. Prot. Dosim.* **2013**, *155*, 335–342. [CrossRef]
41. Sharaf, J.M.; Hamideen, M.S. Measurement of natural radioactivity in Jordanian building materials and their contribution to the public indoor gamma dose rate. *Appl. Radiat. Isot.* **2013**, *80*, 61–66. [CrossRef]
42. UNSCEAR. Exposures from natural radiation sources (Annex B). In *Sources and Effects of Ionizing Radiation*; UNSCEAR: New York, NY, USA, 2000; pp. 84–141.
43. Hanfi, M.Y.; Masoud, M.S.; Ambrosino, F.; Mostafa, M.Y. Natural radiological characterization at the Gabal El Seila region (Egypt). *Appl. Radiat. Isot.* **2021**, *173*, 109705. [CrossRef]
44. Md Sirajul Islam, A.A.; Hossain Miah, M.M.; Ahmed, M.; Hossain, S.; Khandaker, M.U. The presence of terrestrial radionuclides in the Karnaphuli and Halda river sediments and concomitant hazards to the dwellers. *Int. J. Environ. Anal. Chem.* **2021**. [CrossRef]
45. Sivakumar, S.; Chandrasekaran, A.; Senthilkumar, G.; Gandhi, M.S.; Ravisankar, R. Determination of radioactivity levels and associated hazards of coastal sediment from south-east coast of Tamil Nadu with a statistical approach. *Iran. J. Sci. Technol. Trans. A Sci.* **2018**, *42*, 601–614. [CrossRef]
46. Amatullah, S.; Rahman, R.; Ferdous, J.; Siraz, M.M.M.; Khandaker, M.U.; Mahal, S.F. Assessment of radiometric standard and potential health risks from building materials used in Bangladeshi dwellings. *Int. J. Environ. Anal. Chem.* **2021**. [CrossRef]
47. Karim, M.A.; Gafaar, I.; El-Halim, E.A.; Hanfi, M.; El-Dine, N.W. Natural radioactivity and radiological implications of granite rocks, El-Sela area, Southeastern Desert, Egypt. *J. Radioanal. Nucl. Chem.* **2021**, *330*, 707–720. [CrossRef]

materials

Article

A Novel Process to Recover Gypsum from Phosphogypsum

Junhui Xiao [1,2,3,4,*], Tao Lu [2], Yuanfa Zhuang [2] and Huang Jin [2]

1 Sichuan Provincial Engineering Lab of Non-Metallic Mineral Powder Modification and High-Value Utilization, Southwest University of Science and Technology, Mianyang 621010, China
2 Postdoctoral Research Station, Dongfang Boiler Group Co., Ltd., Chengdu 611731, China; lut@dbc.com.cn (T.L.); zhuangyf@dbc.com (Y.Z.); jinh@dbc.com (H.J.)
3 Key Laboratory of Ministry of Education for Solid Waste Treatment and Resource Recycle, Southwest University of Science and Technology, Mianyang 621010, China
4 Institute of Multipurpose Utilization of Mineral Resources, Chinese Academy of Geological Sciences, Chengdu 610041, China
* Correspondence: xiaojunhui33@163.com; Tel.: +86-139-9019-0544

Abstract: In this study, we investigated a coarse phosphogypsum containing 49.63% SO_3, 41.41% CaO, 10.68%, 4.47% SiO_2, 1.28% P_2O_5, 0.11% F, $CaSO_4 \cdot 2H_2O$ purity of 80.65%, and whiteness of 27.68. Phosphogypsum contains calcium sulfate dehydrate as the main mineral, with small amounts of brushite, quartz, muscovite, and zoisite. Harmful elements, such as silicon, phosphorus, and fluorine, are mainly concentrated in the +0.15 mm and −0.025 mm fraction, which can be pre-selected and removed by the grading method to further increase the $CaSO_4 \cdot 2H_2O$ content. Gypsum was recovered using a direct flotation method, which included one roughing, one scavenging, and two cleaning operations, from −0.15 mm to +0.025 mm. The test results show that a gypsum concentrate with a $CaSO_4 \cdot 2H_2O$ purity of 98.94%, $CaSO_4 \cdot 2H_2O$ recovery of 80.02%, and whiteness of 37.05 was achieved. The main mineral in the gypsum concentrate was gypsum, and limited amounts of muscovite and zoisite entered the gypsum concentrate because of the mechanical entrainment of the flotation process.

Keywords: phosphogypsum; gypsum; classification; flotation

Citation: Xiao, J.; Lu, T.; Zhuang, Y.; Jin, H. A Novel Process to Recover Gypsum from Phosphogypsum. *Materials* **2022**, *15*, 1944. https://doi.org/10.3390/ma15051944

Academic Editor: Petrica Vizureanu

Received: 13 February 2022
Accepted: 3 March 2022
Published: 5 March 2022

1. Introduction

Phosphogypsum is mainly obtained from the phosphate fertilizer industry and is a solid waste residue produced by a wet preparation of phosphoric acid. The global cumulative emissions of phosphogypsum total approximately 6 billion tons, which is increasing at a rate of 150 million tons/year. It is projected that the total amount of phosphogypsum will double by 2025–2045 [1,2]. Phosphogypsum is an industrial by-product of gypsum with a low utilization rate owing to its large discharge, complex impurity composition, and difficult treatment. The accumulation of large amounts of phosphogypsum can cause environmental risks. Therefore, effective treatment and efficient utilization of phosphogypsum are urgently needed. The main impurities in phosphogypsum are divided into three categories: phosphorus, fluorine, and organic impurities. Phosphorus impurities in phosphogypsum mainly consist of soluble phosphorus (H_3PO_4, H_2PO^{4-}, and $HPO_4{}^{2-}$), insoluble phosphorus ($Ca_3(PO_4)_2$), and eutectic phosphorus ($CaHPO_4 \cdot 2H_2O$). The fluorine-containing impurities in phosphogypsum mainly consist of soluble fluorine (NaF and KF) and insoluble fluorine (Na_3AlF_6, $CaSiF_6$, and CaF_2). Organic impurities in phosphogypsum mainly consist of organic matter inherent in phosphate rock and organic additives added in the production process [3–5]. The comprehensive utilization rate of phosphogypsum is low, owing to the various complex impurities and the difficult occurrence state. Presently, the main separation and purification methods for phosphogypsum are as follows.

Phosphorus and fluorine in phosphogypsum are largely soluble in water. Owing to their solubility in water, they can be effectively removed by rinsing, filtering, leaching, and dehydration. However, the single water-washing process consumes a large amount of water and a high amount of energy. The newly generated wastewater causes secondary pollution, and the wastewater needs to be treated separately to meet the national discharge standards before discharge, which significantly increases the treatment cost. Therefore, this process has not been applied on a large scale. Moreover, it is necessary to realize stepwise water recycling through technological improvements, reduce water consumption, and recycle wastewater-soluble impurities, such as phosphorus and fluorine, in green and low-cost ways [6–8].

Organic impurities can be removed by conventional flotation. This process includes pouring phosphogypsum and water into flotation equipment in the right proportions and using the natural floating of organic impurities to scrape off the impurities. This process is suitable for treating phosphogypsum with a high organic content and can improve the whiteness of phosphogypsum; however, this method has low efficiency and no significant removal effect on soluble impurities. Due to the fact that the water used in the flotation process can be recycled, it is often combined with the water-washing process. Presently, the purification of phosphogypsum by adding a flotation agent has also been widely studied. This treatment method involves the addition of alkaline-modified materials, such as quick lime, into phosphogypsum; the alkaline-modified material reacts with the soluble phosphorus and soluble fluorine and converts the refractory inert materials to precipitates. This method can homogenize phosphogypsum with high-quality fluctuations and low organic matter content. Lime neutralization treatment is widely used in the production of cement retarders because of its simple process, low investment, obvious effect, and low amount of secondary pollution. However, this method can only temporarily solve the harmful effects of soluble phosphorus and fluorine. Soluble phosphorus and fluorine precipitate after a long time, and this method cannot remove the adverse effects of organic matter on phosphogypsum. In the process of calcination at 800 °C, the P_2O_5 in phosphogypsum is converted into stable inert phosphate, and a small amount of organophosphorus and hydrogen fluoride can be removed by volatilization. Unlike other calcination processes, flash sintering aims to transform soluble phosphorus and fluorine into inert substances without water washing. The flash-burning method is usually combined with the lime neutralization method to avoid the volatilization of fluoride, which pollutes the environment and causes secondary pollution; however, this method has high energy consumption, large initial investment, and its equipment is prone to corrosion [9–13].

In summary, extensive research on phosphogypsum pretreatment has been performed by scientists and technicians. However, each method has some disadvantages, and there are still some gaps in large-scale industrial applications. Thus, it is necessary to combine various pretreatment methods to learn from each of them, which is an important research direction in the future. In addition, the authors believe that flotation is a promising method. Organic impurities surfaced by reverse flotation, $CaSO_4 \cdot 2H_2O$ surfaced by forward flotation, and then a small amount of lime phosphorus fixation and fluorine were added to the forward-flotation-filtered concentrate, such that phosphogypsum could be completely purified at a low cost. However, further experiments are required to verify this hypothesis.

2. Materials and Methods

2.1. Materials

The test samples used in this study were taken from phosphogypsum produced by a phosphorus chemical enterprise in the Deyang Region, Sichuan Province, China. The content of $CaSO_4 \cdot 2H_2O$ in the phosphogypsum was 80.65%, and the whiteness was 27.68. The water content of the phosphogypsum was less than 5%, and the particle size was less than 1 mm. The main chemical compositions of the samples are shown in Table 1.

Table 1. Main chemical composition analysis of phosphogypsum (mass fraction, %).

SO_3	CaO	SiO_2	P_2O_5	Al_2O_3	Fe_2O_3	K_2O	SrO	MgO	F	TiO_2	BaO	Na_2O	Y_2O_3
49.62	41.41	5.47	1.28	0.63	0.31	0.20	0.12	0.12	0.11	0.10	0.07	0.05	0.01

The main chemical reagents used in this test were sulfuric acid, sodium silicate, sodium hexametaphosphate, starch, carboxymethylcellulose, dodecylamine, mixed amine, octadecylamine, dodecyltrimethylammonium chloride, and pine oil. All reagents were of analytical grade and were obtained from a producing area in Tianjin Tianli Chemical Reagent Co., Ltd., Tianjin, China.

2.2. Experiment

Flotation (roughing, scavenging) aimed at improving the recovery of gypsum concentrate was performed using an XFD-1.5 L hanging tank flotation machine (Jinlin Exploration Machinery Plant, Changchun, China) operating at a spindle speed of 1650 r/min. A 500 g mass of phosphogypsum was added to the 1.5 L flotation tank. Flotation (cleaning) aimed at further increasing the purity and whiteness of the gypsum concentrate was performed using an XFD-1.0 L hanging tank flotation machine (Jinlin Exploration Machinery Plant, Changchun, China) operating at a spindle speed of 1,650 r/min. A 300 g mass of gypsum concentrate was added to the 1.0 L flotation tank. Distilled water (1.0 L) was added, and the pulp was stirred and mixed for 3 min, followed by adjustment to the required pH using sulfuric acid. After 5 min of pulping, the depressant was added to the slurry and conditioned for 3 min. The collectors were then added and agitated for 3 min. Before aeration, frothers (pine oil) were added to improve the bubbles and stirred for an additional 3 min. After 3 min of flotation, the froth (gypsum concentrate) and in-tank product (flotation tailings) were separately filtered, the samples were filtered, dried at 40 °C, weighed, and the gypsum recovery was calculated according to Equation (1).

$$R = (Q_1 \times G_1)/(Q_0 \times G_0), \tag{1}$$

where R is the recovery of gypsum (%), Q_1 is the weight of the flotation concentrate (g), G_1 is the calcium sulfate dehydrate of the flotation concentrate, Q_0 is the weight of phosphogypsum (g), and G_0 is the calcium sulfate dehydrate of phosphogypsum (g).

2.3. Analyses

The whiteness of the gypsum concentrate samples was measured using a WSD-3C whiteness instrument manufactured by Beijing Kangguang Optical Instrument Co., Ltd. The equipment was preheated for 30 min before the test, and calibration was subsequently performed using a standard colorimetric plate. After calibration, the samples were tested. The purity of the calcium sulfate dihydrate in the phosphogypsum was determined by the content of crystallized water in the gypsum. The mass fraction of calcium sulfate dihydrate (G) in the sample was calculated using Equation (1). Gypsum recovery was calculated using Equation (2):

$$G = 4.7885 \times H, \tag{2}$$

where G is the purity of calcium sulfate dihydrate (%), and 4.7785 is the coefficient of calcium sulfate dehydrate content converted from crystal water content, which is equal to the molecular weight of the calcium sulfate dihydrate divided by the molecular weight of two water molecules. H represents the crystal water content (%).

The chemical compositions of the solid materials (including phosphogypsum, gypsum concentrate, and flotation tailings) were analyzed using a Z–2000 atomic absorption spectrophotometer (Hitachi Co., Ltd., Tokyo, Japan). The mineral phase compositions of the aforementioned solid substances were examined using an XRD (X' Pert Pro, Panaco, The Netherlands). The microstructures of the solid products were investigated using scanning electron microscopy (SEM, S440, Hirschmann Laborgerate GmbH & Co. KG, Eberstadt,

Germany) equipped with an energy-dispersive X-ray spectroscopy (EDS) detector (Ultra55, CarlzeissNTS GmbH, Hirschmann Laborgerate GmbH & Co. KG, Eberstadt, Germany).

3. Results

3.1. Process Mineralogical Analysis of Phosphogypsum

Figure 1a shows the X-ray diffraction (XRD) patterns of the samples. Phosphogypsum mainly contains calcium sulfate dehydrate ($CaSO_4 \cdot 2H_2O$), along with small amounts of $CaPO_3(OH) \cdot 2H_2O$, SiO_2, $Kal_2SiO_{10}(OH)$, and $Ca_2Al_3[SiO_{12}]OH$. This indicates that the crystalline phase of phosphogypsum is mainly gypsum, and the main mineral impurities are brushite and silicaluminate minerals, which need to be removed in the subsequent purification. The results in Table 2 and Figure 2b show that, compared with the content of SiO_2, F, and P_2O_5 in different fractions, those in the +0.15 mm fraction were 12.89%, 0.41%, and 5.08%, respectively, whereas those in the −0.025 mm fraction were 13.17%, 0.35%, and 1.56%, respectively. This further indicates that the impurity elements, i.e., silicon, fluorine, and phosphorus, are mainly concentrated in the +0.15 mm and −0.025 mm fractions. Therefore, in this study, the +0.15 mm and −0.025 mm products with two particle sizes were pre-selected by a classification method, and gypsum was further recovered from the −0.15 and +0.025 mm raw materials by flotation.

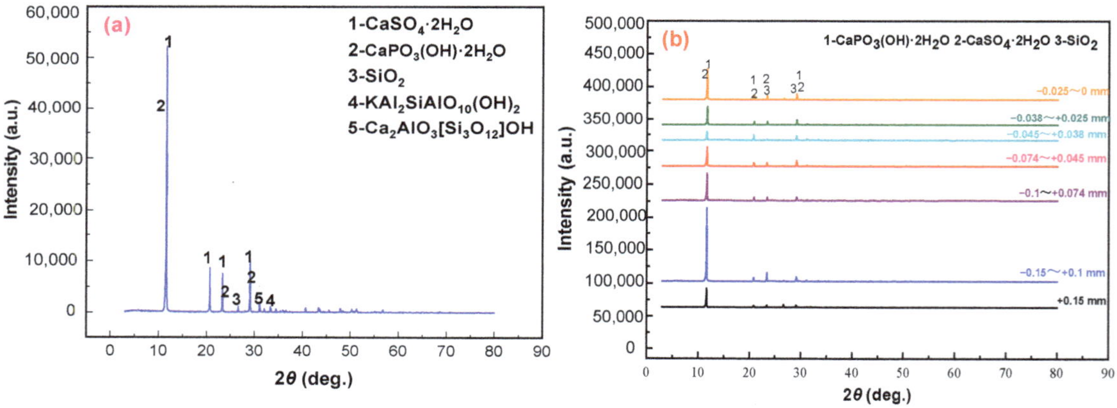

Figure 1. XRD patterns of the phosphogypsum (**a**) and the different grain-sized products (**b**).

Table 2. Grain-sized analysis of phosphogypsum (mass fraction, %).

Fraction (mm)	Yield	SO_3	CaO	SiO_2	P_2O_5	Al_2O_3	Fe_2O_3	K_2O	MgO	F
+0.15	7.60	41.17	38.35	12.89	5.08	1.00	0.57	0.34	0.10	0.41
−0.15~+0.1	11.21	51.89	43.23	2.93	1.15	0.28	0.21	0.08	0.02	0.004
−0.1~0.074	19.01	51.76	44.24	2.25	1.01	0.25	0.18	0.06	0.06	0.005
−0.074~0.045	14.23	52.08	44.09	2.15	1.03	0.19	0.26	0.05	0.02	0.006
−0.045~+0.038	17.35	51.86	43.93	2.34	1.03	0.26	0.28	0.05	0.02	0.005
−0.038~+0.025	10.26	51.09	43.39	3.24	1.16	0.40	0.35	0.08	0.03	0.003
−0.025~0	20.34	44.81	37.28	13.17	1.56	1.36	0.87	0.39	0.12	0.35
Totals	100.00	49.55	42.10	5.46	1.47	0.54	0.39	0.15	0.06	0.11

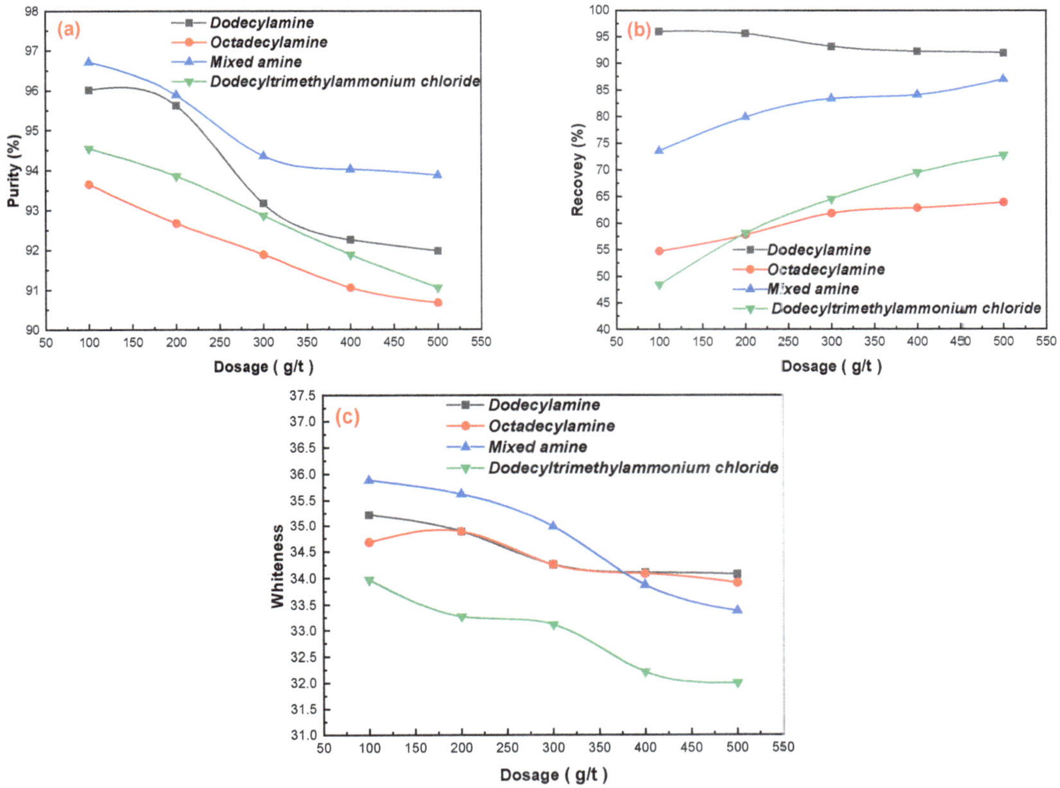

Figure 2. Effects of dosages of dodecylamine, octadecylamine, mixed amine, and dodecyltrimethylammonium chloride on gypsum purification: (**a**) gypsum purity, (**b**) gypsum recovery, and (**c**) gypsum whiteness.

3.2. *Effects of Flotation Flowsheet Parameters on Gypsum Purification*

3.2.1. Collectors Dosages

A study on the surface properties of gypsum showed that the gypsum surface was charged, and the cationic collector was easily adsorbed onto its surface. Amine flotation agents such as dodecylamine and quaternary ammonium salt are common cationic collectors that are widely used in numerous mineral separations [14–17]. The effects of dodecylamine, octadecylamine, mixed amine, and dodecyltrimethylammonium chloride on the separation and purification of gypsum were investigated under the following test conditions: flotation concentration of 25%, flotation pulp pH = 2.5 using sulfuric acid, a pine dosage of 40 g/t; the results are shown in Figure 2a–c). Using the same amount of collector, the order of collecting abilities of gypsum was found to be: mixed amine > dodecylamine > dodecyltrimethylammonium chloride > octadecylamine. With an increase in the dosage of MA, the purity of $CaSO_4{\cdot}2H_2O$ in the concentrate decreased gradually, and the recovery of gypsum increased gradually. When the dosage of mixed amine reached 200 g/t and continued to increase, the purity of $CaSO_4{\cdot}2H_2O$ and whiteness of the concentrate decreased significantly, mainly owing to the increase in the dosage of the reagents, resulting in an increased floatation of the concentrate and the easy inclusion of other impurities. Regarding the cost of the mixing of reagents, a mixed amine dosage of 200 g/t is suitable for flotation. The purity of $CaSO_4{\cdot}2H_2O$ in the concentrate was 95.89%, the recovery of $CaSO_4{\cdot}2H_2O$ was 79.93%, and the whiteness was 35.63.

3.2.2. Depressants Dosages

Sodium silicate, sodium hexametaphosphate, starch, and carboxymethylcellulose were used as flotation agents for gangue minerals [18–20], and the effects of different dosages on gypsum flotation were investigated under the following test conditions: flotation concentration of 25%, flotation pulp pH = 2.5 using sulfuric acid, mixed amine dosage of 200 g/t, and pine dosage of 40 g/t. The results in Figure 3a–c show that sodium silicate, sodium hexametaphosphate, and carboxymethylcellulose influence gangue minerals. However, the addition of ST can reduce the purity of $CaSO_4{\cdot}2H_2O$, the recovery rate of $CaSO_4{\cdot}2H_2O$, and the whiteness of the concentrate, which further indicates that the inhibitory effect of the four inhibitors on gangue minerals is as follows: sodium silicate > carboxymethylcellulose > sodium hexametaphosphate > starch. Sodium silicate can be selectively adsorbed onto the surface of gangue minerals, this results in an increase in the hydrophilicity of gangue minerals and changes to the surface electrical properties of gangue minerals. Thus, the gangue minerals and gypsum surfaces are negative, increasing the electrostatic repulsion between gypsum and gangue minerals, reducing the agglomeration of mineral particles, and improving the dispersion of the flotation system [21–23]. An increase in the dosage of sodium silicate improved the quality of the concentrate; however, when the dosage of sodium silicate exceeded 400 g/t, the purity of $CaSO_4{\cdot}2H_2O$ and the whiteness of the concentrate decreased, and the recovery of $CaSO_4{\cdot}2H_2O$ increased to a certain extent. Therefore, a sodium silicate dosage of 400 g/t is ideal, and the purity of $CaSO_4{\cdot}2H_2O$ in the concentrate increased to 96.38%, the whiteness increased to 36.63, and the recovery of $CaSO_4{\cdot}2H_2O$ increased to 81.51%.

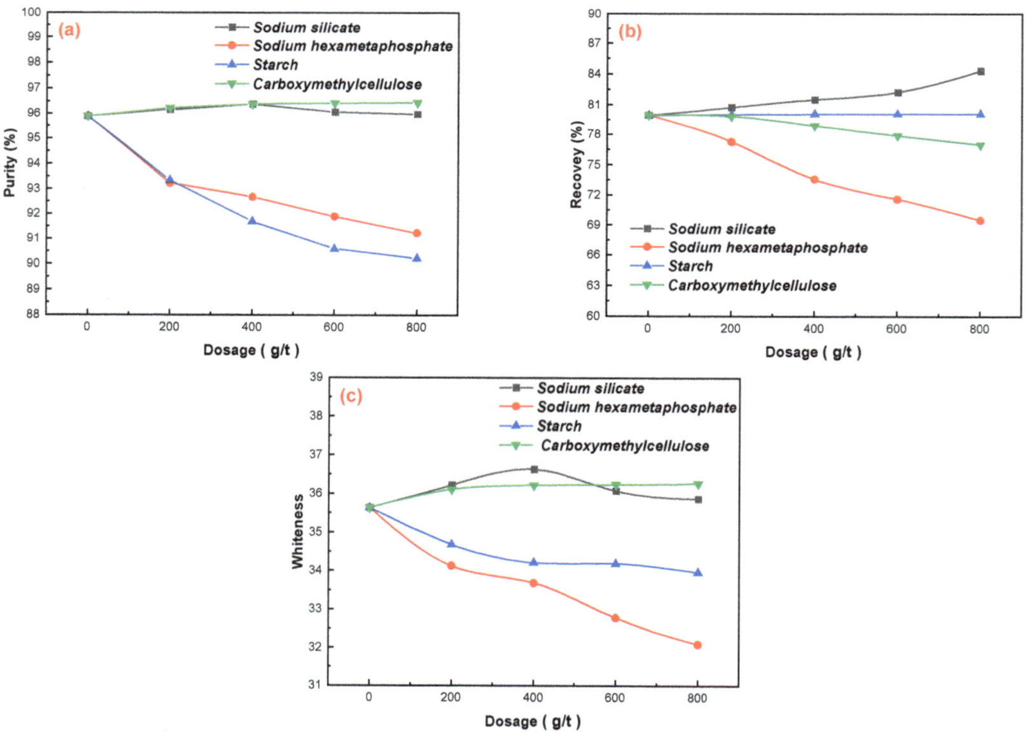

Figure 3. Effects of the dosages of sodium silicate, sodium hexametaphosphate, starch, and carboxymethylcellulose on gypsum purification: (**a**) gypsum purity, (**b**) gypsum recovery, and (**c**) gypsum whiteness.

3.2.3. pH Value of Pulp

Different flotation pH values of the pulp tests were conducted under the following conditions: flotation concentration of 25%, mixed amine dosage of 200 g/t, sodium silicate dosage of 400 g/t, and pine dosage of 40 g/t; the results are shown in Figure 4a–c. When the pH of the pulp is low (pH = 1.5–2.0), the separation of gypsum and quartz is different, and the separation effect of gypsum and quartz is better. When the pH of the pulp is higher (pH > 2.0), the difference in separation between gypsum and quartz becomes small, and the separation effect of gypsum and quartz is poor. This is because the isoelectric point of gypsum is pH = 1–2, and that of quartz is pH = 2.3–3.0. When the pH of the slurry was higher than 2.3, both the gypsum and the quartz minerals were negatively charged on the surface, which could be adsorbed by the cationic collector and showed good floatability; thus, they could not be separated. When the pulp pH is $\leq$2, the gypsum mineral surface with a negative charge can exhibit cationic collector electrostatic adsorption, whereas a quartz surface with positive charge or no charge cannot exhibit cationic collector adsorption, to achieve gypsum and quartz separation. A gypsum surface potential of PZC = 2.3 is negatively charged over a wide pH range, and a cationic collector can easily be adsorbed onto the gypsum surface. The molecular layer of the gypsum crystal water and the Ca–O bond are weak and easy to dissociate from these positions when broken, which also leads to natural hydrophilic gypsum. It is also easy to achieve the separation of hydrophobic organic matter and other impurities from phosphogypsum and gypsum by adding a foaming agent [24–28]. After comprehensive consideration, the purity, whiteness, and recovery of $CaSO_4 \cdot 2H_2O$ in the concentrate were 96.38%, 36.63, and 84.08%, respectively, at pH = 2.0.

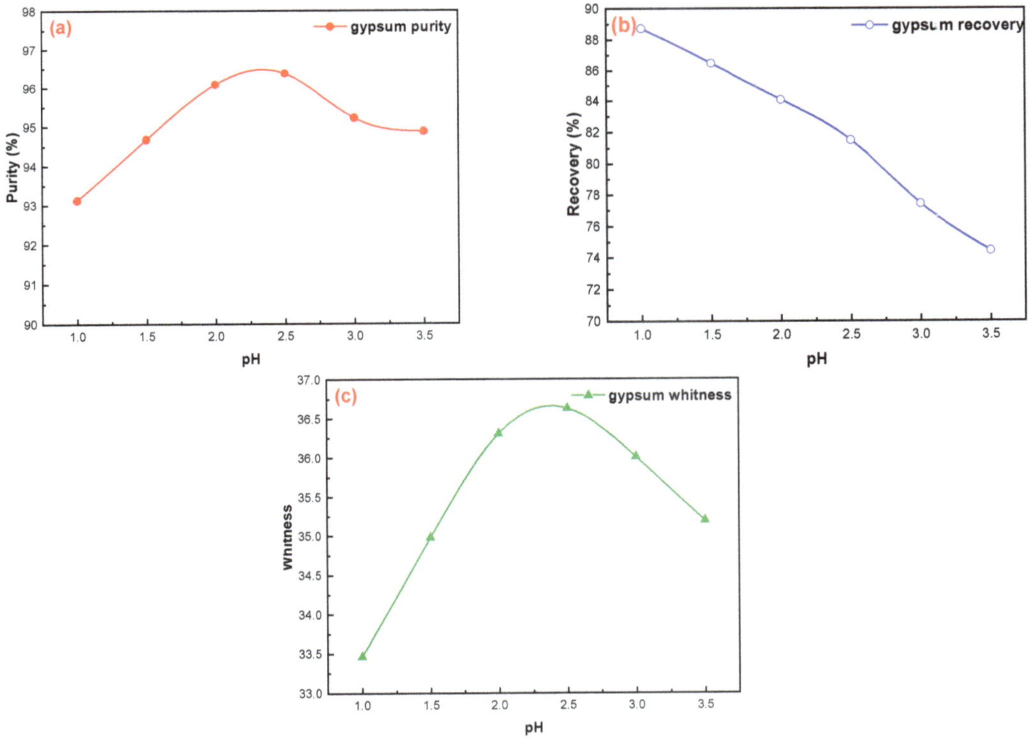

Figure 4. Effects of flotation pH value on gypsum purification: (**a**) gypsum purity, (**b**) gypsum recovery, and (**c**) gypsum whiteness.

3.2.4. Flotation Concentration

The effects of different dosages on gypsum flotation were investigated under the following conditions: pH = 2.0 (H_2SO_4), mixed amine dosage of 200 g/t, sodium silicate dosage of 400 g/t, and pine dosage of 40 g/t; the results are shown in Figure 5a–c. Flotation concentration is an important process parameter that influences the beneficiation index, which mainly influences the pulp filling capacity, the concentration of reagents in the pulp, and the flotation time. Generally, a proper increase in the concentration of the floating material can reduce the probability of particles (especially coarse particles) falling off the bubble, thus improving the recovery of useful minerals from coarse particles. A high concentration of the material plays an important role in weakening the influence of density and particle size on the velocity of the upper floating end [29–34]. However, owing to the fine particle size of the phosphogypsum material, the $CaSO_4{\cdot}2H_2O$ content in the concentrate decreased significantly; the whiteness of the concentrate also decreased when the concentration exceeded 30%, and the recovery of the $CaSO_4{\cdot}2H_2O$ increased slightly. Therefore, 30% is the ideal selection of flotation concentration, and the purity and recovery of $CaSO_4{\cdot}2H_2O$ in the concentrate are 96.18% and 84.16%, respectively, and the whiteness of the concentrate is 36.28.

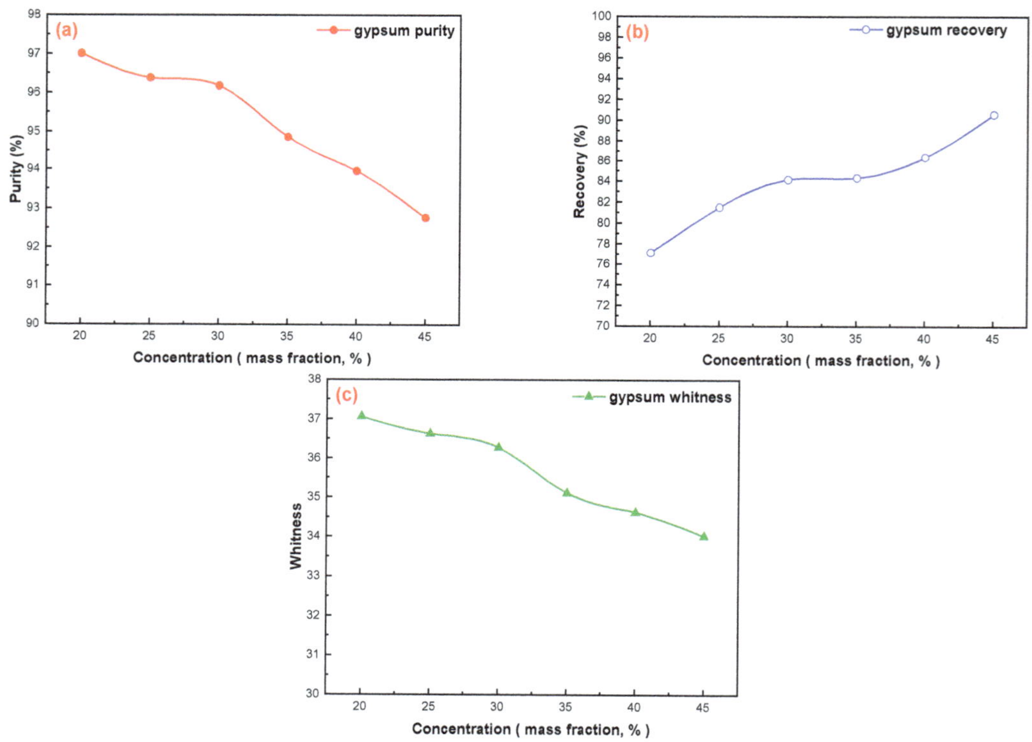

Figure 5. Effects of flotation concentration on gypsum purification: (**a**) gypsum purity, (**b**) gypsum recovery, and (c) gypsum whiteness.

3.2.5. Time of Flotation Scavenging

To further improve the comprehensive recovery rate of gypsum from phosphogypsum, scavenging tests were performed to determine suitable times for the recovery of gypsum. The test conditions for the roughing flotation were a flotation concentration of 30%, pH = 2.0 (H_2SO_4), mixed amine dosage of 400 g/t, sodium silicate dosage of 400 g/t, and pine dosage of 20 g/t. The test conditions for scavenging I were MA dosage of 200 g/t, pine dosage of 20 g/t, and the test conditions of scavenging II were MA dosage of 100 g/t and pine dosage of 10 g/t. The technological process is illustrated in Figure 6, and the results are listed in Table 3.

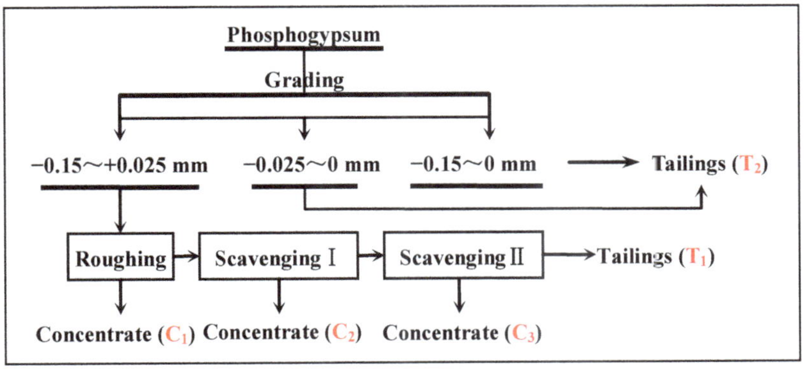

Figure 6. Scavenging test flowsheet of the recovery of gypsum from phosphogypsum.

Table 3. Scavenging test results of the recovery of gypsum from phosphogypsum.

Products	Mass Fraction (%)						Whiteness	
	Yield		Purity		Recovery		Individual	Cumulative
	Individual	Cumulative	Individual	Cumulative	Individual	Cumulative		
C_1	62.54	62.54	96.45	96.45	74.72	74.72	36.28	36.28
C_2	7.55	70.09	96.25	96.43	9.00	83.72	33.16	35.94
C_3	0.76	70.85	92.16	96.38	0.87	84.59	29.63	35.88
T_1	1.18	72.03	76.34	96.05	1.12	85.71	7.35	35.41
T_2	27.97	100	41.23	80.72	14.29	100.00	8.15	27.78
Totals	100.00		80.72		100.00		27.78	

An increase in the time of scavenging can improve the recovery rate of gypsum; however, after two scavenging operations, the whiteness of the gypsum concentrate decreased significantly, and, after two cleaning operations, the whiteness reduced to 29.63%. Owing to the increase in scavenging times, the consumption of flotation reagents increased correspondingly. In the case of the flotation process with one roughing and one scavenging step, a gypsum concentrate with a $CaSO_4{\cdot}2H_2O$ purity of 96.43%, a whiteness of 35.94, and a $CaSO_4{\cdot}2H_2O$ recovery of 83.72% can be obtained.

3.2.6. Time of Flotation Cleaning

Gypsum concentrate with a $CaSO_4{\cdot}2H_2O$ purity of 96.43%, whiteness of 35.94, and $CaSO_4{\cdot}2H_2O$ recovery of 83.72% was obtained from the effects of flotation scavenging, using the flotation process of one roughing and one scavenging step. The technological process of flotation cleaning shown in Figure 7 was used to further purify the gypsum concentrate; the test conditions of the roughing flotation were a flotation concentration of 30%, pH = 2.0 (H_2SO_4), a mixed amine dosage 400 g/t, a sodium silicate dosage of 400 g/t, and a pine dosage of 40 g/t; the test conditions of scavenging were an MA of dosage 200 g/t and pine dosage of 20 g/t. The results in Table 4 indicate that a two-stage concentration process can improve the purity, whiteness, and recovery of the gypsum concentrate. The purity of the $CaSO_4{\cdot}2H_2O$, the recovery of the $CaSO_4{\cdot}2H_2O$, and the whiteness were 98.94%, 79.87%, and 37.05, respectively.

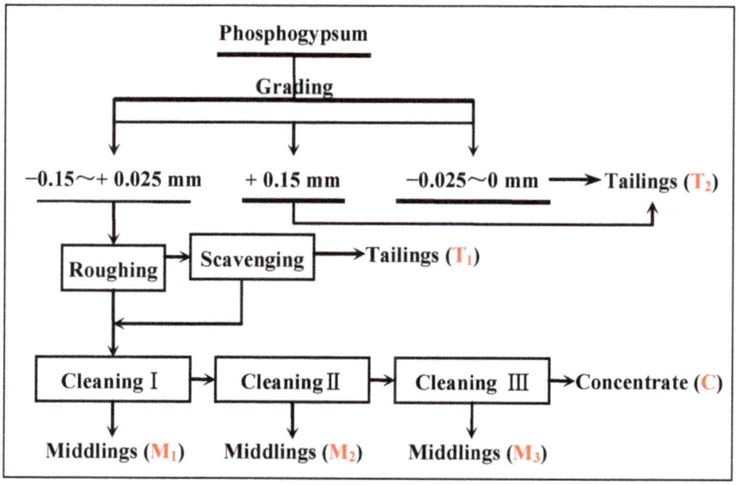

Figure 7. Cleaning test flowsheet of the recovery of gypsum from phosphogypsum.

Table 4. Cleaning test results of recovering gypsum from phosphogypsum.

Products	Mass Fraction (%)						Whiteness	
	Yield		Purity		Recovery			
	Individual	Cumulative	Individual	Cumulative	Individual	Cumulative	Individual	Cumulative
C	64.56	64.56	98.98	98.98	79.11	79.11	37.07	37.07
M_3	0.19	64.75	95.88	98.97	0.22	79.33	35.75	37.07
M_2	0.46	65.21	94.65	98.94	0.54	79.87	35.26	37.05
M_1	0.96	66.17	93.12	98.86	1.11	80.98	32.37	36.98
T_1	5.87	72.04	65.25	91.31	4.74	85.72	6.66	34.52
T_2	27.96	100.00	41.24	80.77	14.28	100.00	8.67	27.72
Totals	100.00		80.77		100.00		27.72	

3.2.7. The Entire Flowsheet Test of Recovering Gypsum from Phosphogypsum

The technological process of recovering gypsum from phosphogypsum was obtained using a single flotation process test, a scavenging test, and a cleaning test. To obtain the product index of the entire process, the process shown in Figure 8 was adopted to perform the entire flowsheet test. The results in Tables 5 and 6 show that gypsum concentrates with a $CaSO_4{\cdot}2H_2O$ purity of 98.94%, a $CaSO_4{\cdot}2H_2O$ recovery of 80.02%, and a whiteness of 37.05 were obtained. The content of silicon, phosphorus, fluorine, and other impurities in the gypsum concentrate was relatively low; therefore, the obtained gypsum concentrate could be used as a high-quality raw material for the preparation of α-hemihydrate high-strength gypsum or β-hemihydrate building gypsum.

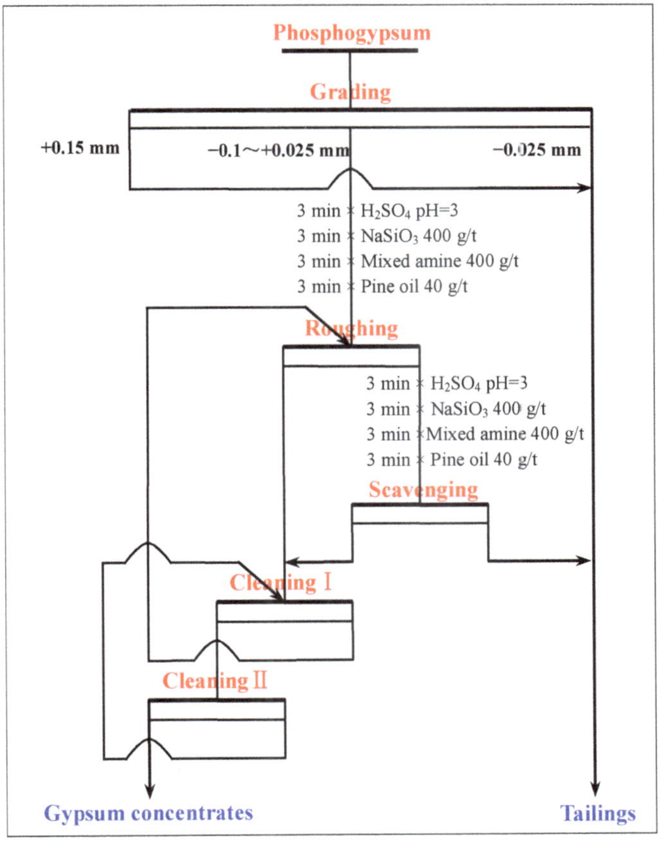

Figure 8. The recovery of gypsum from phosphogypsum using grading and flotation.

Table 5. Entire flowsheet test results of the recovery of gypsum from phosphogypsum.

Products	Mass Fraction (%)			Whiteness
	Yield	Purity	Recovery	
Gypsum concentrates	65.27	98.92	80.02	37.12
Tailings	34.73	46.43	19.98	9.96
Totals	100.00	80.69	100.00	27.69

Table 6. Main chemical composition analysis of gypsum concentrates (mass fraction, %).

SO_3	CaO	SiO_2	P_2O_5	Al_2O_3	Fe_2O_3	K_2O	SrO	MgO	F	TiO_2	BaO	Na_2O	Y_2O_3
52.52	44.88	0.32	0.05	0.05	0.02	0.05	0.01	0.06	0.01	0.02	0.01	0.01	0.01

4. Discussion

Figures 9 and 10 show that gypsum minerals and gangue minerals in untreated phosphogypsum adhere to each other, and fine mineral particles are attached to the surface of the gypsum crystals. The main mineral in the gypsum concentrate was gypsum ($CaSO_4{\cdot}2H_2O$); however, the brushite ($CaPO_3(OH){\cdot}2H_2O$) mineral phase and quartz (SiO_2) were not found. Meanwhile, tiny amounts of muscovite ($KAl_2SiO_{10}(OH)$) and zoisite ($Ca_2Al_3[SiO_{12}]OH$) entered the gypsum concentrate owing to mechanical entrainment of the flotation process.

Figure 11a–d shows further that the phosphogypsum, after grading and flotation of the gypsum concentrate, dispersed and had a smooth crystal surface, which indicates that the phosphogypsum through classification of pre-treatment using direct flotation can effectively remove impurities in the phosphogypsum concentrate, thus significantly improving the product quality of the gypsum concentrate.

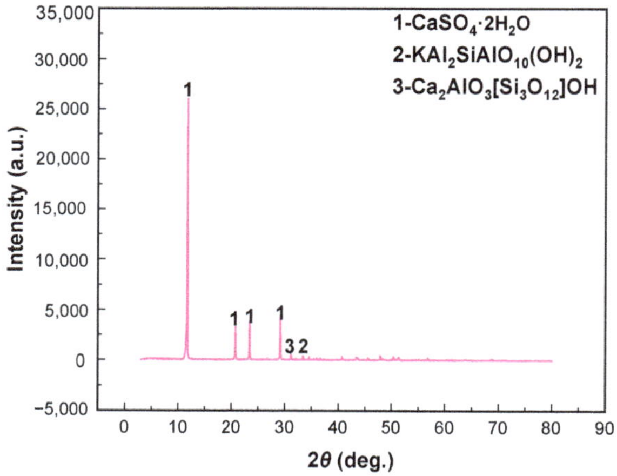

Figure 9. XRD patterns of gypsum concentrates.

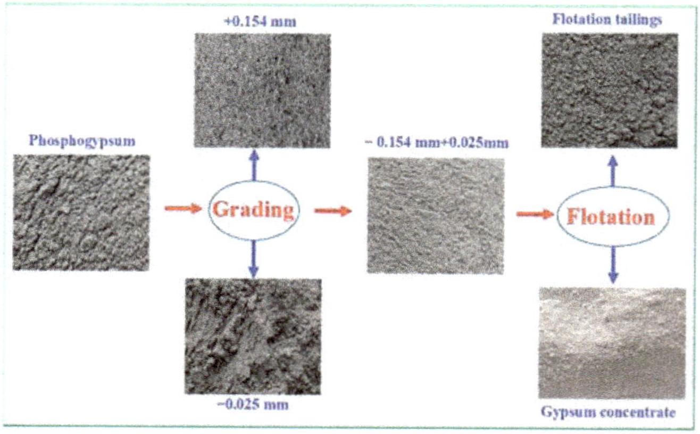

Figure 10. Photographs of characterization analysis of purification process of phosphogypsum.

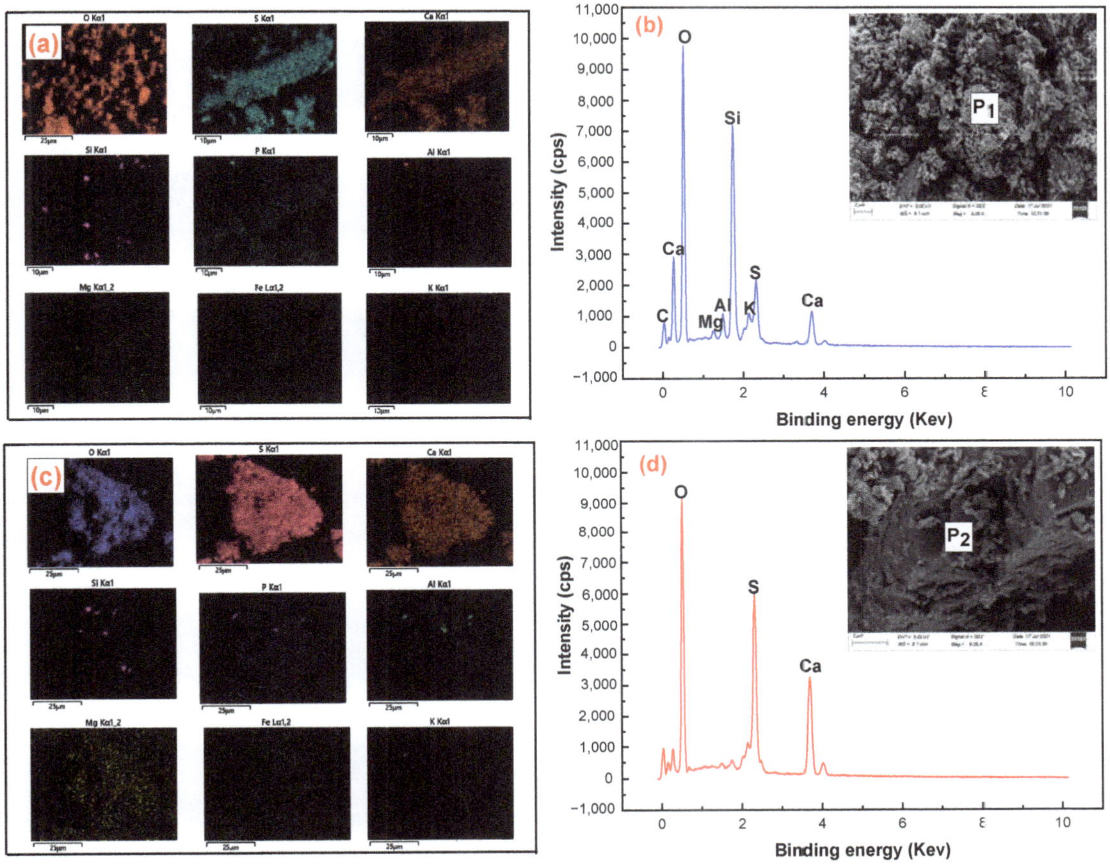

Figure 11. SEM-EDS analysis of (**a**,**b**) phosphogypsum, and (**c**,**d**) gypsum concentrates.

5. Conclusions

The following are the main conclusions based on the results obtained in this study on the recovery of gypsum from phosphogypsum:

(1) Phosphogypsum contained 80.65% $CaSO_4{\cdot}2H_2O$ and whiteness of 27.68. The main mineral is $CaSO_4{\cdot}2H_2O$ with small amounts of brushite, quartz, muscovite, and zoisite in phosphogypsum. Harmful elements, such as silicon, phosphorus, and fluorine, are mainly concentrated in the +0.15 mm and −0.025 mm fraction; these can be pre-selected and removed by the grading method to increase the $CaSO_4{\cdot}2H_2O$ content and reduce the processing cost.

(2) A novel direct flotation process, consisting of one roughing, one scavenging, and two cleaning operations, were employed to recover gypsum from the −0.15 to + 0.025 mm fraction materials. The addition of $NaSiO_3$ enhanced the removal of gangue minerals. Furthermore, the addition of a mixed amine enhanced the capture of $CaSO_4{\cdot}2H_2O$. The test results show that gypsum with a $CaSO_4{\cdot}2H_2O$ content of 98.94%, $CaSO_4{\cdot}2H_2O$ recovery of 80.02%, and whiteness of 37.05 was obtained under the following conditions: flotation roughing concentration of 30%, pH = 2.0 (H_2SO_4), mixed amine dosage of 400 g/t, pine dosage of 40 g/t, flotation scavenging of mixed amine dosage of 200 g/t, sodium silicate dosage of 400 g/t, and pine dosage of 20 g/t. Gypsum concentrate could be used as a high-quality raw material to prepare α-hemihydrate high-strength gypsum or β-hemihydrate-building gypsum.

(3) The results of XRD and SEM-EDS demonstrated that the main mineral in the gypsum concentrate was gypsum; however, brushite mineral phase and quartz were not found. Meanwhile, a tiny amount of muscovite and zoisite entered the gypsum concentrate owing to the mechanical entrainment of the flotation process.

Author Contributions: This is a joint work of the four authors; each author was in charge of their expertise and capability: J.X. of Writing, review, editing, and conceptualization; T.L. of software; Y.Z. of validation and data curation; and H.J. of methodology. All authors have read and agreed to the published version of the manuscript.

Funding: This study was funded by the Sichuan Science and Technology Program (Grant Nos. 22ZDYF1399, 2021YJ0057, 2021YFG0268) and the Research Fund Program of the Key Laboratory of the Guangdong Provincial Key Laboratory of Radioactive and Rare Resource Utilization (Grant Nos. 2018B030322009).

Institutional Review Board Statement: Not applicable.

Informed Consent Statement: Not applicable.

Data Availability Statement: Not applicable.

Conflicts of Interest: The authors declare no conflict of interest. The authors alone are responsible for the content and writing of the article.

References

1. Chernysh, Y.; Yakhnenko, O.; Chubur, V.; Roubík, H. Phosphogypsum Recycling: A Review of Environmental Issues, Current Trends, and Prospects. *Appl. Sci.* **2021**, *11*, 1575. [CrossRef]
2. Bakshi, P.; Pappu, A.; Gupta, M.K. A review on calcium-rich industrial wastes: A sustainable source of raw materials in India for civil infrastructure—opportunities and challenges to bond circular economy. *J. Mater. Cycles Waste Manag.* **2021**, *24*, 49–62. [CrossRef]
3. Bouargane, B.; Marrouche, A.; El Issiouy, S.; Biyoune, M.G.; Mabrouk, A.; Atbir, A.; Bachar, A.; Bellajrou, R.; Boukbir, L.; Bakiz, B. Recovery of $Ca(OH)_2$, $CaCO_3$, and Na_2SO_4 from Moroccan phosphogypsum waste. *J. Mater. Cycles Waste Manag.* **2019**, *21*, 1563–1571. [CrossRef]
4. Attallah, M.; Metwally, S.; Moussa, S.; Soliman, M. Environmental impact assessment of phosphate fertilizers and phosphogypsum waste: Elemental and radiological effects. *Microchem. J.* **2019**, *146*, 789–797. [CrossRef]
5. Garbaya, H.; Jraba, A.; Khadimallah, M.A.; Elaloui, E. The Development of a New Phosphogypsum-Based Construction Material: A Study of the Physicochemical, Mechanical and Thermal Characteristics. *Materials* **2021**, *14*, 7369. [CrossRef]
6. Rashad, A.M. Phosphogypsum as a construction material. *J. Clean. Prod.* **2017**, *166*, 732–743. [CrossRef]
7. Patil, P.P.; Prabhu, M.; Mutnuri, S. A novel and sustainable approach for biotransformation of phosphogypsum to calcium carbonate using urease producing Lysinibacillus sphaericus strain GUMP2. *Environ. Technol.* **2021**, 1–14. [CrossRef]
8. Zhang, D.; Luo, H.; Zheng, L.; Wang, K.; Li, H.; Wang, Y.; Feng, H. Utilization of waste phosphogypsum to prepare hydroxyapatite nanoparticles and its application towards removal of fluoride from aqueous solution. *J. Hazard. Mater.* **2012**, *241–242*, 418–426. [CrossRef]
9. Kuzmanović, P.; Todorović, N.; Forkapić, S.; Petrović, L.F.; Knežević, J.; Nikolov, J.; Miljević, B. Radiological characterization of phosphogypsum produced in Serbia. *Radiat. Phys. Chem.* **2020**, *166*, 108463. [CrossRef]
10. Wei, J.; Gu, Y.; Lv, H.; Wu, X. A zero-emission method for recycling phosphogypsum using Na_2SO_4 electrolysis: Preliminary study. *Sep. Purif. Technol.* **2021**, *259*, 118168. [CrossRef]
11. Kazanskaya, L.; Smirnova, O.; Palomo, Á.; Pidal, I.M.; Romana, M. Supersulfated Cement Applied to Produce Lightweight Concrete. *Materials* **2021**, *14*, 403. [CrossRef] [PubMed]
12. Schaefer, C.O.; Cheriaf, M.; Rocha, J.C. Production of Synthetic Phosphoanhydrite and Its Use as a Binder in Self-Leveling Underlayments (SLU). *Materials* **2017**, *10*, 958. [CrossRef] [PubMed]
13. Hentati, O.; Abrantes, N.; Caetano, A.; Bouguerra, S.; Gonçalves, F.J.M.; Römbke, J.; Pereira, R. Phosphogypsum as a soil fertilizer: Ecotoxicity of amended soil and elutriates to bacteria, invertebrates, algae and plants. *J. Hazard. Mater.* **2015**, *294*, 80–89. [CrossRef] [PubMed]
14. Macías, F.; Cánovas, C.R.; Cruz-Hernández, P.; Carrero, S.; Asta, M.P.; Nieto, J.M.; Pérez-López, R. An anomalous metal-rich phosphogypsum: Characterization and classification according to international regulations. *J. Hazard. Mater.* **2017**, *331*, 99–108. [CrossRef] [PubMed]
15. Zhang, S.; Huang, Z.; Wang, H.; Liu, R.; Cheng, C.; Shuai, S.; Hu, Y.; Guo, Z.; Yu, X.; He, G.; et al. Flotation performance of a novel Gemini collector for kaolinite at low temperature. *Int. J. Min. Sci. Technol.* **2021**, *31*, 1145–1152. [CrossRef]
16. El-Didamony, H.; Gado, H.; Awwad, N.; Fawzy, M.; Attallah, M. Treatment of phosphogypsum waste produced from phosphate ore processing. *J. Hazard. Mater.* **2013**, *244–245*, 596–602. [CrossRef]

17. Yao, W.; Li, M.; Cui, R.; Jiang, X.; Jiang, H.; Deng, X.; Li, Y.; Zhou, S. Flotation Behavior and Mechanism of Anglesite with Salicyl Hydroxamic Acid as Collector. *JOM* **2018**, *70*, 2813–2818. [CrossRef]
18. Marion, C.; Jordens, A.; Li, R.; Rudolph, M.; Waters, K. An evaluation of hydroxamate collectors for malachite flotation. *Sep. Purif. Technol.* **2017**, *183*, 258–269. [CrossRef]
19. Ramirez, A.; Rojas, A.; Gutierrez, L.; Laskowski, J.S. Sodium hexametaphosphate and sodium silicate as dispersants to reduce the negative effect of kaolinite on the flotation of chalcopyrite in seawater. *Miner. Eng.* **2018**, *125*, 10–14. [CrossRef]
20. Hussein, M.; Donaldson, A. Mixing strategy effect on dispersion of amine during reagent production for flotation applications. *Can. J. Chem. Eng.* **2018**, *96*, 360–366. [CrossRef]
21. Özün, S.; Ulutaş, Ş. Interfacial Behavior of Anionic/Cationic Flotation Collectors in Mixed Aqueous Solutions and Their Effect on Flotation Recovery of Quartz. *J. Surfactants Deterg.* **2019**, *22*, 61–71. [CrossRef]
22. Kargupta, W.; Browne, C.; Verdugo, L.; Hunt, I.; Stack, K.; Batchelor, W.; Tanner, J. Flotation as a separation technology for recovering pulp fines and sustainable nanocellulose production. *Sep. Purif. Technol.* **2021**, *270*, 118810. [CrossRef]
23. Bhondayi, C.; Moys, M.; Danha, G. Effect of pulp chemistry and solids on a froth bubble size measurement method. *Powder Technol.* **2016**, *297*, 202–210. [CrossRef]
24. Hosseini, S.H.; Forssberg, E. Physicochemical studies of smithsonite flotation using mixed anionic/cationic collector. *Miner. Eng.* **2007**, *20*, 621–624. [CrossRef]
25. Marion, C.; Jordens, A.; McCarthy, S.; Grammatikopoulos, T.; Waters, K.E. An investigation into the flotation of muscovite with an amine collector and calcium lignin sulfonate depressant. *Sep. Purif. Technol.* **2015**, *149*, 216–227. [CrossRef]
26. Chen, Y.; Hu, S.; Li, J.; Weng, L.; Wu, C.; Liu, K. Improvement on combustible matter recovery in coal slime flotation with the addition of sodium silicate. *Colloids Surfaces A Physicochem. Eng. Asp.* **2020**, *603*, 125220. [CrossRef]
27. Feng, B.; Guo, W.; Xu, H.; Peng, J.; Luo, X.; Zhu, X. The combined effect of lead ion and sodium silicate in the flotation separation of scheelite from calcite. *Sep. Sci. Technol.* **2017**, *52*, 567–573. [CrossRef]
28. Meng, Q.; Yuan, Z.; Xu, Y.; Du, Y. The effect of sodium silicate depressant on the flotation separation of fine wolframite from quartz. *Sep. Sci. Technol.* **2018**, *54*, 1400–1410. [CrossRef]
29. Gao, Z.-Y.; Jiang, Z.-Y.; Sun, W.; Gao, Y.-S. Typical roles of metal ions in mineral flotation: A review. *Trans. Nonferrous Met. Soc. China* **2021**, *31*, 2081–2101. [CrossRef]
30. Solongo, S.K.; Flores, A.G.; You, J.; Choi, S.; Heyes, G.W.; Ilyas, S.; Lee, J.; Kim, H. Cationic collector conformations on an oxide mineral interface: Roles of pH, ionic strength, and ion valence. *Miner. Eng.* **2020**, *150*, 106277. [CrossRef]
31. Mechi, N.; Khiari, R.; Ammar, M.; Elaloui, E.; Belgacem, M.N. Preparation and application of Tunisian phosphogypsum as fillers in papermaking made from *Prunus amygdalus* and *Tamarisk* sp. *Powder Technol.* **2017**, *312*, 287–293. [CrossRef]
32. Jara, A.D.; Woldetinsae, G.; Betemariam, A.; Kim, J.Y. Mineralogical and petrographic analysis on the flake graphite ore from Saba Boru area in Ethiopia. *Int. J. Min. Sci. Technol.* **2020**, *30*, 715–721. [CrossRef]
33. Xiao, J.; Zhang, Y. Recovering Cobalt and Sulfur in Low Grade Cobalt-Bearing V–Ti Magnetite Tailings Using Flotation Process. *Processes* **2019**, *7*, 536. [CrossRef]
34. Xiao, J.; Chen, C.; Ding, W.; Peng, Y.; Chen, T.; Zou, K. Extraction of Phosphorous from a Phosphorous-Containing Vanadium Titano-Magnetite Tailings by Direct Flotation. *Processes* **2020**, *8*, 874. [CrossRef]

materials

Article

The Effects of Niobium and Molybdenum on the Microstructures and Corrosion Properties of CrFeCoNiNbxMoy Alloys

Chun-Huei Tsau *, Yi-Hsuan Chen and Meng-Chi Tsai

Institute of Nanomaterials, Chinese Culture University, Taipei 111, Taiwan; psfohs1130@gmail.com (Y.-H.C.); asd99586@yahoo.com.tw (M.-C.T.)
* Correspondence: chtsau@staff.pccu.edu.tw

Abstract: The present work systematically investigated the effects of niobium and molybdenum on the microstructures and corrosion properties of high-entropy $CrFeCoNiNb_xMo_x$ and $CrFeCoNiNb_xMo_{1-x}$ alloys, the maximum content of (Nb + Mo) was 20 at.%. All of the alloys were prepared by arc melting under an argon atmosphere. In $CrFeCoNiNb_xMo_x$ alloys (x = 0.15, 0.3 and 0.5), increasing Nb and Mo content would change the microstructure of the alloy from a hypoeutectic structure ($x \leq 0.3$) to a hypereutectic one (x = 0.5). All of the $CrFeCoNiNbxMo_{1-x}$ alloys (x = 0.25, 0.5 and 0.75) had a hypereutectic microstructure. Only two phases were analyzed in these alloys, which were face-centered cubic (FCC) and hexagonal close packing (HCP). Increasing the content of Nb and Mo increases the hardness of the alloys by the effects of the solid solution strengthening and formation of the HCP phase. The potentiodynamic polarization curves of these alloys were also measured in 1 M sulfuric acid and 1 M sodium chloride solutions to evaluate the corrosion resistance of these alloys. The $CrFeCoNiNb_{0.3}Mo_{0.3}$ alloy had the smallest corrosion rate (0.0732 mm/yr) in 1 M deaerated H_2SO_4 solution, and the $CrFeCoNiNb_{0.15}Mo_{0.15}$ alloy had the smallest corrosion rate (0.0425 mm/yr) in 1 M deaerated NaCl solution. However, the $CrFeCoNiNb_{0.5}Mo_{0.5}$ alloy still had the best combination of corrosion resistance and hardness in the present study.

Keywords: high-entropy alloy; $CrFeCoNiNb_xMo_x$; $CrFeCoNiNb_xMo_{1-x}$; microstructure; hardness; corrosion

Citation: Tsau, C.-H.; Chen, Y.-H.; Tsai, M.-C. The Effects of Niobium and Molybdenum on the Microstructures and Corrosion Properties of CrFeCoNiNbxMoy Alloys. *Materials* **2022**, *15*, 2262. https://doi.org/10.3390/ma15062262

Academic Editor: Petrica Vizureanu

Received: 28 February 2022
Accepted: 17 March 2022
Published: 18 March 2022

1. Introduction

The alloys used in this study were prepared under the high-entropy alloy concept [1–3]. This high-entropy alloy concept provides researchers to develop new materials with suitable properties for applications. Researchers can smartly select the elements to prepare the materials, and many high-entropy alloys (HEAs) were thus produced with excellent mechanical, physical and chemical properties. Moreover, the shapes of high-entropy materials can be bulk alloys, thin films or coating alloys. For example, the casting, homogenization, cold rolling, recrystallization and deformation mechanism of equiatomic CoCrFeMnNi high-entropy alloy were well investigated [4,5]. The microstructures and compression properties of $CoCrFeNiTiAl_x$ high-entropy alloys were tested, and results indicated that the CoCrFeNiTiAl alloy had good compressive strength and elastic modulus [6]. The elements with a high melting point were selected to produce the refractory alloys, such as NbMoTaW, VNbMoTaW and HfTaTiNbZr-based alloys [7,8]. The high-entropy alloys can be prepared by mechanical alloying (MA) to obtain the alloys with nanocrystalline structures and enhance the properties [9,10].

Cobalt, chromium and nickel are wildly used to produce alloys with good corrosion resistance. Other elements are selected and added into CoCrNi alloy to change the microstructures and enhance the mechanical properties, such as CoCrFeMnNi alloy [11]. Co–CrFeNiSn has good passivation in sodium chloride solution compared with stainless steels [12]. The $AlCoCrFeNiTi_{0.5}$ coating was fabricated by laser cladding coating and showed the optimal performance of corrosion and mechanical properties [13]. Minor additions of molybdenum could improve the corrosion resistance of the $AlCrFe_2Ni_2$ alloy by suppressing pit formation [14]. The addition of molybdenum could increase the corrosion resistance was observed in the $(CoCrFeNi)_{100-x}Mo_x$ high-entropy alloys [15]. The study on the $Al_{0.4}CrFe_{1.5}MnNi_{0.5}Mo_x$ alloys indicated that adding molybdenum can effectively improve the impedance of passive film and reduce the corrosion current density and thus form a more stable passivation film [16]. The non-equimolar $Cr_{19}Fe_{22}Co_{21}Ni_{25}Mo_{13}$ alloy possessed better corrosion resistance compared with 304 stainless steel in both deaerated 1 M HNO_3 and 1 M HCl solutions [17]. The corrosion resistance of FeCuNbSiB and $CrFeCoNiNb_x$ alloys can be improved by adding niobium [18,19].

In our previous study on the corrosion properties of FeCoNi and CrFeCoNi alloys [20], the FeCoNi alloy had a better corrosion resistance by comparing with CrFeCoNi alloy. After adding molybdenum, the corrosion resistance of FeCoNiMo was not as good as that of the CrFeCoNiMo alloy [21]. This indicates that chromium is a very important element in developing a corrosion-resistant alloy. Therefore, the present work studied the effect of adding Nb and Mo on the CrFeCoNi alloy and evaluated the potential of the application.

2. Materials and Methods

The $CrFeCoNiNb_xMo_x$ and $CrFeCoNiNb_xMo_{1-x}$ alloys were prepared by arc-melting under argon atmosphere after accurate weighting. Each melt was about 100 g. Table 1 lists the nominal compositions of the alloys. The microstructures of the alloys were examined by a scanning electron microscope (SEM, JEOL JSM-6335, JEOL Ltd., Tokyo, Japan) after regular metallurgical processes. The compositions of the alloys and the phases existing in the alloys were analyzed by an energy dispersive spectrometer (EDS). An X-ray diffractometer (XRD, Rigaku ME510-FM2, Rigaku Ltd., Tokyo, Japan) was used to identify the phases in the alloys, and the scanning rate was fixed at 0.04 degrees per second. A Vicker's hardness tester (Matsuzawa Seiki MV1, Matsuzawa Ltd., Akita, Japan) was used to measure the hardness of the alloys, and the loading force was 19.61 N (2000 gf).

Table 1. The nominal compositions of the as-cast alloys.

Alloys	Weight Percent					
	Cr	Fe	Co	Ni	Nb	Mo
$CrFeCoNiNb_{0.15}Mo_{0.15}$	20.49	22.00	23.22	23.13	5.49	5.67
$CrFeCoNiNb_{0.3}Mo_{0.3}$	18.43	19.79	20.89	20.81	9.88	10.20
$CrFeCoNiNb_{0.5}Mo_{0.5}$	16.25	17.46	18.42	18.35	14.52	14.99
$CrFeCoNiNb_{0.25}Mo_{0.75}$	16.22	17.42	18.38	18.31	7.24	22.44
$CrFeCoNiNb_{0.75}Mo_{0.25}$	16.29	17.50	18.46	18.40	21.83	7.52

The potentiodynamic polarization curves of the alloys were tested by a three-electrode electrochemical device (Autolab PGSTAT302N, Metrohm Autolab B.V., Utrecht, The Netherlands). One electrode was the specimens mounted in epoxy resin with an exposed area of 0.196 cm^2 (0.5 cm in diameter). The second electrode was a counter (a platinum wire). The third electrode was a reference one, which was a saturated silver chloride electrode (Ag/AgCl, SSE). This potential of the reference Ag/AgCl electrode was 0.197 V higher than the standard hydrogen electrode (SHE) at 25 °C [22]. All of the potentiodynamic polarization curves were tested at 30 °C, and the scanning rate of the potentiodynamic polarization test was 1 mV per second. Nitrogen bubbled through the process to degas the oxygen in the solutions. The solutions of 1 M sulfuric acid and 1 M sodium chloric solutions were prepared by reagent-grade acids and deionized water.

3. Results and Discussion

This work was divided into two parts to investigate the effect of Nb and Mo content on the effects of $CrFeCoNiNb_xMo_y$ alloys. Part 1 was the $CrFeCoNiNb_xMo_x$ alloys, and x was 0.15, 0.3 and 0.5. The same amount of Nb and Mo was added to the CrFeCoNi alloy. The microstructure revolution of $CrFeCoNiNb_xMo_x$ alloys were studied in this part; the relationships between the properties and the microstructures were also investigated. Part 2 studied the properties of $CrFeCoNiNb_xMo_{1-x}$ alloys, where x was 0.25, 0.5 and 0.75. This part studied the effect of different ratios of Nb and Mo on the microstructures and properties of the alloys. The total amount of Nb and Mo was fixed at one part (20 at.%) because the alloys would easily crack during solidification if an excess amount of Nb and Mo was added.

3.1. $CrFeCoNiNb_xMo_x$ Alloys

The microstructures of as-cast $CrFeCoNiNb_xMo_x$ alloys, x = 0.15, 0.3 and 0.5, are shown in Figure 1. In our previous study, the CrFeCoNi alloy had an FCC-structured granular microstructure and some Cr-rich precipitates with HCP structure [20]. After adding Nb and Mo, the microstructures of as-cast $CrFeCoNiNb_xMo_x$ alloys became dendritic ones. The dendrites of $CrFeCoNiNb_xMo_x$ alloys showed a single phase, and the interdendrities of $CrFeCoNiNb_xMo_x$ alloys showed a eutectic structure. Table 2 shows the chemical compositions of the alloys and the phases existing in the alloys. According to our previous study [23], the HCP phase had higher Nb and Mo content, and FCC had higher Cr- Fe and Ni content. Therefore, the FCC and HCP phases were easy to identify by detecting the compositions. The dendrites of $CrFeCoNiNb_{0.15}Mo_{0.15}$ and $CrFeCoNiNb_{0.3}Mo_{0.3}$ alloys were an FCC phase and the dendrites of $CrFeCoNiNb_{0.5}Mo_{0.5}$ alloy were an HCP-structured laves phase. All of the interdendrites of $CrFeCoNiNb_xMo_x$ alloys were a eutectic structure with two phases which were FCC and HCP phases (laves phase). This indicated that the alloy changed from a hypoeutectic alloy to a hypereutectic one. That is, $CrFeCoNiNb_{0.15}Mo_{0.15}$ and $CrFeCoNiNb_{0.3}Mo_{0.3}$ alloys were hypoeutectic alloys, and $CrFeCoNiNb_{0.5}Mo_{0.5}$ alloy was a hypereutectic alloy.

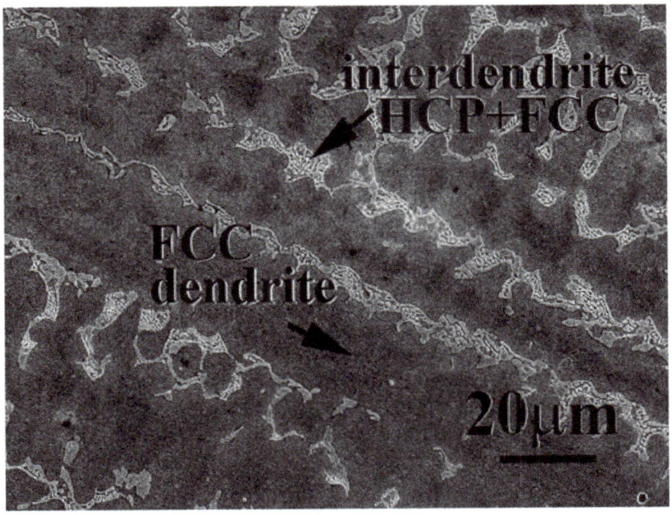

(a)

Figure 1. *Cont.*

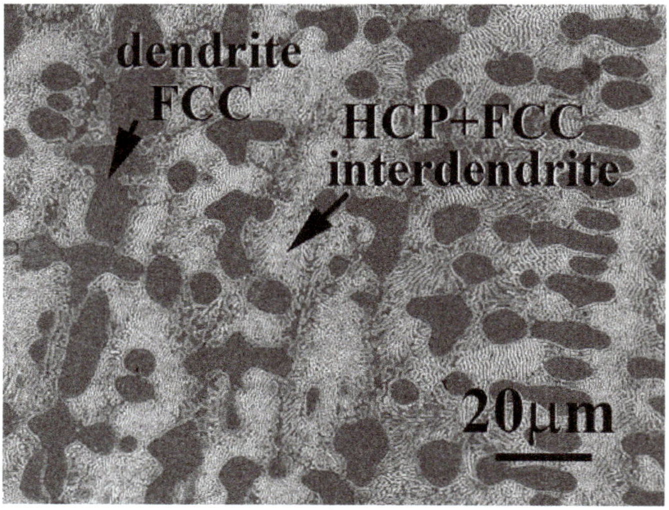

(**b**)

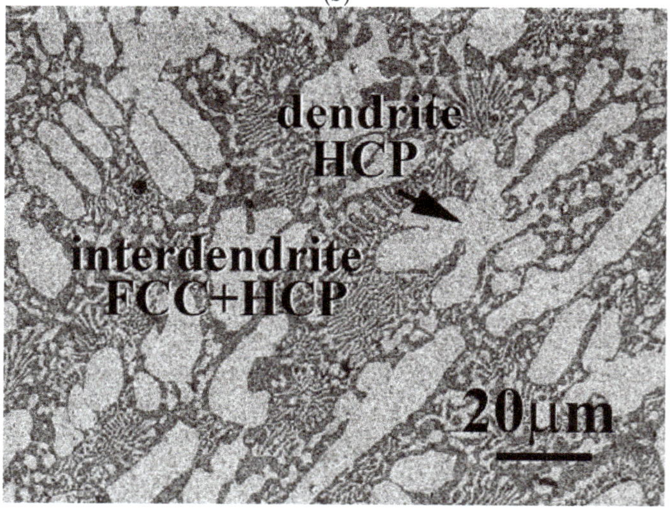

(**c**)

Figure 1. SEM micrographs of as-cast $CrFeCoNiNb_xMo_x$ alloys: (**a**) $CrFeCoNiNb_{0.15}Mo_{0.15}$ alloy; (**b**) $CrFeCoNiNb_{0.3}Mo_{0.3}$ alloy; (**c**) $CrFeCoNiNb_{0.5}Mo_{0.5}$ alloy.

Figure 2 shows the XRD patterns of the $CrFeCoNiNb_xMo_x$ alloys. Two phases of FCC and laves phases (HCP structure) were identified in these alloys. The FCC phase was the main phase in the $CrFeCoNiNb_{0.15}Mo_{0.15}$ alloy. Only a small amount of laves phase was in this alloy. Increasing the Nb and Mo content resulted in increasing the amount of laves phase. The laves phase became the major phase in the $CrFeCoNiNb_{0.5}Mo_{0.5}$ alloy. This result quite matches with the SEM observation.

Figure 3 plots the hardness of $CrFeCoNiNb_xMo_x$ alloys as a function of Nb and Mo content. Increasing Nb and Mo content would almost linearly increase the hardness of the $CrFeCoNiNb_xMo_x$ alloys. The hardness of $CrFeCoNiNb_{0.15}Mo_{0.15}$ alloy was only 215 HV, and the hardness of $CrFeCoNiNb_{0.5}Mo_{0.5}$ reached 553 HV. Adding Nb and Mo into CrFeCoNi alloy increases the hardness because the atomic radiuses of niobium and molybdenum are larger than those of the other elements. The atomic radiuses of niobium and molybdenum are 1.43 and 1.40 Å, respectively; the atomic radiuses of cobalt, chromium, iron and nickel are 1.253, 1.249, 1.241 and 1.243 Å, respectively [24]. Therefore, the hardness of the $CrFeCoNiNb_xMo_x$ alloys increased due to the solid solution strengthening effect. Our previous work indicated that increasing niobium and molybdenum content increases the density of dislocation in the FCC phase [23]. This was another effect enhancing the hardness of the alloys. Additionally, the hardness of the HCP-structured laves phase was higher than that of the FCC phase because the slip system of the HCP phase was less than that of the FCC phase. The hardness of $CrFeCoNiNb_xMo_x$ alloys increased after adding more niobium and molybdenum elements due to the increase in the HCP phase and solid solution strengthening effect.

Table 2. The chemical compositions of the as-cast $CrFeCoNiNb_xMo_x$ alloys and the phases in each alloy analyzed by SEM/EDS.

Alloys	Weight Percent					
	Cr	Fe	Co	Ni	Nb	Mo
$CrFeCoNiNb_{0.15}Mo_{0.15}$						
overall	23.9 ± 0.9	24.2 ± 4.9	25.6 ± 0.4	17.2 ± 4.8	5.0 ± 0.2	4.1 ± 0.2
FCC	24.9 ± 1.3	25.3 ± 0.3	22.3 ± 1.1	20.5 ± 1.2	3.1 ± 0.7	3.9 ± 0.6
HCP	19.8 ± 0.5	16.0 ± 1.4	26.6 ± 1.3	15.6 ± 0.5	10.4 ± 0.3	10.6 ± 0.2
$CrFeCoNiNb_{0.3}Mo_{0.3}$						
overall	23.1 ± 1.3	20.4 ± 1.5	18.1 ± 2.1	21.7 ± 1.6	9.2 ± 0.5	7.5 ± 0.8
FCC	23.6 ± 2.0	23.3 ± 2.2	20.4 ± 2.8	25.8 ± 3.0	2.0 ± 0.4	5.1 ± 0.6
HCP	18.7 ± 0.6	14.6 ± 0.2	17.9 ± 2.6	17.8 ± 0.2	19.2 ± 0.2	11.8 ± 2.5
$CrFeCoNiNb_{0.5}Mo_{0.5}$						
overall	21.4 ± 0.4	19.4 ± 1.6	19.8 ± 0.3	17.6 ± 2.1	11.0 ± 0.2	10.8 ± 0.7
FCC	22.1 ± 0.9	20.4 ± 1.4	18.4 ± 0.3	22.8 ± 0.4	7.1 ± 1.2	9.2 ± 0.5
HCP	17.3 ± 0.3	23.6 ± 5.2	19.1 ± 0.8	14.2 ± 0.8	16.9 ± 0.1	15.6 ± 0.9

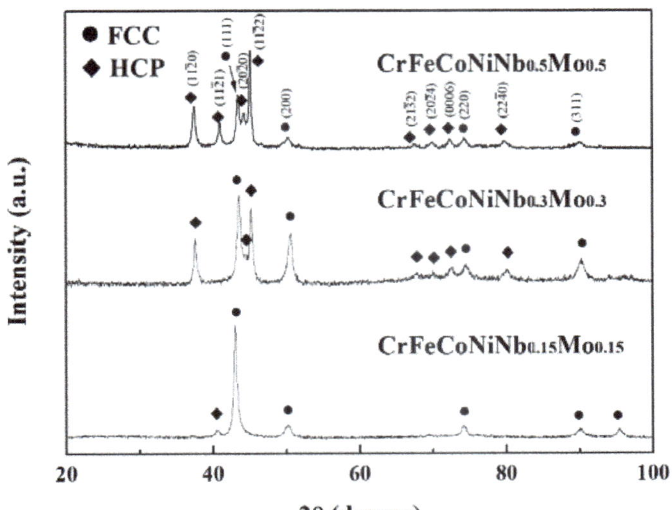

Figure 2. XRD patterns of as-cast $CrFeCoNiNb_xMo_x$ alloys.

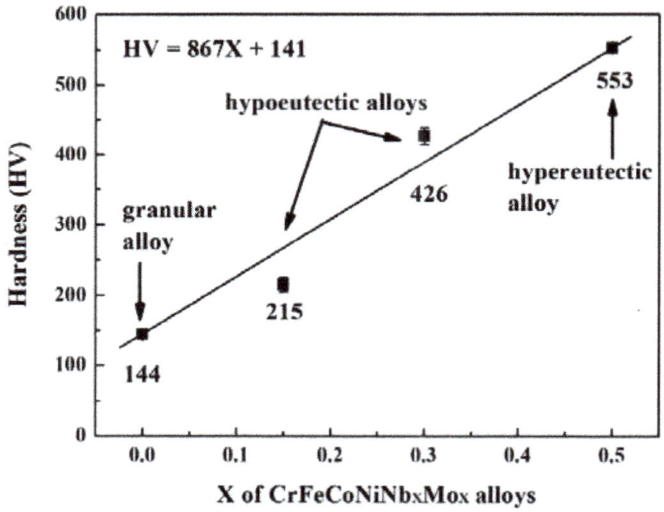

Figure 3. Hardness of as-cast $CrFeCoNiNb_xMo_x$ alloys. Each value is the average hardness of the alloy.

Figure 4 shows the potentiodynamic polarization curves of the as-cast $CrFeCoNiNb_xMo_x$ alloys in 1 M deaerated H_2SO_4 solution at 30 °C. The potentiodynamic polarization data of these polarization curves are listed in Table 3. The potentiodynamic polarization curves of the alloys with potential negative than corrosion potential (E_{corr}) was the cathode. The potentiodynamic polarization curves of the alloys with potential positive than corrosion potential was the anode. The corrosion potential (E_{corr}) of $CrFeCoNiNb_xMo_x$ alloys was very close. The standard electrode potential of the elements used in present work is listed in Table 4 [25]. The niobium is more active than the other elements because the standard electrode potential of niobium is more negative. Therefore, the corrosion potential of $CrFeCoNiNb_{0.5}Mo_{0.5}$ alloy had the most negative corrosion potential (E_{corr}). The corrosion current densities (i_{corr}) of $CrFeCoNiNb_xMo_x$ alloys were all around 10 μA/cm^2. The potentiodynamic polarization curve of $CrFeCoNiNb_{0.15}Mo_{0.15}$ alloy had an apparent anodic peak, and the other alloys had no anodic peak. The passivation potential (E_{pp}) and critical current density (i_{crit}) of the anodic peak of $CrFeCoNiNb_{0.15}Mo_{0.15}$ alloy is listed in Table 3. Thus, the $CrFeCoNiNb_{0.3}Mo_{0.3}$ and $CrFeCoNiNb_{0.5}Mo_{0.5}$ alloys easily entered into the passivation regions and formed the passive films during corrosion in H_2SO_4 solution. The lowest passivation current densities (i_{pass}) of these alloys were about 12 A/cm^2. All the passivation regions of these alloys were breakdown at a potential (E_b) of about 1.2 V (SHE) due to oxygen evolution reaction [26].

Table 3. Potentiodynamic polarization data of the as-cast $CrFeCoNiNb_xMo_x$ alloys in 1 M deaerated H_2SO_4 solution at 30 °C.

	$CrFeCoNiNb_{0.15}Mo_{0.15}$	$CrFeCoNiNb_{0.3}Mo_{0.3}$	$CrFeCoNiNb_{0.5}Mo_{0.5}$
E_{corr} (V vs. SHE)	−0.03	−0.05	−0.06
i_{corr} (μA/cm^2)	11.9	7.0	10.0
E_{pp} (V vs. SHE)	0.18	N/A	N/A
i_{crit} (μA/cm^2)	67.5	N/A	N/A
i_{pass} (μA/cm^2)	12.5	13.1	12.2
E_b (V vs. SHE)	1.21	1.19	1.19

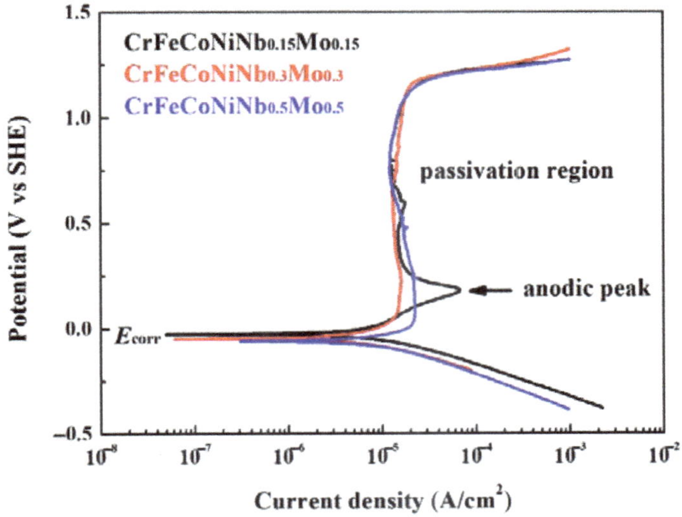

Figure 4. Potentiodynamic polarization curves of as-cast $CrFeCoNiNb_xMo_x$ alloys tested in the 1 M deaerated sulfuric acid solution at 30 °C.

Table 4. Standard electrode potential at 25 °C [25].

Reaction	Electrode Potential (E° vs. SSE)
Cr, Cr^{3+}	−0.74
Fe, Fe^{2+}	−0.44
Co, Co^{2+}	−0.227
Ni, Ni^{2+}	−0.25
Nb, Nb^{3+}	−1.10
Mo, Mo^{3+}	−0.20

Figure 5 displays the potentiodynamic polarization curves of the as-cast $CrFeCoNiNb_xMo_x$ alloys in 1 M deaerated NaCl solution at 30 °C. The potentiodynamic polarization data of these polarization curves are listed in Table 5. The corrosion potential (E_{corr}) of CrFeCoNi $Nb_{0.15}Mo_{0.15}$ alloy was more negative than the other alloys, and $CrFeCoNiNb_{0.15}Mo_{0.15}$ alloy also had the smallest corrosion current density (i_{corr}). All of the potentiodynamic polarization curves of $CrFeCoNiNb_xMo_x$ alloys in 1 M deaerated NaCl solution had apparent anodic peaks. Additionally, the potentiodynamic polarization curves of Cr–$FeCoNiNb_{0.15}Mo_{0.15}$ and $CrFeCoNiNb_{0.3}Mo_{0.3}$ alloys had small secondary anodic peaks. The $CrFeCoNiNb_{0.15}Mo_{0.15}$ alloy had the lowest passivation current density. The passivation current densities (i_{pass}) of these alloys were about 9–15 μA/cm^2.

3.2. $CrFeCoNiNb_xMo_{1-x}$ Alloys

The microstructures and properties of $CrFeCoNiNb_xMo_{1-x}$ alloys were studied in this part. The total amount of Nb and Mo of these alloys was kept as one part; the amount of Nb and Mo thus equaled 20 at.%. The microstructures of as-cast $CrFeCoNiNb_xMo_{1-x}$ alloys, x = 0.25 and 0.75, are shown in Figure 6. All of the $CrFeCoNiNb_xMo_{1-x}$ alloys, x = 0.25, 0.5 and 0.75, had a hypereutectic microstructure because a large amount of Nb and Mo were added into these alloys. The dendrites of $CrFeCoNiNb_xMo_{1-x}$ alloys were an HCP phase (laves phase), and the interdendrites of $CrFeCoNiNb_xMo_{1-x}$ alloys were a eutectic structure with FCC and HCP phases (laves phase). Table 6 lists the chemical compositions of the alloys and the phases existing in the alloys. As described above, the Nb and Mo content in the HCP phase of the alloys is higher than that in the FCC phase.

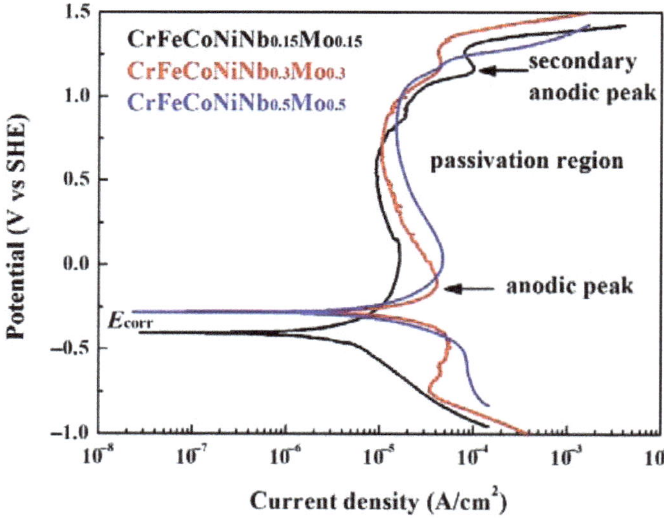

Figure 5. Potentiodynamic polarization curves of as-cast CrFeCoNiNb$_x$Mo$_x$ alloys tested in the 1 M deaerated sodium chloric solution at 30 °C.

Table 5. Potentiodynamic polarization data of the as-cast CrFeCoNiNb$_x$Mo$_x$ alloys in 1 M deaerated NaCl solution at 30 °C.

	CrFeCoNiNb$_{0.15}$Mo$_{0.15}$	**CrFeCoNiNb$_{0.3}$Mo$_{0.3}$**	**CrFeCoNiNb$_{0.5}$Mo$_{0.5}$**
E_{corr} (V vs. SHE)	−0.40	−0.28	−0.28
i_{corr} (μA/cm^2)	4.2	10.0	10.7
E_{pp} (V vs. SHE)	0.13	−0.10	0.03
i_{crit} (μA/cm^2)	16.2	40.9	47.0
i_{pass} (μA/cm^2)	9.1	10.5	15.0
E_b (V vs. SHE)	1.31	1.33	1.19

Table 6. The chemical compositions of the as-cast CrFeCoNiNb$_x$Mo$_{1-x}$ alloys and the phases in each alloy analyzed by SEM/EDS.

Alloys	**Weight Percent**					
	Cr	**Fe**	**Co**	**Ni**	**Nb**	**Mo**
CrFeCoNiNb$_{0.25}$Mo$_{0.75}$						
overall	16.3 ± 3.4	17.6 ± 1.6	18.5 ± 1.0	18.1 ± 1.6	7.8 ± 1.3	21.7 ± 2.5
FCC	21.7 ± 2.2	21.2 ± 0.5	18.1 ± 1.9	25.6 ± 2.4	4.4 ± 3.3	9.0 ± 0.1
HCP	14.8 ± 2.3	15.2 ± 2.8	20.2 ± 1.5	15.4 ± 1.3	8.7 ± 0.2	25.7 ± 0.6
CrFeCoNiNb$_{0.75}$Mo$_{0.25}$						
overall	16.9 ± 0.6	17.5 ± 2.6	18.8 ± 2.0	18.9 ± 1.2	20.2 ± 0.2	7.7 ± 0.2
FCC	23.2 ± 2.5	23.8 ± 1.0	19.3 ± 0.6	27.7 ± 2.7	3.6 ± 0.1	2.4 ± 1.7
HCP	14.4 ± 0.6	15.5 ± 1.4	21.0 ± 0.4	14.1 ± 0.7	26.5 ± 1.1	8.5 ± 0.5

Figure 7 shows the XRD patterns of the CrFeCoNiNb$_x$Mo$_{1-x}$ alloys. Two phases existing in these alloys were FCC and HCP-structured laves phases. The HCP phase was the main phase in these CrFeCoNiNb$_x$Mo$_{1-x}$ alloys, and the FCC phase became the minor phase in these alloys. Figure 8 displays the hardness of CrFeCoNiNb$_x$Mo$_{1-x}$ alloys, some of the data were from our previous study [23]. According to our previous study, the overall hardness of the alloy was contributed by the hard HCP dendrites and the soft interdendrities (HCP + HCC). The hardness of both HCP and FCC phases increased with increasing the content of Nb and Mo due to solid solution strengthening. However, the hardness of the FCC phase was still softer than that of the HCP phase. Because the hardness of interdendrities was softer than that of dendrites, increasing the volume fraction of interdendrite would decrease the overall hardness of the alloy. The $CrFeCoNiNb_{0.5}Mo_{0.5}$ alloy had the lowest hardness among these alloys because of the largest ratio of interdendrities in this alloy. The $CrFeCoNiNb_{0.75}Mo_{0.25}$ alloy had the highest hardness of 625 HV among the alloys in the present study.

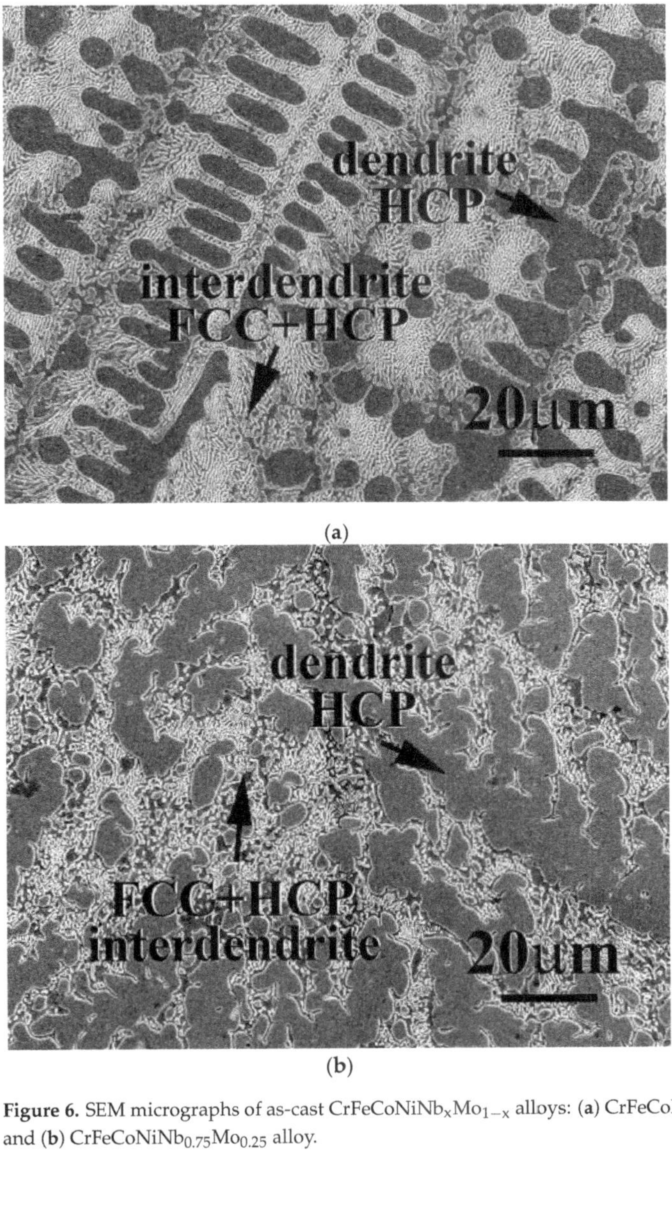

Figure 6. SEM micrographs of as-cast $CrFeCoNiNb_xMo_{1-x}$ alloys: (**a**) $CrFeCoNiNb_{0.25}Mo_{0.75}$ alloy and (**b**) $CrFeCoNiNb_{0.75}Mo_{0.25}$ alloy.

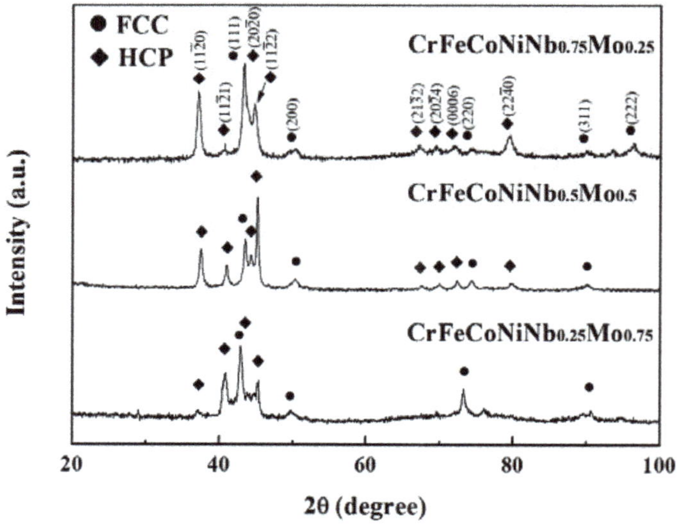

Figure 7. XRD patterns of as-cast $CrFeCoNiNb_xMo_{1-x}$ alloys.

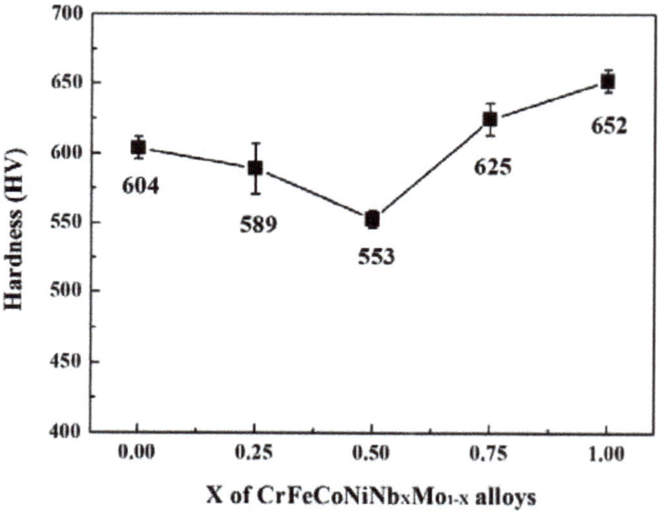

Figure 8. Hardness of as-cast $CrFeCoNiNb_xMo_{1-x}$ alloys. Each value is the average hardness of the alloy. Some of the data are from our previous study [23].

Figure 9 shows the potentiodynamic polarization curves of the as-cast $CrFeCoNiNb_xMo_{1-x}$ alloys in 1 M deaerated H_2SO_4 solution at 30 °C. The potentiodynamic polarization data of these polarization curves are listed in Table 7. The corrosion potential (E_{corr}) and the corrosion current densities (i_{corr}) of $CrFeCoNiNb_xMo_{1-x}$ alloys were very close; the $CrFeCoNiNb_{0.25}Mo_{0.75}$ alloy had the lowest i_{corr} of about 5 $\mu A/cm^2$. The potentiodynamic polarization curve of $CrFeCoNiNb_{0.25}Mo_{0.75}$ and $CrFeCoNiNb_{0.75}Mo_{0.25}$ alloys had apparent anodic peaks, but the $CrFeCoNiNb_{0.5}Mo_{0.5}$ alloy had no anodic peak. The passivation potential (E_{pp}) and critical current density (i_{crit}) of the anodic peaks of the alloys are listed in Table 7. This indicated that the $CrFeCoNiNb_{0.5}Mo_{0.5}$ alloys easily entered the passivation regions and formed the passive films during corrosion in H_2SO_4 solution among these alloys. The current densities of the passivation regions (i_{pass}) of $CrFeCoNiNb_{0.5}Mo_{0.5}$

and $CrFeCoNiNb_{0.75}Mo_{0.25}$ alloys were about 12 A/cm^2, but the $CrFeCoNiNb_{0.25}Mo_{0.75}$ alloy had a larger passivation current density (i_{pass}). The passivation regions of these alloys were all breakdown at a potential (E_b) of about 1.2 V (SHE) because of oxygen evolution reaction [26]. In order to compare the properties of corrosion potential, corrosion current density, anodic peak and passivation region of each alloy in the present study, the $CrFeCoNiNb_{0.5}Mo_{0.5}$ alloy had the best corrosion resistance among the alloys in H_2SO_4 solution.

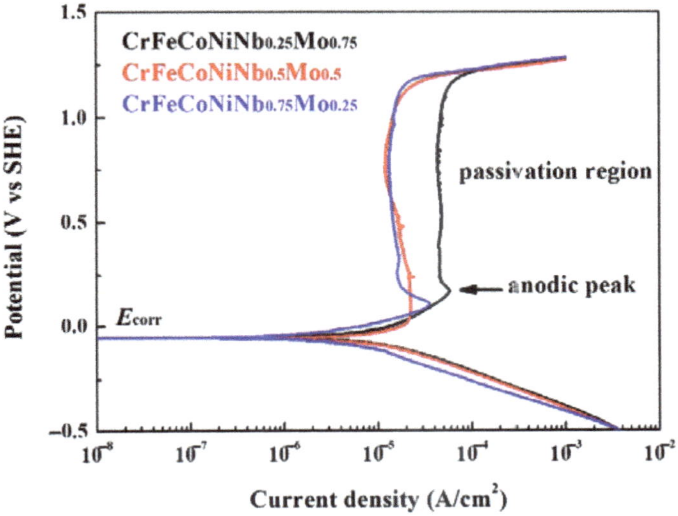

Figure 9. Potentiodynamic polarization curves of as-cast $CrFeCoNiNb_xMo_{1-x}$ alloys tested in the 1 M deaerated sulfuric acid solution at 30 °C.

Table 7. Potentiodynamic polarization data of the as-cast $CrFeCoNiNb_xMo_{1-x}$ alloys in 1 M deaerated H_2SO_4 solution at 30 °C.

	$CrFeCoNiNb_{0.25}Mo_{0.75}$	$CrFeCoNiNb_{0.5}Mo_{0.5}$	$CrFeCoNiNb_{0.75}Mo_{0.25}$
E_{corr} (V vs. SHE)	−0.06	−0.06	−0.06
i_{corr} ($\mu A/cm^2$)	10.2	10.0	4.9
E_{pp} (V vs. SHE)	0.17	N/A	0.11
i_{crit} ($\mu A/cm^2$)	58.3	N/A	35.7
i_{pass} ($\mu A/cm^2$)	44.1	12.2	11.9
E_b (V vs. SHE)	1.21	1.19	1.19

Figure 10 shows the potentiodynamic polarization curves of the as-cast $CrFeCoNiNb_xMo_{1-x}$ alloys in 1 M deaerated NaCl solution at 30 °C. The potentiodynamic polarization data of these polarization curves are listed in Table 8. The cathodic limit current density (i_L) was found in these $CrFeCoNiNb_xMo_{1-x}$ alloys. The cathodic limit current density meant that the maximum reaction rate was limited because of the diffusion rate of hydroxyl ions (OH^-) in the solution [26]. The corrosion potential (E_{corr}) of $CrFeCoNiNb_{0.25}Mo_{0.75}$ alloy was more positive than the other alloys; the corrosion potential (E_{corr}) of $CrFeCoNiNb_{0.5}Mo_{0.5}$ and $CrFeCoNiNb_{0.75}Mo_{0.25}$ alloy was very close. Moreover, the corrosion current density (i_{corr}) of $CrFeCoNiNb_{0.25}Mo_{0.75}$ alloy was larger than the other alloys. The corrosion current density (i_{corr}) of $CrFeCoNiNb_{0.5}Mo_{0.5}$ and $CrFeCoNiNb_{0.75}Mo_{0.25}$ alloys were about 10 $\mu A/cm^2$. All of the $CrFeCoNiNb_xMo_{1-x}$ alloys had apparent anodic peaks in 1 M deaerated NaCl solution. The passivation regions of $CrFeCoNiNb_{0.5}Mo_{0.5}$ and $CrFeCoNiNb_{0.75}Mo_{0.25}$ alloys were better than that of $CrFeCoNiNb_{0.25}Mo_{0.75}$ alloy. The minimum passivation current densities (i_{pass}) of $CrFeCoNiNb_{0.5}Mo_{0.5}$ and $CrFeCoNiNb_{0.75}Mo_{0.25}$

alloys were about 15 μA/cm². When comparing the properties of corrosion potential, corrosion current density, anodic peak and passivation region of each alloy in the present study, the $CrFeCoNiNb_{0.5}Mo_{0.5}$ alloy had the best corrosion resistance among the alloys in NaCl solution.

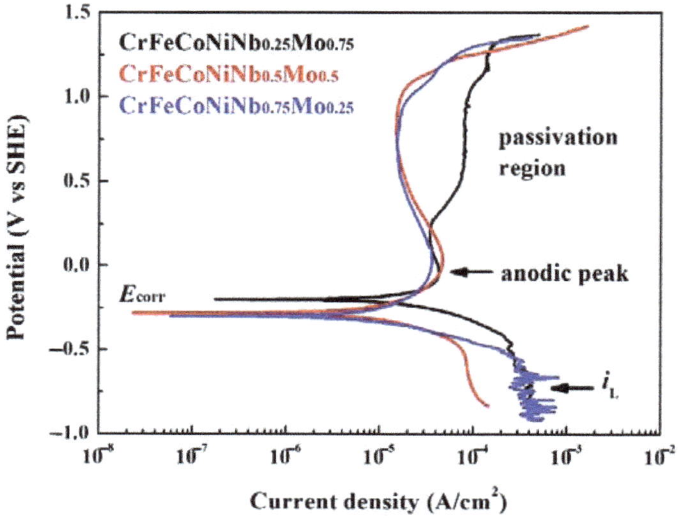

Figure 10. Potentiodynamic polarization curves of as-cast $CrFeCoNiNb_xMo_{1-x}$ alloys tested in the 1 M deaerated sodium chloric solution at 30 °C.

Table 8. Potentiodynamic polarization data of the as-cast $CrFeCoNiNb_xMo_{1-x}$ alloys in 1 M deaerated NaCl solution at 30 °C.

	$CrFeCoNiNb_{0.25}Mo_{0.75}$	$CrFeCoNiNb_{0.5}Mo_{0.5}$	$CrFeCoNiNb_{0.75}Mo_{0.25}$
E_{corr} (V vs. SHE)	−0.20	−0.28	−0.30
i_{corr} (μA/cm²)	31.0	10.7	10.8
E_{pp} (V vs. SHE)	−0.04	0.03	0.03
i_{crit} (μA/cm²)	43.2	47.0	36.0
i_{pass} (μA/cm²)	34.4	15.0	15.6
E_b (V vs. SHE)	1.32	1.19	1.26

The corrosion rate of the alloys in deaerated 1 M H_2SO_4 and 1 M NaCl solutions can be calculated by assuming that all of the corrosion types of the alloys in these solutions are a type of uniform corrosion. Therefore, the relationship between corrosion depth of one year and corrosion current density is listed as the following Equation [27]:

$$\text{corrosiondepth} = \frac{M \cdot i_{corr} \cdot t}{\rho \cdot n \cdot F} \quad (1)$$

where M is the average atomic mass (g/mol), i_{corr} is the corrosion current density (A/cm²), t is the corrosion time (31,536,000 s, 1 yr), ρ is the average density (g/cm³), n is the number of average valence electron and F is the Faraday constant (96,500 C/mol). This study assumes that the average density of an alloy is $\rho = \sum X_i \rho_i$, where X_i and ρ_i are the molar fraction and density of element i. The corrosion rates (mm per year) of the alloys in deaerated 1 M H_2SO_4 and 1 M NaCl solutions are listed in Table 9. The $CrFeCoNiNb_{0.3}Mo_{0.3}$ alloy had the smallest corrosion rate (0.0732 mm/yr) in 1 M deaerated H_2SO_4 solution, and alloy $CrFeCoNiNb_{0.75}Mo_{0.25}$ had the largest corrosion rate (0.3315 mm/yr) in this solution. The $CrFeCoNiNb_{0.15}Mo_{0.15}$ alloy had the smallest corrosion rate (0.0425 mm/yr) in 1 M deaerated NaCl solution, and $CrFeCoNiNb_{0.5}Mo_{0.5}$ and $CrFeCoNiNb_{0.75}Mo_{0.25}$ alloys had larger corrosion rate in this solution. The corrosion rate of $CrFeCoNiNb_{0.5}Mo_{0.5}$ alloy (0.1152 mm/yr) in 1 M deaerated NaCl solution was higher than that of $CrFeCoNiNb_{0.15}Mo_{0.15}$ alloy in the same solution. However, the $CrFeCoNiNb_{0.5}Mo_{0.5}$ alloy still possessed the best combination of corrosion resistance and hardness among these alloys.

Table 9. Corrosion rates of the as-cast alloys in deaerated 1 M H_2SO_4 and 1 M NaCl solutions at 30 °C.

Alloys	1 M H_2SO_4 (mm/yr)	1 M NaCl (mm/yr)
$CrFeCoNiNb_{0.15}Mo_{0.15}$	0.1204	0.0425
$CrFeCoNiNb_{0.3}Mo_{0.3}$	0.0732	0.1043
$CrFeCoNiNb_{0.25}Mo_{0.75}$	0.1091	0.0532
$CrFeCoNiNb_{0.5}Mo_{0.5}$	0.1077	0.1152
$CrFeCoNiNb_{0.75}Mo_{0.25}$	0.3315	0.1172

4. Conclusions

The microstructures, hardness and corrosion properties of hypoeutectic $CrFeCoNiNb_xMo_x$ alloys (x = 0.15 and 0.3) and hypereutectic $CrFeCoNiNb_xMo_{1-x}$ alloys (x = 0.25, 0.5 and 0.75) were studied. There were two phases (FCC and HCP) in these alloys. The dendrites of hypoeutectic $CrFeCoNiNb_xMo_x$ alloys (x = 0.15 and 0.3) were an FCC phase; the interdendrities of these alloys were a eutectic structure which the phases were HCP and FCC phases. The dendrites of hypereutectic $CrFeCoNiNb_xMo_{1-x}$ alloys (x = 0.25, 0.5 and 0.75) were an HCP-structured laves phase, and the interdendrities of these alloys were a eutectic structure with HCP and FCC phases. Increasing the content of Nb and Mo would increase the hardness of the alloys because of the formation of the hard HCP phase and the effect of solid solution strengthening. The $CrFeCoNiNb_{0.75}Mo_{0.25}$ alloy had the highest hardness of 625 HV among the alloys in present study. After potentiodynamic polarization test in deaerated 1 M H_2SO_4 and 1 M NaCl solutions at 30 °C, the $CrFeCoNiNb_{0.5}Mo_{0.5}$ alloy had the best corrosion resistance among these alloys by comparing the properties of corrosion potential, corrosion current density, anodic peak and passivation region of the alloys. Therefore, $CrFeCoNiNb_{0.5}Mo_{0.5}$ alloy was the best alloy among these alloys by comparing the corrosion properties and hardness.

Author Contributions: Conceptualization, C.-H.T.; methodology, C.-H.T., Y.-H.C. and M.-C.T.; formal analysis, C.-H.T., Y.-H.C. and M.-C.T.; investigation, C.-H.T., Y.-H.C. and M.-C.T.; data curation, C.-H.T., Y.-H.C. and M.-C.T.; writing—original draft preparation, C.-H.T.; writing—review and editing, C.-H.T.; supervision, C.-H.T. All authors have read and agreed to the published version of the manuscript.

Funding: This research received no external funding.

Institutional Review Board Statement: Not applicable.

Informed Consent Statement: Not applicable.

Data Availability Statement: Not applicable.

Acknowledgments: The authors are grateful to the Ministry of Science and Technology of the Republic of China for its financial support under the project MOST 110-2221-E-034-007.

Conflicts of Interest: The funders had no role in the design of the study; in the collection, analyses, or interpretation of data; in the writing of the manuscript, or in the decision to publish the results.

References

1. Murty, B.S.; Yeh, J.W.; Ranganathan, S.; Bhattacharjee, P.P. *High-Entropy Alloys*, 2nd ed.; Elsevier Co.: Amsterdam, The Nederlands, 2019; pp. 13–30.
2. Yeh, J.W. Alloy Design Strategies and Future Trends in High-Entropy Alloys. *JOM* **2013**, *65*, 1759–1771. [CrossRef]

3. George, E.P.; Raabe, D.; Ritchie, R.O. High-entropy alloys. *Nat. Rev. Mater.* **2019**, *4*, 515–534. [CrossRef]
4. Otto, F.; Dlouhy, A.; Somsen, C.; Bei, H.; Eggeler, G.; George, E.P. The influences of temperature and microstructure on the tensile properties of a CoCrFeMnNi high-entropy alloy. *Acta Mater.* **2013**, *61*, 5743–5755. [CrossRef]
5. Lu, K.; Chauhan, A.; Tirunilai, A.S.; Freudenberger, J.; Kauffmann, A.; Heilmaier, M.; Aktaa, J. Deformation mechanisms of CoCrFeMnNi high-entropy alloy under low-cycle-fatigue loading. *Acta Mater.* **2021**, *215*, 117089. [CrossRef]
6. Zhang, K.B.; Fu, Z.Y.; Zhang, J.Y.; Wang, W.M.; Wang, H.; Wang, Y.C.; Zhang, Q.J.; Shib, J. Microstructure and mechanical properties of $CoCrFeNiTiAl_x$ high-entropy alloys. *Mater. Sci. Eng. A* **2009**, *508*, 214–219. [CrossRef]
7. Senkov, O.N.; Wilks, G.B.; Scott, J.M.; Miracle, D.B. Mechanical properties of $Nb_{25}Mo_{25}Ta_{25}W_{25}$ and $V_{20}Nb_{20}Mo_{20}Ta_{20}W_{20}$ refractory high entropy alloys. *Intermetallics* **2011**, *19*, 698–706. [CrossRef]
8. Shkodich, N.; Sedegov, A.; Kuskov, K.; Busurin, S.; Scheck, Y.; Vadchenko, S.; Moskovskikh, D. Refractory High-Entropy HfTaTiNbZr-Based Alloys by Combined Use of Ball Milling and Spark Plasma Sintering: Effect of Milling Intensity. *Metals* **2020**, *10*, 1268. [CrossRef]
9. Vaidya, M.; Gariapati, M.M.; Murty, B.S. High-entropy alloys by mechanical alloying: A review. *J. Mater. Res.* **2019**, *34*, 664–686. [CrossRef]
10. Zhang, Y.; Li, R. New Advances in High-Entropy Alloys. *Entropy* **2020**, *22*, 1158. [CrossRef]
11. Xu, D.; Wang, M.; Li, T.; Wei, X.; Lu, Y. A critical review of the mechanical properties of CoCrNi-based medium-entropy alloys. *Microstructures* **2022**, *2*, 2022001. [CrossRef]
12. Muangtong, P.; Rodchanarowan, A.; Chaysuwan, D.; Chanlek, N.; Goodall, R. The corrosion behaviour of CoCrFeNi-x (x = Cu, Al, Sn) high entropy alloy systems in chloride solution. *Corros. Sci.* **2020**, *172*, 108740. [CrossRef]
13. Ma, G.; Zhao, Y.; Cui, H.; Song, X.; Wang, M.; Lee, K.; Gao, X.; Song, Q.; Wang, C. Addition Al and/or Ti Induced Modifications of Microstructures, Mechanical Properties, and Corrosion Properties in CoCrFeNi High-Entropy Alloy Coatings. *Acta Metall. Sin.* **2021**, *34*, 1087–1102. [CrossRef]
14. Godlewska, E.M.; Mitoraj-Królikowska, M.; Czerski, J.; Jawańska, M.; Gein, S.; Hecht, U. Corrosion of Al(Co)CrFeNi High-Entropy Alloys. *Front. Mater.* **2020**, *7*, 566336. [CrossRef]
15. Wang, W.; Wang, J.; Yi, H.; Qi, W.; Peng, Q. Effect of Molybdenum Additives on Corrosion Behavior of $(CoCrFeNi)_{100-x}Mo_x$High-Entropy Alloys. *Entropy* **2018**, *20*, 908. [CrossRef]
16. Hsu, W.C.; Kao, W.P.; Yeh, J.W.; Tsai, C.W. Effect of Mo on the Mechanical and Corrosion Behaviors in Non-Equal Molar AlCrFeMnNi BCC High-Entropy Alloys. *Materials* **2022**, *15*, 751. [CrossRef] [PubMed]
17. Tsau, C.H.; Chen, P.M. Corrosion Behavior of $Cr_{19}Fe_{22}Co_{21}Ni_{25}Mo_{13}$ Alloy in 1 M Nitric Acid and 1 M Hydrochloric Acid Solutions. *Crystal* **2021**, *11*, 1289. [CrossRef]
18. Mariano, N.A.; Souza, C.A.C.; May, J.E.; Kuri, S.E. Influence of Nb content on the corrosion resistance and saturation magnetic density of FeCuNbSiB alloys. *Mater. Sci. Eng. A* **2003**, *354*, 1–5. [CrossRef]
19. Tsau, C.H.; Yeh, C.Y.; Tsai, M.C. The Effect of Nb-Content on the Microstructures and Corrosion Properties of $CrFeCoNiNb_x$ High-Entropy Alloys. *Materials* **2019**, *12*, 3716. [CrossRef]
20. Tsau, C.H.; Lin, S.X.; Fang, C.H. Microstructures and corrosion behaviors of FeCoNi and CrFeCoNi equimolar alloys. *Mater. Chem. Phys.* **2017**, *186*, 534–540. [CrossRef]
21. Tsau, C.H.; Tsai, M.C.; Wang, W.L. Microstructures of FeCoNiMo and CrFeCoNiMo Alloys, and the Corrosion Properties in 1 M Nitric Acid and 1 M Sodium Chloride Solutions. *Materials* **2022**, *15*, 888. [CrossRef]
22. Bard, A.J.; Faulkner, L.R. *Electrochemical Methods: Fundamentals and Applications*, 2nd ed.; The Table on Inside Back Cover; John Wiley & Sons: New York, NY, USA, 2000.
23. Tsau, C.H.; Tsai, M.C. The Effects of Mo and Nb on the Microstructures and Properties of CrFeCoNi(Nb,Mo) Alloys. *Entropy* **2018**, *20*, 648. [CrossRef] [PubMed]
24. Smith, W.F. *Foundations of Materials Science and Engineering*, 3rd ed.; McGraw–Hill Co.: Singapore, 2004; pp. 877–878.
25. Chawla, S.L. *Materials Selection for Corrosion Control*; ASM International: Materials Park, OH, USA, 1993; p. 18.
26. Revie, R.W.; Uhlig, H.H. *Corrosion and Corrosion Control, an Introduction to Corrosion Science and Engineering*, 4th ed.; John Wiley & Sons: New York, NY, USA, 2008; p. 476.
27. Andrade, C.; Alonso, C. Corrosion rate monitoring in the laboratory and on-site. *Constr. Build. Mater.* **1996**, *10*, 315–328. [CrossRef]

materials

MDPI

Article

Re-Melting Behaviour and Wear Resistance of Vanadium Carbide Precipitating $Cr_{27.5}Co_{14}Fe_{22}Mo_{22}Ni_{11.65}V_{2.85}$ High Entropy Alloy

Kai Treutler *, Swenja Lorenz and Volker Wesling

Institute of Welding and Machining, Clausthal University of Technology, 38678 Clausthal-Zellerfeld, Germany; swenja.lorenz@tu-clausthal.de (S.L.); office@isaf.tu-clausthal.de (V.W.)
* Correspondence: treutler@isaf.tu-clausthal.de

Abstract: High entropy alloys (HEAs) are among of the most promising new metal material groups. The achievable properties can exceed those of common alloys in different ways. Due to the mixture of five or more alloying elements, the variety of high entropy alloys is fairly huge. The presented work will focus on some first insights on the weldability and the wear behavior of vanadium carbide precipitation $Cr_{27.5}Co_{14}Fe_{22}Mo_{22}Ni_{11.65}V_{2.85}$ HEA. The weldability should always be addressed in an early stage of any alloy design to avoid welding-related problems afterwards. The cast $Cr_{27.5}Co_{14}Fe_{22}Mo_{22}Ni_{11.65}V_{2.85}$ HEA has been remelted using a TIG welding process and the resulting microstructure has been examined. The changes in the microstructure due to the remelting process showed little influence of the welding process and no welding-related problems like hot cracks have been observed. It will be shown that vanadium carbides or vanadium-rich phases precipitate after casting and remelting in a two phased HEA matrix. The hardness of the as cast alloy is 324HV0.2 and after remelting the hardness rises to 339HV0.2. The wear behavior can be considered as comparable to a Stellite 6 cobalt base alloy as determined in an ASTM G75 test. Overall, the basic HEA design is promising due to the precipitation of vanadium carbides and should be further investigated.

Keywords: high entropy alloy (HEA); wear; weldability; microstructure; ASTM G75; precipitating; vanadium carbide

Citation: Treutler, K.; Lorenz, S.; Wesling, V. Re-Melting Behaviour and Wear Resistance of Vanadium Carbide Precipitating $Cr_{27.5}Co_{14}Fe_{22}Mo_{22}Ni_{11.65}V_{2.85}$ High Entropy Alloy. *Materials* **2021**, *14*, 1871. https://doi.org/10.3390/ma14081871

Academic Editor: Petrica Vizureanu

Received: 19 March 2021
Accepted: 7 April 2021
Published: 9 April 2021

1. Introduction

Highly entropy alloys have been the subject of increasing scientific research in the last 20 years and are considered to be the next generation of materials in some areas because of their properties. Due to their composition of a larger number of different elements with approximately equal distribution as well as the possible addition of various other alloying elements, the number of potentially applicable alloying systems is immeasurable. Nevertheless, on the basis of various experiments and procedures, such as CALPHAD [1–3], it has been possible to identify and investigate promising systems for different applications. The fields of application range from high-temperature or cryogenic materials to corrosion-resistant and wear-resistant materials [4–7]. In addition to the single-phase high entropy alloys, other alloys have been developed which form precipitates in order to further influence the properties of these alloys [8]. Precipitated phases can be used to increase strength [9] and wear resistance, as in conventional non-high-entropy alloys [10]. One famous high entropy alloy is the so called Cantor alloy [11] which is investigated in various studies for example references [12,13] including its derivatives. This high entropy alloy and its derivatives consist of near equiatomic CrCoFeMnNi.

In addition to this alloy with its partly known properties, other systems are also being investigated for their properties. The results suggest a wide range of applications, ranging from cryogenic conditions [14,15] to high-temperature applications [16]. Furthermore, some of the high entropy alloys have very good corrosion resistance and wear protection properties [2,8]. Especially in wear protection, applied coatings of different materials are used, so that a change to high entropy alloys requires only minor adjustments on the process side.

The welding of high entropy alloys becomes more and more important due to the increased development in the field of alloy designs. This is reflected in the increasing number of publications in this field [17]. One aspect that is considered in the welding related HEA research so far are the mechanical and microstructural properties of the welds. Besides of the mechanical properties and the weldability the machinability becomes more important for HEA as well. Richter et al. showed that modern machining process have a beneficial influence on the surface integrity of a HEA [18].

The basic weldability of HEA is currently been seen as good with different welding processes [17,19,20]. Especially, for the CrCoFeMnNi basis the welding with laser or friction stir welding has been successfully performed [21,22]. The change in one alloying element in the concept can change its properties. So for each alloy the general welding technological occurring phenomena regarding the weldability and needs to be addressed besides metallurgical effects. These four main phenomena are:

- Hot-cracking Hot-cracking occurs due to shrinkage phenomena near the solidification of material and the lack of present melt to close the occurring gaps [23]. This phenomena is supported by the occurrence of a second low melting phase which solidifies in the interdendritic region of the first phase [23].
- Cold-cracking Cold cracking occurs due to high hardness and/or low toughness of the material [24]. This can be intensified due to external restraints or other phenomena like the hydrogen assisted cracking [25,26].
- Stress relief cracking Cracking which occurs due to high strength and low ductility during post weld heat treatment [27–29].
- Pore formation Pore formation occurs mostly due to a gap in the solubility of gases between the solid material and the melt [30].

For each of this phenomena there are designated welding test methods to verify the susceptibility to these phenomena. The basis for these tests is the remeltability of the material regarding the microstructural ability to form a suitable microstructure and that none of the weldability related phenomena occur without any additional constraints on the melt with for example free shrinkage.

In this study, the basic properties of a newly CrCoFeMoNiV alloy with small additions of Nb and Si is presented. This type of alloy exceeds in the studies on the wear resistance and the weldability/remeltability due to the exchange of Mn with Mo in the CrCoFeMnNi HEA and the addition of vanadium to the alloy to increase the wear resistance of the alloy by precipitating vanadium carbides. The possibility of precipitating hard phases for a similar alloy ($CoCrFeNi_{2.1}Nb_{0.2}$) have been shown by Sunkari et al., for Nb-containing hard phases [31]. The alloy used in this study can be called a complex concentrated alloy due to the intentional precipitation of carbides [2], but due to the fact that the matrix should be a single solution, the term high entropy alloy is used. The precipitation of carbides in similar HEAs has been reported by Huang et al., and the amount of carbides and the hardness of the alloy is directly linked to the carbon content [32]. The weldability and wear protection properties under abrasive wear are presented for CrCoFeMoNiV and are compared with conventional wear protection alloys such as Stellite 6 and Stellite 12. These materials are typically used in three-body abrasion, so the ASTM G75 "Miller Test" is used to characterize their wear properties. The wear resistance of these kinds of HEA are already of high interest and subject to studies in various alloy compositions [33–36]. The creation of an alloy-side basis for weldable wear-resistant high-entropy alloys is the aim of the investigations and the focus is on weldability. The integration of weldability into the alloy development represents a special approach that differs from the normally chosen path in alloy development, where the mechanical properties are in the foreground.

2. Materials and Methods

The methods used and the production of the alloy are described below. First the material and the casting process used will be described, secondly information on the applied wear test will be given followed by a description of the microstructural analysis methods and a short info about the used weldability test.

2.1. Used Material and Casting

The high entropy alloy was produced by casting small test pieces. A high-temperature furnace was used for this. The educts were therefore available in powder form in various master alloys. The mixing ratios of the different master alloys are shown in Table 1. These were mixed and melted in a graphite crucible at approx. 1600 °C under argon atmosphere. A total of 300 g was initially mixed and melted. The melt was kept at 1600 °C for 10 min under sequential stirring to ensure complete melting. The casting was done in a mould made of cast iron with the dimensions 60 mm × 30 mm × 10 mm. Cooling was done on air without any acceleration.

Table 1. Used pre alloys for casting.

Weight-%							
FeNb	**FeMo**	**Fe**	**FeSi**	**FeV**	**Ni**	**Cr**	**Co**
0.14	24.4	19.25	0.24	3,47	17.5	17.5	17.5

The targeted chemical composition of the alloy is shown in Table 2. The target alloy has an increased proportion of iron compared to the other alloy constituents. The proportion of carbon and vanadium should lead to the formation of vanadium carbide, which should precipitate in the matrix.

Table 2. Intended chemical composition of the used HEA in advance of casting.

Element in at.-%								
Fe	**C/B**	**SI**	**Cr**	**Mo**	**Co**	**V**	**Nb**	**Ni**
25.65	0.3	0.5	17.5	18.1	17.5	2.85	0.1	17.5

After casting, the surface was smoothed by grinding and samples were taken for metallographic examinations, welding and wear testing by wire electric discharge machining (EDM). These samples were freed from any erosion layer by grinding.

2.2. ASTM G75 Test

In order to map the three-body abrasive wear, the Miller test according to ASTM G75 is used [37]. This was originally developed to determine the wear in pipes that transport suspensions. With the help of the so-called Miller number, the abrasiveness of different suspensions can be determined and thus compared with each other [37–39]. For this purpose, [37] gives the chemical composition of a steel that is used for pumps in pipelines. Furthermore, the SAR number can be calculated with the Miller test. It is an indicator for the wear behaviour of a material in a certain suspension. If the suspension is kept constant and the material is changed, the materials can be compared in terms of their wear resistance using the SAR number. If the determined Miller number or SAR number is less than 50, the suspension is likely to cause only minor wear, while higher Miller numbers cause greater abrasive wear attack. With the help of these numbers, for example, the design of pumps can be carried out if the abrasiveness of the suspension used is known [10].

In this test, according to the standard, a sample is moved under a defined load of 22.24 N oscillating through an abrasive suspension against a counter body of chloroprene rubber at a speed of 20 m/min (Figure 1)).

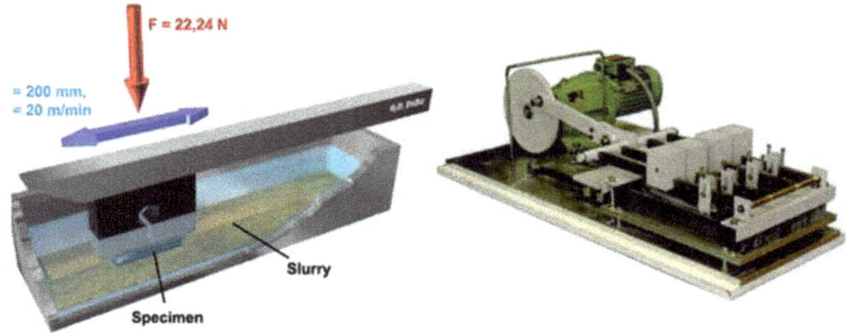

Figure 1. ASTM G75 (Miller Test) [10].

At the end of each stroke, the sample is lifted so that there is abrasive material mixture in the gap between the sample and the counterbody at any time. The path length of one stroke is 200 mm. The total duration of the test is six hours, whereby the loss of mass of the samples is determined gravimetrically after every two hours [10].

According to [37], the suspension used consists of 150 g solid and 150 g liquid. The standard material used for the tests at the Institute of Welding Technology and Separating Production Processes is a high-grade corundum of grain size F220 (d50 = 57.5 µm), (Figure 2) [10].

Figure 2. SEM Image of corundum F220 [10].

2.3. *Micrstructural Analysis*

For the metallographic analysis of the samples, they were embedded in a polymer and then automatically ground and polished. Subsequently, images were taken using an optical microscope and a scanning electron microscope. In order to emphasize the structure of the samples more clearly, they were etched. For this purpose, an etching solution according to Murakami with Murakami etchant (10 g KOH or NaOH, 10 g potassium ferricyanide, 100 mL water) was used. The preparation and evaluation of the micrographs has been conducted in accordance with the common standards including but not limited to: DIN CEN ISO/TR 16,060 "Etchants for macroscopic and microscopic examination" [40], DIN EN ISO 6520-1 "Classification of geometric imperfections in metallic materials" [41] and DIN EN ISO 17,639 "Macroscopic and microscopic examination of welds" [42].

2.4. Re-Melting Behaviour

In order to obtain a statement about the weldability at this stage of the alloy development, the cast samples were re-melted using a TIG welding process on one side of the sample. This then allowed statements to be made about the microstructure resulting from welding and welding-related problems such as hot or cold cracks and any pore formation to be identified.

3. Results

No casting-related issues arose during processing. The melt was sufficiently fluid and filled the mould well. A slag formed on the surface of the melt, but this was not examined more closely. Significant outgassing of the melt was also not observed. There were no cracks or shrinkage cavities. After grinding the specimen showed a metallic blank surface without any defects. Due to the forming slag an additional determination of the chemical composition has been done.

3.1. Chemical Composition

To determine the losses of alloying elements an EDX analysis has been carried out, Table 3. This showed that the intended increased iron content was reduced during melting and casting, Table 3. In addition, the amount of dissolved nickel was about 6 at.-% below the target. Correspondingly, the Cr and Mo contents are significantly higher. This suggests that the slag formed consists largely of iron and nickel.

Table 3. Chemical composition of the used HEA in advance of casting (EDX).

Element in at.-%					
Fe	Cr	Mo	Co	V	Ni
22.21	27.51	22.06	13.82	2.85	11.66

Nevertheless, the microstructure exhibits the desired effects.

3.2. Microstructure

Figure 3 shows the microstructure of the high entropy alloy not etched. Finely distributed yellowish precipitates with an average size between 1 μm and 2 μm are visible. The distribution is uniform over the entire cross-section. Next to the precipitates, two different phases are visible, Figure 3.

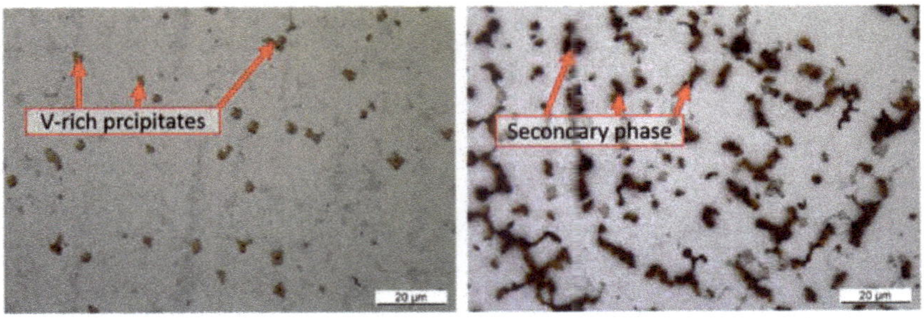

(a) **(b)**

Figure 3. Microstructure of the as cast alloy (**a**) not etched, (**b**) Murakami solution.

The secondary phase can be made visible by etching with Murakami solution (Figure 3). Here, the dendrites can be seen. This indicates that the second phase is a phase with reduced solidification temperature. Since the precipitates occur almost exclusively in the first phase, it can be concluded that they only form after solidification at lower temperatures.

A detailed analysis of the chemical composition of the three different phases shows that the precipitates are vanadium, chromium and nickel rich phases (Figure 4 and Table 4). In addition, significant amounts of cobalt and iron are present as well. Due to the large difference in the carbon content, it can be stated that this alloy has two different precipitates forming. One is a chromium-vanadium carbide and one is a chromium-vanadium intermetallic.

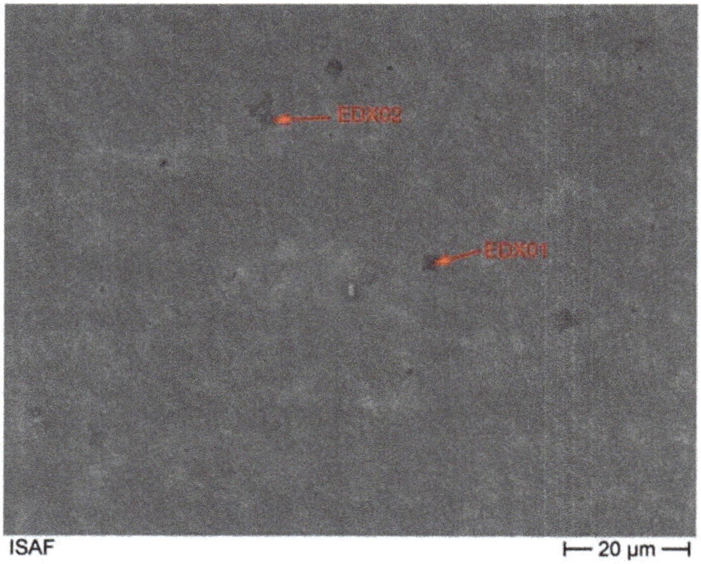

Figure 4. SEM image of the alloy with marks for EDX measurements of precipitates.

Table 4. Chemical composition of the precipitates (EDX).

Position	Element in at.-%						
-	Fe	Cr	Mo	Co	V	Ni	C
EDX1	11.68	22.63	1.59	8.41	38.89	9.39	7.13
EDX2	5.57	25.88	0.89	3.57	58.14	5.02	0.7

The corresponding spectra of the presented measurements are given in Figures A1 and A2, which can be found in the Appendix A. The chemical composition of the second phase, measured by EDX, is shown in Table 5. The corresponding measuring points are marked in the SEM image (Figure 5). The second phase of the matrix shows a reduced molybdenum content compared to the surface measurement of 6 at.-% and 12 at.-% with simultaneously high iron, cobalt and especially nickel contents. This corresponds to an average reduction of the molybdenum content of 13 at.-%. The nickel content in this phase is twice as high as in the entire alloy. The chromium content is missing by 7 at.-% and the iron content is increased by 6 at.-%. This, in conjunction with the area analysis to determine the overall chemical composition, shows that the main phase is a molybdenum-rich phase and the second phase is a molybdenum-poor phase. The increased nickel content in the second phase leads to the assumption that the solubility of nickel in the molybdenum rich matrix alloy is limited. And vice versa that there is a corresponding alloy nickel rich alloy composition with limited molybdenum solubility. Incomplete melting can be dismissed due to the clear presence of chromium cobalt and iron. In addition, the larger areas of the second phase no longer have a particle-like shape.

Table 5. Chemical composition of the used HEA in advance of casting (EDX).

Position	Element in at.-%					
	Fe	Cr	Mo	Co	V	Ni
EDX3	27.55	22.19	6.64	20.91	1.7	21.77
EDX4	26.23	18.75	12.32	19.64	1.03	18.39
EDX5	30.27	17.99	2.97	21.43	1.13	26.21
EDX6	28.90	18.42	3.8	21.31	1.09	26.24

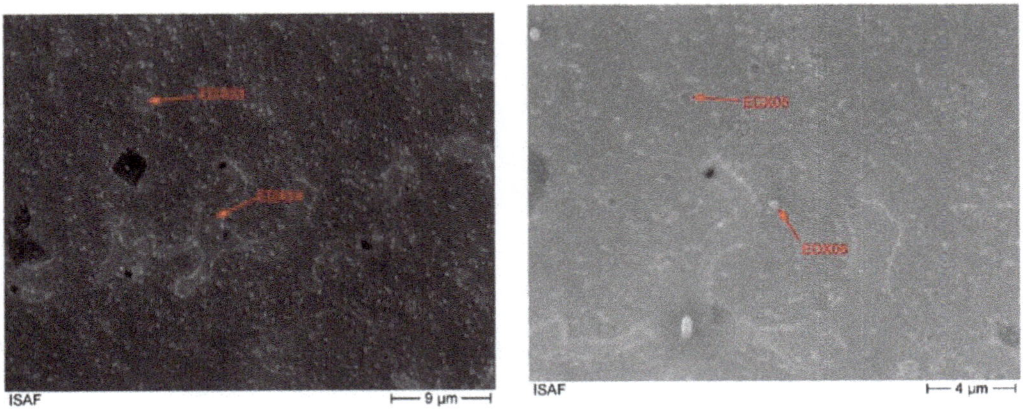

(**a**) (**b**)

Figure 5. SEM images of different magnifications (**a**) 2500×, (**b**) 5000× of the alloy with marks for EDX measurements of two main phases.

Furthermore, Figure 5 shows that over the entire cross-sectional area there are further precipitates with a size of a few hundred nanometres. The composition of these fine precipitates indicates that they are not precipitates but the second molybdenum-poor phase, Table 5. The corresponding EDX spectra are shown in the Appendix A in Figures A1–A4.

The formation of two phases is rather unfavourable for a possible weldability of the alloy, since two-phase systems with a dendritic growth as shown in Figure 4 and a low-melting phase within the interdendritic space have a greater risk of hot cracking, [23]. However, there are approaches to modify the original alloy design to a single-phase system. For this purpose, the molybdenum content could be adjusted. Detailed investigations are needed here.

3.3. Remelting Behaviour

In order to be able to give a basic assessment of the weldability of the alloy, the samples were locally remelted with the help of a TIG welding process with argon as shielding gas and examined for the resulting microstructure. A current of 100 A with a tip to workpiece distance of 10 mm has been used. The specimen was 10 mm × 10 mm × 50 mm and was melted on the smaller surface at on end. The heating/melting was done until the full face of the specimen was molten. Figure 6 shows the microstructure of the remelted area.

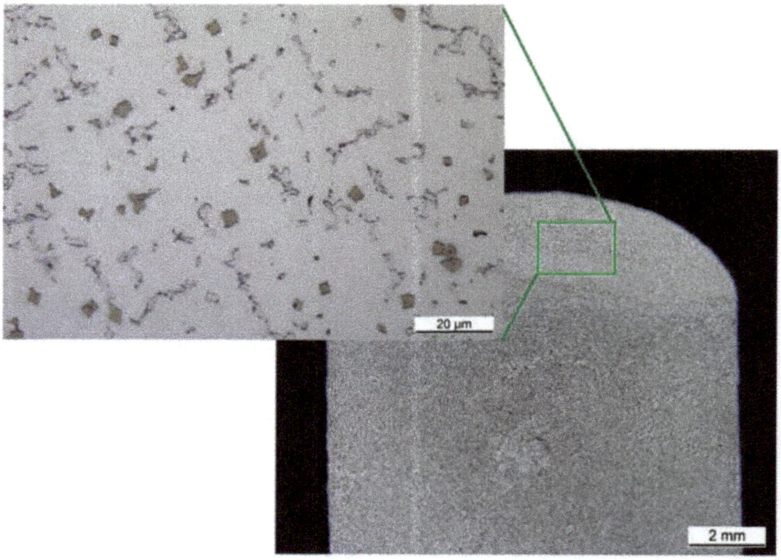

Figure 6. Microstructure of the re-melted sample.

The lower right part of Figure 6 shows an overview of the remelted area and the base material. As can be seen, there is no distinct heat-affected zone and no clear difference in precipitation distribution between the re-melted area and the base material. Neither hot cracks nor pore formation were observed. At most, the remelted area is slightly finer in terms of its phase distribution and size. This is due to the accelerated cooling during welding compared to the casting tests. In general, the accelerated cooling and the change in the distribution of the phases should result in a change in the hardness of the material. Overall, it can be stated that the used high entropy alloy has a basic remeltability.

3.4. *Hardness*

The hardness measurement across the cross-section of the sample showed that the base material has a hardness of 324HV0.2 and the remelted area has a hardness of 339HV0.2 on average, Table 6. The series measurement carried out to identify different areas also below the re-melted material is shown in Figure 7. The variations that occur within the measurements are within the normal range for hardness measurements on welded and materials. The hardness measurements also show that there is no distinct heat-affected zone with changed properties, as is typical for other material groups, like high strength low alloyed steels. Due to the considerable high hardness of the material a good wear resistance under three body abrasive wear conditions can be anticipated.

Table 6. Hardness of the used alloy.

Material Condition	Hardness HV0.2	Deviation
As Cast	324	7.2
Re-melted	339	7.3

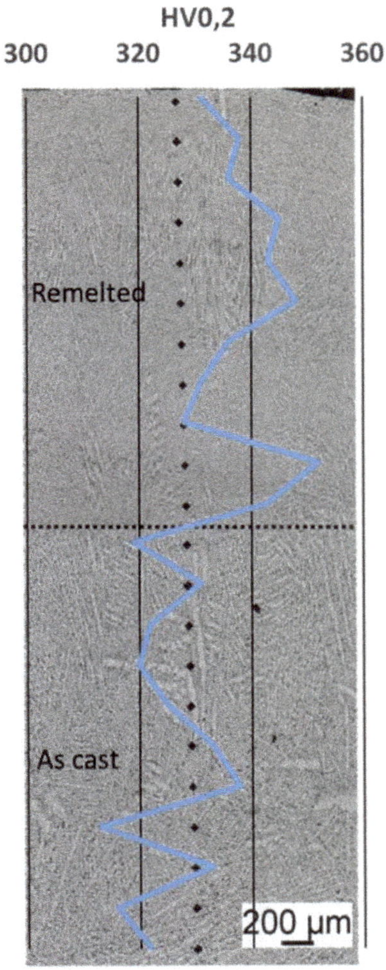

Figure 7. Hardnessprofile of re-melted alloy and base material.

3.5. Waer Resitance

In order to be able to compare the properties of the high-entropy alloy used with other wear protection materials used, the wear resistance of cladded Stellite 6 and Stellite 12 was determined. Stellites are hard alloys based on cobalt-chromium [10].

These two alloys are currently widely used as protection in abrasive three-body wear. In the ASTM G57 "Miller Test", these show an ablation of 91 mm^3 for the Stellite 12 and 122 mm^3 for the Stellite 6. In comparison, the CrCoFeMoNi + VC high entropy alloy used shows an ablation of 126 mm^3 (Figure 8).

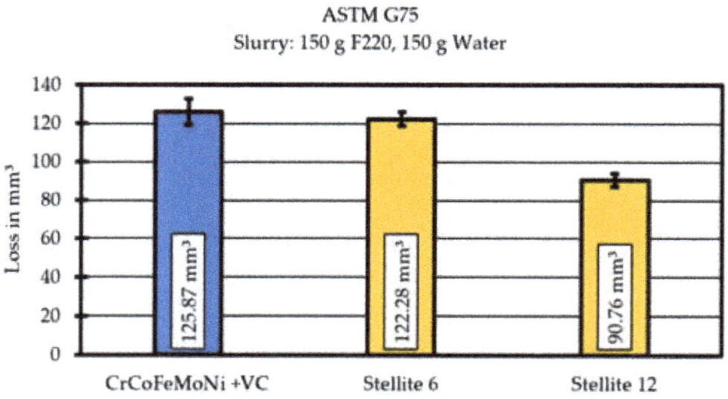

Figure 8. Mass loss for CrCoFeMoNi + VC and Stellite 6 and Stellite 12.

These results show that the wear resistance of the CrCoFeMoNi + VC alloy is within the range of conventionally used wear protection alloys and clearly shows potential for a further increase in wear resistance by, for example, increasing the proportion of carbides by increasing the carbon content of the starting alloy.

4. Discussion and Summary

Due to the possible casting of the material the presented alloy is in line with comparable results from the literature for casted or cladded CrCoFe-containing HEAs [43]. The intended chemical composition could not be reached through casting and some alloying elements were lost in a slag. In the as-cast materials two phases besides precipitates are present. One a molybdenum rich and one a molybdenum poor and nickel ridge phase. A similar phase distribution has been reported for an AlCrFeNiMo HEA produced by laser-cladding and there a similar reduction of molybdenum in the second phase has been reported [43]. Furthermore, the microstructure contains vanadium-rich precipitates, carbides and another type of precipitation. The possibility of precipitating other phases during a heat treatment has been reported for Cu-rich phases in [9,32]. Vanadium-rich precipitates also occur after a conducted remelting as well as the two main phases. A specific analysis of the precipitated phases and the main phases with regard to their atomic structure and composition has not yet been carried out, but will be aimed at after further adaptation of the alloy system. The occurrence of a secondary matrix phase in the selected alloy system can lead to the formation of hot cracks during welding processes. The formation of hot cracks through different phases is a known phenomenon for many material systems [23]. However, the results presented for the remelting tests show no evidence of such behaviour without external restraints, It should be noted, however, that these effects usually occur more frequently when external residual stress-inducing restrictions, such as restraints or distortion, are present. The same applies to the consideration of cold cracks or stress relief cracking. These mostly occur in large components with the potential for the formation of high residual stresses and in connection with embrittlement of the material. In future work adequate welding test for determining the cold cracking susceptibility have to be applied, which are well summarized by Kannengiesser et al., in [44]. The found absence of an pronounced HAZ is reported in the literature for AlCrFeNiMo and low molybdenum contents as well [43]. The modification of the presented alloy design can lead to a pronounced HAZ, which is common for other HEA steels and aluminium alloys.

The results for the hardness (of 324HV0.2) of the high entropy alloy are in line with some similar alloys without molybdenum from the literature [32] and furthermore show the influence of the cooling time on the hardness. A faster cooling increases the hardness to 339HV0.2. Here, an approach for the modification of the presented alloy-design can be derived from. The increase of the carbon content should increase the hardness of the material and, hopefully, increase the wear resistance against three body abrasive wear. The effect of the carbon content on the precipitation behaviour in HEA has been demonstrated by Huang et al. and can be transferred [32]. Furthermore, it has been shown that the addition of other alloying elements can have a significant influence on the phase morphology, the hardness and the wear morphology. The addition of small amounts of Titanium may have a positive effect like shown by Löbe et al. [35].

Nevertheless, it should be noted that the methods used for determination of the chemical composition, the EDX measurements, suffer from some restrictions in their accuracy. The measured values have been reported as given in the measurement results.

The wear test results the show that the chosen alloy design can compete with commonly used hardfacings like Stellite 6.

Overall, it can be stated that the results presented herein are in line with other HEA studies and the usage of Mo as alloying element in CrCoFeNi-based alloy systems is possible. This type of alloy can be further strengthened by the addition of vanadium due to the precipitation of vanadium-rich phases.

5. Conclusions

The presented study shows that the used CrCoFeMoNi + VC high entropy alloy has the following features:

- Castable

The small samples produced showed no casting defects and a homogeneous microstructural morphology. The microstructure present in the as cast specimen show precipitations and two common main phases. One molybdenum rich and one molybdenum poor phase.

- Contains vanadium carbide precipitates

As intended, precipitates were found in the microstructure of the high entropy alloy, which have an increased vanadium and carbon content.

- Hardness and heat affected zone

The presented alloy with its vanadium rich precipitated reaches a hardness of 324HV0.2. The TIG-remelted material has a slightly higher hardness of 339HV0.2. A pronounced HAZ with an increase or drop in the hardness has not been observed. The microstructure shows no evidence for a pronounced HAZ as well.

- Remeltable

The remeltability investigations carried out show no indications of welding-related failures which need to be supported with corresponding welding tests as there is a possibility of hot cracking due to the second phase in the matrix of this alloy. Pore formation or cold cracks could not be found during and after remelting.

- Wear resistant

Under three-body abrasion wear testing in the G75 "Miller test" the alloy showed similar behaviour to conventional cobalt-chromium hard (Stellite 6 and Stellite 12) alloys.

Furthermore, it can be stated that the given $Cr_{27.5}Co_{14}Fe_{22}Mo_{22}Ni_{11.65}V_{2.8}$ high entropy alloy show the potential of generating a remeltable wear resistant alloy, but further work on the alloy design and the determination of its properties is necessary. Especially, the dependence of the precipitation behaviour from the carbon content is of great interest for this promising new type of alloy.

Author Contributions: Conceptualization, K.T.; methodology, K.T.; investigation, K.T. and S.L.; resources, V.W.; writing—original draft preparation, K.T.; writing—review and editing, K.T., S.L. and V.W.; visualization, K.T. and S.L.; supervision, V.W. All authors have read and agreed to the published version of the manuscript.

Funding: We acknowledge support by Open Access Publishing Fund of Clausthal University of Technology.

Institutional Review Board Statement: Not applicable.

Informed Consent Statement: Not applicable.

Data Availability Statement: The used data is available on request from the corresponding author.

Conflicts of Interest: The authors declare no conflict of interest.

Appendix A

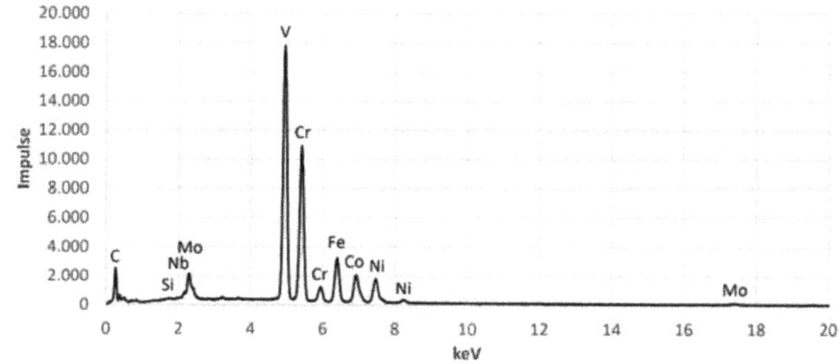

Figure A1. EDX-Spectrum for EDX Point 1.

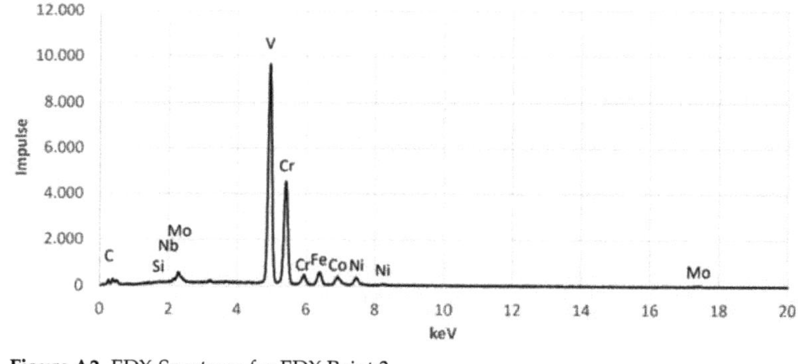

Figure A2. EDX-Spectrum for EDX Point 2.

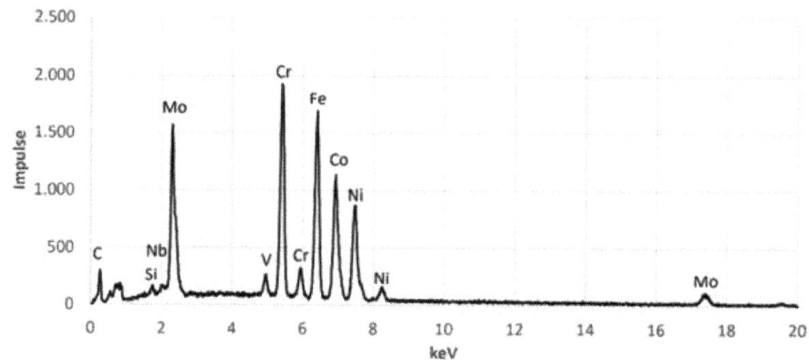

Figure A3. EDX-Spectrum for EDX Point 3.

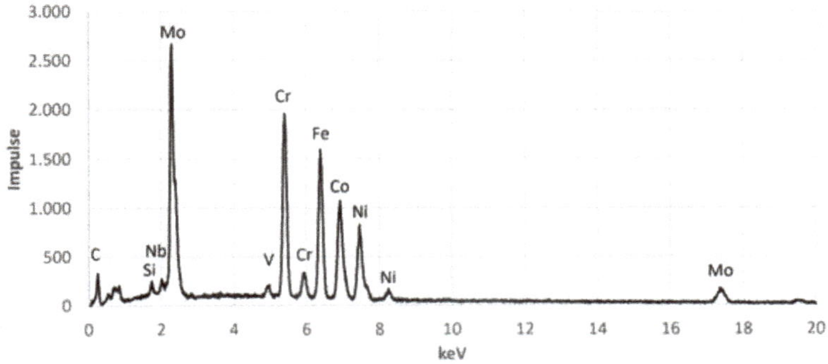

Figure A4. EDX-Spectrum for EDX Point 4.

References

1. Durga, A.; Kumar, K.C.H.; Murty, B.S. Phase Formation in Equiatomic High Entropy Alloys: Calphad Approach and Experimental Studies. *Trans. Indian Inst. Met.* **2012**, *65*, 375–380. [CrossRef]
2. Miracle, D.; Senkov, O. A critical review of high entropy alloys and related concepts. *Acta Mater.* **2017**, *122*, 448–511. [CrossRef]
3. Zhang, W.; Liaw, P.K.; Zhang, Y. Science and technology in high-entropy alloys. *Sci. China Mater.* **2018**, *61*, 2–22. [CrossRef]
4. Jones, M.R.; Nation, B.L.; Wellington-Johnson, J.A.; Curry, J.F.; Kustas, A.B.; Lu, P.; Chandross, M.; Argibay, N. Evidence of Inverse Hall-Petch Behavior and Low Friction and Wear in High Entropy Alloys. *Sci. Rep.* **2020**, *10*, 10151. [CrossRef]
5. Huo, W.-Y.; Shi, H.-F.; Ren, X.; Zhang, J.-Y. Microstructure and Wear Behavior of CoCrFeMnNbNi High-Entropy Alloy Coating by TIG Cladding. *Adv. Mater. Sci. Eng.* **2015**, *2015*, 647351. [CrossRef]
6. Shi, Y.; Yang, B.; Liaw, P.K. Corrosion-Resistant High-Entropy Alloys: A Review. *Metals* **2017**, *7*, 43. [CrossRef]
7. Tian, L.-H.; Xiong, W.; Liu, C.; Lu, S.; Fu, M. Microstructure and Wear Behavior of Atmospheric Plasma-Sprayed AlCoCrFeNiTi High-Entropy Alloy Coating. *J. Mater. Eng. Perform.* **2016**, *25*, 5513–5521. [CrossRef]
8. Chen, M.-R.; Lin, S.-J.; Yeh, J.-W.; Chuang, M.-H.; Chen, S.-K.; Huang, Y.-S. Effect of vanadium addition on the microstructure, hardness, and wear resistance of $Al_{0.5}CoCrCuFeNi$ high-entropy alloy. *Met. Mater. Trans. A* **2006**, *37*, 1363–1369. [CrossRef]
9. Ayyagari, A.V.; Gwalani, B.; Muskeri, S.; Mukherjee, S.; Banerjee, R. Surface degradation mechanisms in precipitation-hardened high-entropy alloys. *NPJ Mater. Degrad.* **2018**, *2*, 33. [CrossRef]
10. Kamper, S. *Eisenbasierte Intermetallische Hartlegierungen für den Verschleißschutz am Beispiel von Einschneckenextrudern*; Universitätsbibliothek Der TU Clausthal: Clausthal Zellerfeld, Germany, 2019.
11. Cantor, B.; Chang, I.; Knight, P.; Vincent, A. Microstructural development in equiatomic multicomponent alloys. *Mater. Sci. Eng. A* **2004**, *375–377*, 213–218. [CrossRef]
12. Oh, H.S.; Ma, D.; Leyson, G.P.; Grabowski, B.; Park, E.S.; Körmann, F.; Raabe, D. Lattice Distortions in the FeCoNiCrMn High Entropy Alloy Studied by Theory and Experiment. *Entropy* **2016**, *18*, 321. [CrossRef]
13. Chen, S.; Oh, H.S.; Gludovatz, B.; Kim, S.J.; Park, E.S.; Zhang, Z.; Ritchie, R.O.; Yu, Q. Real-time observations of TRIP-induced ultrahigh strain hardening in a dual-phase CrMnFeCoNi high-entropy alloy. *Nat. Commun.* **2020**, *11*, 826. [CrossRef]
14. Haglund, A.; Koehler, M.; Catoor, D.; George, E.; Keppens, V. Polycrystalline elastic moduli of a high-entropy alloy at cryogenic temperatures. *Intermetallics* **2015**, *58*, 62–64. [CrossRef]
15. Gludovatz, B.; Hohenwarter, A.; Catoor, D.; Chang, E.H.; George, E.P.; Ritchie, R.O. A fracture-resistant high-entropy alloy for cryogenic applications. *Science* **2014**, *345*, 1153–1158. [CrossRef]
16. Stepanov, N.; Shaysultanov, D.; Yurchenko, N.; Zherebtsov, S.; Ladygin, A.; Salishchev, G.; Tikhonovsky, M. High temperature deformation behavior and dynamic recrystallization in CoCrFeNiMn high entropy alloy. *Mater. Sci. Eng. A* **2015**, *636*, 188–195. [CrossRef]
17. Lopes, J.G.; Oliveira, J.P. A Short Review on Welding and Joining of High Entropy Alloys. *Metals* **2020**, *10*, 212 [CrossRef]
18. Richter, T.; Schröpfer, D.; Rhode, M.; Börner, A. Influence of modern machining processes on the surface integrity of high-entropy alloys. *IOP Conf. Ser. Mater. Sci. Eng.* **2020**, *882*, 12016. [CrossRef]
19. Guo, J.; Tang, C.; Rothwell, G.; Li, L.; Wang, Y.-C.; Yang, Q.; Ren, X. Welding of High Entropy Alloys—A Review *Entropy* **2019**, *21*, 431. [CrossRef]
20. Filho, F.C.G.; Monteiro, S.N. Welding Joints in High Entropy Alloys: A Short-Review on Recent Trends. *Materials* **2020**, *13*, 1411. [CrossRef]
21. Zherebtsov, S.; Stepanov, N.; Shaysultanov, D.; Malopheyev, S.; Vysotskiy, I.; Sanin, V.; Kashaev, N.; Kaibyshev, R. Use of Novel Welding Technologies for High-Entropy Alloys Joining. *Mater. Sci. Forum* **2018**, *941*, 919–924. [CrossRef]
22. Zhou, Z.; Zhang, F.; Wu, J.; Pi, J.; Chen, F. Laser Beam Welding of Feconicrmn High-Entropy Alloys with Preplaced Powders. *Metals* **2020**, *10*, 1402. [CrossRef]
23. Böllinghaus, T.; Herold, H. *Hot Cracking Phenomena in Welds*; Springer: Berlin/Heidelberg, Germany, 2005; ISBN 3540223320.

24. Lalpoor, M.; Eskin, D.; Ruvalcaba, D.; Fjær, H.; Cate, A.T.; Ontijt, N.; Katgerman, L. Cold cracking in DC-cast high strength aluminum alloy ingots: An intrinsic problem intensified by casting process parameters. *Mater. Sci. Eng. A* **2011**, *528*, 2831–2842. [CrossRef]
25. Alexandrov, B.; Theis, K.; Streitenberger, M.; Herold, H.; Martinek, I. Cold Cracking in Weldments of Steel S 690 QT. *Weld. World* **2005**, *49*, 64–73. [CrossRef]
26. Schaupp, T.; Rhode, M.; Yahyaoui, H.; Kannengiesser, T. Hydrogen-assisted cracking in GMA welding of high-strength structural steels using the modified spray arc process. *Weld. World* **2020**, *64*, 1997–2009. [CrossRef]
27. Kromm, A.; Lausch, T.; Schroepfer, D.; Rhode, M.; Kannengiesser, T. Influence of welding stresses on relief cracking during heat treatment of a creep-resistant $_{13}$CrMoV steel: Part I—Effect of heat control on welding stresses and stress relief cracking. *Weld. World* **2020**, *64*, 807–817. [CrossRef]
28. Kromm, A.; Lausch, T.; Schroepfer, D.; Rhode, M.; Kannengiesser, T. Influence of welding stresses on relief cracking during heat treatment of a Creep-Resistant $_{13}$CrMoV steel Part II: Mechanisms of stress relief cracking during post weld heat treatment. *Weld. World* **2020**, *64*, 819–829. [CrossRef]
29. Schroepfer, D.; Kromm, A.; Lausch, T.; Rhode, M.; Wimpory, R.C.; Kannengiesser, T. Influence of welding stresses on relief cracking during heat treatment of a creep-resistant $_{13}$CrMoV steel Part III: Assessment of residual stresses from Small-Scale to real component welds. *Weld. World* **2021**, *64*, 1–15. [CrossRef]
30. Treutler, K.; Brechelt, S.; Wiche, H.; Wesling, V. Beneficial use of hyperbaric process conditions for welding of aluminium and copper alloys. *Weld. World* **2021**, 1–9. [CrossRef]
31. Sunkari, U.; Reddy, S.R.; Rathod, B.D.S.; Kumar, S.S.S.; Saha, R.; Chatterjee, S.; Bhattacharjee, P.P. Heterogeneous precipitation mediated heterogeneous nanostructure enhances strength-ductility synergy in severely cryo-rolled and annealed $CoCrFeNi_{2.1}Nb_{0.2}$ high entropy alloy. *Sci. Rep.* **2020**, *10*, 1–9. [CrossRef]
32. Huang, T.; Jiang, L.; Zhang, C.; Jiang, H.; Lu, Y.; Li, T. Effect of carbon addition on the microstructure and mechanical properties of CoCrFeNi high entropy alloy. *Sci. China Ser. E Technol. Sci.* **2017**, *61*, 117–123. [CrossRef]
33. Lindner, T.; Löbel, M.; Saborowski, E.; Rymer, L.-M.; Lampke, T. Wear and Corrosion Behaviour of Supersaturated Surface Layers in the High-Entropy Alloy Systems CrMnFeCoNi and CrFeCoNi. *Crystals* **2020**, *10*, 110. [CrossRef]
34. Liu, Y.; Ma, S.; Gao, M.C.; Zhang, C.; Zhang, T.; Yang, H.; Wang, Z.; Qiao, J. Tribological Properties of AlCrCuFeNi2 High-Entropy Alloy in Different Conditions. *Met. Mater. Trans. A* **2016**, *47*, 3312–3321. [CrossRef]
35. Löbel, M.; Lindner, T.; Mehner, T.; Lampke, T. Influence of Titanium on Microstructure, Phase Formation and Wear Behaviour of $AlCoCrFeNiTi_x$ High-Entropy Alloy. *Entropy* **2018**, *20*, 505. [CrossRef]
36. Löbel, M.; Lindner, T.; Pippig, R.; Lampke, T. High-Temperature Wear Behaviour of Spark Plasma Sintered $AlCoCrFeNiTi_{0.5}$ High-Entropy Alloy. *Entropy* **2019**, *21*, 582. [CrossRef]
37. G02 Committee. *Test Method for Determination of Slurry Abrasivity (Miller Number) and Slurry Abrasion Response of Materials (SAR Number)*; ASTM International: West Conshohocken, PA, USA, 2019.
38. Bayer, R.J. *Mechanical Wear Fundamentals and Testing, Revised and Expanded*; Apple Academic Press: Palm Bay, FL, USA, 2004.
39. Uetz, H.; Berndgen, W. (Eds.) *Abrasion und Erosion: Grundlagen, Betriebliche Erfahrungen, Verminderung*; Hanser: München, Germany, 1986; ISBN 3446142150.
40. *Destructive Tests on Welds in Metallic Materials—Etchants for Macroscopic and Microscopic Examination (ISO/TR 16060:2003)*, German version CEN ISO/TR 16060:2014; Beuth Verlag GmbH: Berlin, Germany, 2014.
41. *Schweißen und Verwandte Prozesse—Einteilung von Geometrischen Unregelmäßigkeiten an Metallischen Werkstoffen—Teil 1: Schmelzschweißen (ISO 6520-1:2007)*; Dreisprachige Fassung EN ISO 6520-1:2007; Beuth Verlag GmbH: Berlin, Germany, 2007.
42. *Zerstörende Prüfung von Schweißverbindungen an Metallischen Werkstoffen—Makroskopische und Mikroskopische Untersuchungen von Schweißnähten (ISO 17639:2003)*; Deutsche Fassung EN ISO 17639:2013; Beuth Verlag GmbH: Berlin, Germany, 2013.
43. Wu, W.; Jiang, L.; Jiang, H.; Pan, X.; Cao, Z.; Deng, D.; Wang, T.; Li, T. Phase Evolution and Properties of $Al_2CrFeNiMo \times$ High-Entropy Alloys Coatings by Laser Cladding. *J. Therm. Spray Technol.* **2015**, *24*, 1333–1340. [CrossRef]
44. Kannengiesser, T.; Boellinghaus, T. Cold cracking tests—An overview of present technologies and applications. *Weld. World* **2013**, *57*, 3–37. [CrossRef]

materials

MDPI

Article

Thermal Shock Resistance and Thermal Insulation Capability of Laser-Glazed Functionally Graded Lanthanum Magnesium Hexaluminate/Yttria-Stabilised Zirconia Thermal Barrier Coating

Muhammed Anaz Khan [1], Annakodi Vivek Anand [2], Muthukannan Duraiselvam [3], Koppula Srinivas Rao [4], Ramachandra Arvind Singh [5,*] and Subramanian Jayalakshmi [5,*]

1 Department of Mechanical Engineering, MLR Institute of Technology, Hyderabad 500043, India; muhammedanazkhan@gmail.com
2 Department of Aeronautical Engineering, MLR Institute of Technology, Hyderabad 500043, India; vivekanandbit@gmail.com
3 Department of Production Engineering, National Institute of Technology, Tiruchirappalli 620015, India; durai@nitt.edu
4 Department of Computer Science and Engineering, MLR Institute of Technology, Hyderabad 500043, India; ksreenu2k@gmail.com
5 Institute of Laser Optoelectronics and Intelligent Manufacturing, College of Mechanical and Electrical Engineering, Wenzhou University, Wenzhou 325035, China
* Correspondence: r.arvindsingh@gmail.com (R.A.S.); jayalakshmi.subramanian@gmail.com (S.J.)

Citation: Anaz Khan, M.; Vivek Anand, A.; Duraiselvam, M.; Srinivas Rao, K.; Arvind Singh, R.; Jayalakshmi, S. Thermal Shock Resistance and Thermal Insulation Capability of Laser-Glazed Functionally Graded Lanthanum Magnesium Hexaluminate/Yttria-Stabilised Zirconia Thermal Barrier Coating. *Materials* **2021**, *14*, 3865. https://doi.org/10.3390/ma14143865

Academic Editor: Petrica Vizureanu

Received: 31 May 2021
Accepted: 7 July 2021
Published: 10 July 2021

Abstract: In this work, functionally graded lanthanum magnesium hexaluminate ($LaMgAl_{11}O_{19}$)/yttria-stabilised zirconia (YSZ) thermal barrier coating (FG-TBC), in as-sprayed and laser-glazed conditions, were investigated for their thermal shock resistance and thermal insulation properties. Results were compared with those of a dual-layered coating of $LaMgAl_{11}O_{19}$ and YSZ (DC-TBC). Thermal shock tests at 1100 °C revealed that the as-sprayed FG-TBC had improved thermal stability, i.e., higher cycle lifetime than the as-sprayed DC-TBC due to its gradient architecture, which minimised stress concentration across its thickness. In contrast, DC-TBC spalled at the interface due to the difference in the coefficient of thermal expansion between the $LaMgAl_{11}O_{19}$ and YSZ layers. Laser glazing improved cycle lifetimes of both the types of coatings. Microstructural changes, mainly the formation of segmentation cracks in the laser-glazed surfaces, provided strain tolerance during thermal cycles. Infrared rapid heating of the coatings up to 1000 °C showed that the laser-glazed FG-TBC had better thermal insulation capability, as interlamellar pores entrapped gas and constrained heat transfer across its thickness. From the investigation, it is inferred that (i) FG-TBC has better thermal shock resistance and thermal insulation capability than DC-TBC and (ii) laser glazing can significantly enhance the overall thermal performance of the coatings. Laser-glazed FG-TBC provides the best heat management, and has good potential for applications that require effective heat management, such as in gas turbines.

Keywords: thermal barrier coating; yttria-stabilised zirconia (YSZ); lanthanum magnesium hexaluminate ($LaMgAl_{11}O_{19}$); thermal shock resistance; thermal insulation; laser glazing

1. Introduction

Thermal barrier coatings (TBC) are multi-layered ceramic coatings, usually used in gas turbines to impart thermal insulation to turbine components from hot combustion gases [1,2]. Typically, a TBC consists of two distinctive layers, namely (i) metallic bond coat and (ii) ceramic top coat. The metallic bond coat is coated over turbine components to provide better compliance with the ceramic top coat. The two layers of a TBC have distinct physical, thermal and mechanical properties. Thermal loading conditions are a major factor that determines the material selection for these two layers [3]. Turbine components such as combustor liners, blades, vanes and nozzles coated with TBCs are required to withstand high thermal loads and render thermal insulation, so as to achieve (i) higher engine efficiency, (ii) emission reduction, and (iii) cooling requirements. Myoung et al. [4] observed improvement in thermal durability upon air cooling thick ZrO_2-8% Y_2O_3 TBCs coated on Ni-superalloy. The magnitude of thermal drop is influenced by factors such as heat transfer coefficients, heat flux, internal cooling, coating thickness and thermal conductivity. Ceramic top coats are expected to impart (a) low thermal conductivity, to enhance thermal insulation, (b) high strain tolerance under cyclic loading, to improve lifetime, and (c) stable microstructure, to minimise deleterious temperature effects such as phase transformations, grain growth and sintering. Yttria-stabilised zirconia (YSZ) is a widely used thermal barrier coating material. However, YSZ as a material has severe limitations, such as (i) ageing, (ii) post-sintering, and (iii) detrimental phase transformation (at temperatures >1200 °C) [5,6]. These limitations cause early failure of YSZ coatings. In YSZ, tetragonal to monoclinic phase transformation occurs during service and is the major reason for coating failure. Gu et al. [5] reported that YSZ-$Y_3Al_5O_{12}$ (YAG) composite coatings can suppress monoclinic phase transformation. Freidrich et al. [6], in their work on YSZ, observed that above 1100 °C, the high oxygen ion conducting nature of zirconia caused increased diffusion of oxygen through the dense ceramic coating, resulting in oxidation of metallurgical interlayer. This consequently led to chipping of the ceramic coating, which limited its long-term high-temperature application [6]. To overcome the limitations of YSZ, (a) doping it with oxide stabilisers (e.g., MgO, Y_2O_3, Sc_2O_3, In_2O_3, CeO_2, SnO_2 and TiO_2) has been investigated [7], and (b) other new materials such as those containing pyrochlore [8], fluorite [9], and perovskite [10] have been developed. Among the new materials, the hexaluminates ($MMeAl_{11}O_{19}$, M = La, Pr, Nd, Sm, Eu, Gd, Ca, Sr; Me = Mg, Mn, Fe, Co, Ni, Cu, Zn), which have a magnetoplumbite structure, exhibit improved structural and thermal stability up to 1400 °C. Hexaluminates have low thermal conductivity [6]. Among hexaluminates, lanthanum magnesium hexaluminate, LaMA ($LaMgAl_{11}O_{19}$) has good thermo-chemical stability [11], and also has an identical cyclic lifetime similar to that of YSZ [12]. The composition of LaMA is able to prevent post-sintering densification, as was reported by Freidrich et al. [6]. Additionally, high-temperature ageing in LaMA occurs more slower than other commercial zirconia-based TBCs, as was reported in [6]. This makes LaMA a promising material for TBC applications.

Conventional double-layer coatings are susceptible to cracking due to thermal stress mismatch and lower fracture toughness, which reduce their lifetime. Functionally graded thermal barrier coatings that have a multi-layered architecture are designed and developed with the aim of enhancing coating compliance and reducing thermal stress mismatch between the two layers, namely, the ceramic layer and metallic bond coat [13]. Functionally graded thermal barrier coatings have composite layers of two different ceramic materials. These coatings are designed such that their top layer is made from ceramic material that has a lower coefficient of thermal expansion, and its weight ratio with the other selected ceramic materials decreases in the subsequent underlying layers. As a consequence of such an architecture in functionally graded thermal barrier coatings, their physical and mechanical properties vary gradually across their coating thickness. Functionally graded thermal barrier coatings have improved thermal cycle lifetime and adhesion strength compared to conventional double-layer structures [12,14–18]. Kim et al. [15] investigated thermoelastic characteristics in TBCs with a graded layer between the top coat and bond coat. By using the finite element method (FEM), they identified that the functionally graded layer can considerably improve cycle lifetime. Kirbiyik et al. [16] synthesised multi-layered ceria and yttria stabilised zirconia (CYSZ)/Al_2O_3 ceramic TBCs, both in double-layered and functionally graded architectures. It was observed that the functionally graded architecture improved bonding strength between layers, and provided better thermal cycle performance than single-layered and double-layered coatings. Gok et al. [17] conducted thermal cycling experiments on multi-layered and functionally graded $Gd_2Zr_2O_7$/CYSZ thermal barrier coatings. It was found that the functionally graded coating had lifetimes almost double those of the single-layered coatings.

Surface modification techniques have also been developed to increase the lifetime of TBCs. As an example, by optimising the coating parameters (such as material feed rate, spray distance, etc.), segmentation cracks can be induced in the top coat to provide better coating compliance [19]. However, in this case, the crack geometries become irregular and cannot be controlled during coating process. Several post-treatment processes have been developed to create segmentation cracks in the top coat. Laser glazing is one such advanced process, in which the top coat is remelted to a depth of few hundred microns by scanning a laser beam over the top coat. As a consequence, rapid resolidification within the treated depth induces a controlled network of segmentation cracks, in the direction perpendicular to the coating surface [20,21]. Glazing also reduces surface roughness [20,21]. Segmentation cracks are vertical cracks in the top coat, with a length at least half that of the thickness of the top coat [19]. The formation of segmentation cracks in a controlled manner relieves thermal stress (which is usually induced during thermal cycles) and improves coating compliance. Reduction in residual stress due to presence of segmentation cracks improves strain tolerance of coatings (i.e., accommodation of large thermal strains without failure [19]). In addition, they act as barrier for the propagation of delamination cracks (i.e., parallel cracks [19]). Segmentation cracks thus enhance thermal shock resistance [19–24]. Laser glazing thus improves the structural integrity of coatings [21,25–27]. Ghasemi et al. [27] laser glazed the top coat of YSZ-based nanostructured TBCs. They found that laser glazing eliminates surface porosities and reduces surface roughness of the coatings. Lee et al. [25] laser glazed plasma-sprayed CYSZ thermal barrier coatings. Their tests on thermal cyclic performance revealed a twofold increase in the lifetime of laser-glazed coatings when compared to their as-sprayed counterparts. The thermo-mechanical behaviour of laser-glazed TBCs is hence an important aspect that greatly influences the heat management performance of TBCs, and so requires detailed investigation.

The present investigation examines (i) the performance of two types of TBC architecture, namely, dual-layered and functionally graded architectures on the heat management capability of as-sprayed $LaMgAl_{11}O_{19}$/YSZ coatings. Thermal shock resistance and thermal insulation capability of the two types of archtiectures are evaluated to decipher the difference in their heat management capabilities; and (ii) the effect of laser glazing on the heat management capability of the coatings. The coatings were applied on Hastealloy (a nickel superalloy) surfaces.

2. Experimental Procedure

2.1. Test Substrate and Coating Materials

Hastelloy C-263 superalloy (Ni-Co-Cr-Mo alloy), used for combustion liners of gas turbines, was selected as the test substrate. The test coupons of 25 mm × 25 mm × 5 mm were machined and grit blasted (average surface roughness, R_a~3 to 4 microns). Prior to coating, the coupons were degreased in an acetone bath. To synthesise $LaMgAl_{11}O_{19}$ (LaMA), a high-temperature solid-state reaction strategy was followed (Equation (1)). La_2O_3 oxide powder was preheated at 973 °C for 2 h, as it absorbs moisture and converts to lanthanum hydroxide [5,28]. Commercially available La_2O_3, Al_2O_3 and MgO were blended in a ball mill with 2:11:1 molar ratio.

$$2MgO + 11Al_2O_3 + La_2O_3 \rightarrow 2LaMgAl_{11}O_{19} \quad (1)$$

Next, the blended powder was ball milled for 5 h and was subsequently heated in a ceramic tubular furnace at 1000 °C for 7 h. Heating temperature was progressively increased to 1650 °C, for 10 h. Eventually, the synthesised LaMA powder was ball milled and dried to obtain free-flowing powder (average particle size: 45–130 µm). Commercially available 8 wt.% YSZ was used for preparing the composite coatings.

2.2. Coating Architecture

Two different coating architectures were followed: (i) double-layer structure (DC-TBC) having two separate layers of YSZ and LaMA above the bond coat and (ii) five ceramic layers above the bond coat with varying weight fraction of YSZ and LaMA (FG-TBC). Bond coat was made of NiCrAlY (composition: Ni-22Cr-10Al-1.0Y (wt.%)). Coatings were deposited via atmospheric plasma spray (APS) process (machine model: Sulzer Metco 9MP, Sulzer Metco India Limited). Optimised parameters of atomic spray process used to deposit the coatings are given in Table 1. The process parameters were selected based on initial trials and from information based on previous studies [29–33]. Total thickness of the deposited coatings was 480 μm. Architectures of DC-TBC and FG-TBC are shown as schematics in Figure 1.

Table 1. Optimised parameters of atomic spray process used to deposit the coatings.

Coating Type	Current (A)	Voltage (V)	Stand-Off Distance (mm)	Primary Gas, Ar (slpm)	Secondary Gas, H_2 (slpm)	Carrier Gas, Ar (slpm)
Bond Coat	550	75	110	35	14	2.3
Ceramic Top Coat	650	61	120	65	12	2.6

slpm: standard litres per minute.

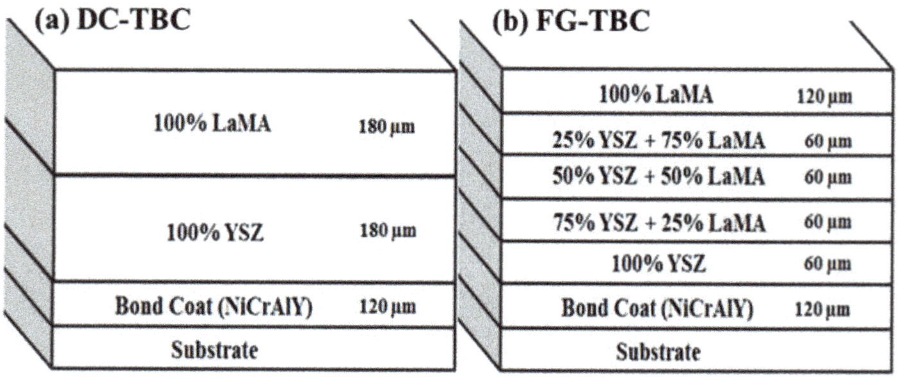

Figure 1. Coating architectures: (**a**) dual-layered architecture, DC-TBC and (**b**) functionally graded architecture, FG-TBC.

2.3. Laser Glazing

DC-TBC and FG-TBC surfaces were laser glazed using a ytterbium-doped fibre laser (wavelength: 1080 nm). The laser was operated in continuous wave (CW) mode and the beam was kept at a slightly defocused position. Diameter of the circular laser beam was d_{spot} = 0.4 mm. Defocused position of laser was used so as to control the delivery of concentrated energy density and to eliminate the deterioration of ceramic layers upon interaction with the focussed beam. For this purpose, initial trials were conducted on single tracks on the developed coatings, with laser power setting at 500 W, 700 W and 900 W. Scanning speed was kept constant at 150 mm/min.

From post-deposition surface analysis, the optimal laser power setting was identified as 700 W and this laser power setting was used to glaze the coatings for thermal tests. The percentage of overlap was selected after measuring the glazed layer width of a single track. In the present work, the coated surfaces were glazed through 30% overlapped parallel tracks to achieve uniform remelting across the surface.

2.4. Surface Characterisation and Phase Analysis

Surface roughness of the TBCs was measured using a 3-D surface profilometer (Rtec instruments, San Jose, CA, USA) with vertical resolution less than 0.1 nm and lateral resolution of 100 nm. Presence of defects such as cracks and pores was identified using scanning electron microscope (SEM). Energy dispersive spectroscopy (EDS) was used for elemental analysis on the TBC surfaces. Phase analysis of the as-synthesised, as-sprayed and laser-glazed surfaces was conducted using X-ray diffraction (XRD, Rigaku ULTIMA-IV, Tokyo, Japan) with Cu-kα radiation. After the thermal shock resistance tests, the failure mechanism of TBCs were identified using SEM.

2.5. Thermal Shock Test

Thermal shock tests were conducted to determine the resistance of the coatings to thermal spalling, using a high-temperature muffle furnace at 1100 °C. Samples were heated for 10 min and were subseqeuntly quenched in water that was maintained at 20 to 25 °C. This heating–quenching cycle was repeated to determine the thermal shock resistance. Surfaces of the coated samples were monitored after every test, and the heating–quenching cycles were repeated until 20% of spallation was observed [34,35].

2.6. Infrared Rapid Heating Test

Thermal insulation capability of the ceramic coated samples was evaluated using a 150 kW infrared (IR) rapid heater (Figure 2). Samples were mounted on sample holder, such that their coated surfaces were exposed towards IR heater. Ni-superalloy substrate was taken as the reference sample to characterise the thermal insulation of the coated test coupons. A gap of about 75 mm was maintained between the heater and the test coupons.

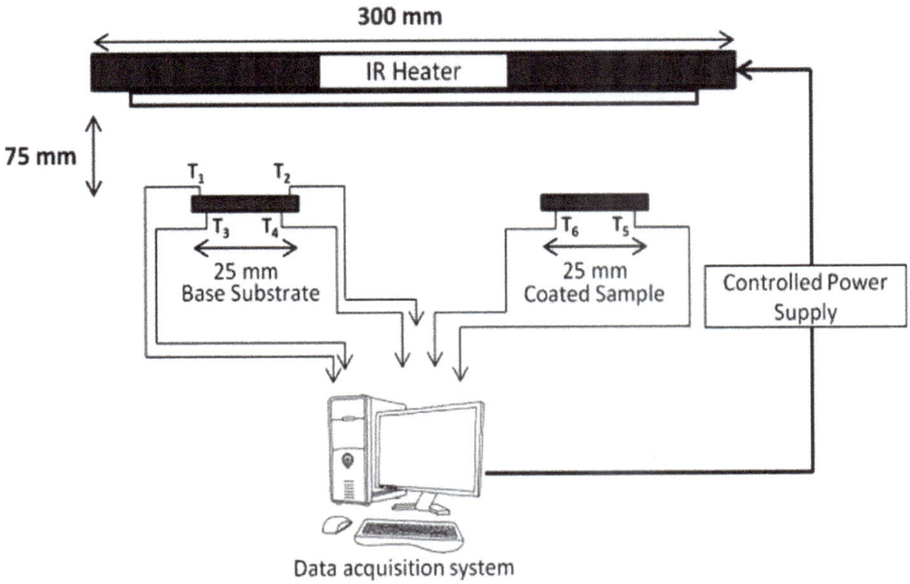

Figure 2. Schematic of infrared (IR) rapid heater.

Type-R thermocouples were used to measure surface temperatures. Back wall temperature drop was measured with time. Thermocouples T_1 and T_2 were attached to the front side of the base reference sample facing the IR heater. Thermocouples T_3 and T_4 were attached on the back side of the base reference sample. T_3 and T_4 were the controller and redundant thermocouples. Thermocouples T_5 and T_6 were attached to the back side of the coated test coupon. Thermal insulation provided by ceramic layers in the coated samples was analysed by measuring the difference in temperature recorded by the thermocouples attached to the back side of the uncoated base reference sample and the coated test coupons. Test specimens were heated to 1000 °C at the rate of 25 °C/s. Peak temperature was attained in 100 s.

3. Results and Discussion

3.1. Surface Topography

The plasma spray process involves accelerating ceramic powders towards a target surface using high-energy plasma. In the case of TBCs, the size of the ceramic particles and surface roughness of bond coat influence their microstructure and surface roughness. During the APS process, molten and semi-molten particles impinge on the targeted substrate and/or previously deposited ceramic layers at higher temperature and pressure. This causes flattening and solidification of thin splats and results in the formation of anisotropic lamellar structure. The coatings consist of various types of defects, which include globular pores. Process parameters influence the adhesion strength of the APS coatings [36,37]. In the present work, the absence of microcracks between the layers indicates that the selected process parameters were optimal for producing good coatings.

Rougher bond coat surfaces facilitate better wettability of the molten splats and improve the adhesion of ceramic material to bond coat. This increases the coating lifetime [38,39]. During the spray process, as the molten splats solidify, the surface becomes rougher. The molten splats, as they reach the target surface, spread over the previously deposited solidified splats [22,23]. This imparts higher surface roughness to the as-sprayed coatings (Figure 3a). Laser glazing reduces the roughness of the ceramic coating (Figure 3b).

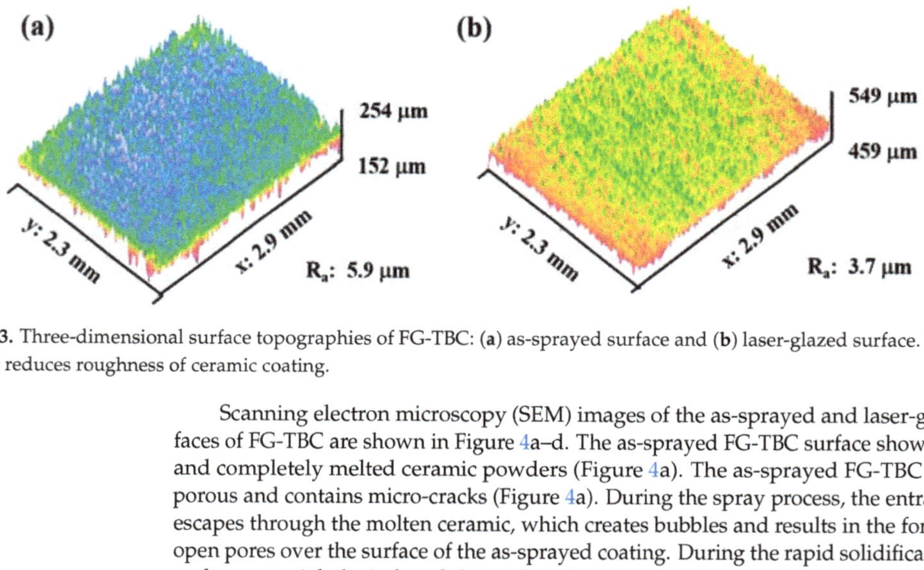

Figure 3. Three-dimensional surface topographies of FG-TBC: (**a**) as-sprayed surface and (**b**) laser-glazed surface. Laser glazing reduces roughness of ceramic coating.

Scanning electron microscopy (SEM) images of the as-sprayed and laser-glazed surfaces of FG-TBC are shown in Figure 4a–d. The as-sprayed FG-TBC surface shows partially and completely melted ceramic powders (Figure 4a). The as-sprayed FG-TBC surface is porous and contains micro-cracks (Figure 4a). During the spray process, the entrapped gas escapes through the molten ceramic, which creates bubbles and results in the formation of open pores over the surface of the as-sprayed coating. During the rapid solidification of the molten material, the induced thermal strain across the coating thickness and the relieving strain due to solidification cause micro-cracks in the as-sprayed coating [27]. However, the propagation of micro-cracks across the coating thickness is restricted by the mechanical interlocking of the overlapped resolidified splats. A significant difference between the surface topography of as-sprayed and laser-glazed surfaces can be observed (Figure 4a,b). Due to the laser glazing, the coarser and rougher surface of the as-sprayed ceramic is remelted and densifies (Figure 4b–d).

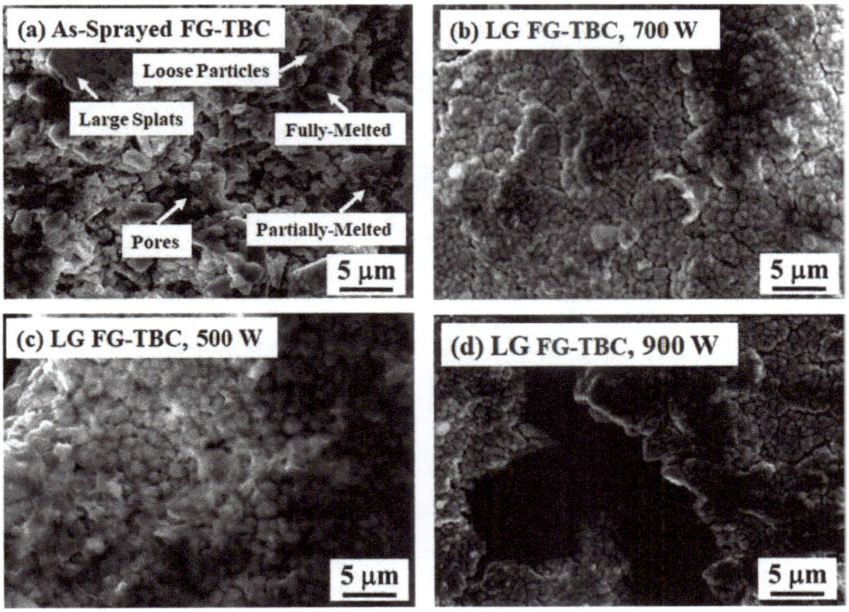

Figure 4. SEM images: (**a**) as-sprayed FG-TBC. Laser-glazed (LG) surfaces obtained using laser power setting of (**b**) 700 W, (**c**) 500 W and (**d**) 900 W.

Laser glazing improves the surface characteristics of TBC by increasing microhardness, sealing surface porosity, reducing surface roughness, minimising the bending modulus of coatings, and by creating a controlled network of segmented cracks over the coatings [21,40,41]. The laser glazing parameters can be varied to obtain significant variation in surface morphology over the glazed surfaces [42]. Both pulsed wave and continuous lasers can be used for surface glazing. Important laser parameters include pulse power, peak power, pulse length, pulse shape, laser beam wavelength, laser scanning speed and the geometry of the laser beam (i.e., depth of focus, spot size) [43]. In the present work, the laser scanning speed was kept at 150 mm/min, which was selected on the basis of preliminary trials. Upon visual inspection of the laser-glazed surface, it was seen that the colour of the coatings changed from a pale grey to a light yellowish glossy surface. This change in colour is known to occur during laser glazing, and indicates optimal laser glazing conditions. Scanning speeds greater than 150 mm/min cause a higher thermal gradient across the coating thickness and a higher rate of thermal stress [44].

Upon interaction with the laser beam, the pores and micro-cracks heal significantly, leading to a homogeneously resolidified net-shaped structure (Figure 4b, laser glazed at 700 W). Segmentation micro-cracks occur due to the higher solidification rate imparted by the raster scanning of the laser source at the optimal laser power settings of 700 W. Segmentation cracks are known to influence thermal shock resistance and thermal cycle lifetime in TBCs [19,22,25,45–47]. The partially dense surface topography of FG-TBC glazed at 500 W (Figure 4c) indicates that this laser power level was not sufficient to glaze the surface effectively. The presence of macro-cracks on the surface of FG-TBC glazed at 900 W (Figure 4d) indicates that this laser power level was not optimal for laser glazing. Based on these observations, a laser power level of 700 W was selected to laser glaze the coatings for their investigation.

Figure 5a,b show the surface roughness and surface porosity of the as-sprayed and laser-glazed DC-TBC and FG-TBC, respectively. The DC-TBC surface has higher roughness compared to FG-TBC surface, both in the as-sprayed and laser-glazed conditions. For both coating architectures, the as-sprayed surfaces are rougher than the laser-glazed surfaces. Partially melted particles (Figure 4a) cause higher roughness (Figure 5a). DC-TBC and FG-TBC both have higher porosity levels in their as-sprayed conditions (Figure 5b). Porosity in the coatings appears in the form of open pores, interlamellar pores (i.e., interlamellar spaces between splats), and globular pores, causing coating failure [48,49]. Open pores permit diffusion of oxygen ions from flue gas into the metallic bond coat, causing oxide formation (thermally grown oxides, TGOs [49]). TGOs cause coating failure at the bond coat interface. The interlamellar pores that form due to rapid solidification lead to delamination of coating [48]. Globular pores, a result of improper filling of the coating material, stacking inconsistencies, incomplete contact between the splats, and the presence of unmelted particles, are initiation sites for coating failures [39]. Post-processing treatments can reduce surface porosity, as is evident from the low porosity levels in the laser-glazed TBCs (Figures 5b and 6).

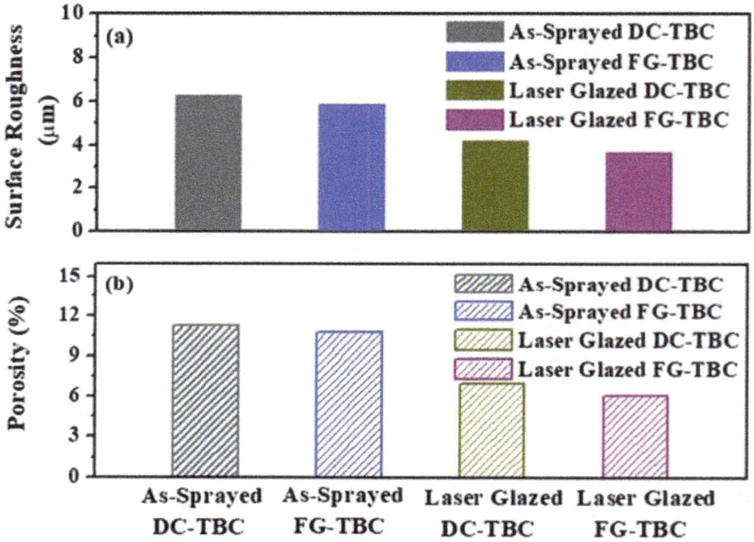

Figure 5. (**a**) Surface roughness (μm) and (**b**) surface porosity (%) of as-sprayed and laser-glazed coating surfaces.

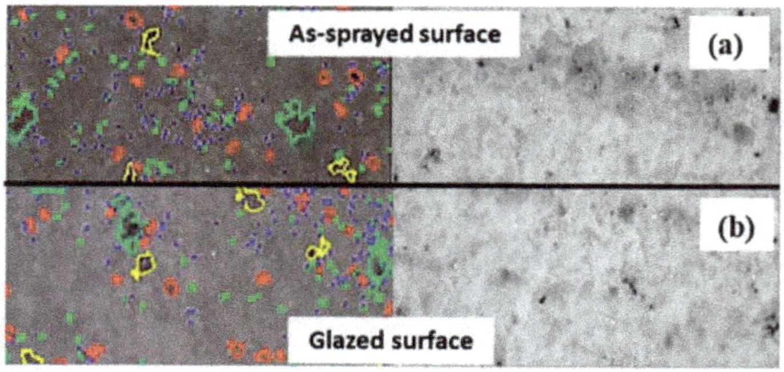

Figure 6. Optical microscopic images of surfaces of: (**a**) as-sprayed and (**b**) laser-glazed FG-TBC. In the right-hand side images, pores appear in grey/black colours. The left-hand side pictures are from the image analysis of the respective optical microscopic images, in which different colours represent pores of different size.

During laser glazing, owing to the rapid melting and resolidification, the open pores on surface close; in other words, they get patched. Additionally, remelting and resolidification densifies the coating material and induces segmentation cracks in the coating [19,49]. The glazed FG-TBC surface thus has a lower surface roughness (3.7 µm) and a lower porosity level (6.1%). Similar behaviour was reported by Ghasemi et al. [27] when nanostructured TBCs containing a YSZ ceramic top coat were laser glazed. A significant reduction in surface roughness after laser glazing was observed. They reported that the surface roughness (Ra) of the as-sprayed coating was 9.2 µm, which upon laser-glazing was reduced to 2.5 µm. Furthermore, they observed that the as-sprayed surface had cracks, voids and pores. Upon laser glazing, they observed an absence of defects, complete resolidification, and a dense microstructure with segmentation cracks [27].

3.2. Microstructure and Phase Analysis

The interface between the YSZ and LaMA ceramic layers in the as-sprayed DC-TBC is shown in Figure 7a. Conformal deposition of LaMA over YSZ is evident. EDS analysis of the region (Figure 7a) confirms the presence of LaMA and YSZ elements (Figure 7b,c). The SEM cross-section image of FG-TBC and the corresponding EDS spectra are shown in Figure 8a,b. Elemental mapping of La and Zr taken across the FG-TBC (Figure 8c,d) shows gradual variation of the La and Zr elements across the coating thickness. This confirms the formation of the graded layer across FG-TBC.

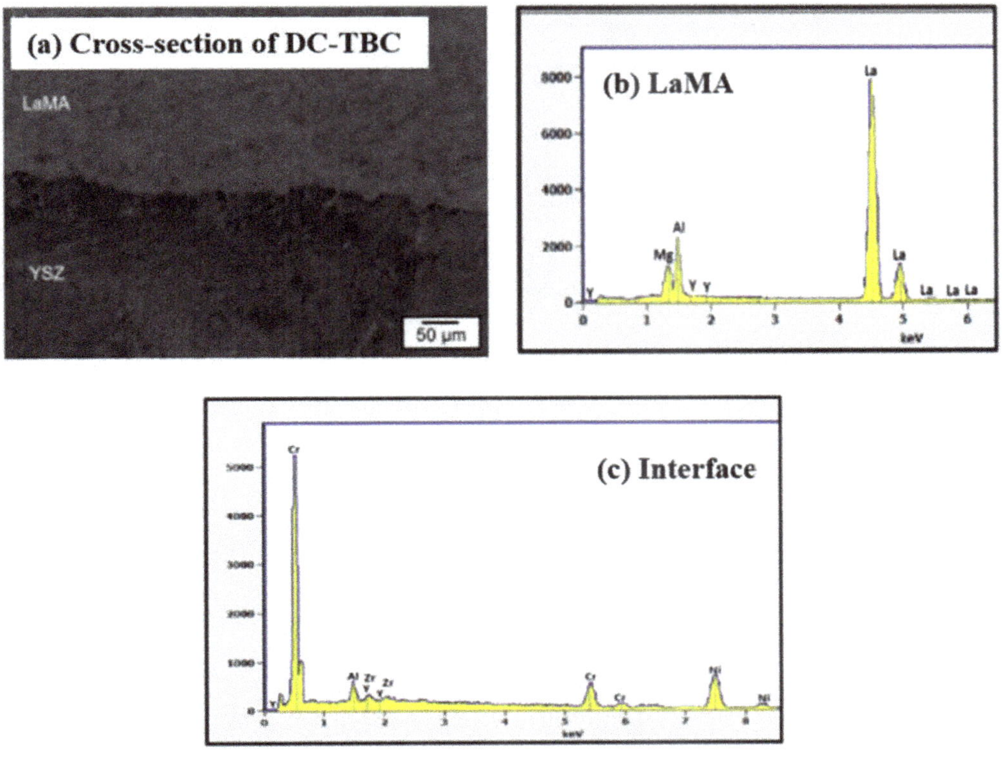

Figure 7. (**a**) SEM image of DC-TBC (cross-section). (**b**) EDS spectrum on LaMA. Peaks corresponding to La, Mg, Al and Y are present. (**c**) EDS spectrum of the YSZ/metallic bond coat interface. Peaks corresponding to Y, Zr, La, Al, Cr and Ni are present.

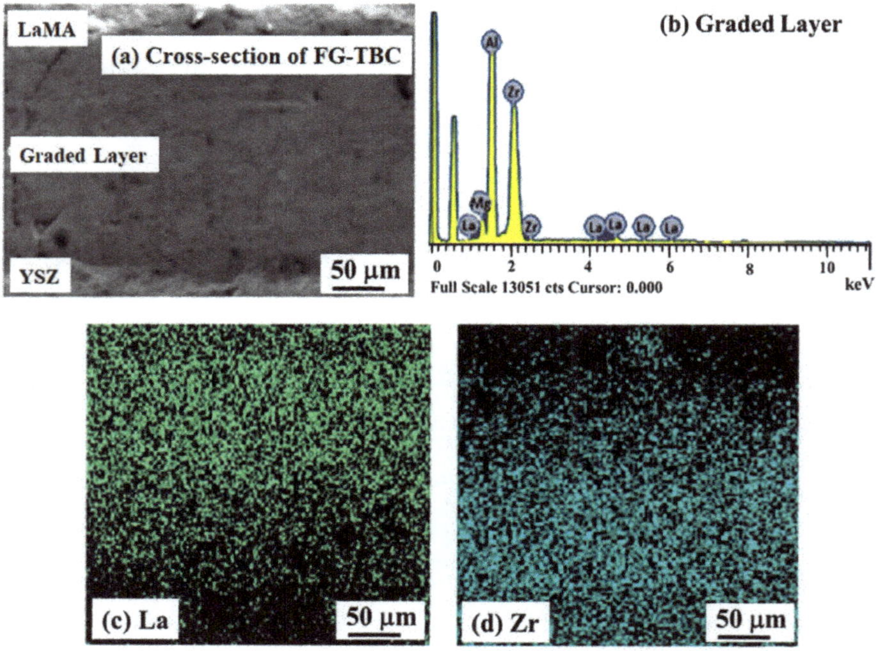

Figure 8. (**a**) SEM image of the cross-section of FG-TBC. (**b**) EDS spectrum across the graded layer showing the presence of La, Mg, Zr and Al elements. (**c**,**d**) Elemental maps showing the gradual variation of La and Zr across the coating thickness.

The XRD patterns for the as-synthesised LaMA powder, as-sprayed and laser-glazed surfaces are shown in Figure 9. Over the as-sprayed surface, the LaMA amorphous phase can be observed as a major phase with broader peaks. Peaks of $LaAlO_3$ are present in the as-sprayed and laser-glazed samples due to partial decomposition of LaMA oxides during the spray process. Other volatile intermetallic peaks are not observed. Partial decomposition of LaMA oxides along with volatilisation during high-temperature synthesis reduced the percentage of volatile intermetallics [11]. For the laser-glazed surface, the narrow peaks indicate the crystallisation of LaMA oxides. The presence of α-Al_2O_3 peaks in the XRD pattern of the laser-glazed surface indicates the partial decomposition of LaMA oxides.

3.3. Thermal Shock Resistance

Thermal shock resistance is an important property of TBCs. The reliability of TBCs under extreme operating conditions is a critical factor. Gas turbine components operate with repeated run–stop cycles, inducing large fluctuations in temperature (i.e., cyclic thermal loads) to the TBCs used for the components. This causes thermal stresses across coatings [34,50–53]. As a consequence, degradation mechanisms manifest, such as sintering effect, thermal expansion, and high temperature friction. Evaluation of the thermal shock resistance of TBCs is therefore vital for their screening and selection for gas turbine components.

The number of cycles to failure of DC-TBC and FG-TBC coatings are shown in Figure 10. In the as-sprayed condition, FG-TBC has a higher cycle lifetime than DC-TBC, by 30 cycles. In the laser-glazed condition, FG-TBC has a higher cycle lifetime than DC-TBC, by 77 cycles. Among the FG-TBC, the laser-glazed coating has a higher cycle lifetime than its as-sprayed counterpart, by 65 cycles. These results indicate that the laser-glazed FG-TBC has the best thermal shock resistance.

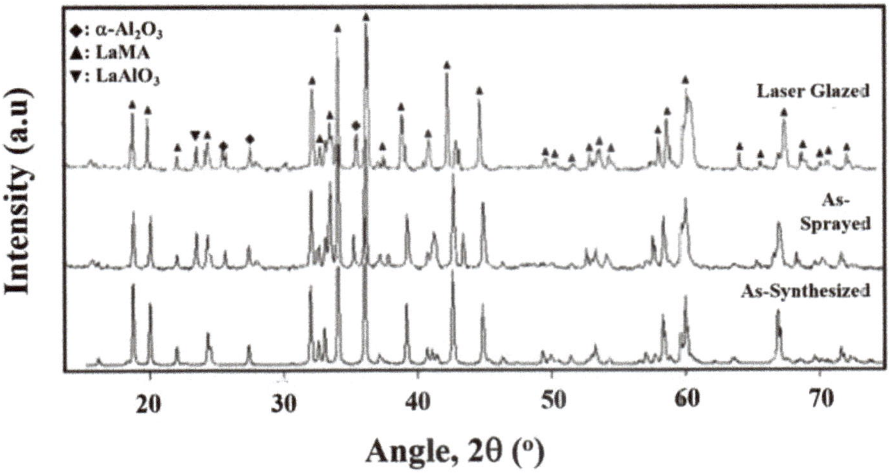

Figure 9. X-ray diffraction patterns of as-synthesised, as-sprayed and laser-glazed TBCs.

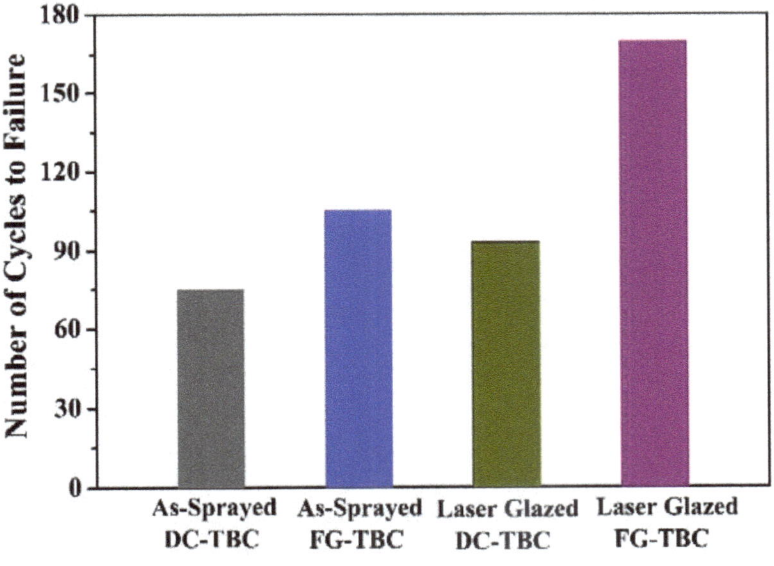

Figure 10. Thermal shock resistance, i.e., number of cycles to failure of DC-TBC and FG-TBC coatings in their as-sprayed and laser-glazed conditions.

3.3.1. As-Sprayed TBCs

Different failure mechanisms were observed for as-sprayed DC-TBC and FG-TBC. In the as-sprayed DC-TBC, horizontal cracks form and propagate at the interface between YSZ and $LaMgAl_{11}O_{19}$ layers due to the difference in their coefficient of thermal expansion (YSZ CTE: 10.2×10^{-6} K^{-1}, room temperature to 877 °C; $LaMgAl_{11}O_{19}$ CTE: 5.13×10^{-6} K^{-1} [11]). Mismatch in the thermal expansion coefficient causes thermal stress mismatch at the interface, which changes the local volume along the interface. With the increase in the number of thermal cycles, spallation of the coating occurs due to the propagation of horizontal cracks.

SEM images of as-sprayed DC-TBC, taken after different numbers of thermal cycles, are shown in Figure 11a–e. Horizontal cracks are initiated along the YSZ/$LaMgAl_{11}O_{19}$ interface after 22 cycles. The intensity of these cracks increases with increasing numbers of thermal cycles. The high thickness of both YSZ and $LaMgAl_{11}O_{19}$ (thickness: 180 μm each) induces a lower thermal gradient across the coating thickness and favours the accumulation of stress. Branching of the micro-cracks after 65 cycles can be observed in Figure 11d. The coalescence of the micro-cracks that can be seen in the sample after 74 cycles (Figure 11e) is caused by the accumulated thermal elastic strain, which is relieved rapidly upon quenching during the thermal shock test.

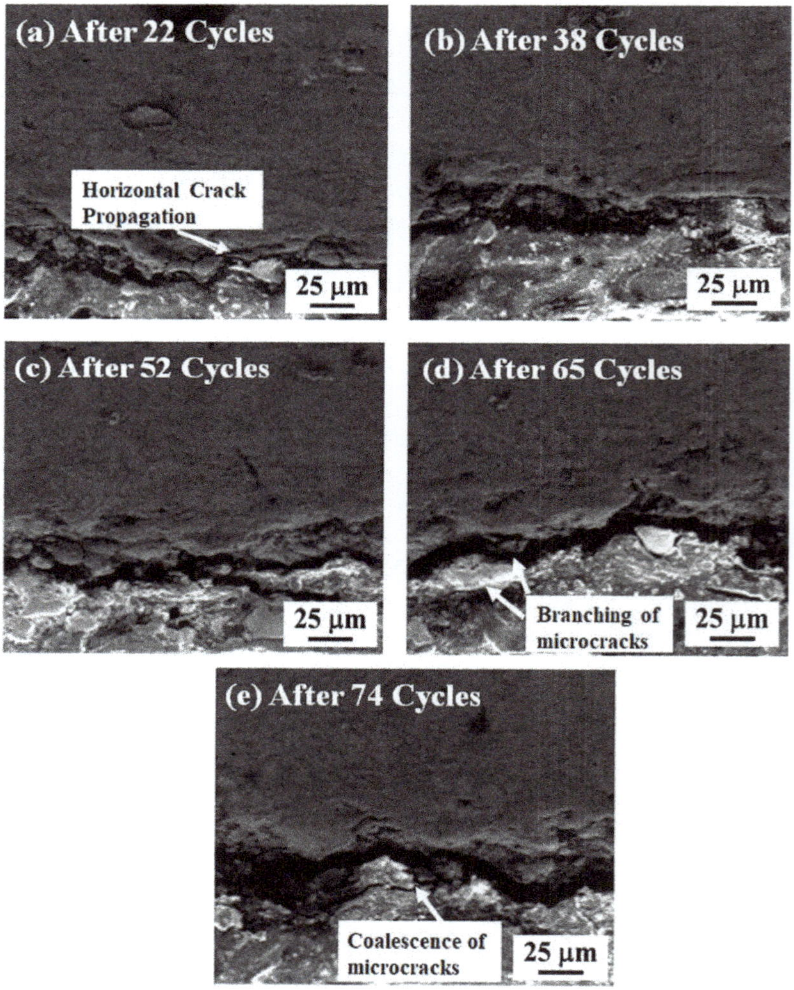

Figure 11. SEM images of as-sprayed DC-TBC after (**a**) 22, (**b**) 38, (**c**) 52, (**d**) 65 and (**e**) 74 heating–quenching cycles.

Compared to DC-TBC, the as-sprayed FG-TBC showed a different failure mechanism (Figure 12a–c). The graded layers of FG-TBC effectively prevent the accumulation of stress and provide better thermal insulation across the coating. The thermal insulation of the coatings is influenced by their microstructure and crystal structure [54]. FG-TBC has a top layer that consists of 100% $LaMgAl_{11}O_{19}$ (thickness: 120 μm), which has lower

CTE, and its weight percentage decreases with subsequent underlying layers. The graded architecture reduces the propensity for the propagation of defects (such as micro-cracks) to the subsequent layers and to the substrate [55]. Due to this gradient architecture, stress accumulation in FG-TBC is lower than that in DC-TBC, which has a dual-layered architecture; consequently, FG-TBC has higher thermal shock resistance than DC-TBC (Figure 10).

Figure 12. SEM images of as-sprayed FG-TBC after (**a**) 53, (**b**) 78 and (**c**) 106 cycles.

In TBCs, there are two different crack propagation mechanisms: (i) cracks that propagate along the coated surface, i.e., parallel to the coated surface, termed inter-splat cracks; and (ii) cracks that are oriented across the thickness of coatings, i.e., perpendicular to the coated surface, termed intra-splat cracks [27,56]. Reports have shown that parallel cracks provide better thermal compliance and insulation than intra-splat cracks [57]. In the as-sprayed FG-TBC, the cracks initiate along the highly stressed brim region and propagate across the coating thickness. The intensity of these cracks increases with the number of thermal cycles (Figure 12). Coatings were partly purged along with the top coat and spalled within the ceramic layer.

3.3.2. Laser-Glazed TBCs

Laser-glazed coatings have higher thermal shock resistance than their as-sprayed counterparts (Figure 10). Laser glazing causes the remelting and resolidification of the coating surfaces and induces segmentation cracks. During laser glazing, the higher thermal gradients and non-uniform resolidification (i.e., rapid solidification ~10^7 K/s) favour accumulation of thermal stress across the treated depth [20,46]. The shear force across the molten layer is accumulated due to the induced surface tension [20,47]. Gravitational force stabilises the induced shear force in the remelted zone. Therefore, the accumulated thermal stress to which segmentation cracks are subjected will be relatively lower than that experience by non-segmentation cracks. Thus, it can be observed that laser-glazed coatings have a higher cycle lifetime than the as-sprayed coatings (Figure 10). Similar observations have been reported previously, when plasma-sprayed ceria-yttira-stabilised TBCs were laser glazed [25]. The results showed that the thermal cycling lifetime of laser-glazed TBCs increased twofold [25]. Kadhim [33] treated surfaces of yttria partially stabilised zirconia (YPSZ) by laser sealing. Compared to the as-sprayed surface, laser surface processing effectively modified the surface layer by sealing the porosity and reducing surface roughness, and enhanced the thermal shock resistance due to the presence of segmentation cracks [33].

SEM images of laser-glazed DC-TBC after 54, 68, 85 and 93 heating–quenching cycles are shown in Figure 13a–d. The segmentation cracks in the laser-glazed coatings accommodate thermal stresses and improve the strain tolerance. The difference in CTE between the dual layers induces thermal strain along the coating, causing formation and propagation of delamination cracks. The glazed coating surface becomes densified after 54 cycles (Figure 13), and the coating spalls. Guo et al. [23] reported significant improvement in thermal shock resistance of plasma-sprayed YSZ due to the presence of segmentation cracks. They observed that the coatings failed by spalling and delamination [23].

Laser-glazed FG-TBC has a higher thermal shock resistance, i.e., a higher cycle lifetime, than its as-sprayed counterpart and laser-glazed DC-TBC (Figure 10). The better shock resistance of laser-glazed FG-TBC is due to (i) the gradient architecture and (ii) the beneficial effect of laser glazing. To elucidate, (i) gradient architecture prevents the accumulation of stress and reduces the crack propagation rate across the graded thickness of the coating [33]. In addition, it provides strain tolerance during heating–quenching cycles. (ii) Laser glazing densifies the structure, eliminates open pores, and thus prevents the diffusion of oxygen ions into the metallic bond coat. This reduces the propensity of oxide formation and coating failure thereof. In addition, the formation of segmentation cracks in the top coat upon laser glazing provides strain tolerance during heating–quenching cycles. With prolonged test cycles, the cracks propagated across the coating thickness and spalled the glazed layer, as shown in Figure 14a–c. The gradient ceramic layer is below the spalled region (not shown here).

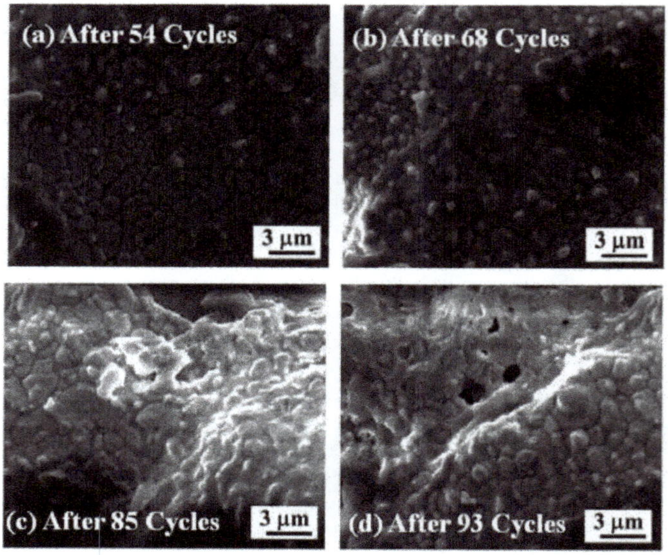

Figure 13. High-magnification SEM images of laser-glazed DC-TBC after (**a**) 54, (**b**) 68, (**c**) 85, and (**d**) 93 heating–quenching cycles.

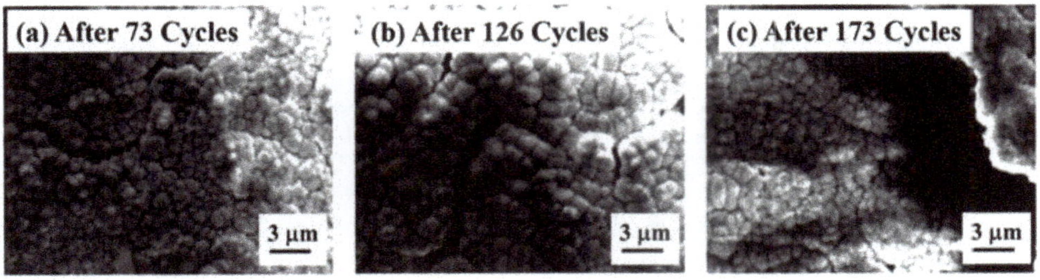

Figure 14. High-magnification SEM images of laser-glazed FG-TBC after (**a**) 73, (**b**) 126, and (**c**) 173 heating–quenching cycles.

3.4. Thermal Insulation Capability

TBCs with both the types of architectures, in both as-sprayed and laser-glazed conditions, sustained the IR rapid heating test (1000 °C at 25 °C/s) without spallation. Acquired back wall temperature of the TBCs is shown in Figure 15. All traces have three distinct regions, namely, (i) the commencement region, wherein the temperature increases at a lower rate during the beginning of test (until 25 s), (ii) incubation time, wherein the temperature increases linearly (25 s to 65 s), and (iii) stabilised time, wherein the temperature reaches stable values. Two base reference samples (Hastealloy) were tested to study the accuracy of the data analyser in recording the drop in temperature. An identical back wall temperature of 998 °C was observed for both the samples. This means that the drop in the back wall temperature was 2 °C, which is considered to be negligible.

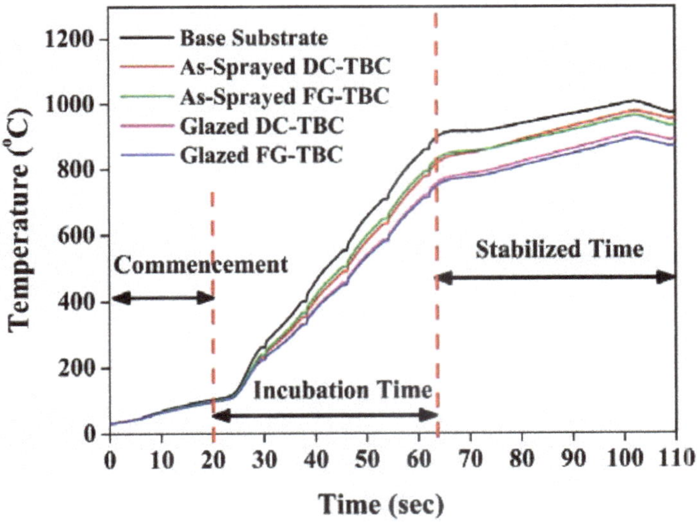

Figure 15. Back wall temperature plot of infrared rapid heating of DC-TBC and FG-TBC in their as-sprayed and laser-glazed conditions.

Back wall temperature drops of 67 °C and 83 °C were observed for the as-sprayed DC-TBC and FG-TBC coatings, respectively. Pores are known to impede thermal conduction and thereby enhance the thermal insulation of coatings [58,59]. In FG-TBC, which has multiple layers of YSZ-LaMA, mechanical interlocking of splats increases the roughness across the ceramic layers and imparts increased levels of porosity between the layers (i.e., interlamellar pores) [59]. Increased porosity induces increased thermal insulation. For this reason, FG-TBC has a higher back wall temperature drop. Reduction in the back wall temperature of the as-sprayed TBCs compared with the base reference material shows the thermal insulation capability of the ceramic coatings. Back wall temperature drops of 102 °C and 117 °C were observed for the laser-glazed DC-TBC and FG-TBC coatings, respectively. Under laser-glazed conditions, i.e., due to the densification of top coat, the interlamellar pores entrap gas and contribute towards enhancing thermal insulation [59]. Guo et al. [60] studied the effect of splat interfaces on the thermal conductivity of YSZ coatings, by finite element simulations and experiments. They identified that interlamellar pores cause lowering of thermal conductivity of TBCs, i.e., the interlamellar pores promoted thermal insulation. On similar lines, Wei et al. [61] conducted simulations and studied the effect of lamellar interspaces on thermal conductivity of TBCs. They opined that interlamellar pores trap gas molecules and limit the conduction of heat flow [60,61]. They reported that the presence of interlamellar pores can contribute up to 70% of reduction in thermal conductivity [61].

In the present case, laser-glazed FG-TBC showed better thermal insulation capability due to (i) increased formation of interlamellar pores owing to the multiple layers of YSZ-LaMA [5] and (ii) densification of the top coat, which leads to the entrapment of gas by interlamellar pores, preventing heat conduction across the coating thickness (schematically shown in Figure 16). Thus, by this mechanism, heat transfer is suppressed, and thermal insulation is enhanced in laser-glazed FG-TBC (Figure 15).

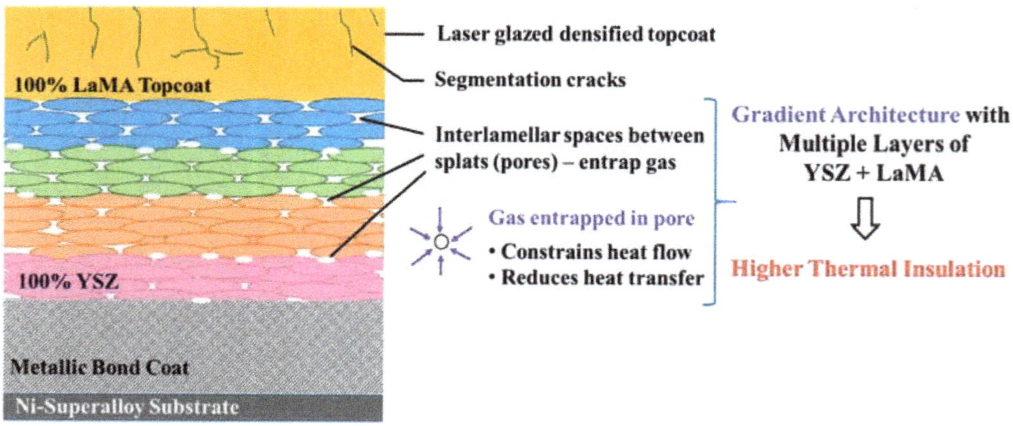

Figure 16. Schematic representation of the heat insulation mechanism in laser-glazed FG-TBC. Laser glazing densifies the top coat. Interlamellar space between splats (pores) entraps gas, constrains heat flow and prevents heat transfer across the coating thickness.

In thermal barrier coatings, thermal conduction is known to occur by phonon transmission. Ceramic oxides in thermal barrier coatings have lattice imperfections that scatter phonons [60]. Scattering of phonons hinders their free flow across coating thickness and consequently lowers thermal conductivity. In YSZ, the addition of yttria to zirconia requires O^{2-} vacancies in order to retain the electrical neutrality of ionic lattice. The O^{2-} vacancy and yttria scatter the incoming phonons across the lattice structure [27,60,62–66]. This phenomenon of phonon scattering induces thermal insulation, as there are insufficient free electrons (phonons are less effective in conducting heat energy compared to free electrons). YSZ and hexaluminate (LaMA) have low thermal conductivities (YSZ: 1.3 W/mK [66]; hexaluminates: 0.8 to 2.6 W/mK [6]). These ceramics, taken in combination to synthesise thermal barrier coatings with a functionally graded architecture, can provide good thermal insulation capability, as is evident from the results in the present investigation.

4. Conclusions

Lanthanum magnesium hexaluminate ($LaMgAl_{11}O_{19}$)/yttria-stabilised zirconia (YSZ) thermal barrier coatings were prepared with two architectures: (i) functionally graded coating (FG-TBC) and (ii) dual-layered coating (DC-TBC). The influence of the architecture type on the thermal shock resistance and thermal insulation capability of the TBCs was examined. TBCs were subjected to laser glazing, in order to determine the effect of glazing on their thermal shock resistance and thermal insulation capability. The main conclusions drawn from the investigation are as follows:

- Surface topography: Laser glazing significantly altered the surface topography of both coating architectures, such that the roughness and porosity of DC-TBC and FG-TBC on their surfaces was reduced. Densification of the top coat material due to laser glazing caused these reductions.
- Thermal shock resistance: FG-TBC has better thermal shock resistance, i.e., higher cycle lifetime, than DC-TBC, in both the as-sprayed and laser-glazed conditions.
    (a) As-sprayed DC-TBC spalled along the YSZ/LaMA interface due to thick dual layers and lower thermal gradient that caused stress accumulation along the interface. In as-sprayed FG-TBC, the functionally graded architecture reduced stress concentration, which increased cycle lifetime.
    (b) Laser-glazed FG-TBC has higher thermal shock resistance than laser-glazed DC-TBC due to (i) the formation of segmentation cracks, (ii) improved strain tolerance, and (iii) closure of surface pores.
- Thermal insulation capability: FG-TBC has better thermal insulation capability, i.e., higher back wall temperature drop, than DC-TBC, in both the as-sprayed and laser-glazed conditions. The multiple layers in FG-TBC cause increased formation of interlamellar pores. Laser-glazed FG-TBC showed better thermal insulation capability due to densification of the top coat, causing the entrapment of gas in interlamellar pores, which constrains heat transfer across the coating thickness.
- Laser-glazed FG-TBC has the best heat management, in terms of both thermal shock resistance and thermal insulation capability. It has good potential for applications that require effective heat management, such as in gas turbines.

Author Contributions: Conceptualisation, M.A.K.; methodology, M.A.K., A.V.A., R.A.S. and S.J.; software, M.A.K. and S.J.; validation, S.J. and R.A.S.; formal analysis, M.A.K., R.A.S., and S.J.; investigation, M.A.K.; resources, M.A.K., M.D. and K.S.R.; data curation, M.A.K., R.A.S. and S.J.; writing—original draft preparation, M.A.K., S.J. and R.A.S.; writing—review and editing, S.J. and R.A.S.; visualisation, S.J. and R.A.S.; supervision, M.D.; project administration, M.D. and K.S.R.; funding acquisition, M.D. All authors have read and agreed to the published version of the manuscript.

Funding: The work was financially supported by Defence Research and Development Laboratory (DRDL), Government of India. Project number: DRDL/24/08P/12/0513/41812.

Institutional Review Board Statement: Not applicable.

Informed Consent Statement: Not applicable.

Data Availability Statement: Data is contained within the article.

Conflicts of Interest: The authors declare no conflict of interest.

References

1. Feuerstein, A.; Knapp, J.; Taylor, T.; Ashary, A.; Bolcavage, A.; Hitchman, N. Technical and economical aspects of current thermal barrier coating systems for gas turbine engines by thermal spray and EB-PVD: A review. *J. Therm. Spray Technol.* **2008**, *17*, 199–213. [CrossRef]
2. Kumar, V.; Balasubramanian, K. Progress update on failure mechanisms of advanced thermal barrier coatings: A review. *Prog. Org. Coat.* **2016**, *90*, 54–82. [CrossRef]
3. Karaoglanli, A.C.; Ogawa, K.; Ozdemir, I. Thermal shock and cycling behavior of thermal barrier coatings (TBCs) used in gas turbines. In *Progress in Gas Turbine Performance*; Benini, E., Ed.; IntechOpen: London, UK, 2013; pp. 237–268.

4. Myoung, S.W.; Lee, S.S.; Kim, H.S.; Kim, M.S.; Jung, Y.G.; Jung, S.I.; Woo, T.K.; Park, U. Effect of post heat treatment on thermal durability of thermal barrier coatings in thermal fatigue tests. *Surf. Coat. Technol.* **2013**, *215*, 46–51. [CrossRef]
5. Gu, L.; Zhao, S.; Xu, J.; Hui, Y.; Fan, X.; Zou, B.; Wang, Y.; Cao, X. Phase stability of plasma sprayed YAG–YSZ composite beads/coatings at high temperature. *J. Eur. Ceram. Soc.* **2013**, *33*, 3325–3333. [CrossRef]
6. Friedrich, R.; Gadow, T.; Schirmer, T. Lanthanum hexaaluminate—A new material for atmospheric plasma spraying of advanced thermal barrier coatings. *J. Therm. Spray. Technol.* **2001**, *10*, 592–598. [CrossRef]
7. Pitek, F.M.; Levi, C.G. Opportunities for TBCs in the ZrO_2–$YO_{1.5}$–$TaO_{2.5}$ system. *Surf. Coat. Technol.* **2007**, *201*, 6044–6050. [CrossRef]
8. Vaben, R.; Traeger, F.; Stöver, D. New thermal barrier coatings based on pyrochlore/YSZ double-layer systems. *Int. J. Appl. Ceram. Technol.* **2004**, *1*, 351–361.
9. Cao, X.Q.; Vassen, R.; Tietz, F.; Stöver, D. Lanthanum-cerium oxide as a thermal barrier-coating material for high-temperature applications. *Adv. Mater.* **2003**, *15*, 1438–1442. [CrossRef]
10. Ma, W.; Mack, D.E.; Vassen, R. Perovskite-type strontium zirconate as a new material for thermal barrier coating. *J. Am. Ceram. Soc.* **2008**, *91*, 2630–2635. [CrossRef]
11. Chen, X.; Gu, L.; Zou, B.; Wang, Y.; Cao, X. New functionally graded thermal barrier coating system based on $LaMgAl_{11}O_{19}$/YSZ prepared by air plasma spraying. *Surf. Coat. Technol.* **2012**, *206*, 2265–2274. [CrossRef]
12. Cao, X.Q.; Zhang, Y.F.; Zhang, J.F.; Zhong, X.H.; Wang, Y.; Ma, H.M.; Xu, Z.H.; He, L.M.; Lu, F. Failure of the plasma-sprayed coating of lanthanum hexaluminate. *J. Eur. Ceram. Soc.* **2008**, *28*, 1979–1986. [CrossRef]
13. Łatka, L.; Pawłowski, L.; Winnicki, M.; Sokołowski, P.; Małachowska, A.; Kozerski, S. Review of functionally graded thermal sprayed coatings. *Appl. Sci.* **2020**, *10*, 5153. [CrossRef]
14. Xie, L.; Ma, X.; Jordan, E.H.; Padture, N.P.; Xiao, T.D.; Gell, M. Identification of coating deposition mechanisms in the solution-precursor plasma-spray process using model spray experiments. *Mater. Sci. Eng. A* **2003**, *362*, 204–212. [CrossRef]
15. Kim, S.; Go, J.; Jung, Y.G.; Lee, J.H. Thermoelastic characteristics in thermal barrier coatings with a graded layer between the top and bond coats. *Math. Prob. Eng.* **2013**, *2013*, 515792. [CrossRef]
16. Gok, M.G.; Goller, G. Production and characterisation of GZ/CYSZ alternative thermal barrier coatings with multilayered and functionally graded designs. *J. Eur. Ceram. Soc.* **2016**, *36*, 1755–1764.
17. Kirbiyik, F.; Gok, M.G.; Goller, G. Microstructural, mechanical and thermal properties of Al_2O_3/CYSZ functionally graded thermal barrier coatings. *Surf. Coat. Technol.* **2017**, *329*, 193–201. [CrossRef]
18. Schlichting, K.W.; Padture, N.P.; Jordan, E.H.; Gell, M. Failure modes in plasma-sprayed thermal barrier coatings. *Mater. Sci. Eng. A* **2003**, *342*, 120–130. [CrossRef]
19. Shinde, S.V.; Curtis, E.J.G.V.; Johnson, A.; Sampath, S. Segmentation crack formation dynamics during air plasma spraying of zirconia. *Acta Mater.* **2020**, *183*, 196–206. [CrossRef]
20. Zhang, G.; Liang, Y.; Wu, Y.; Feng, Z.; Zhang, B.; Liu, F. Laser remelting of plasma sprayed thermal barrier coatings. *J. Mater. Sci. Technol.* **2001**, *17*, 105–110.
21. Batista, A.; Ribeiro, R.M.; Teixeira, V.; Costa, M.F.; Oliveira, C.R. Surface laser-glazing of plasma-sprayed thermal barrier coatings. *Appl. Surf. Sci.* **2005**, *247*, 313–319. [CrossRef]
22. Ito, K.; Kuriki, H.; Araki, H.; Kuroda, S.; Enoki, M. Detection of segmentation cracks in top coat of thermal barrier coatings during plasma spraying by non-contact acoustic emission method. *Sci. Technol. Adv. Mater.* **2014**, *15*, 035007. [CrossRef]
23. Guo, H.B.; Vaßen, R.; Stöver, D. Thermophysical properties and thermal cycling behaviour of plasma sprayed thick thermal barrier coatings. *Surf. Coat. Technol.* **2005**, *192*, 48–56. [CrossRef]
24. Guo, H.; Murakami, H.; Kuroda, S. Thermal Cycling Behavior of Plasma Sprayed Segmentation Thermal Barrier Coatings. *Mater. Trans.* **2006**, *47*, 306–309. [CrossRef]
25. Lee, J.H.; Tsai, P.C.; Chang, C.L. Microstructure and thermal cyclic performance of laser-glazed plasma-sprayed ceria–yttria-stabilized zirconia thermal barrier coatings. *Surf. Coat. Technol.* **2005**, *202*, 5607–5612. [CrossRef]
26. Taha-al, Z.Y.; Hashmi, M.S.; Yilbas, B.S. Laser treatment of HVOF coating: Model study and characterization. *J. Mech. Sci. Technol.* **2007**, *21*, 1439–1444. [CrossRef]
27. Ghasemi, R.; Shoja-Razavi, R.; Mozafarinia, R.; Jamali, H. Laser glazing of plasma-sprayed nanostructured yttria stabilized zirconia thermal barrier coatings. *Ceram. Int.* **2013**, *39*, 9483–9490. [CrossRef]
28. Gell, M.; Wang, J.; Kumar, R.; Roth, J.; Jordan, E.H. Higher Temperature Thermal Barrier Coatings with the Combined Use of Yttrium Aluminum Garnet and the Solution Precursor Plasma Spray Process. *J. Therm. Spray Technol.* **2018**, *27*, 543–555. [CrossRef]
29. Dhineshkumar, S.R.; Duraiselvam, M.; Natarajan, S.; Panwar, S.S.; Jana, T.; Khan, M.A. Enhanced ablation resistance through laser glazing of plasma sprayed $LaTi_2Al_9O_{19}$-based functionally graded thermal barrier coating. *Ceram. Int.* **2016**, *42*, 10184–10190. [CrossRef]
30. Dhineshkumar, S.R.; Duraiselvam, M.; Natarajan, S.; Panwar, S.S.; Jana, T.; Khan, M.A. Effect of laser glazing on the thermo-mechanical properties of plasma-sprayed $LaTi_2Al_9O_{19}$ thermal barrier coatings. *Mater. Manuf. Process.* **2017**, *32*, 1573–1580. [CrossRef]
31. Dhineshkumar, S.R.; Duraiselvam, M.; Natarajan, S.; Panwar, S.S.; Jena, T.; Khan, M.A. Enhancement of strain tolerance of functionally graded $LaTi_2Al_9O_{19}$ thermal barrier coating through ultra-short pulse based laser texturing. *Surf. Coat. Technol.* **2016**, *304*, 263–271. [CrossRef]

32. Khan, M.A.; Duraiselvam, M.; Panwar, S.S.; Jena, T.; Dhineshkumar, S.R. Thermo-mechanical characterization of laser textured $LaMgAl_{11}O_{19}$/YSZ functionally graded thermal barrier coating. *Surf. Coat. Technol.* **2017**, *321*, 146–155. [CrossRef]
33. Khan, M.A.; Duraiselvam, M.; Panwar, S.S.; Jena, T.; Dhineshkumar, S.R. Improvement of strain tolerance of functionally graded TBCs through laser surface micro-texturing. *J. Mater. Sci.* **2016**, *52*, 696–708. [CrossRef]
34. Lia, M.; Sun, X.; Hu, W.; Guan, H. Thermal shock behaviour of EB-PVD thermal barrier coatings. *Surf. Coat. Technol.* **2007**, *201*, 7387–7391. [CrossRef]
35. Ahmadi-Pidani, R.; Shoja-Razavi, R.; Mozafarinia, R.; Jamali, H. Improving the thermal shock resistance of plasma sprayed CYSZ thermal barrier coatings by laser surface modification. *Opt. Lasers Eng.* **2012**, *50*, 780–786. [CrossRef]
36. Bai, Y.; Zhao, L.; Qu, Y.M.; Fu, Q.Q.; Wang, Y.; Liu, K.; Tang, J.J.; Li, B.Q.; Han, Z.H. Particle In-flight behavior and its influence on the microstructure and properties of supersonic-atmospheric-plasma-sprayed nanostructured thermal barrier coatings. *J. Alloys Compd.* **2015**, *644*, 873–882. [CrossRef]
37. Mostaghimi, J.; Chandra, S. Droplet Impact and Solidification in Plasma Spraying. In *Handbook of Thermal Science and Engineering*; Springer: Cham, Switzerland, 2018. [CrossRef]
38. Traeger, F.; Ahrens, M.; Vaben, R.; Stöver, D. A life time model for ceramic thermal barrier coatings. *Mater. Sci. Eng. A* **2003**, *358*, 255–265. [CrossRef]
39. Rajasekaran, B.; Mauer, G.; Vaben, R. Enhanced Characteristics of HVOF-sprayed MCrAlY Bond Coats for TBC Applications. *J. Therm. Spray Technol.* **2011**, *20*, 1209–1216. [CrossRef]
40. Tsai, P.C.; Lee, J.H.; Chang, C.L. Improving the erosion resistance of plasma-sprayed zirconia thermal barrier coatings by laser glazing. *Surf. Coat. Technol.* **2007**, *202*, 719–724. [CrossRef]
41. Ahmaniemi, S.; Vuoristo, P.; Mäntylä, T.; Gualco, C.; Bonadei, A.; di Maggio, R. Thermal cycling resistance of modified thick thermal barrier coatings. *Surf. Coat. Technol.* **2005**, *190*, 378–387. [CrossRef]
42. Chwa, S.O.; Ohmori, A. Microstructures of ZrO_2-8wt.%Y_2O_3 coatings prepared by a plasma laser hybrid spraying technique. *Surf. Coat. Technol.* **2002**, *153*, 304–312. [CrossRef]
43. Olsen, F.O.; Alting, L. Pulsed laser materials processing, ND-YAG versus CO_2 lasers. *CIRP Ann. Manuf. Technol.* **1995**, *44*, 141–145. [CrossRef]
44. Yang, Y.; Song, Y.; Wu, W.; Wang, M. Multi-pass overlapping laser glazing of FeCrPC and CoNiSiB alloys. *Thin Solid Films* **1998**, *323*, 199–202. [CrossRef]
45. Kadhim, M.J. Laser sealing and thermal shock resistance of 6.5 wt% yttria partially stabilized zirconia plasma sprayed layers. *Eng. Technol. J.* **2009**, *27*, 1038–1045.
46. Wang, D.; Tian, Z.; Shen, L.; Liu, Z.; Huang, Y. Effects of laser remelting on microstructure and solid particle erosion characteristics of ZrO_2–7wt%Y_2O_3 thermal barrier coating prepared by plasma spraying. *Ceram. Int.* **2014**, *40*, 8791–8799. [CrossRef]
47. Wang, L.; Wang, Y.; Sun, X.G.; He, J.Q.; Pan, Z.Y.; Zhou, Y.; Wu, P.L. Influence of pores on the thermal insulation behavior of thermal barrier coatings prepared by atmospheric plasma spray. *Mater. Des.* **2011**, *32*, 36–47. [CrossRef]
48. Tillmann, W.; Khalil, O.; Abdulgader, M. Porosity Characterization and Its Effect on Thermal Properties of APS-Sprayed Alumina Coatings. *Coatings* **2019**, *9*, 601. [CrossRef]
49. Ekberg, J.; Ganvir, A.; Klement, U.; Creci, S.; Nordstierna, L. The Influence of Heat Treatments on the Porosity of Suspension Plasma-Sprayed Yttria-Stabilized Zirconia Coating. *Intl. J. Therm. Spray Technol.* **2018**, *27*, 391–401. [CrossRef]
50. Siebert, B.; Funke, C.; Vaßen, R.; Stover, D. Changes in porosity and Young's Modulus due to sintering of plasma sprayed thermal barrier coatings. *J. Mater. Process. Technol.* **1999**, *92–93*, 217–223. [CrossRef]
51. Fry, A.T.; Banks, P.J.; Nunn, J.; Brown, J.L. Comparison of the Thermal Cycling Performance of Thermal Barrier Coatings under Isothermal and Heat Flux Conditions. *Mater. Sci. Forum* **2008**, *595–598*, 77–85. [CrossRef]
52. Balint, D.S.; Hutchinson, J.W. An analytical model of rumplingin thermal barrier coatings. *J. Mech. Phys. Solids* **2005**, *53*, 949–973. [CrossRef]
53. Srinivasa Rao, L.; Jürgen, R.; Brian R., L. Cyclic Fatigue from Frictional Degradation at Bridging Grains in Alumina. *J. Am. Ceram. Soc.* **1991**, *74*, 1340–1348.
54. Jamali, H.; Mozafarinia, R.; Razavi, R.S.; Ahmadi-Pidani, R.; Loghman-Estarki, M.R. Fabrication and evaluation of plasma-sprayed nanostructured and conventional YSZ thermal barrier coatings. *Curr. Nanosci.* **2012**, *8*, 402–409. [CrossRef]
55. Odhiambo, J.G.; Li, W.G.; Zhao, Y.T.; Li, C.L. Porosity and Its Significance in Plasma-Sprayed Coatings. *Coatings* **2019**, *9*, 460. [CrossRef]
56. Schlichting, K.W.; Padture, N.P.; Klemens, P.G. Thermal conductivity of dense and porous yttria-stabilized zirconia. *J. Mater. Sci.* **2001**, *36*, 3003–3010. [CrossRef]
57. Padture, N.P.; Gell, M.; Eric, L.; Jordan, H. Thermal barrier coatings for gas turbine engine applications. *Mater. Sci.* **2002**, *296*, 280–285. [CrossRef] [PubMed]
58. Rai, A.K.; Schmitt, M.P.; Zhu, D.; Wolfe, D.E. *Sintering Characteristics of Multilayered Thermal Barrier Coatings under Thermal Gradient and Isothermal High Temperature Annealing Conditions*; NASA Report; NASA: Washington, DC, USA, 2014; pp. 1–16.
59. Zhao, M.; Pan, W.; Wan, C.; Qu, Z.; Li, Z.; Yang, J. Defect engineering in development of low thermal conductivity materials: A review. *J. Eur. Ceram. Soc.* **2017**, *37*, 1–13. [CrossRef]
60. Guo, X.; Zhao, W.; Zeng, Y.; Lin, C.; Zhang, J. Effects of splat Interfaces, Monoclinic Phase and Grain Boundaries on the Thermal Conductivity of Plasma Sprayed Yttria-Stabilized Zirconia Coatings. *Coatings* **2019**, *9*, 26. [CrossRef]

61. Wei, S.; Fu-chi, W.; Qun-bo, F.; Dan, H.; Zhuang, M. Proposal of new expressions for effects of splat interfaces and defects on effective properties of thermal barrier coatings. *Surf. Coat. Technol.* **2010**, *204*, 3376–3381. [CrossRef]
62. Li, X.; Yang, R. Effect of lattice mismatch on phonon transmission and interface thermal conductance across dissimilar material interfaces. *Phys. Rev. B* **2012**, *86*, 054305. [CrossRef]
63. Klemens, P.G.; Gell, M. Thermal conductivity of thermal barrier coatings. *Mater. Sci. Eng. A* **1998**, *245*, 143–149. [CrossRef]
64. Nicholls, J.R.; Lawson, K.J.; Johnstone, A.; Rickerby, D.S. Methods to reduce the thermal conductivity of EB-PVD TBCs. *Surf. Coat. Technol.* **2002**, *151–152*, 383–391. [CrossRef]
65. Jiang, B.; Fang, M.H.; Huang, Z.H.; Liu, Y.G.; Peng, P.; Zhang, J. Mechanical and thermal properties of $LaMgAl_{11}O_{19}$. *Mater. Res. Bull.* **2010**, *45*, 1506–1508. [CrossRef]
66. Hu, N.; Khan, M.; Wang, Y.; Song, X.; Lin, C.; Chang, C.; Zeng, Y. Effect of Microstructure on the Thermal Conductivity of Plasma Sprayed Y_2O_3 Stabilized Zirconia (8% YSZ). *Coatings* **2017**, *7*, 198. [CrossRef]

materials

MDPI

Article

Transmission, Reflection and Dissipation of Microwaves in Magnetic Composites with Nanocrystalline Finemet-Type Flakes

Anatoly B. Rinkevich [1,*], Dmitry V. Perov [1] and Yuriy I. Ryabkov [2]

1 M.N. Miheev Institute of Metal Physics UB RAS, Sofia Kovalevskaya St., 18, Ekaterinburg 620108, Russia; peroff@imp.uran.ru
2 Institute of Chemistry UB RAS, Pervomayskaya St., 48, Syktyvkar 167000, Russia; ryabkov-yi@chemi.komisc.ru
* Correspondence: rin@imp.uran.ru

Abstract: The microwave properties of a composite material containing flakes of finemet-type nanocrystalline alloy placed in the epoxy matrix have been investigated. Two compositions have been studied: with 15% and 30% flakes. Frequency dependences of transmission and reflection coefficients are measured in the frequency range from 12 to 38 GHz. The dielectric permittivity and magnetic permeability are obtained, and the microwave losses are calculated. The dependences of transmission and reflection coefficients have been drawn as functions of wave frequency and thickness of the composite material, taking into account the frequency dependences of permittivity and permeability. The regions of maximal and minimal microwave absorption have been defined. The influence of wave interference on the frequency dependence of microwave absorption is studied.

Keywords: magnetic composites; microwaves; absorption; transmission and reflection coefficients

Citation: Rinkevich, A.B.; Perov, D.V.; Ryabkov, Y.I. Transmission, Reflection and Dissipation of Microwaves in Magnetic Composites with Nanocrystalline Finemet-Type Flakes. *Materials* **2021**, *14*, 3499. https://doi.org/10.3390/ma14133499

Academic Editor: Petrica Vizureanu

Received: 23 April 2021
Accepted: 21 June 2021
Published: 23 June 2021

1. Introduction

The study of nanocomposites consisting of metallic nanoparticles and polymer matrices has become a significant active field due to their physical properties attractive for applications [1]. There is large variety of mixing formulas allowing determine the permittivity of a composite media if concentrations and permittivities of its components are assumed known [2]. The microwave magnetic properties of composites and the mixing rules are briefly reviewed in [3], and the laws governing the magnetic frequency dispersion in magnetic composites are discussed. The composites containing magnetic metallic particles are regarded as electromagnetic wave absorbers and as materials for other devices such as microwave antennas, materials for mobile communications, etc. A method of studying microwave magnetic properties of metal particles based on swept frequency measurements under magnetic bias in a coaxial line is developed [4]. Polymer media, ceramics and other dielectrics with high-quality factors are often be chosen as matrices for composites [5].

The best microwave absorbing materials should fulfil such requirements as lightweight, more effective and broader bandwidth absorption. The types of fillers and polymer matrices, as well as the number of layers and thickness, have to be chosen to improve the absorbing capability [6]. The frequency dispersion of flake-shaped and spherical-like $Fe_{16}Ni_{82}Mo_2$ alloy particles has been studied [7]. The composite that contains Fe-50 wt% Ni alloy particles with spherical form has been prepared to improve the microwave absorbing properties in the range of 1–4 GHz [8]. The complex permittivity, complex permeability and reflection loss of the microwave absorbing material are also studied. Improved electromagnetic shielding and absorption properties are obtained for a polymer–metal composite based on polyvinylidene fluoride dispersed with varying concentration of nanocrystalline iron [9]. The possibilities of improving the shielding and absorptive properties of carbon-based admixtures, nickel powder, iron powder, ferrites, magnetite and other materials from the pulsed, high power microwave irradiation have been reviewed in [10].

The dielectric and magnetic losses in the granular structures constituted by ferromagnetic nanoparticles (Co, Fe, B) in an insulating amorphous SiO_2 matrix are investigated at microwave frequencies. The magnetic losses are caused mainly by the fast spin-polarized relaxation mechanism [11]. The microwave refraction coefficient of a composite consisting of Fe-Si-Nb-Cu-B alloy flakes placed into an epoxy resin matrix is investigated. It has been shown that the material under consideration behaves as a dielectric at direct current (DC) and as a lossy dielectric at microwave frequencies in the absence of a magnetic field. Near the field of ferromagnetic resonance, the real and imaginary parts of the complex refraction coefficient are of the same order as in a conductive medium [12]. Systematic permeability measurements of magnetically structured granular systems using the transmission/reflection waveguide method are carried out. The effective complex permeability is measured in the frequency range of 0.01–10 GHz [13]. The distribution of the ferromagnetic Ni in the ceramic matrix of ZrO_2 is studied [14]. The gradient nanocomposite films are promising for better matching between air and metal in the microwave shielding problem.

Due to high magnetic permeability, the finemet-type alloys are regarded as suitable materials for radio- and microwave engineering. The frequency response of magnetic cores produced from pulverized FeNbBSiCu-based nanophase alloy ribbon is studied [15]. The measurements of the magnetic properties of powdered $Fe_{73.5}Cu_1Nb_3Si_{13.5}B_9$ nanostructured alloy have been carried out in the frequency range 0.2–10.2 GHz [16]. Optimization of magnetic properties and magnetic anisotropy of thin, soft magnetic films of finemet alloys is accomplished [17]. The 100–200 nm films are found to be appropriate materials for sensor and actuator devices. Coupling between microstructure and the magnetic properties of FeNbBSiCu alloys is discussed systematically in the paper [18]. The best soft magnetic performance can be obtained when the average nanocrystal size is 16 nm. The FeNbBSiCu thin films have been deposited using RF sputtering [19]. The magnetic field microsensors based on the magneto-impedance effect have been fabricated by stacking up finemet/copper/finemet films. Thus, the thin film and composite materials with finemet alloys are applicable in many microelectronic devices and magnetic field sensors.

To build up the composite materials, which have optimal properties, it is necessary to elaborate on the methods of calculation of microwave dielectric and magnetic losses. The complexity of this problem lies, among other factors, in the fact that the shape, the dimensions and the spatial orientation of ferromagnetic particles have to be considered. The problems arising in the calculation of the dynamic magnetic permeability of composites have been discussed in [20–22]. A possible method of introduction of the magnetic permeability tensor based on the Maxwell–Garnet model is presented in the paper [20]. The difference between a magnetic field inside of a ferromagnetic particle and a given magnetic field outside it can be defined using the effective demagnetizing tensor, which depends on the portion of the ferromagnetic phase in the composite [21,22].

The calculation of magnetic losses is important in order to estimate microwave absorption. The computation of the effective magnetic permeability is worked out in this paper for composite materials containing ferromagnetic particles. The calculation is performed on the frequency dependences of the microwave transmission and reflection coefficients. The measurements of these coefficients are carried out for the composite material with the flakes made of finemet-like alloy in the frequency range from 12 to 38 GHz. The performed calculation allows one to choose the optimal conditions, namely the thickness of the plate and the wave frequency, for maximal or, in opposite, minimal reflection coefficient and dissipated power inside the composite. The comparison is drawn between the experimental data on microwave dissipation and computed ones. The role of wave interference inside the composite plate is clarified.

2. Materials and Methods

Particles of Fe-Si-Nb-Cu-B alloy in the form of flakes have been used for the preparation of the composite. The chemical analysis is carried out with the X-ray fluorescence analyzer Horiba 500 as well as with the method of atomic emission spectroscopy with the inductive-coupled plasma. The elemental composition of the flakes is as follows: Fe—80.1%; Si—8.5%; Nb—8.4%; Cu—1.1%; B—1.2%; Cr—0.2%; Mn—0.1%; Ni—0.1%; Co—0.3%.

It is a nanocrystalline finemet-like alloy with high magnetic permeability. The distribution of the mean radii of the flakes is shown in Figure 1a. The mean radius, which is the mean distance from the center of mass to the border, is 20.1 µm. The maximal Martin diameter, which is the maximal length of a line fit into a particle, is 51 µm on average, and the minimal Martin diameter is 27 µm. The ratio of these diameters equals 1.9, which characterizes the relation between the lateral sizes.

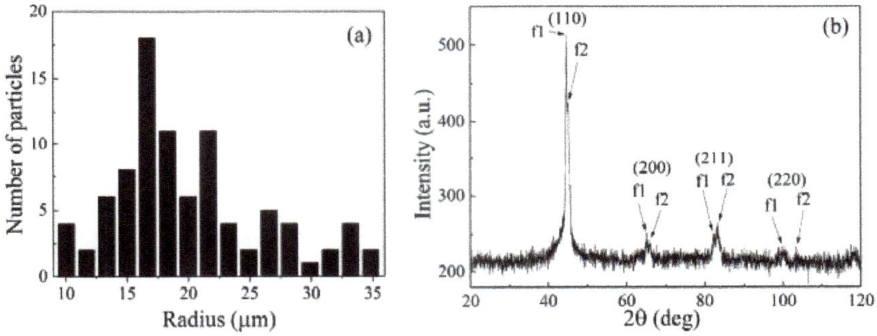

Figure 1. The distribution of mean radiuses of flakes (**a**); the X-ray diffraction image of the composite medium (**b**).

The composite material is prepared by mechanical mixing of flake particles in an epoxy oligomer to matrix polymerization. The epoxy matrix is chosen because of its moderate dielectric permittivity as well as the simplicity of preparation of the composite. After mechanical mixing, the treatment in an ultrasonic bath is performed. Further, the liquid mixture of epoxy and particles is sealed in the metallic mounts, which cavities have similar dimensions as the lateral dimensions of the rectangular waveguides in which the microwave measurements are carried out. Hardening of the mixture lasts several hours. Two sequences of composite samples have been prepared with 15 wt.% and 30 wt.% of flakes. The X-ray phase analysis with a "Pananalytical" spectrometer shows that the main phases are two phases of *bcc* lattice of α-Fe type (f1 and f2), distinguished only by the lattice parameters, which equal to 2.871 Å and 2.841 Å. This difference can be explained as follows. Because the significant amounts of Si and Nb are present in the finemet-like alloy, two phases are formed during the crystallization from a homogeneous melt; these are α-Fe–Nb and α-Fe–Si. The atomic radii of these solid solutions differ from the atomic radius of Fe. The phases cause the splitting of the main picks of an X-ray diffraction pattern into the doublets. The X-ray diffraction image for the sample with 15% flakes is shown in Figure 1b.

The structure of the composite is studied with the Vega3 from Tescan, Brno, Czech republic electron microscope at an accelerating voltage of 30 kV. The structure of the composite with 15% flakes is shown, and the samples are prepared from the chip, i.e., from the inner part of the sample (Figure 2a) and the top surface (Figure 2b). It should be noted that the particles in the inner part of the sample are oriented stochastically, but on the top surface, there is a preferred orientation of the particles in parallel to the surface. The preferred orientation of flakes is seen also for the top view of the composite with 30% flakes (Figure 2c). The preferred orientation in the top part of the sample is formed during the hardening process because of the influence of surface tension forces. From the microscopy data, one can conclude that approximately 20% of flakes have preferable orientation in parallel to the surface, and 80% of flakes are oriented stochastically. Let us also point out that electric contact between the flakes is absent, so the DC conductivity is negligible.

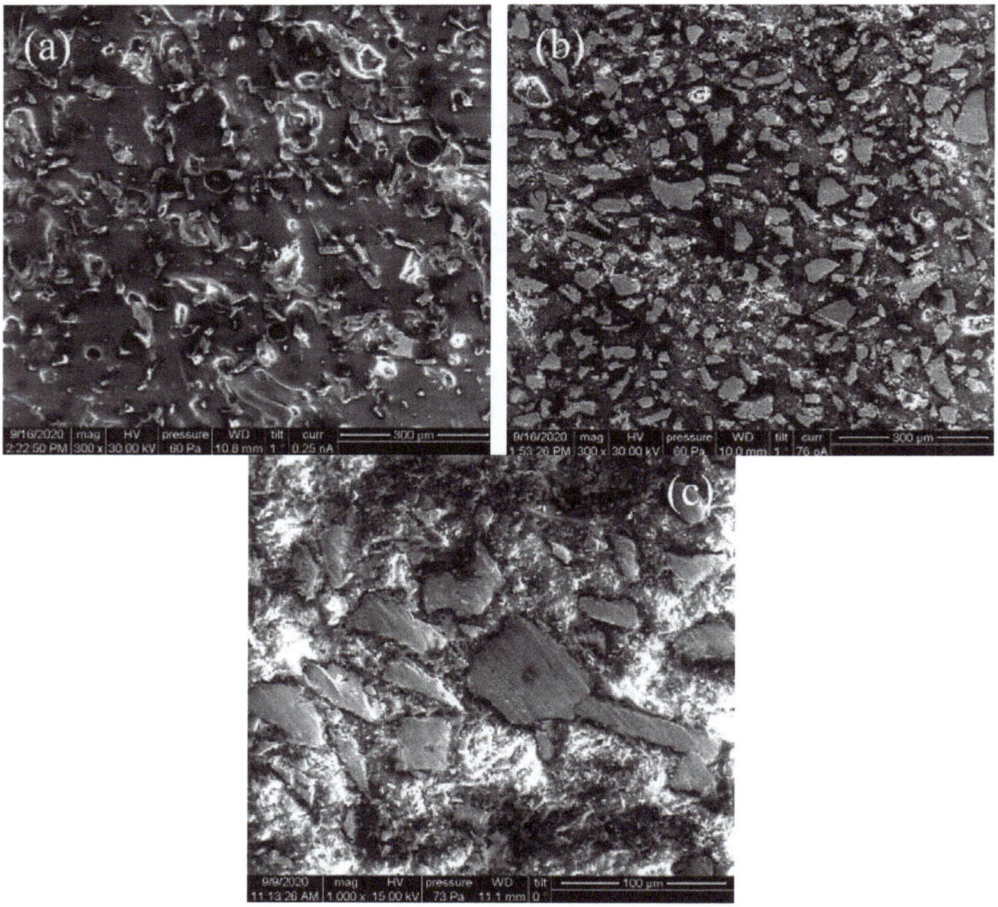

Figure 2. The structure of the composite medium with 15% flakes: chip (**a**); its top surface (**b**); the structure of the composite medium with 30% flakes obtained from its top surface (**c**).

The microwave measurements are performed at frequencies from 12 to 38 GHz according to the method described in [23]. The scheme of the experiment is shown in Figure 3. The sample is placed into a rectangular waveguide *1* to completely overlap its cross-section. The thickness of the sample is from 1.5 to 2 mm. The waveguide operates at TE_{10} mode, and its dimensions are defined by the frequency range: 16 mm × 8 mm for 12–17 GHz (WR-62); 11.5 mm × 5 mm for 17–26 GHz (WR-42); 7.2 mm × 3.4 mm for 27–38 GHz (WR-28). The wave impinges upon the surface of the sample normally. The measurements are carried out with the scalar network analyzer. The amplitudes of transmitted and reflected waves are measured with directional couplers *3*. The modules of transmission *T* and reflection *R* coefficients have been measured as well as their frequency dependences. The measurements of the coefficients are used to determine the complex dielectric permittivity $\dot{\varepsilon}$.

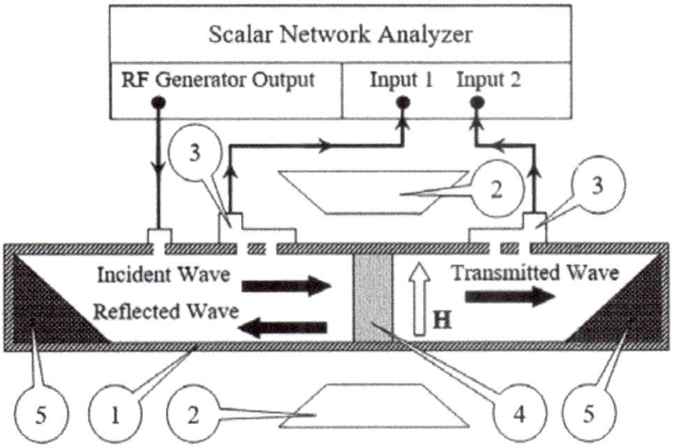

Figure 3. The scheme of microwave measurements: 1—the waveguide; 2—the electrcmagnet; 3—the directional couplers; 4—the sample; 5—the microwave absorber.

Let us shortly describe the procedure of measurements and calculations of the complex dielectric permittivity. The complex transmission $\dot{T}$ and reflection $\dot{R}$ coefficients can be calculated via the formulas [24,25]:

$$\dot{T} = \frac{2Z_1\dot{Z}_2}{2Z_1\dot{Z}_2\cos\left(\dot{k}_2d_2\right) + i\left(Z_1^2 + \dot{Z}_2^2\right)\sin\left(\dot{k}_2d_2\right)} \quad (1)$$

$$\dot{R} = \frac{i\left(\dot{Z}_2^2 - Z_1^2\right)\sin\left(\dot{k}_2d_2\right)}{2Z_1\dot{Z}_2\cos\left(\dot{k}_2d_2\right) + i\left(Z_1^2 + \dot{Z}_2^2\right)\sin\left(\dot{k}_2d_2\right)} \quad (2)$$

In Equations (1) and (2), the medium "1" is either the inner space of the waveguide or free space. The medium "2" is the sample, i.e., imperfect dielectric ferromagnet with the thickness d_2.

The complex impedance $\dot{Z}_2$ for this medium can be written as follows:

$$\dot{Z}_2 = \mathrm{Re}\dot{Z}_2 + i\mathrm{Im}\dot{Z}_2 = \sqrt{\frac{\mu_0\dot{\mu}}{\varepsilon_0\dot{\varepsilon}}} = \sqrt{\frac{\mu_0|\dot{\mu}|}{\varepsilon_0|\dot{\varepsilon}|}}\cdot\exp\left[-i\frac{(\arg\dot{\mu} - \arg\dot{\varepsilon})}{2}\right] \quad (3)$$

$$\left|\dot{Z}_2\right| = \sqrt{\frac{\mu_0|\dot{\mu}|}{\varepsilon_0|\dot{\varepsilon}|}}$$

$$\mathrm{Re}\dot{Z}_2 = \sqrt{\frac{\mu_0|\dot{\mu}|}{\varepsilon_0|\dot{\varepsilon}|}}\cdot\cos\left[\frac{(\arg\dot{\mu} - \arg\dot{\varepsilon})}{2}\right]$$

$$\mathrm{Im}\dot{Z}_2 = \sqrt{\frac{\mu_0|\dot{\mu}|}{\varepsilon_0|\dot{\varepsilon}|}}\cdot\sin\left[-\frac{(\arg\dot{\mu} - \arg\dot{\varepsilon})}{2}\right]$$

$$\mathrm{Re}\left(\dot{Z}_2^2\right) = \frac{\mu_0|\dot{\mu}|}{\varepsilon_0|\dot{\varepsilon}|}\cdot\cos(\arg\dot{\mu} - \arg\dot{\varepsilon})$$

The impedance of the space "1" for the waveguide is

$$Z_1 = \sqrt{\frac{\mu_0}{\varepsilon_0}}\frac{1}{\sqrt{1 - \left(\frac{\pi c}{\omega a}\right)^2}} \quad (4)$$

and the same for the free space

$$Z_1 = \sqrt{\frac{\mu_0}{\varepsilon_0}}$$

where μ_0 and ε_0 are the magnetic permeability and dielectric permittivity of vacuum, $c = \frac{1}{\sqrt{\varepsilon_0 \mu_0}}$ is the speed of light, a is the greater size of cross section of the rectangular waveguide, and $\omega = 2\pi f$ is the cyclic frequency.

The complex constitutive parameters of the sample, namely, the magnetic permeability $\dot{\mu} = \mu' - i\mu''$ and dielectric permittivity $\dot{\varepsilon} = \varepsilon' - i\varepsilon''$ enter the Equation (3). The designations μ', ε' and μ'', ε'' correspond, respectively, to the real and imaginary parts of the permeability and permittivity. Similar marking for the components of complex values will be used throughout the article. The complex wavenumber $\dot{k}_2 = k_2' - ik_2''$ for an imperfect dielectric ferromagnet in Equations (1) and (2) is calculated in the following manner. For TE_{10} mode of a rectangular waveguide, the wavenumber's components are equal to

$$\begin{aligned} k_2' &= \sqrt{\frac{\sqrt{\Re^4 + \Im^4} + \Re^2}{2}}, \\ k_2'' &= \sqrt{\frac{\sqrt{\Re^4 + \Im^4} - \Re^2}{2}}, \end{aligned} \tag{5}$$

where $\Re = \sqrt{\left(\frac{\omega}{c}\right)^2 (\varepsilon'\mu' - \varepsilon''\mu'') - \left(\frac{\pi}{a}\right)^2}$ and $\Im = \frac{\omega}{c}\sqrt{(\varepsilon''\mu' + \varepsilon'\mu'')}$, and for the free space

$$\begin{aligned} k_2' &= \frac{\omega}{c}\sqrt{\frac{|\dot{\varepsilon}||\dot{\mu}| + \varepsilon'\mu' - \varepsilon''\mu''}{2}}, \\ k_2'' &= \frac{\omega}{c}\sqrt{\frac{|\dot{\varepsilon}||\dot{\mu}| - \varepsilon'\mu' + \varepsilon''\mu''}{2}}. \end{aligned}$$

Additionally, the complex impedance $\dot{Z}_2$, which is defined by Equation (3) for an unbounded medium, for the TE_{10} mode of the waveguide takes the view:

$$\dot{Z}_2 = \frac{\omega \mu_0 \dot{\mu}}{\dot{k}_2}$$

where the wavenumber $\dot{k}_2$ is determined by equation for free space.

The power transmission T_P and reflection R_P coefficients are formally given by the equations

$$\begin{aligned} T_P &= \dot{T} \cdot T^* = \\ &= \frac{4Z_1^2 Z_2^{*2}}{\left[2Z_1\dot{Z}_2\cos\left(\dot{k}_2 d_2\right) + i\left(Z_1^2 + \dot{Z}_2^2\right)\sin\left(\dot{k}_2 d_2\right)\right]\left[2Z_1 Z_2^*\cos(k_2^* d_2) - i\left(Z_1^2 + Z_2^{*2}\right)\sin(k_2^* d_2)\right]} \end{aligned} \tag{6}$$

$$\begin{aligned} R_P &= \dot{R} \cdot R^* = \\ &= \frac{\left(\dot{Z}_2^2 - Z_1^2\right)\left(Z_2^{*2} - Z_1^2\right)\sin\left(\dot{k}_2 d_2\right)\sin\left(k_2^* d_2\right)}{\left[2Z_1\dot{Z}_2\cos\left(\dot{k}_2 d_2\right) + i\left(Z_1^2 + \dot{Z}_2^2\right)\sin\left(\dot{k}_2 d_2\right)\right]\left[2Z_1 Z_2^*\cos(k_2^* d_2) - i\left(Z_1^2 + Z_2^{*2}\right)\sin(k_2^* d_2)\right]} \end{aligned} \tag{7}$$

where the asterisk means the complex conjugation.

If one knows the coefficients from Equations (6) and (7), the dissipation D can be calculated, which presents the portion of the microwave power dissipated inside the sample:

$$D = 1 - T_P - R_P \tag{8}$$

Dissipation of the microwave power occurs for the following reasons: absorption in the sample, scattering on the inner heterogeneities and a transformation into the evanescent modes of the waveguide at the boundaries of the sample.

Let us first accept the magnetic permeability as a known value and discuss the procedure how to extract the complex dielectric permittivity from the measured frequency dependences of transmission and reflection coefficients moduli [23]. Denote the measured experimentally frequency dependence of transmission coefficient modulus as $\left|\widetilde{T}(\omega;\dot{\varepsilon},\dot{\mu})\right|$ and reflection coefficient modulus as $\left|\widetilde{R}(\omega;\dot{\varepsilon},\dot{\mu})\right|$. Let us write the difference between calculated $|T|$ and measured $\left|\widetilde{T}\right|$ values of the transmission coefficient modulus as $\Delta_T = \left|\widetilde{T}_p(\omega,\dot{\varepsilon},\dot{\mu})\right| - |T_p(\omega,\dot{\varepsilon},\dot{\mu})|$, and in a similar manner, for the reflection coefficient, $\Delta_R = \left|\widetilde{R}_p(\omega,\dot{\varepsilon},\dot{\mu})\right| - |R_p(\omega,\dot{\varepsilon},\dot{\mu})|$. Here, the complex dielectric permittivity $\dot{\varepsilon}$ is an unknown value. To find it, one chooses $\dot{\varepsilon}$ so that the full difference

$$\Delta = \min_{\dot{\varepsilon}}\left[\sqrt{\left(\Delta_R(\omega,\dot{\varepsilon},\dot{\mu})\right)^2 + \left(\Delta_T(\omega,\dot{\varepsilon},\dot{\mu})\right)^2}\right] \tag{9}$$

has a minimum. The obtained value of $\dot{\varepsilon}$ are accepted as an estimation of the dielectric permittivity. To fulfil the minimization procedure, the frequency range should be chosen in which the amplitude-frequency characteristics of T and R coefficients are measured. The full working frequency range of the waveguide can be initially accepted.

If for none of the frequency belonging to this range, the difference between $|T|$ and $\left|\widetilde{T}\right|$, as well as between $|R|$ and $\left|\widetilde{R}\right|$, does not exceed a preassigned value (this assigned value defines the accuracy of estimation of $\dot{\varepsilon}$), then one can regard the obtained value of $\dot{\varepsilon}$ is constant within the chosen frequency range. The latter is true if the frequency dispersion of $\dot{\varepsilon}$ is weak. If, however, the difference exceeds the preassigned value, determination of $\dot{\varepsilon}$ and the minimization procedure repeats using the sliding frequency window. The minimization procedure (9) is performed at every position of the sliding window, and the frequency dependence of $\dot{\varepsilon}(\omega)$ is obtained as a result. From the obtained $\dot{\varepsilon}$, the microwave conductivity can be calculated: $\sigma = \omega\varepsilon_0\varepsilon''$.

3. Results

The frequency dependences of transmission and reflection coefficient modules have been measured for the composite samples with 15% and 30% ferromagnetic particles. The measurements are carried out within three frequency ranges: from 12 to 17 GHz, from 17 to 26 GHz, from 26 to 38 GHz, with the corresponding waveguide in every frequency range. The results of measurements are presented partially in Figure 4. It is found that for the composite with 30% flakes in the frequency range from 26 to 38 GHz, the dependences calculated with one optimally chosen $\dot{\varepsilon}$ value fairly good approximate the measured dependences, see Figure 4a. For the composite with 15% flakes, the frequency dependences of transmission and reflection coefficients moduli have an essential difference between the measured and calculated dependences at frequencies above 20.5 GHz. The method with the sliding window is used in this case, and the $\dot{\varepsilon}(\omega)$ dependence has been obtained, which is presented in the next section. Depending on the frequency dispersion of $\dot{\varepsilon}(\omega)$ function, the width of the sliding window is chosen from 0.4 to 4 GHz. At that, the difference between approximated and measured dependences of T and R does not exceed 0.02.

Let us present the data on magnetic permeability and dielectric permittivity of the composite samples. The method of how the dielectric permittivity has been obtained is described above. For magnetic permeability, the problem lies in the fact that there is no measured frequency dependence of magnetic permeability for so high frequencies in literature. Let us, at first, obtain the approximation of frequency dependency of the material from which the flakes are produced and then calculate the permeability of the composite material. The frequency dependence of complex magnetic permeability of Hitachi Finemet alloys in the frequency range from 1 kHz to 10 MHz is presented in [26], see Figure 5. The values of magnetic permeability for the material of flakes at frequency f = 10 GHz are given in the patent [27]. In Figure 5, besides the experimental data on permeability, the approximation using the Cole–Cole formula [28] is also shown. This formula represents a typical view of the frequency dependence

$$\dot{\chi}(\omega) = \chi'(\omega) - i\chi''(\omega) = \chi_{\infty} + \frac{4\pi(\chi_0 - \chi_{\infty})}{1 + (i\omega\tau)^{1-\alpha}} \quad (10)$$

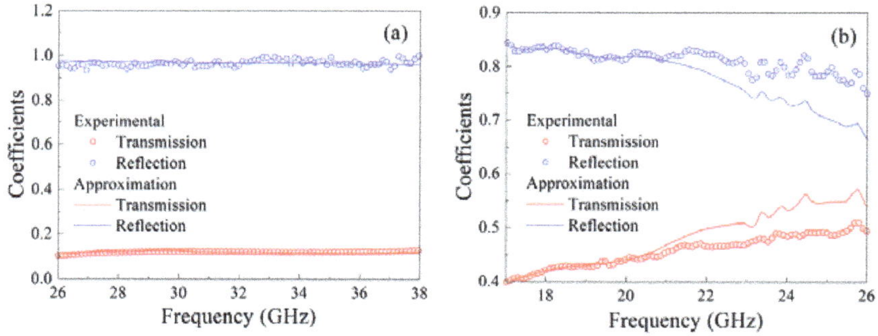

Figure 4. Frequency dependences of transmission and reflection coefficients: for the composite with 30% flakes in the frequency range from 26 to 38 GHz (**a**); for the composite with 15% flakes in the frequency range from 17 to 26 GHz (**b**).

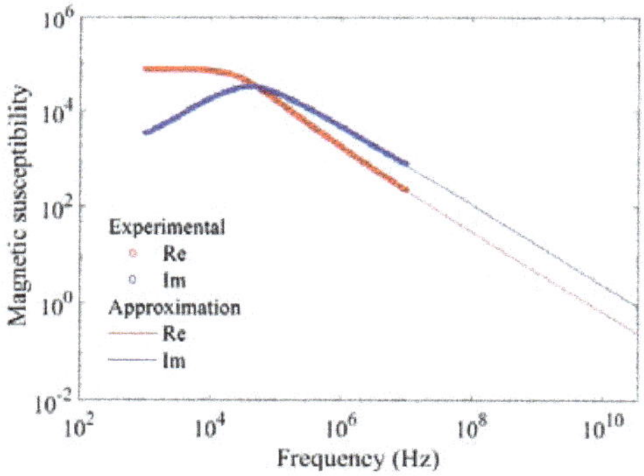

Figure 5. A comparison between measured and approximated frequency dependences of the real and the imaginary parts of magnetic permeability of Finemet alloy.

In Equation (10), χ_0 is the permeability at $\omega \to 0$, χ_{∞} is the permeability at $\omega \to \infty$, τ is the mean relaxation time of the magnetic moment, which is an adjustable parameter, and α is the parameter chosen for better approximation, $0 < \alpha < 1$.

From the view of dependences shown in Figure 5, let us accept $\chi_{\infty} = 0$ because, with the increase in frequency, the monotonic and strong decrease in permeability is fixed. The best results of approximation have been obtained with $\chi_0 = 1.78 \cdot 10^3$, $\tau = 3.84 \cdot 10^{-6}$ s, $\alpha = 0.17$. At these values of the approximation parameters at frequency f = 10 GHz $\mathrm{Re}(\dot{\mu}) \approx 1.2$, in accordance with data [27], and at frequency f = 30 GHz $\mathrm{Re}(\dot{\mu}) \approx 1.08$. Equation (10) permits calculating the magnetic susceptibility and permeability for the material of flakes in a wide frequency range. After that, it is necessary to calculate the magnetic permeability of the composite.

The calculation of magnetic permeability is carried out for the ensemble of ferromagnetic particles flakes-type composed of 1600 randomly oriented flakes and 400 flakes parallel to the top surface of the sample.

The complex effective magnetic permeability of the composite $\dot{\mu}_{eff}$ is calculated using the formula

$$\dot{\mu}_{eff} = \left\langle \dot{\mu}^{m}(\boldsymbol{\Theta}) \right\rangle \tag{11}$$

where index m means belonging to a separate ferromagnetic particle, and the angle brackets mean averaging over the ensemble, taking into consideration the angle $\boldsymbol{\Theta}$ of the orientation of every flake. The spatial orientation of a flake can be characterized by the direction of the normal to its flat surface. The orientation of a ferromagnetic particle is given by the vector $\boldsymbol{\Theta} = (\alpha \;\; \beta \;\; \gamma)$, where α, β and γ are the elements of the independent sets of random numbers with the uniform laws of distribution and belonging to the following intervals: $\alpha \in [-\pi;\ \pi]$, $\beta \in [-\pi;\ \pi]$ and $\gamma \in [-\pi;\ \pi]$.

In computer modelling of $\dot{\mu}_{eff}$, the discrete sets of numbers α_p, β_p and γ_p are formed by a random-number generator; they are combined into the corresponding set of the vectors $\boldsymbol{\Theta}_p = (\alpha_p \;\; \beta_p \;\; \gamma_p)$. These vector sets specify an ensemble of the randomly oriented flakes. Take into consideration that some part of ferromagnetic particles inside the composite material is oriented in a definite manner, namely, the normal vector $\mathbf{n} = (0 \;\; 1 \;\; 0)$ is directed along the y-axis; that is, these flakes lie in the top plane of the sample. This orientation is defined by the vector $\boldsymbol{\Theta}_0$. If there are L_1 randomly oriented flakes and L_2 flakes with orientation $\boldsymbol{\Theta}_0$, then Equation (11) can be rewritten in the following way:

$$\dot{\mu}_{eff} = \frac{L_2 \dot{\mu}^{m}(\boldsymbol{\Theta}_0) + \sum\limits_{p=1}^{L_1} \dot{\mu}^{m}(\boldsymbol{\Theta}_p)}{L_1 + L_2} \tag{12}$$

In the case under consideration, we accept L_1 = 1600 and L_2 = 400. Equation (12) assumes that particles interact only with the external field that is correct only for low concentrations of particles. The ensemble averaging of Equations (11) and (12) over orientations can be performed using the formulas shown in the monograph [2]. Specifically, if there are the particles oriented with the tensor of demagnetizing factors $\overleftrightarrow{\mathbf{N}} = 1$ and $\mathrm{tr}(\overleftrightarrow{\mathbf{N}}) = 1$, then the tensor of magnetic permeability $\overleftrightarrow{\boldsymbol{\mu}}^{m}$ of the composite medium takes the form

$$\begin{gathered} \overleftrightarrow{\boldsymbol{\mu}}^{m} = \begin{pmatrix} \mu^{m}_{xx} & 0 & 0 \\ 0 & \mu^{m}_{yy} & 0 \\ 0 & 0 & \mu^{m}_{zz} \end{pmatrix}, \\ \mu^{m}_{ii} = 1 + \theta_v \frac{\mu - 1}{1 + (1-\theta_v) N_{ii}(\mu - 1)}, \end{gathered} \tag{13}$$

where μ is the magnetic permeability of an isotropic magnetic medium that satisfies the formula $\dot{\mu}(\omega) = 1 + \theta_v \dot{\chi}(\omega)$. In Equation (13) and further, the summation is realized by the repeating indices. If the particles of the same sort are oriented randomly, then, according to [2], it is possible to introduce a scalar effective permeability of the composite medium the following view

$$\mu_{eff} = 1 + \frac{\frac{\theta_v}{3} \cdot \frac{\mu - 1}{1 + N_{ii}(\mu - 1)}}{1 - \frac{\theta_v}{3} \cdot \frac{N_{ii}(\mu - 1)}{1 + N_{ii}(\mu - 1)}} \tag{14}$$

As soon as it can be assumed for flakes that $\overleftrightarrow{\mathbf{N}} = \delta_{i2}\delta_{2j}$, where δ_{ij} is the Kronecker delta function, Equation (14) can be transformed as

$$\mu_{eff} = 1 + \theta_v(\mu - 1)\frac{2\mu + 1}{(3 - \theta_v)\mu + \theta_v} \tag{15}$$

The results of calculation of the frequency dependences of the real and imaginary parts of magnetic permeability, for the composite with 15% flakes, carried out following Equation (12) and taking into account the results of approximation of the magnetic permeability of the material of flakes and the following averaging following Equation (15), are shown in Figure 6a.

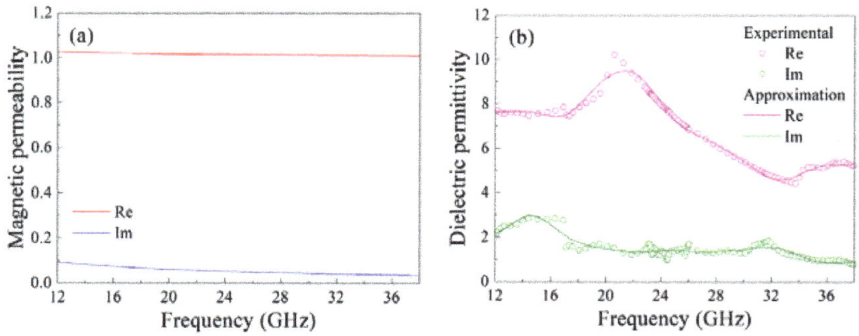

Figure 6. Frequency dependences of magnetic permeability (**a**) and dielectric permittivity (**b**) for the composite with 15% flakes.

In Figure 6b, the frequency dependence of the dielectric permittivity of the composite is presented, which is restored from the measurements of the transmission and reflection coefficients. This dependence is obtained according to the method described above with the sliding frequency window. The frequency dependences of the permeability, which are smoothed using the fast wavelet transform-based wavelet filtering algorithm, are shown as solid lines. This algorithm is widely used to process the measurement results of various physical quantities [29,30]. We have used the following parameters of the wavelet filtering here: the wavelet function is sym10 from the symlets family, the number of decomposition levels is equal to 8, and the thresholding strategy is to zero out all the detail coefficients. The mean values of the real and imaginary parts of the dielectric permittivity for three frequency ranges, as well as the values of the microwave conductivity for the composite with 15% flakes, are listed in Table 1. In this table, the values of dielectric permittivity of the epoxy matrix are also presented. It is seen that added metallic particles essentially increase both the real and the imaginary parts of the permittivity. For the composite with 30% flakes in the frequency range from 26 to 38 GHz the mean values are the following: $\varepsilon' = 38$, $\varepsilon'' = 13$ and $\sigma = 22$ S/m.

Table 1. Estimations of complex dielectric permittivity and microwave conductivity, averaged over the frequency ranges.

Frequency Range (GHz)	Sample	ε'	ε''	σ (S/m)
12–18	Epoxy matrix	3.22	0.23	
	Composite 15%	7.50	3.13	2.45
18–26	Epoxy matrix	2.94	0.11	
	Composite 15%	8.23	1.52	1.90
26–38	Epoxy matrix	2.61	0.33	
	Composite 15%	5.41	1.10	2.01

4. Discussion

Equations (6)–(8) give the opportunity to calculate the transmission and reflection coefficients and the portion of dissipated power as functions of the wave frequency and the thickness of the composite plate. Of course, the frequency dependence calculated for a fixed thickness equal to the thickness of the sample in the experiment coincides with the experimental dependency. The goal of the calculations is to obtain the frequency and thickness dependences for a wide range of parameters. For more wide applicability of the results of calculations presented below, Equation (4b) for the impedance of the free space is used. The results of calculations of the transmission T_P and reflection R_P coefficients, as well as the portions of dissipated power D, are presented in graphical form, and the values of the coefficients are represented by colors. In Figure 7, the dependences of transmission (a), reflection (b) coefficients and dissipation (c) on the frequency and the thickness of the plate are shown for the composite with 15% flakes. If one observes these dependences as the functions of either the frequency or the thickness, it is seen that these dependences are nonmonotonic, as a rule. The presence of the extrema is linked with the fact that a quarter or half of the wavelength fits the thickness of the plate.

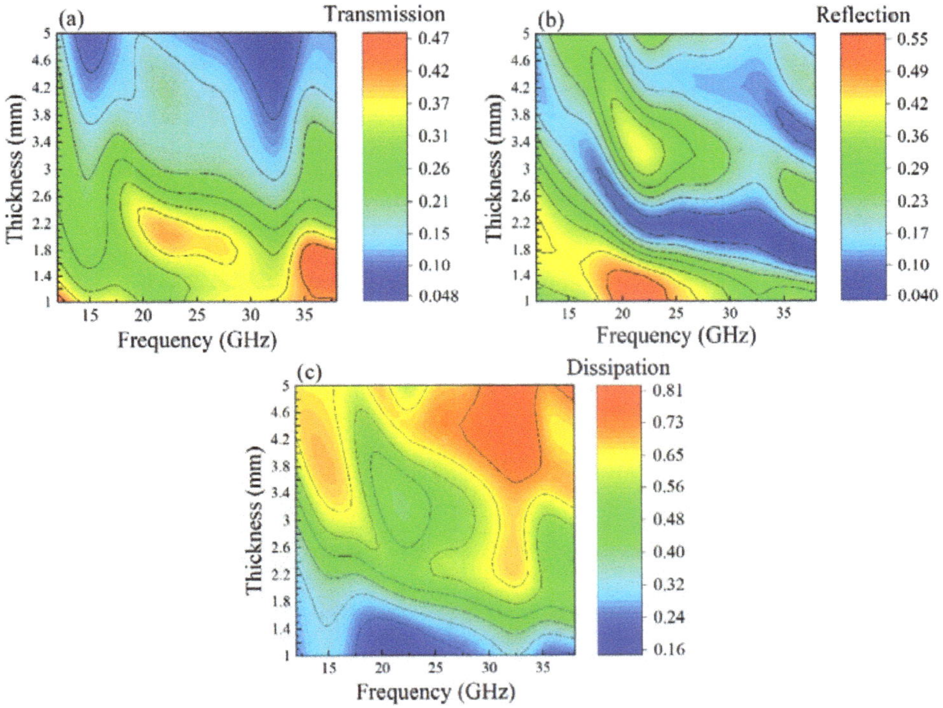

Figure 7. The dependences of transmission coefficient (**a**), reflection coefficient (**b**) and dissipation (**c**) on wave frequency and plate thickness for composite with 15% flakes.

Because these dependences are nonmonotonic, so, it is reasonable to examine the regions where the coefficients become maximal or, in the opposite, minimal values. Knowing these regions is crucial for applications. Certainly, the transmission coefficient T_P decrease with a substantial increase in the plate thickness, as a whole, owing to absorption. The coefficient T_P assumes the maximal values at frequencies 35–38 GHz when the plate thickness is small and the inequality $\lambda_2 >> d_2$ is valid, where λ_2 is the wavelength in the composite. One can see in Figure 7a also two regions of weak transmission for the plates with the thickness exceeding 3–4 mm. These regions are located near the frequencies 15 and 32 GHz. The conducted analysis shows that at frequencies 12–15 GHz, the condition of a quarter-wavelength plate is fulfilled $d_2/\lambda_2 = 1/4$. Near the frequency of 32 GHz, one wavelength fits the plate thickness. Because the impedances of the plate and the surrounding space differ drastically, therefore the standing wave arises in these frequency ranges. In consequence, the amplitude of oscillations increases, and that is why dissipation grows essentially, see Figure 7c. The module of transmission coefficient is small as a result: $T_P \approx 0.05$.

The distribution of the regions of small reflection in Figure 7b has a rather complicated view. Among other things, reflection is high in the region f = 19–24 GHz but drops down to $R_P \approx 0.04$ in the same frequency range for thicker plates d_2 ~ 2 mm. The region of small values of reflection in Figure 7b moves to the lower thickness of the plate when frequency increases. Compare Figure 7b,c, one can conclude that the regions of small reflection do not always coincide with the regions of high dissipation. For example, the maximal dissipation realizes for frequencies f = 30–34 GHz, where reflection is small but not minimal. Of course, the presence of the local extrema in Figure 7 is caused by interference of forward and backward waves. The frequency-dependent absorption, which is defined by the imaginary parts of dielectric permittivity and magnetic permeability, also influences.

Let us now consider the results obtained for the composite with 30% flakes. For this composite, the averaged imaginary part of dielectric permittivity equals ε'' = 13, and the microwave conductivity is σ = 22 S/m, that is much higher than for the composite with 15% flakes. Therefore, the effect from the interference of forward and backward waves inside the plate has to be essentially lower. In Figure 8, the dependences have been shown of the transmission coefficient on the wave frequency and the thickness of the plate for the composite with 30% flakes. There are no local maximums and minimums in this case. The transmission coefficient decreases rapidly when the thickness arises. Let us perform the analysis of Equations (6) and (7), under strong absorption, when $k_2'' d_2 >> 1$. Given what if $x >> 1$ then the approximate equalities $\sinh(x) \approx \frac{\exp(x)}{2}$, $\cosh(x) \approx \frac{\exp(x)}{2}$ are valid, we can obtain the following equations:

$$T_P = \frac{16Z_1^2\left|\dot{Z}_2\right|^2 \exp\left(-k_2'' d_2\right)}{4Z_1^2\left|\dot{Z}_2\right|^2 + Z_1^4 + \left|\dot{Z}_2\right|^4 + 2Z_1^2\mathrm{Re}\left(\dot{Z}_2^2\right) + 4Z_1\mathrm{Re}\left(\dot{Z}_2\right)\left(Z_1^2 + \left|\dot{Z}_2\right|^2\right)} \tag{16}$$

$$R_P \approx \frac{Z_1^4 + \left|\dot{Z}_2\right|^4 - 2Z_1^2\mathrm{Re}\left(\dot{Z}_2^2\right)}{4Z_1^2\left|\dot{Z}_2\right|^2 + Z_1^4 + \left|\dot{Z}_2\right|^4 + 2Z_1^2\mathrm{Re}\left(\dot{Z}_2^2\right) + 4Z_1\mathrm{Re}\left(\dot{Z}_2\right)\left(Z_1^2 + \left|\dot{Z}_2\right|^2\right)} \tag{17}$$

From Equations (16) and (17), the expression can be obtained for the microwave dissipation

$$D = \frac{4Z_1\left[Z_1\left[\left|\dot{Z}_2\right|^2 + \mathrm{Re}\left(\dot{Z}_2^2\right)\right] + \mathrm{Re}\left(\dot{Z}_2\right)\left(Z_1^2 + \left|\dot{Z}_2\right|^2\right)\right]}{4Z_1^2\left|\dot{Z}_2\right|^2 + Z_1^4 + \left|\dot{Z}_2\right|^4 + 2Z_1^2\mathrm{Re}\left(\dot{Z}_2^2\right) + 4Z_1\mathrm{Re}\left(\dot{Z}_2\right)\left(Z_1^2 + \left|\dot{Z}_2\right|^2\right)} \tag{18}$$

It follows from Equation (16) that a monotonic decreasing thickness dependence of the transmission coefficient should exist. Curiously, it is found that, in this case of strong absorption, the reflection coefficient almost does not depend on the thickness of the plate that declares Equation (17).

At last, let us discuss the dependence of dissipation on the thickness of the plate. In Figure 9, these dependences are shown drawn for the composites with 15% and 30% flakes at several frequencies. For the composite with 15% flakes, the dependences are oscillating at all frequencies and in the whole range of the plate thicknesses; there is a trend of dissipation increasing while the thickness increases. For the composite with 30% flakes, the oscillations present only for thin plates, so far as the influence of backward waves weakens if the thickness becomes higher. It seems surprising at first glance that the dissipation at high thicknesses of 4–5 mm for the composite with 15% flakes is higher than for the composite with 30% flakes. This peculiarity is caused by the high reflection of waves from the composite with the high content of metallic particles.

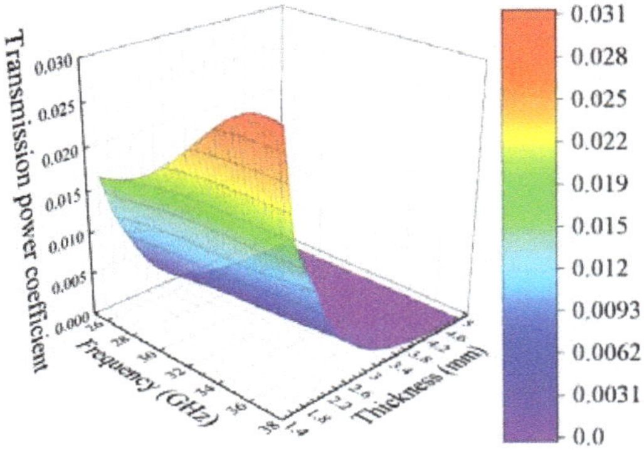

Figure 8. The dependences of the transmission coefficient on wave frequency and plate thickness for the composite with 30% flakes.

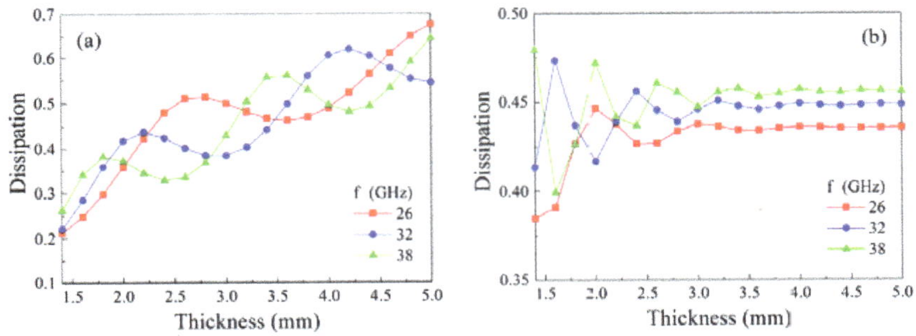

Figure 9. The dependences of microwave dissipation on the thickness of the plate for the composite with 15% (**a**) and 30% (**b**) flakes calculated for several frequencies.

5. Conclusions

The investigation of the microwave properties of the composites prepared from nanocrystalline flakes of finemet-type alloy has been made. The mean size of flakes is 50 μm. The transmission and reflection coefficients, as well as the portion of dissipated power, have been studied both experimentally and theoretically. The measurements and calculations are carried out in the frequency range from 12 to 38 GHz for the composites with 15% and 30% flakes.

The method of restoring the frequency dependence of the complex magnetic permeability is worked up, taking into consideration the distribution of spatial orientation of flakes. The frequency dependence of the dielectric permittivity is obtained from the measured frequency dependences of the transmission and reflection coefficients. The frequency dependences of the transmission and reflection coefficients have been calculated for the plate thickness from 1.4 to 5 mm, taking into account the frequency dependences of the magnetic permeability and dielectric permittivity.

It has been found that the frequency and the thickness dependences of the coefficients are nonmonotonic ones owing to the standing waves. The regions of frequencies and thicknesses where the reflection minimum is observed do not precisely coincide with the maxima of the dissipation. For the composite with 30% flakes, the oscillations in the dependences are weak because of strong absorption.

The obtained results can be useful for the calculations of microwave devices containing magnetic composites.

Author Contributions: Conceptualization, A.B.R.; methodology, A.B.R. and D.V.P.; software, D.V.P.; validation, A.B.R., D.V.P. and Y.I.R.; formal analysis, D.V.P.; investigation, A.B.R. and Y.I.R.; resources, Y.I.R.; writing—original draft preparation, A.B.R.; writing—review and editing, A.B.R. and Y.I.R. All authors have read and agreed to the published version of the manuscript.

Funding: This research was funded by the Russian Ministry of Science and Education, the theme "Function" No AAAA-A19-119012990095-0. The calculation of the transmission and reflection coefficients and discussion was carried out with the support of the Russian Science Foundation, grant No 17-12-01002.

Institutional Review Board Statement: Not applicable.

Informed Consent Statement: Not applicable.

Data Availability Statement: Not applicable.

Acknowledgments: The authors are grateful to Yu.V. Korkh for the analysis of the distribution of flakes over dimensions and E.A. Kuznetsov for help in the microwave measurements.

Conflicts of Interest: The authors declare no conflict of interest.

References

1. Kumar, V.; Kalia, S.; Swart, H.C. *Conducting Polymer Hybrids*; Springer Series on Polymer and Composite Materials; Springer: Cham, Switzerland, 2017.
2. Sihvola, A. *Electromagnetic Mixing Formulas and Applications*; The Institution of Electrical Engineers: London, UK, 1999; 284p.
3. Lagarkov, A.N.; Rozanov, K.N. High-frequency behavior of magnetic composites. *JMMM* **2009**, *321*, 2082–2092. [CrossRef]
4. Shiryaev, A.O.; Rozanov, K.N.; Starostenko, S.N.; Bobrovskii, S.Y.; Osipov, A.V.; Petrov, D.A. The bias effect on the frequency dispersion of microwave permeability of composites filled with metal films or flakes. *JMMM* **2019**, *470*, 139–142. [CrossRef]
5. Yang, S.; Liang, B.; Liu, C.; Liu, J.; Fang, C.; Ai, Y. Microwave sintering and microwave dielectric properties of $(1-x)Ca_{0.61}La_{0.26}TiO_3$-$xNd(Mg_{0.5}Ti_{0.5})O_3$ Ceramics. *Materials* **2021**, *14*, 438. [CrossRef] [PubMed]
6. Idris, F.M.; Hashim, M.; Abbas, Z.; Ismail, I.; Nazlan, R.; Ibrahim, I.R. Recent developments of smart electromagnetic absorbers based polymer-composites at gigahertz frequencies. *JMMM* **2016**, *405*, 197–208. [CrossRef]
7. Yang, R.B.; Liang, W.F. Microwave absorbing characteristics of flake-shaped FeNiMo/epoxy composites. *J. Appl. Phys.* **2013**, *113*, 17A315. [CrossRef]
8. Feng, Y.; Qiu, T. Enhancement of electromagnetic and microwave absorbing properties of gas atomized Fe-50 wt% Ni alloy by shape modification. *JMMM* **2012**, *324*, 2528–2533. [CrossRef]
9. Gargama, H.; Thakur, A.K.; Chaturvedi, S.K. Polyvinylidene fluoride/nanocrystalline iron composite materials for EMI shielding and absorption applications. *J. Alloys Compd.* **2015**, *117*, 224903. [CrossRef]
10. Majcher, K.; Musiał, M.; Pakos, W.; Rózanski, A.; Sobótka, M.; Trapko, T. Methods of protecting buildings against HPM radiation—A review of materials absorbing the energy of electromagnetic waves. *Materials* **2020**, *13*, 5509. [CrossRef]
11. Lutsev, L.V.; Kazantseva, N.E.; Tchmutin, I.A.; Ryvkina, N.G.; Kalinin, Y.E.; Sitnikoff, A.V. Dielectric and magnetic losses of microwave electromagnetic radiation in granular structures with ferromagnetic nanoparticles. *J. Phys. Condens. Matter.* **2003**, *15*, 3665–3681. [CrossRef]
12. Rinkevich, A.B.; Ryabkov, Y.I.; Perov, D.V.; Nemytova, O.V. Microwave refraction coefficient of composite with flakes of Fe-Si -Nb-Cu-B alloy. *JMMM* **2021**, *529*, 167901. [CrossRef]
13. Brosseau, C.; Talbot, P. Effective magnetic permeability of Ni and Co micro- and nanoparticles embedded in a ZnO matrix. *J. Appl. Phys.* **2005**, *97*, 104325. [CrossRef]
14. Zygmuntowicz, J.J.; Wachowski, M.; Zielant, D.; Kaszuwara, W. Influence of magnetic field on the distribution of the ferromagnetic component in centrifugally cast ceramic-metal gradient composites. *Materials* **2021**, *14*, 955. [CrossRef] [PubMed]
15. Iqbal, Y.; Davies, H.A.; Gibbs, M.R.J.; Woodcock, T.R.; Todd, I.; Major, R.V. Nanocrystalline powder cores for high frequency applications. *JMMM* **2002**, *242–245*, 282–284. [CrossRef]
16. Kubacki, R.; Nowosielski, L.; Przesmycki, R.; Ferenc, J.; Frender, R. Magnetic properties of the finemet alloys in the microwave frequency range. In Proceedings of the 18th International Conference on Microwave Radar and Wireless Communications, Vilnius, Lithuania, 14–16 June 2010; pp. 103–106, ISBN 978-1-4244-5288-0.
17. Mikhalitsyna, E.A.; Kataev, V.A.; Larrañaga, A.; Lepalovskij, V.N.; Kurlyandskaya, G.V. Nanocrystallization in FINEMET-type $Fe_{73.5}Nb_3Cu_1Si_{13.5}B_9$ and $Fe_{72.5}Nb_{1.5}Mo_2Cu_{1.1}Si_{14.2}B_{8.7}$ thin films. *Materials* **2020**, *13*, 348. [CrossRef] [PubMed]
18. Bai, F.; Dong, Y.; Xie, L.; Li, Q.; He, A.; Jia, X.; Li, J.; Wang, X. Effect of pre-existing nuclei on microstructure and magnetic properties of high Bs FINEMET-like nanocrystalline alloys. *J. Mater. Sci.* **2021**, *56*, 9254–9262. [CrossRef]
19. Moulin, J.; Shahosseini, I.; Alves, F.; Mazaleyrat, F. Ultrasoft Finemet thin films for magneto-impedance microsensors. *J. Micromech. Microeng.* **2011**, *21*, 074010. [CrossRef]
20. Ramprasad, R.; Zurcher, P.; Petras, M.; Miller, M. Magnetic properties of metallic ferromagnetic nanoparticle composites. *J. Appl. Phys.* **2004**, *96*, 519–529. [CrossRef]

21. Chevalier, A.; Mattei, J.-L.; Le Floc'h, M. Ferromagnetic resonance of isotropic heterogeneous magnetic materials: Theory and experiments. *JMMM* **2000**, *215–216*, 66–68. [CrossRef]
22. Mattei, J.-L.; Le Floc'h, M. A numerical approach of the inner demagnetizing effects in soft magnetic composites. *JMMM* **2000**, *215–216*, 589–591. [CrossRef]
23. Rinkevich, A.B.; Perov, D.V.; Vaskovsky, V.O.; Gorkovenko, A.N.; Kuznetsov, E.A. Millimeter wave resistance of metal-dielectric $Co_X(SiO_2)_{1-X}$ and $Co_X(Al_2O_3)_{1-X}$ films. *IEEE Trans. Nano.* **2017**, *16*, 1067–1072. [CrossRef]
24. Collin, R.E. *Field Theory of Guided Waves*; IEEE Press: Piscataway, NJ, USA; John Wiley & Sons: New York, NY, USA, 1991; 852p.
25. Ustinov, V.V.; Rinkevich, A.B.; Perov, D.V.; Burkhanov, A.M.; Samoilovich, M.I.; Kleshcheva, S.M.; Kuznetsov, E.A. Giant antiresonance in electromagnetic wave reflection from a 3D structure with ferrite spinel nanoparticles. *Tech. Phys.* **2013**, *58*, 568–577. [CrossRef]
26. Nanocrystalline Soft Magnetic Material. Hitachi Metals. Available online: https://www.hitachi-metals.co.jp/e/product/finemet/cmc/characteristic/index.html (accessed on 22 April 2021).
27. Zajtseva, N.V.; Korobejnikov, G.V.; Kokhnjuk, D.D.; Ivanova, L.N.; Slavin, V.V.; Kuznetsov, P.A.; Marennikov, N.V.; Semenenko, V.N. Electromagnetic Wave Absorber. Russian Patent No RU2 414 029 C1, 10 March 2011. Russian Bulletin of Patents 2011, No 7.
28. Cole, K.S.; Cole, R.H. Dispersion and absorption in dielectrics. *J. Chem. Phys.* **1941**, *9*, 341. [CrossRef]
29. Perov, D.V.; Rinkevich, A.B. Using wavelets for analyzing ultrasonic fields detected by a laser interferometer. Basic concepts of the wavelet analysis. *Russ. J. Nondest. Test.* **2001**, *37*, 879–888. [CrossRef]
30. Perov, D.V.; Rinkevich, A.B.; Smorodinskii, Y.G. Wavelet filtering of signals from ultrasonic flaw detector. *Russ. J. Nondest. Test.* **2002**, *38*, 869–882. [CrossRef]

materials

MDPI

Article

A Mixing Model for Describing Electrical Conductivity of a Woven Structure

Magdalena Tokarska

Faculty of Material Technologies and Textile Design, Institute of Architecture of Textiles, Lodz University of Technology, 90-924 Lodz, Poland; magdalena.tokarska@p.lodz.pl

Abstract: The main aim of the research was to describe electro-conductive woven structures by specifying the phases' exponents using the generalised Archie's law. Special woven structures were designed to transfer Archie's model to the textile object. The woven structure was treated as a complex multiphase mixture. The structure was composed of two conducting phases (strips and strip contacts) and one non-conducting phase (pore space). It was found that the designed structures were characterised by the phases' exponents that exceeded the value of 2, which denoted low connectivity in the conductive phases. A qualitative and quantitative description of the woven structure was feasible, i.e., the connectedness and the connectivity, respectively. The connectedness of both of the phases was dependent on the material from which the structure was designed. The fraction of each of the phases involved in the current conductivity was important. The connectivity connected with structure density, in varying degrees, affected the electro-conductive properties of the woven structure. It was important how the phases were arranged in the whole composite. It was found that the strips' contact phases played an important role in the structure of the composite.

Keywords: Archie's law; mixing model; electrical conductivity; woven structure; composite

Citation: Tokarska, M. A Mixing Model for Describing Electrical Conductivity of a Woven Structure. *Materials* **2022**, *15*, 2512. https://doi.org/10.3390/ma15072512

Academic Editor: Petrica Vizureanu

Received: 4 February 2022
Accepted: 23 March 2022
Published: 29 March 2022

1. Introduction

Due to their electrical properties, woven and knitted fabrics belonging to flat textile objects can be used as textile sensors [1–4]. Most textile-based sensors rely on a change in electrical resistance. To predict their conductivity, models based on equivalent resistance schemes are used [5–8]. The fabric is seen from an electrical point of view as an electrical circuit composed of connected resistors and a battery. Yarns are ideal resistors of known resistance. The simulated structure was regular so the resistive model was a simplified model of the textile object, which was much more complex and showed the anisotropy of the electrical properties [9–12]. The electrical conductivity of flat textile materials results from the electrical conductivity of their components, i.e., fibres and yarns and contact resistances resulting from interlaced yarns [8,13–15]. Woven fabric can be compared to metal-dielectric composites where conductive linear components (yarns) create a system of empty spaces filled with dielectric air [10,14]. To predict the resistivity of fabrics, the McLachlan equation described in [16,17] was modified [14]. A prediction level in the range of 83–88% for fabrics with a surface percentage covered with yarns above 96% was obtained [14]. Contact resistance resulting from the interlaced yarns was not taken into consideration.

Many types of mixing models are used to determine the electrical properties of porous materials. One of them is Archie's law [18] extended for n phases and known as the generalised Archie's law [19]. Archie developed an empirical quantitative relationship between the porosity, electrical conductivity, and brine saturation of rocks. The law describes the relationship of the conductivity of a clean reservoir rock to its porosity and the conductivity of phase (e.g., fluid) that completely saturates the pore space, and is given as follows [19]:

$$\sigma = \sigma_f \phi^m \quad (1)$$

where σ is the bulk effective conductivity of the rock, σ_f is the conductivity of the fluid occupying the pores, ϕ is the volume fraction of the fluid phase, and m is the cementation exponent. Due to the fluid occupying the pores, and the pore space being fully saturated, ϕ is identical to the porosity.

The parameter $G = \sigma/\sigma_f = \phi^m$, called the connectedness of the porous medium, was defined by Glover and Walker [20]. The connectedness of a given phase is a measure of the availability of pathways for conduction through that phase. The connectivity defined as $\chi = \phi^{m-1}$ is a measure of how the pore space is arranged [19]. A small exponent m (below 2) occurs for high connectivity phases [21]. Pores that are well connected provide an efficient pathway for the medium flow. A large exponent m (equal to or above 2) occurs for low connectivity phases. It was noticed that a cementation exponent m depends on the shape and type of the sedimentary rock grains, the shape and type of pores, specific surface area, tortuosity, anisotropy, and compaction [22]. Archie's law was used to predict the conductivity-brine volume trend for sandstone [23], and to characterise sedimentary rock formations, i.e., carbonate rocks, which are prone to develop a wide variety of pore structures [22]. The law modifications were presented for tight and clay-rich reservoirs to describe electrical conductivity in such reservoirs [24]. The generalised Archie's law was also used for the modelling of electro-conductive properties of woven structures [25]. Parameters such as connectedness and connectivity, determined for conducting phases, enabled an explanation of the phenomenon of current conduction in a woven structure. It was found that a decrease in the connectedness of strips and strip contacts' phases could be obtained by adding another component to the woven structure, which would reduce the conductivity of the whole structure. The larger values of connectivity for the strips' phase compared to the connectivity of the strip contacts' phase meant that the strips' phase (in terms of their quantity) had a greater effect on the conductivity of the woven structure than the strip contacts' phase.

Archie's law is not applicable if there are two or more conducting phases, or if the conducting phase does not fill the pore space. Due to the occurrence of porous media consisting of solid and fluid phases of known conductivities, volume fractions, and distributions, Archie's law was modified by Glover et al. for two [21], and n conducting phases [19]. The generalised Archie's law is as follows:

$$\sigma = \sum_{i=1}^{n} \sigma_i \phi_i^{m_i} \quad (2)$$

where σ is the mixing model conductivity, σ_i is the i-th phase conductivity, ϕ_i is the i-th phase volume fraction, m_i is the i-th phase exponent, and n is the number of conducting phases.

It was stated that the sum of the volume fractions of all of the phases in a porous medium (a rock) must equal unity. Each of the n phases that occupy the rock must share the same total space. This means that an increase in the connectedness of one of the phases must lead to a reduction in the connectedness of at least one of the other phases [19].

In the generalised Archie's law, phase exponent $m < 1$ represents a phase with a high degree of connectivity [19,26]. A value of $m \approx 1$ can be observed for rocks with a low porosity but a well-developed fracture network, the network then has fairly direct flow paths. A phase exponent $m \approx 2$ means that the phase is partially connected in a similar way as in sandstone [23,24,27]. A higher value of m represents lower phase connectivity as in the case of vuggy limestone [22]. The classical and generalised Archie's laws share the property that the exponents modify the volume fraction of the relevant phase concerning the total volume of the rock.

The main aim of this research was to describe electro-conductive woven structures by specifying the phases' exponents using the generalised Archie's law. This meant that by designing the same structure, but from other components, you could predict the conductivity of the new structure based on parameters determined from the mixing model. The law can be applied to objects containing pores filled with phases. Therefore, special woven structures were designed to transfer Archie's model to the textile object. New structures were composed of interlaced strips cut from fabrics. Such a structure was treated as a complex multiphase mixture containing two phases and a matrix. The strips and the strip contacts were phases that allowed current to be conducted in the composite. The remaining part was pore space, corresponding to the non-conducting matrix. So, the pores of the woven structure were not identical with the pores of Archie's model, but were opposite to the matrix. This comparison was an original and innovative approach to the woven structure for modelling its electro-conductive properties. The features of the fabric's linear elements and the number of their contacts seemed to be important in controlling the electrical conductivity of the woven structure.

2. Materials

Three commercially available electro-conductive woven fabrics were chosen to construct new woven structures. The raw material composition of the fabrics is presented in Table 1. The fabric denoted as S1 was purchased from Laird™ (ABC Elektronik Sp.z o.o., Gorlice, Poland), and the fabrics denoted as S2 and S3 were purchased from Soliani™ (Como, Italy). According to datasheets, the surface resistivity of S1 was below 0.07 Ω/sq, and for S2 and S3 did not exceed 0.40 Ω/sq.

Table 1. The raw material composition of woven fabrics.

Woven Fabric	S1	S2	S3
Raw material composition	100% polyamide woven fabric; nickel and copper metalised	100% polyester woven fabric; nickel metalised	100% polyester woven fabric; nickel metalised
Weave	Plain	Plain	Twill
Microscopic image with total visual magnification 30× ↓ the warp direction → the weft direction	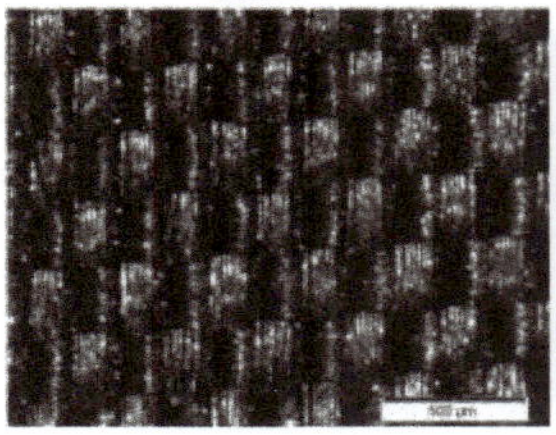	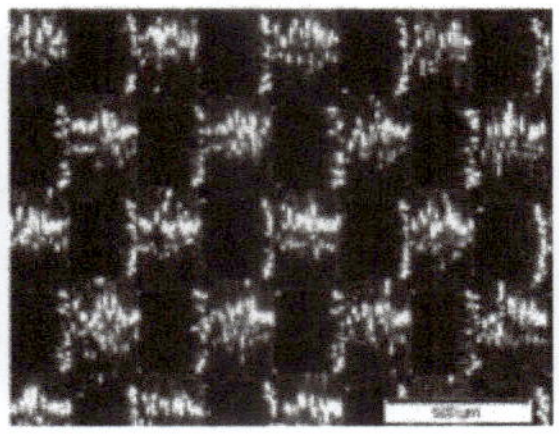	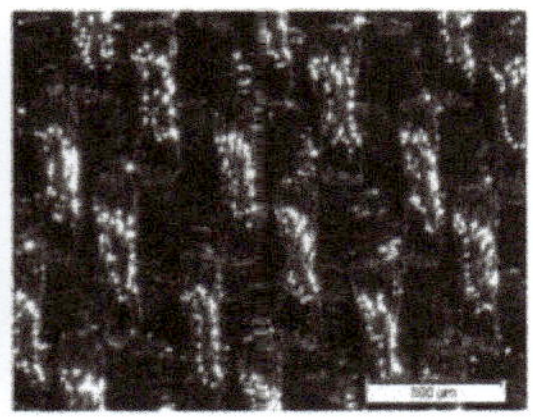

Parameters of the textile materials (woven fabrics) are presented in Table 2.

Table 2. Parameters of woven fabrics.

Woven Fabric	Thickness (mm)	Areal Density (g/m^2)	Bulk Density (kg/m^3)	Warp Density (Yarns/1 cm)	Weft Density (Yarns/1 cm)
S1	0.124 (10.5%)	86 (2.3%)	694 (10.8%)	57.0 (1.7%)	41.0 (1.2%)
S2	0.078 (15.4%)	75 (2.7%)	966 (15.8%)	40.0 (1.2%)	30.0 (1.7%)
S3	0.270 (5.6%)	152 (2.6%)	564 (5.8%)	47.5 (1.9%)	34.0 (1.7%)

The mean values of quantities such as thickness, areal density, and bulk density, and their relative expanded uncertainties U (given in parentheses in Table 2) for confidence level equal to 0.95, were calculated according to the following equation [28]:

$$U = k_p\, u_c(y) \tag{3}$$

wherein

$$u_C^2(y) = \sum_{i=1}^{N}\left[\left(\frac{\partial f}{\partial x_i}\right)^2\left({u_A}^2(x_i) + {u_B}^2(x_i)\right)\right] \tag{4}$$

where k_p is the coverage factor (for confidence level equal to 0.95, the coverage factor equals 2), $u_C(y)$ is the combined variance, $u_A(x_i)$ is the Type A standard uncertainty estimated from independent repeated observations, $u_B(x_i)$ is the Type B standard uncertainty evaluated by scientific judgment based on all of the available information on the possible variability of input quantity, N is the number of independent input quantities x_i with $i = 1, 2, \ldots, N$, y is the estimate of an output quantity, and f is the functional relationship between input and output quantities (for direct measurements).

Measurements of thickness and mass were repeated five times. Measurements of yarn densities were repeated three times. A rectangular distribution of possible values for the calculation of the Type B uncertainty was assumed [29].

Strips of length 15 cm and two different widths (1.0 cm and 1.5 cm) were cut from fabrics S1, S2, and S3. New plain weave structures composed of the same strips and the same fabric were designed for research purposes. Another woven structure was handmade from interlaced strips based on a specific report. The three structures for strip width equal to 1.5 cm are shown in Figure 1. The fabric components are presented in Figure 1c. Three more structures differing in the width of the strips (1.0 cm) were also designed. The number k of strips in the weft and warp directions was the same for the chosen structure but all the structures differed in strip density (strips per unit of length). The structure dimension $l \times w$ was assumed for the target research, where l was the structure length (wherein l = 9 cm), and w was the structure width (wherein w = 13 cm) (see Figure 1a).

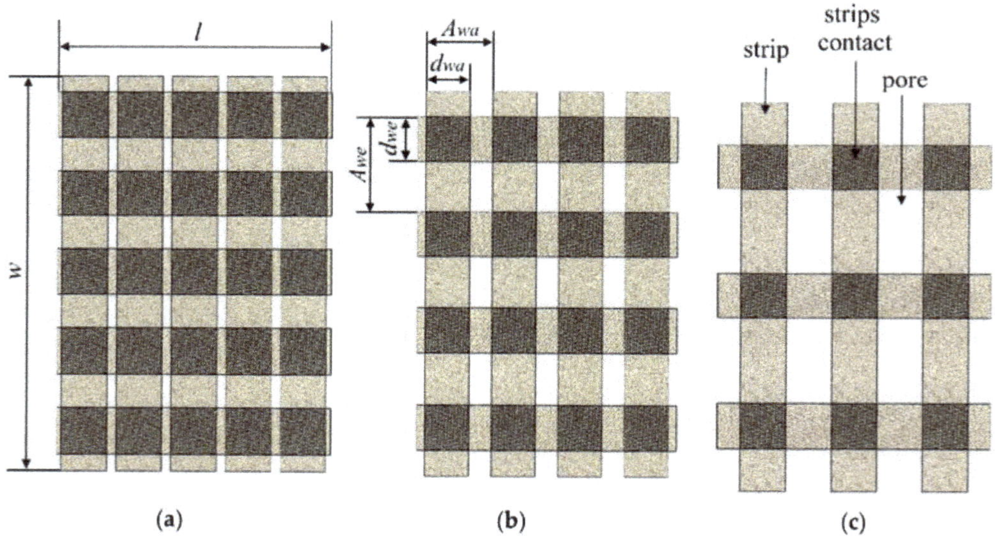

Figure 1. Woven structure types designed from strips: (**a**) 5 × 5; (**b**) 4 × 4; (**c**) 3 × 3.

The designed woven structures were characterised by parameters such as the spacing of warp and weft strips A_{wa} and A_{we}, respectively, and the width of warp and weft strips d_{wa} and d_{we}, respectively [10,30]. The parameters were determined and are shown in Figure 1b for the second structure. Parameters of six designed woven structures are presented in Table 3.

Table 3. Parameters of designed structures.

d_{wa} = d_{we}	$k \times k$	A_{wa} (cm)	Pore Width [1] (cm)	A_{we} (cm)	Pore Length [2] (cm)
1.5 cm	5 × 5	1.80	0.30	2.60	1.10
	4 × 4	2.25	0.75	3.25	1.75
	3 × 3	3.00	1.50	4.33	2.83
1.0 cm	5 × 5	1.80	0.80	2.60	1.60
	4 × 4	2.25	1.25	3.25	2.25
	3 × 3	3.00	2.00	4.33	3.33

[1] Pore width equals A_{wa}–d_{wa}. [2] Pore length equals A_{we}–d_{we}.

Based on parameters of the designed woven structures (Table 3), fractions of components (i.e., strips, strips contacts, and pores) in the whole woven structure and percentage surface cover were determined and are presented in Table 4. The fraction of the component was calculated as a quotient of the particular component area and the whole area of the woven structure ($l \times w$), wherein C_{th} + C_{cont} + C_p = 1. The fraction of strips applied only to strips, excluding their contact surfaces.

Table 4. The fraction share of components in the whole woven structure.

d_{wa} = d_{we}	$k \times k$	Fraction of Strips C_{th} (-)	Fraction of Strip Contacts C_{cont} (-)	Fraction of Pores C_p (-)	Percentage Surface Cover C_{str} (%) [1]
1.5 cm	5 × 5	0.449	0.481	0.070	93
	4 × 4	0.513	0.308	0.179	82
	3 × 3	0.500	0.173	0.327	67
1.0 cm	5 × 5	0.513	0.214	0.273	73
	4 × 4	0.479	0.137	0.384	62
	3 × 3	0.410	0.077	0.513	49

[1] Values calculated according to the equation: $C_{str} = \frac{A_{we}d_{wa}+A_{wa}d_{we}-d_{wa}d_{we}}{A_{wa}A_{we}}100$ [10].

Eighteen woven structures were prepared using three different woven fabrics S1, S2, and S3.

3. Methods

The generalised Archie's law (see Equation (2)) for the woven structure can be rewritten as:

$$\sigma_{str} = \sigma_{th}\phi_{th}^{m_{th}} + \sigma_{cont}\phi_{cont}^{m_{cont}} \quad (5)$$

where σ_{str} is the conductivity of the woven structure, σ_{th} is the conductivity of the strips phase, σ_{cont} is the conductivity of strip contacts' phase, ϕ_{th} is the area fraction of the strips' phase, ϕ_{cont} is the area fraction of the strip contacts' phase, m_{th} is the strips' phase exponent, and m_{cont} is the strip contacts' phase exponent. Equation (5) is valid for σ_{th}, σ_{cont}, m_{th}, m_{cont} > 0, and ϕ_{th}, $\phi_{cont} \in (0,1)$, wherein $\phi_{th} + \phi_{cont} = 1$. Due to the thickness of strips being very small compared to the dimensions of strips and contact strips, instead of the volume fraction of each phase, the area fraction was taken into consideration in Equation (5). Based on the assumption that each conducting phase was fully saturated, i.e., identical to the area fraction of the phase, and detailed analysis for two phases considered by Glover [19], the following equality holds:

$$\left(-\frac{\phi_{th}^2}{2}\right)m_{cont}^2 + \left(\phi_{th} - \frac{\phi_{th}^2}{2}\right)m_{cont} - \phi_{th}^{m_{th}} = 0 \quad (6)$$

If the conductivities of the individual phases and their fractions are known, the application of simultaneous Equations (5) and (6) enables the determination of the strips' phase exponent and the strip contacts' phase exponent. As it was stated that the sum of the area fractions of all of the phases in the composite must equal unity, the same assumption was adopted for the sum of the connectedness of all of the phases.

To describe the woven structure, connectedness and connectivity were adopted. The connectedness of the strips' phase is given by

$$G_{th} = \frac{\sigma_{str}}{\sigma_{th}} \tag{7}$$

While the connectivity is given by

$$\chi_{th} = \phi_{th}^{m_{th}-1} \tag{8}$$

The connectedness of the strip contacts' phase is given by

$$G_{cont} = \frac{\sigma_{str}}{\sigma_{cont}} \tag{9}$$

While the connectivity is given by

$$\chi_{cont} = \phi_{cont}^{m_{cont}-1} \tag{10}$$

The resistance of the strips was determined based on the four-electrode method [31]. Parallel brass plates were used as electrodes. Current I was injected through the two outer electrodes (1 and 4) and voltage drop UI between the two inner electrodes (2 and 3) was measured (Figure 2). The resistance could then be calculated.

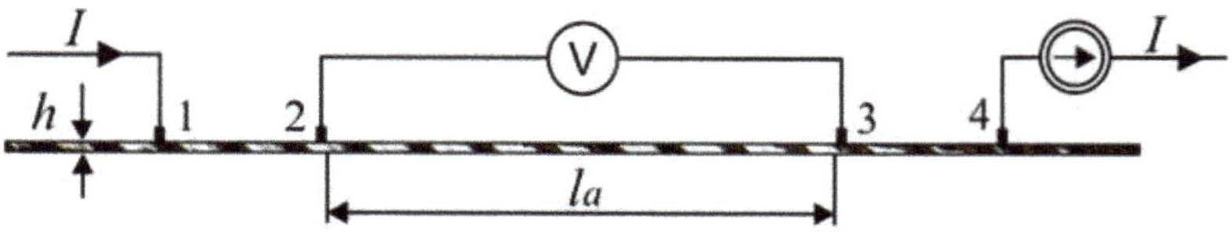

Figure 2. Strip resistance measurements (the side view).

Conductivity σ_{th} of the strip could be determined using the following equation:

$$\sigma_{th} = \frac{l_a}{dhR} \tag{11}$$

where R is the strip resistance, l_a is the voltage electrodes spacing (l_a = 5 cm), d is the strip width (d = 1.0 cm or d = 1.5 cm), and h is the strip thickness (corresponding to the fabric thickness).

Measurements were conducted for 10 strips cut from the same woven fabric. All measurements were repeated three times.

The resistance of strip contacts was determined using the four-electrode method described in detail in [13,32]. Brass plates were used as electrodes. The idea of the measurement method is presented in Figure 3. An initial load of 5 cN was applied to avoid the strips moving relative to each other. Based on the indirect method, resistance could be determined using Ohm's law.

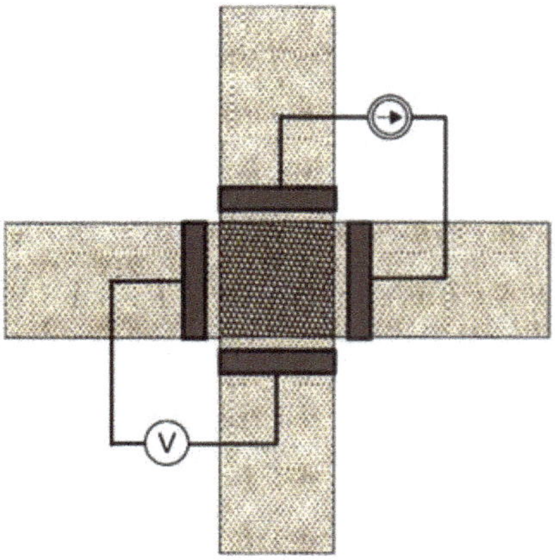

Figure 3. Strip contact resistance measurements (the top view).

As shown in Figure 3, the contact surface was in the shape of a square with each side being 1.0 cm or 1.5 cm depending on the width of the strips. Measurements were conducted for three pairs of strips cut from the same woven fabric. All measurements were repeated three times.

Conductivity σ_{cont} of the strip contact could be determined using the following equation:

$$\sigma_{cont} = \frac{1}{2Rh} \quad (12)$$

where R is the resistance of the strips contact, and h is the strip thickness (double fabric thickness was assumed in Equation (12)).

Resistance measurements of the designed woven structure were performed by the four-wire method using two electrodes [33]. Brass plates were used as electrodes. The direct measurement method is presented in Figure 4. Measurements were carried out in the weft direction.

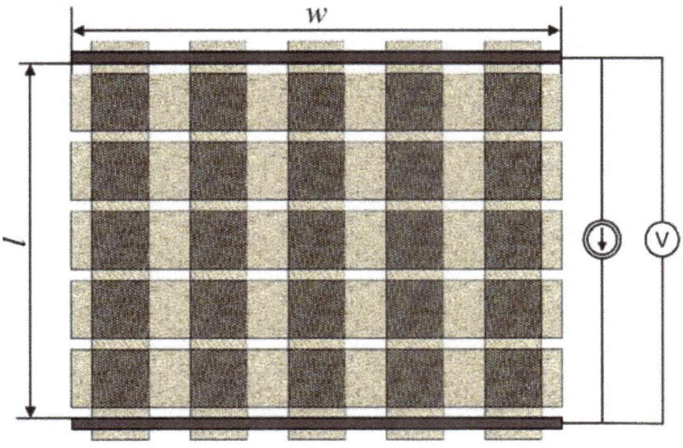

Figure 4. Woven structure resistance measurements (the top view).

It was assumed, that the yarn deformations in the strips forming the woven structure had a negligible effect on the resistance measurements of the structure. Conductivity σ_{str} of the woven structure could be determined using the following equation:

$$\sigma_{str} = \frac{l}{wh'R} \tag{13}$$

wherein

$$h' = 2\,hC_{cont} + 1\,hC_{th} + 0\,hC_p = 2\,hC_{cont} + hC_{th} \tag{14}$$

where R is the woven structure resistance, l is the spacing of the electrodes (l = 9 cm), w is the structure width (w = 13 cm), h' is the resultant fabric thickness, C_{cont} is the fraction of strip contacts in the whole woven structure, C_{th} is the fraction of strips in the structure, and C_p is the fraction of pores in the structure, called the porosity.

A DC power supply Agilent E3644A (Agilent, Santa Clara, CA, USA) was used as an ammeter. The resolution of the ammeter was 0.001 A. A multimeter Agilent 34410A (Agilent) was used as a voltmeter. The resolution of the voltmeter was 0.0001 V.

4. Results and Discussion

Measurements of resistance of strips and strip contacts were carried out in standard atmospheric conditions according to the standard [34]. Conductivities were calculated according to Equations (11)–(14). Received results are given in Table 5. The coefficient of variation determined for conductivities is given in parentheses. Area fractions of phases ϕ_{th} and ϕ_{cont} were also determined and are given in Table 5.

Table 5. Characterisation of woven structures and their components.

Group	Sample	ϕ_{th} (-)	σ_{th} $(\Omega\ \mathrm{cm})^{-1}$	ϕ_{cont} (-)	σ_{cont} $(\Omega\ \mathrm{cm})^{-1}$	σ_{str} $(\Omega\ \mathrm{cm})^{-1}$
A1	S1 5 × 5_1.5	0.483	2909.3 (6%)	0.517	2489.0 (30%)	1170.7 (9%)
	S2 5 × 5_1.5	0.483	617.5 (16%)	0.517	2202.8 (24%)	263.6 (8%)
	S3 5 × 5_1.5	0.483	140.7 (6%)	0.517	138.2 (38%)	61.3 (4%)
A2	S1 5 × 5_1.0	0.706	2291.1 (4%)	0.294	3200.2 (17%)	911.4 (5%)
	S2 5 × 5_1.0	0.706	712.1 (6%)	0.294	3338.7 (13%)	286.1 (7%)
	S3 5 × 5_1.0	0.706	150.3 (9%)	0.294	145.8 (11%)	67.9 (4%)
B1	S1 4 × 4_1.5	0.625	2909.3 (6%)	0.375	2489.0 (30%)	1203.2 (3%)
	S2 4 × 4_1.5	0.625	617.5 (16%)	0.375	2202.8 (24%)	270.3 (2%)
	S3 4 × 4_1.5	0.625	140.7 (6%)	0.375	138.2 (38%)	60.9 (3%)
B2	S1 4 × 4_1.0	0.778	2291.1 (4%)	0.222	3200.2 (17%)	966.7 (4%)
	S2 4 × 4_1.0	0.778	712.1 (6%)	0.222	3338.7 (13%)	285.7 (7%)
	S3 4 × 4_1.0	0.778	150.3 (9%)	0.222	145.8 (11%)	69.6 (9%)
C1	S1 3 × 3_1.5	0.743	2909.3 (6%)	0.257	2489.0 (30%)	1271.6 (2%)
	S2 3 × 3_1.5	0.743	617.5 (16%)	0.257	2202.8 (24%)	268.3 (3%)
	S3 3 × 3_1.5	0.743	140.7 (6%)	0.257	138.2 (38%)	65.1 (1%)
C2	S1 3 × 3_1.0	0.842	2291.1 (4%)	0.158	3200.2 (17%)	1032.2 (7%)
	S2 3 × 3_1.0	0.842	712.1 (6%)	0.158	3338.7 (13%)	302.3 (2%)
	S3 3 × 3_1.0	0.842	150.3 (9%)	0.158	145.8 (11%)	69.4 (1%)

Solving the simultaneous Equations (5) and (6) in Mathematica® 8, the exponents m_{th} and m_{cont} of phases were determined. Based on Equations (7)–(10) the connectedness and connectivity for phases of the woven structures were determined. The results are juxtaposed in Table 6.

Table 6. Characterisation of phases of woven structures.

Group	Sample	m_{th} (-)	G_{th} (-)	χ_{th} (-)	m_{cont} (-)	G_{cont} (-)	χ_{cont} (-)
A1	S1 5 × 5_1.5	1.414	0.402	0.740	4.453	0.470	0.103
	S2 5 × 5_1.5	1.840	0.427	0.542	4.659	0.120	0.089
	S3 5 × 5_1.5	1.297	0.436	0.806	4.379	0.521	0.108
A2	S1 5 × 5_1.0	2.814	0.398	0.532	3.388	0.285	0.054
	S2 5 × 5_1.0	3.161	0.402	0.471	3.445	0.086	0.050
	S3 5 × 5_1.0	2.393	0.452	0.616	3.305	0.466	0.059
B1	S1 4 × 4_1.5	2.002	0.414	0.624	3.653	0.483	0.074
	S2 4 × 4_1.5	2.262	0.438	0.553	3.725	0.123	0.069
	S3 4 × 4_1.5	1.904	0.433	0.654	3.622	0.517	0.076
B2	S1 4 × 4_1.0	3.559	0.422	0.526	3.140	0.302	0.040
	S2 4 × 4_1.0	4.037	0.401	0.467	3.195	0.086	0.037
	S3 4 × 4_1.0	3.147	0.463	0.583	3.085	0.477	0.043
C1	S1 3 × 3_1.5	2.867	0.437	0.574	3.210	0.511	0.050
	S2 3 × 3_1.5	3.154	0.434	0.527	3.256	0.122	0.047
	S3 3 × 3_1.5	2.678	0.462	0.607	3.177	0.553	0.052
C2	S1 3 × 3_1.0	4.722	0.451	0.527	2.951	0.323	0.027
	S2 3 × 3_1.0	5.245	0.425	0.482	2.993	0.091	0.025
	S3 3 × 3_1.0	4.546	0.462	0.543	2.936	0.476	0.028

First of all, linear regression analysis using Statistica® 13 was performed assuming a significance level $\alpha = 0.10$ and Pearson's correlation coefficient R_P was calculated (Table 7). A significance level equal to 0.10 meant that we were willing to make 10 mistakes out of 100 tests. In this situation, it was easier to reject the null hypothesis. This approach made it possible to detect potential relationships and analyze them, and determine whether they made sense from the point of view of designing the woven structure and its electro-conductive properties.

Table 7. Results of regression analysis [1].

	G_{th}	G_{cont}	m_{th}	ϕ_{th}	χ_{th}	m_{cont}	ϕ_{cont}	χ_{cont}	C_{str}	σ_{th}	σ_{cont}	σ_{str}
G_{th}	1.000	**0.459**	0.160	0.318	0.134	−0.342	−0.318	−0.243	−0.350	**−0.473**	**−0.717**	**−0.435**
G_{cont}	**0.459**	1.000	−0.374	−0.141	**0.716**	0.027	0.141	0.284	0.141	0.105	**−0.684**	0.110
m_{th}	0.160	−0.374	1.000	**0.903**	**−0.738**	**−0.804**	**−0.903**	**−0.933**	**−0.953**$^{(3)}$	−0.043	0.352	−0.020
m_{cont}	−0.342	0.027	**−0.804**	**−0.979**	**0.602**	1.000	**0.979**	**0.938**	**0.924**$^{(3)}$	0.054	−0.074	0.030
χ_{th}	0.134	**0.716**	**−0.738**$^{(2)}$	**−0.688**$^{(2)}$	1.000	**0.602**	**0.688**	**0.797**	**0.646**$^{(3)}$	0.025	**−0.568**	0.014
χ_{cont}	−0.243	0.284	**−0.933**	**−0.985**	**0.797**	**0.938**$^{(2)}$	**0.985**$^{(2)}$	1.000	**0.967**$^{(3)}$	0.048	−0.267	0.024
ϕ_{th}	0.318	−0.141	**0.903**	1.000	**−0.688**	**−0.979**	**−1.000**	**−0.985**	**−0.978**$^{(3)}$	−0.055	0.162	−0.029
ϕ_{cont}	−0.318	0.141	**−0.903**	**−1.000**	**0.688**	**0.979**	1.000	**0.985**	**0.978**$^{(3)}$	0.055	−0.162	0.029
C_{str}	−0.350	0.141	**−0.953**	**−0.978**	**0.646**	**0.924**	**0.978**	**0.967**	1.000	0.055	−0.163	0.027
σ_{th}	**−0.473**	0.105	−0.043	−0.055	0.025	0.054	0.055	0.048	0.055	1.000	**0.623**	**0.998**
σ_{cont}	**−0.717**	**−0.684**	0.352	0.162	**−0.568**	−0.074	−0.162	−0.267	−0.163	**0.623**	1.000	**0.619**
σ_{str}	**−0.435**	0.110	−0.020	−0.029	0.014	0.030	0.029	0.024	0.027	**0.998**$^{(1)}$	**0.619**$^{(1)}$	1.000

[1] Significant statistical correlation coefficients are in bold.

The initial analysis indicated relationships$^{(1)}$ between the conductivity of the whole structure σ_{str} and conductivities of phases σ_{th} and σ_{cont}, wherein the relationships σ_{str} and σ_{th} were stronger ($R_P = 0.998$), while σ_{str} and σ_{th} were weaker ($R_P = 0.619$), assuming a significance level $\alpha = 0.10$. No significant dependences were observed between the parameter C_{str} connected with the structure and parameters G_{th} and G_{cont} connected with electro-conductive features of the used materials.

There were significant correlations$^{(2)}$ between the parameters ϕ_{th}, m_{th}, χ_{th}, and ϕ_{cont}, m_{cont}, χ_{cont} resulted from the fact that they met Equation (10). The parameters were directly related$^{(3)}$ to the woven structure; no relation to the conductivities of phases was observed.

Next, statistical analysis was performed using Statistica® 13 based on the Kruskal–Wallis (K–W) test [35]. The determined p-value was compared with the critical value α for rejecting the null hypothesis. If α was less than the p-value, the null hypothesis was not rejected. When the K–W test led to significant results, at the assumed significance level α, at least one group was different from the other groups. To identify the particular differences between pairs of groups, a post hoc test was used.

The nonparametric statistical procedure was used for comparing chosen parameters (G_{th} and G_{cont}) in three independent groups, i.e., woven structures designed from the same electro-conductive woven fabric (S1, S2, or S3). The test was performed assuming $\alpha = 0.10$. Results of statistical analysis are presented in Table 8.

Table 8. Results of the statistical analysis for three groups: S1, S2, and S3 [1].

Parameter	K–W Test		Post Hoc Test	*p*-Value
G_{th}	Value of test statistic	7.3801	S1–S2 [2]	1.0000
	p-value	**0.0250**	S1–S3 [2]	**0.0520**
			S2–S3 [2]	**0.0602**
G_{cont}	Value of test statistic	12.7836	S1–S2 [2]	**0.0602**
	p-value	**0.0017**	S1–S3 [2]	0.7026
			S2–S3 [2]	**0.0013**

[1] Significant difference in groups is in bold. [2] Compared pairs of groups.

It was found that there were significant differences in groups for both the G_{th} and G_{cont} parameters. This meant that the connectedness of both of the phases was dependent on the material (electro-conductive woven fabric) from which the structure wa s designed (see Figure 5). It was noted that the range of connectedness change G_{th} was not large compared to G_{cont}. This was due to the design assumptions of woven structures. A greater variation (59%) of the area fraction of the strip contacts' phase ϕ_{cont} compared to the variation (8%) of the area fraction of the strips' phase ϕ_{th} was observed (Table 5). Therefore, each phase fraction was important in the conductivity of the current through the woven structure. The connectedness could be considered in terms of woven structure quality. Electrical conduction of a woven structure depends on the phases, which are pathways enabling current conduction. Phases, in varying degrees, affect the electro-conductive properties of a woven structure as shown in Figure 6. It was noticed that the contribution of the strip contacts' phase in the conductivity of woven structure for fabric S2 was lower than that in the case of fabrics S1 and S3. The surface of woven fabric S2 seemed to be much smoother than the surfaces of the remaining fabrics. The resistivity of strips cut from fabric S2 was higher than the resistivity of the strip contacts.

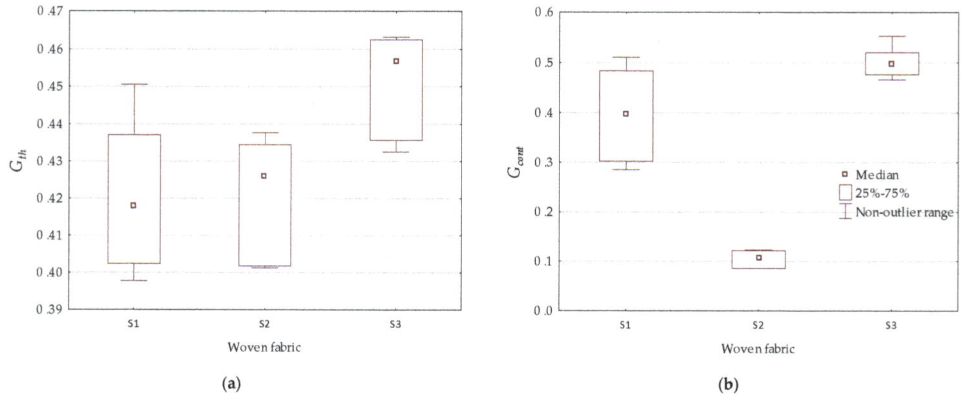

Figure 5. Influence of material on phase connectedness: (**a**) strips' phase; (**b**) strip contacts' phase.

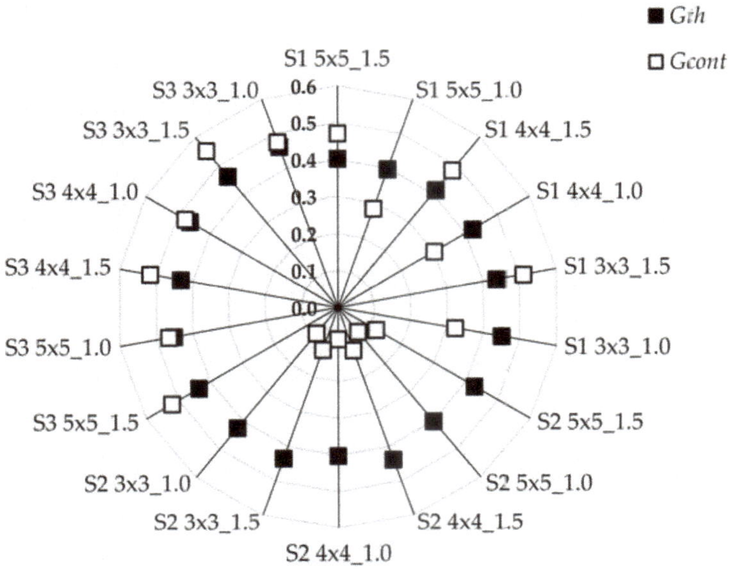

Figure 6. Connectedness of strips' and strip contacts' phases.

The nonparametric statistical procedure was used for comparing chosen parameters (m_{th}, m_{cont}, χ_{th}, and χ_{cont}) in three independent groups, i.e., woven structures designed with the same density of strips (5 × 5, 4 × 4 or 3 × 3). Results of statistical analysis conducted for significance level α = 0.10 are presented in Table 9.

Table 9. Results of the statistical analysis for three groups: 5 × 5, 4 × 4, and 3 × 3 [1].

Parameter	K–W Test		Post Hoc Test	*p*-Value
m_{th}	Value of test statistic *p*-value	6.2222 **0.0446**	3 × 3–4 × 4 [2] 3 × 3–5 × 5 [2] 4 × 4–5 × 5 [2]	0.8384 **0.0386** 0.4792
m_{cont}	Value of test statistic *p*-value	8.8538 **0.0120**	3 × 3–4 × 4 [2] 3 × 3–5 × 5 [2] 4 × 4–5 × 5 [2]	0.4792 **0.0088** 0.3505
χ_{th}	Value of test statistic *p*-value	1.0643 0.5873	3 × 3–4 × 4 [2] 3 × 3–5 × 5 [2] 4 × 4–5 × 5 [2]	– – –
χ_{cont}	Value of test statistic *p*-value	7.9064 **0.0192**	3 × 3–4 × 4 [2] 3 × 3–5 × 5 [2] 4 × 4–5 × 5 [2]	0.4792 **0.0148** 0.4792

[1] Significant difference in groups is in bold. [2] Compared pairs of groups.

It was found that there were significant differences in groups for three of the four parameters. The phases' exponents m_{th}, m_{cont}, differed significantly in two groups of structures: 3 × 3 and 5 × 5. The relationship of the connectivity χ_{th} with the designed three different structures was not found in contrast to χ_{cont}. As mentioned earlier, this might be due to the low variation of the area fraction of the strips' phase ϕ_{th}.

The connectivity could be considered in terms of woven structure quantity. Structure density, in varying degrees, affects the electro-conductive properties of a woven structure. It is important how phases are arranged in the whole composite. Connectedness for both phases that characterise woven structures is shown in Figure 7. Larger values of χ_{th}

compared to χ_{cont} were observed for all woven structures. The results show that the strip contacts' phase played an important role in the structure of the composite.

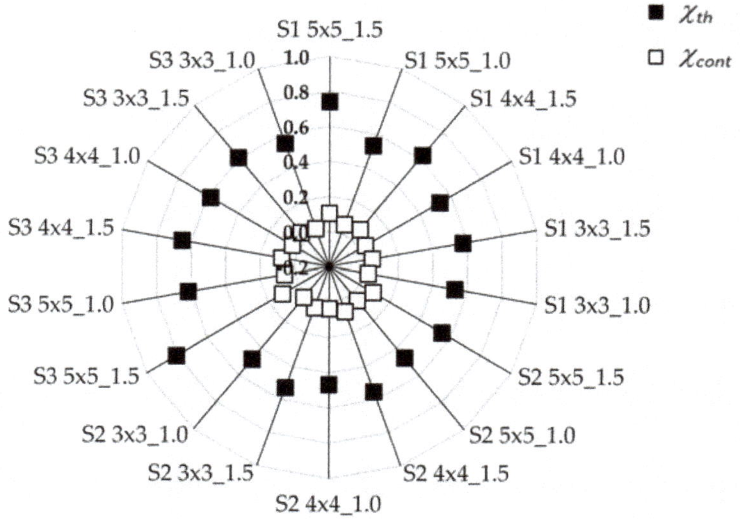

Figure 7. Connectivities of strips' and strip contacts' phases.

Parameters m_{th}, m_{cont}, χ_{th}, and χ_{cont} were connected with the woven structure and especially with percentage surface cover C_{str}, as shown in Table 7. Some dependences were observed and are presented in Figure 8.

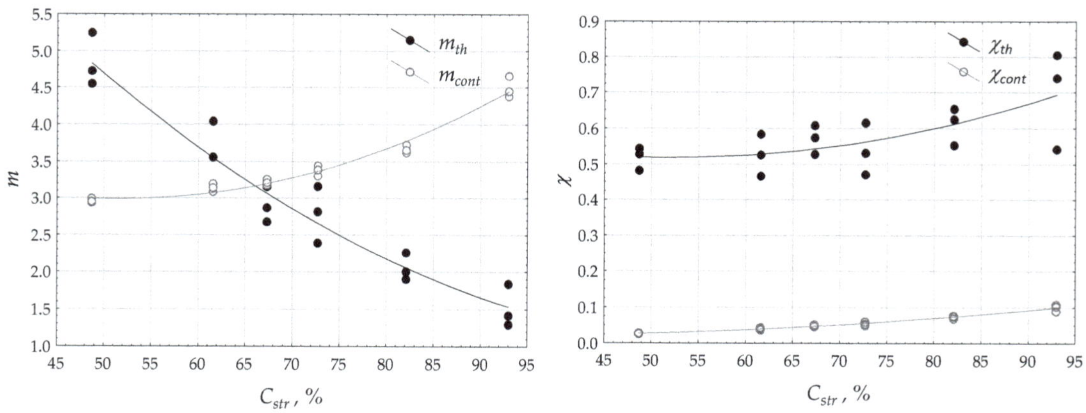

(a) (b)

Figure 8. Influence of percentage surface cover on: (**a**) phase exponent; (**b**) phase connectivity.

The results of the measurements were approximated by quadratic *polynomials*. Coefficients of determination R^2 were as follows for $m_{th} = f(C_{str}) - R^2 = 0.908$, and for $m_{cont} = f(C_{str}) - R^2 = 0.854$ (Figure 8a); for $\chi_{th} = f(C_{str}) - R^2 = 0.417$, and for $\chi_{cont} = f(C_{str}) - R^2 = 0.936$ (Figure 8b). All the coefficients were significant at the 0.10 significance level. It was found that the percentage surface cover increase caused an increase in connectivity of each phase. An increase in percentage surface cover caused a decrease in the strips' phase exponent and an increase in the strip contacts' phase exponent. Each of the phases that occupy the composite must share the same total space. This meant that the increase in the fraction of one of the phases must lead to a reduction in the fraction of the second phase.

The mean values of phases' exponents are presented in Table 10. Types of designed woven structures and the width of strips were taken into account. Variation coefficients are given in parentheses.

Table 10. The phases' exponents for designed woven structures.

Phase Exponent		m_{th} (-)	m_{th} (-)	m_{cont} (-)	m_{cont} (-)
Width of strip		1.0 cm	1.5 cm	1.0 cm	1.5 cm
Structure type	3 × 3	4.84 (8%)	2.90 (8%)	2.96 (1%)	3.21 (1%)
	4 × 4	3.58 (12%)	2.06 (9%)	3.14 (2%)	3.67 (1%)
	5 × 5	2.79 (14%)	1.52 (19%)	3.38 (2%)	4.50 (3%)

A wider range of phase exponent values was observed for the strips' phase than for the strip contacts' phase. The vast majority of phases' exponents exceeded 2. It could therefore be concluded that the low connectivity of conductive phases in the composites occurred. The exception was the strips' phase in the woven structure of which the percentage surface cover was C_{str} = 93%. The percentage surface cover for the remaining woven structures was in the range of 49–82%. However, further research is needed in this area.

5. Conclusions

The generalised Archie's law can be used for describing the electrical conductivity of a woven structure. Every phase occurring in a composite has a well-defined exponent. In general, it could be concluded that the designed structures were characterised by the phases' exponents exceeding 2, denoting low connectivity of the conductive phases. A qualitative and quantitative description of the woven structure was feasible. The connectedness could be considered in terms of woven structure quality. The connectedness of both of the phases was dependent on the material from which the structure was designed. Electrical conduction of woven structures depends on the phases, which are pathways enabling current conduction. The fraction of each of the phases involved in the current conductivity is also important. The fraction of phase is connected with its connectivity. The connectivity can be considered in terms of woven structure quantity. Structure density, in varying degrees, affects the electro-conductive properties of a woven structure. It is important how phases are arranged in the whole composite. This meant that the strip contact's phase plays an important role in the structure of the composite. By designing the same structure but from other components, you could predict the conductivity of the new structure based on parameters determined from the mixing model.

Funding: This research received no external funding.

Institutional Review Board Statement: Not applicable.

Informed Consent Statement: Not applicable.

Data Availability Statement: Experimental methods and results are available from the author.

Conflicts of Interest: The author declares no conflict of interest.

References

1. Tyurin, I.N.; Getmantseva, V.V.; Andreeva, E.G. Van der Pauw method for measuring the electrical conductivity of smart textiles. *Fibre Chem.* **2019**, *51*, 139–146. [CrossRef]
2. Castano, L.M.; Flatau, A.B. Smart fabric sensors and e-textile technologies: A review. *Smart Mater. Struct.* **2014**, *23*, 053001–053027. [CrossRef]
3. Gonçalves, C.; da Silva, A.F.; Gomes, J.; Simoes, R. Wearable e-textile technologies: A review on sensors, actuators and control elements. *Inventions* **2018**, *3*, 14. [CrossRef]
4. Korzeniewska, E.; Krawczyk, A.; Mróz, J.; Wyszyńska, E.; Zawiślak, R. Applications of smart textiles in post-stroke rehabilitation. *Sensors* **2020**, *20*, 2370. [CrossRef]

5. Tokarska, M.; Frydrysiak, M.; Zięba, J. Electrical properties of flat textile material as inhomogeneous and anisotropic structure. *J. Mater. Sci. Mater. Electron.* **2013**, *24*, 5061–5068. [CrossRef]
6. Neruda, M.; Vojtech, L. Verification of surface conductance model of textile material. *J. Appl. Res. Technol.* **2012**, *10*, 578–584. [CrossRef]
7. Liu, S.; Tong, J.; Yang, C.; Li, L. Smart e-textile: Resistance properties of conductive knitted fabric—Single pique. *Text. Res. J.* **2017**, *87*, 1669–1684. [CrossRef]
8. Gunnarsson, E.; Karlsteen, M.; Berglin, L.; Stray, J. A novel technique for direct measurements of contact resistance between interlaced conductive yarns in a plain weave. *Text. Res. J.* **2015**, *85*, 499–511. [CrossRef]
9. Kazani, I.; De Mey, G.; Hertleer, C.; Banaszczyk, J.; Schwarz, A.; Guxho, G.; Van Langenhove, L. Van Der Pauw method for measuring resistivities of anisotropic layers printed on textile substrates. *Text. Res. J.* **2011**, *81*, 2117–2124. [CrossRef]
10. Tokarska, M. New concept in assessing compactness of woven structure in terms of its resistivity. *J. Mater. Sci. Mater. Electron.* **2016**, *27*, 7335–7341. [CrossRef]
11. Jiyong, H.; Xiaofeng, Z.; Guohao, L.; Xudong, Y.; Xin, D. Electrical properties of PPy-coated conductive fabrics for human joint motion monitoring. *Autex Res. J.* **2016**, *16*, 7–12. [CrossRef]
12. Tokarska, M.; Gniotek, K. Anisotropy of the electroconductive properties of flat textiles. *J. Text. Inst.* **2015**, *106*, 9–18. [CrossRef]
13. Vasile, S.; Deruck, F.; Hertleer, C.; De Raeve, A.; Ellegiers, T.; De Mey, G. Study of the contact resistance of interlaced stainless steel yarns embedded in hybrid woven fabrics. *Autex Res. J.* **2017**, *17*, 170–176. [CrossRef]
14. Tokarska, M. Mathematical model for predicting the resistivity of an electroconductive woven structure. *J. Electron. Mater.* **2017**, *46*, 1497–1503. [CrossRef]
15. Hertleer, C.; Meul, J.; De Mey, G.; Vasile, S.; Odhiambo, S.A.; Van Langenhove, L. Mathematical model predicting the heat and power dissipated in an electroconductive contact in a hybrid woven fabric. *Autex Res. J.* **2020**, *20*, 133–139. [CrossRef]
16. McLachlan, D.S. Equations for the conductivity of macroscopic mixtures. *J. Phys. C Solid State Phys.* **1986**, *19*, 1339–1354. [CrossRef]
17. McLachlan, D. An equation for the conductivity of binary mixtures with anisotropic grain structures. *J. Phys. C Solid State Phys.* **1987**, *20*, 865–877. [CrossRef]
18. Archie, G.E. Electrical resistivity log as an aid in determining some reservoir characteristics. *Trans. AIME* **1942**, *146*, 54–62. [CrossRef]
19. Glover, P.W.J. A generalized Archie's law for n phases. *Geophysics* **2010**, *75*, E247–E265. [CrossRef]
20. Glover, P.W.J.; Walker, E. Grain-size to effective pore-size transformation derived from an electrokinetic theory. *Geophysics* **2009**, *74*, E17–E29. [CrossRef]
21. Glover, P.W.J.; Hole, M.J.; Pous, J. A modified Archie's law for two conducting phases. *Earth Planet. Sci. Lett.* **2000**, *180*, 369–383. [CrossRef]
22. Verwer, K.; Eberli, G.P.; Weger, R.J. Effect of pore structure on electrical resistivity in carbonates. *AAGP Bull.* **2011**, *95*, 175–190. [CrossRef]
23. Kennedy, W.D.; Herrick, D.C. Conductivity models for Archie rocks. *Geophysics* **2012**, *77*, WA109–WA128. [CrossRef]
24. Cai, J.; Wei, W.; Hu, X.; Wood, D.A. Electrical conductivity models in saturated porous media: A review. *Earth Sci. Rev.* **2017**, *171*, 419–433. [CrossRef]
25. Tokarska, M. Modeling of electro-conductive properties of woven structure based on mixing model. *Commun. Dev. Assem. Text. Prod.* **2020**, *1*, 12–19.
26. Glover, P.W.J. A new theoretical interpretation of Archie's saturation exponent. *Solid Earth* **2017**, *8*, 805–816. [CrossRef]
27. Berrezueta, E.; González-Menéndez, L.; Ordóñez-Casado, B.; Olaya, P. Pore network quantification of sandstones under experimental CO_2 injection using image analysis. *Comput. Geosci.* **2015**, *77*, 97–110. [CrossRef]
28. ISO/IEC Guide 98-3:2008(En). Uncertainty of Measurement—Part 3: Guide to the Expression of Uncertainty in Measurement (GUM:1995). Available online: https://www.iso.org/obp/ui/#iso:std:iso-iec:guide:98:-3:ed-1:v2:en (accessed on 14 December 2021).
29. Tokarska, M. Characterization of electro-conductive textile materials by its biaxial anisotropy coefficient and resistivity. *J. Mater. Sci. Mater. Electron.* **2019**, *30*, 4093–4103. [CrossRef]
30. Kostajnšek, K.; Urbas, R.; Dimitrovski, K. A new simplified model for predicting the UV-protective properties of monofilament pet fabrics. *Autex Res. J.* **2019**, *19*, 263–270. [CrossRef]
31. *EN 16812:2016*; Textiles and Textile Products—Electrically Conductive Textiles—Determination of the Linear Electrical Resistance of Conductive Tracks. European Committee for Standardization: Brussels, Belgium. Available online: https://standards.iteh.ai/catalog/standards/cen/6a8bc45e-d439-493d-ba92-58698b5ce97b/en-16812-2016 (accessed on 14 December 2021).
32. Meul, J. Study of Electro-Conductive Contacts in Hybrid Woven Fabrics. Master's Thesis, Ghent University, Ghent, Belgium, 2015. Available online: https://lib.ugent.be/catalog/rug01:002224697 (accessed on 14 December 2021).
33. *AATCC 76*; Test Method for Electrical Surface Resistivity of Fabrics. American Association of Textile Chemists and Colorists (AATCC): Research Triangle Park, NC, USA. Available online: https://global.ihs.com/doc_detail.cfm?document_name=AATCC%2076&item_s_key=00156638 (accessed on 14 December 2021).
34. ISO 139:2005. Textiles—Standard Atmospheres for Conditioning and Testing. Available online: https://www.iso.org/obp/ui/#iso:std:iso:139:ed-2:v1:en (accessed on 14 December 2021).
35. Corder, G.W.; Foreman, D.I. *Nonparametric Statistics for Non-Statisticians: A Step-By-Step Approach*; Wiley: Hoboken, NJ, USA, 2009.

materials

MDPI

Review

Recent Developments in Steelmaking Industry and Potential Alkali Activated Based Steel Waste: A Comprehensive Review

Ikmal Hakem Aziz [1,2,*], Mohd Mustafa Al Bakri Abdullah [1,2,*], Mohd Arif Anuar Mohd Salleh [1,2], Liew Yun Ming [1,2], Long Yuan Li [3], Andrei Victor Sandu [4,*], Petrica Vizureanu [4,5], Ovidiu Nemes [5,*] and Shaik Numan Mahdi [6]

1 Faculty of Chemical Engineering Technology, Universiti Malaysia Perlis (UniMAP), Arau 02600, Perlis, Malaysia; arifanuar@unimap.edu.my (M.A.A.M.S.); ymliew@unimap.edu.my (L.Y.M.)
2 Geopolymer & Green Technology, Center of Excellence (CEGeoGTech), Universiti Malaysia Perlis (UniMAP), Arau 02600, Perlis, Malaysia
3 School of Engineering, University of Plymouth, Plymouth PL4 8AA, UK; long-yuan.li@plymouth.ac.uk
4 Faculty of Materials Science and Engineering, Gheorghe Asachi Technical University of Iasi, D. Mangeron 41, 700050 Iasi, Romania; peviz@tuiasi.ro
5 Department of Environmental Engineering and Sustainable Development Entrepreneurship, Faculty of Materials and Environmental Engineering, Technical University of Cluj-Napoca, B-dul Muncii 103-105, 400641 Cluj-Napoca, Romania
6 School of Civil Engineering, CERSSE-JAIN (Deemed to be University), Bangalore 560069, Karnataka, India; shaik.mahdi@gmail.com
* Correspondence: ikmalhakem@unimap.edu.my (I.H.A.); mustafa_albakri@unimap.edu.my (M.M.A.B.A.); sav@tuiasi.ro (A.V.S.); ovidiu.nemes@sim.utcluj.ro (O.N.)

Citation: Aziz, I.H.; Abdullah, M.M.A.B.; Salleh, M.A.A.M.; Ming, L.Y.; Li, L.Y.; Sandu, A.V.; Vizureanu, P.; Nemes, O.; Mahdi, S.N. Recent Developments in Steelmaking Industry and Potential Alkali Activated Based Steel Waste: A Comprehensive Review. *Materials* **2022**, *15*, 1948. https://doi.org/10.3390/ma15051948

Academic Editors: Hansang Kwon and Irina Hussainova

Received: 10 February 2022
Accepted: 3 March 2022
Published: 6 March 2022

Abstract: The steel industry is responsible for one-third of all global industrial CO_2 emissions, putting pressure on the industry to shift forward towards more environmentally friendly production methods. The metallurgical industry is under enormous pressure to reduce CO_2 emissions as a result of growing environmental concerns about global warming. The reduction in CO_2 emissions is normally fulfilled by recycling steel waste into alkali-activated cement. Numerous types of steel waste have been produced via three main production routes, including blast furnace, electric arc furnace, and basic oxygen furnace. To date, all of the steel waste has been incorporated into alkali activation system to enhance the properties. This review focuses on the current developments over the last ten years in the steelmaking industry. This work also summarizes the utilization of steel waste for improving cement properties through an alkali activation system. Finally, this work presents some future research opportunities with regard to the potential of steel waste to be utilized as an alkali-activated material.

Keywords: steel waste; steelmaking; steel waste; alkali-activated cement

1. Introduction

Steel is a globally traded commodity that is manufactured all over the world. It is worth noting that, in 2019, 88% of steel produced in the EU (139 million tonnes) was traded outside of the country of origin, with 111 million tonnes (70% of production) traded on the EU internal market and 28 million tonnes (18% of production) exported outside of the EU, primarily to other European countries (9 million tonnes) and North America (6 million tonnes). Despite producing half of the world's steel (996 million tonnes), China only export 6% of its output (64 million tonnes), mainly to other Asian countries in 2019 [1]. In comparison, China's crude steel production reached 627 million tonnes in 2010, demonstrating that steel is in great demand as a result of growing industrialization and urbanization [2].

The steelmaking industry has become the second-largest energy consuming process in global industrial sectors and emits huge amounts of environmentally harmful substances, such as dust, sulphur dioxide (SO_2), nitrogen oxides (NO_x), and carbon dioxide (CO_2). Since the majority of steelmaking operations are still coal-based and heavily reliant on fossils fuels, such as oil and diesel, significant volumes of CO_2 emissions are emitted. As a result, the steel industry contributes about 6.7 percent of the total global CO_2 emissions [3]. The manufacturing sector in the EU is responsible for 4.7% of total CO_2 emission (182 million tonnes) and about 27% of CO_2 emissions from the worldwide manufacturing sector [4,5]. Steel production emits roughly 1.8 t CO_2 per tonne, while the total energy demand of steel production is 21.0–35.4 GJ/t steel [6].

Steel and iron are manufactured from the metallurgical industry, which is classified into three major routes namely blast furnace (BF), basic oxygen furnace (BOF) and electric arc furnace (EAF). The integrated steel production BF-BOF route is the most crucial steel production route, accounting for roughly 70% of global steel production. The mini-mill approach, which accounts for 25% of global steel production, comprises of EAF in which recycled steel crap is melted and then cast into semi-finished slab, billet or bloom form. As shown in Figure 1 [7], the BF-BOF routes produces one tonne of hot-rolled coil, while emitting approximately 1.8 tonnes of CO_2. The iron-making processes of blast furnace, sintering, and coke making account for about 90% of the total. The steelmaking pathway has the largest energy consumption and associated CO_2 emissions, with 12.31 Gj/tHM and 1.22 t CO_2/tHM [8].

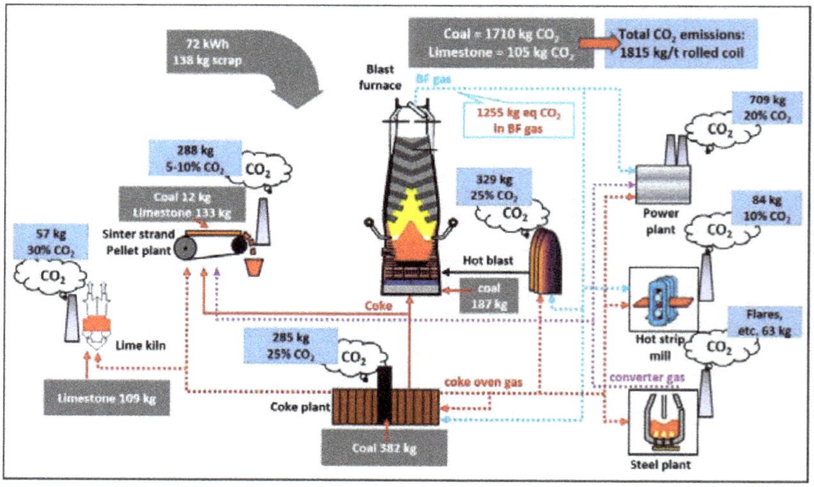

Figure 1. CO_2 emissions in an integrated steel mill [7].

Pollution is produced by all industrial activity, and the steel industry is no exception. Steelmaking and galvanizing processes generate a wide range of waste with varying class, volume, and toxicity. Despite substantial attempts by these industrial sectors to reduce global environmental impacts, there is a constant demand for new technology fresh technology to reduce CO_2, boost efficiency in recycling waste, and produce clean gaseous and liquid effluents. In fact, environmental control has become a study and technique in the metallurgical engineering industry. According to a study by Pardo and Moya [8], the CO_2 reduction achievable by 2030, while maintaining the competitiveness of the European steel sector, is in the range of 14–21%, compared to 2010. This necessitates both the incremental development of existing technologies and the incorporation of new, cutting-edge technologies. The future prices of fuels, energy, and other resources, as well as carbon pricing, will have a significant impact on the adoption of these revolutionary technologies. In the economic scenario, the steel sector's specific, CO_2 emissions in 2050 would be about 15% lower than in 2010. The European steel industry's highest specific

CO_2 emission reduction potential in 2050 compared to 2010 is roughly 57%. For this to be realized, all blast furnaces (BFs) would have to be retrofitted with top-gas recycling and carbon capture and storage (CCS) [9].

The solid waste management of steelmaking entails a more complex procedure that aims to limit the quantity of waste that ends up in landfill or is incinerated. It comprises environmentally sound strategies for preventing, reusing, and recycling garbage, as well as the recovery of resources and energy whenever possible. It is preferable to include solid waste in the steelmaking process itself, and then sell it as a raw material for other industrial processes. Instead, the waste could be processed to minimize toxicity and recover commercially valuable materials. Only a small portion of the material is used in this case, with the rest being sent for final disposal as tailing or incineration as required by environmental regulations.

The alkali activation system can be brought by recycling the steel waste to manufacture eco-friendly cement with more exceptional properties than conventional cement. The utilization of steel waste into alkali-activated cement enhances the mechanical and chemical properties. It is noteworthy to highlight that various types of steel waste can be categorized into blast furnace steelmaking waste, electric arc furnace steelmaking waste, and basic oxygen furnace waste (Table 1).

Table 1. Type of steel waste from different types of steelmaking production.

Steelmaking Production	Type of Steel Waste	Description
Blast furnace	Ground granulated blast furnace slag (GGBFS)	Cement replacement [10], high-performance concrete [11], electromagnetic performance [12], and steel reinforcement material [13]
	Ladle slag	Supplementary material [14], High-strength cement [15], cement replacement [16]. soft clay stabilization [17]
Electric arc furnace	Electric arc furnace slag	One part hybrid cement [18], cement mortar [19], concrete pavement [20], self-compacting concrete [21]
	Steel slag	Alkali-activated cement [22], high-strength cement [23], cement-based composite binders [24]
Basic oxygen furnace	Basic oxygen furnace slag	Cement replacement [25], bacterial community succession [26], cement mortar [27]

This review focuses on the previous research works on the utilization of varied steel waste in the steelmaking industry. Moreover, cost analysis and energy consumption will be discussed in this paper. In particular, sintering returns approximately 80–90% of mill scale steelmaking processes, while 85–90% of slags are commercialized to other industrial process [28]. Previous papers addressed a wide range of steel manufacturing pathways, from economic and environmental aspects to technological highlights [3,29–31]. This review, in contrast, will concentrate on alkali-activated cement application in the most often used integrated steel production pathway. Steel with a high recycling potential necessitates the implementation of long-term management techniques. Meanwhile, there are no recycling options in the cement sector, but cutting energy use and employing alternative fuel sources that produce fewer emissions can assist the industry in becoming more sustainable. The alkali activation method with varied steel waste from the steelmaking sector can be implemented for a reduction in greenhouse gases in the environment. Eventually, additional research opportunities have been offered based on the gaps discovered in the previous literature.

2. Steel Production Routes

2.1. Steel Waste

Blast furnace or basic oxygen furnace route, melting of scrap in electric arc furnace, direct reduction iron/electric arc furnace, and smelting reduction/basic oxygen furnace are the four primary routes for iron steelmaking. Integrated steel production is the most important steel production route, accounting for roughly 70% of global steel production. The mini-mill approach, which accounts for 25% of global steel production, comprises of EAF in which recycled steel scrap is melted and then cast into semi-finished slab, bloom, or billet forms. The reduction iron/electric arc furnace pathway, which produces around 5% of the world's steel, primarily uses natural gas as an energy reducing agent. The smelting reduction/basic oxygen furnace approach relies on the burning of coal fines to reduce iron ore fines without agglomeration, and it accounts for only 0.4% of global steel production [32].

The following is a brief overview of the production process in an integrated steel factory, as well as the main forms of solid waste generated at each stage. In order to produce pig iron, the raw material is fed into a blast furnace. Slag forms in the blast furnace, as well as sludge and dust collected in the reactor gas system treatment, are the main type of solid waste generated in this production routes. Meanwhile, the ladle slag and sludge from the gas handling system are two of the common types of solid waste produced in the pig iron refinement process. The refined steel is subsequently delivered to the continuous casting phase for solidification after undergoing a secondary refining step, which take place in the ladle furnace. The steel then acquires the appropriate thickness and mechanical properties during the rolling steel. Additionally, mill scale is a solid waste produced by the oxidation of the metal surface during continuous rolling and casting operations [28]. The numerous residues evaluated in this review, as well as the typical amount produced are listed in Table 2.

Table 2. The average of steel waste generated from steelmaking production routes.

Type of Steel Waste	The Average Amount Generated
Blast furnace slag	150 up to 300 kg per tonne of pig iron (blast furnace powered by charcoal) and 200 up to 400 kg per tonne of pig iron (blast furnace fuelled by mineral coal) [33,34]
Ladle slag	Each tonne of liquid steel weighs around 200 kg [35]
Electric arc slag	Approximately 130 up to 180 kg per tonne of [36]
Blast furnace sludge	Precisely 6 kg per tonne of pig iron [37,38]
Ladle sludge	15 up to 16 kg per tonne of hot metals [39,40]
Electric arc dust	15 up to 20 kg per tonne of steel [41]
Mill scale	34 up to 40 kg per tonne of steel [28,42]

2.2. Cost Analysis

In the next sections, we evaluate the main technology of blast furnace/basic oxygen furnace regarding process-emission-free and process-emission-intensive technologies in terms of economic costs and process-emission intensities. The statistics are derived from and cross-verified by a variety of sources [29,43–46], as well as a stakeholder discussion, and refer to a European viewpoint, particularly in terms of resource and energy cost (Table 3).

Table 3. Cost analysis of different iron and steel production routes (net of taxes).

Technology (EUR/t Steel)	Blast Furnace/Basic Oxygen Furnace	Direct Reduction Iron/Electric Arc Furnace
Electricity	0	219
Iron pellets	0	84
Coke	84	0
Iron ore	189	189
Services	45	40
Skilled labour	44	40
Unskilled labour	5	4
OPEX (EUR/t steel)	415	624
Process emission (t CO_2/t steel)	1.5	-
Investment cost (EUR/t steel)	-	1113

The most important aspect is that, in terms of operational expenditures (OPEX), direct reduction iron/electric arc furnace steel is around 50% more expensive than blast furnace/basic oxygen furnace steel for the given prices of major elements (labour and capital) and intermediate inputs. Even though usage of direct reduction iron/electric arc furnace reduces costs associated with coke, the iron and steel industry in the analysis provided here converts to hydrogen via water electrolysis using polymer electrolyte membranes. Considering that industrial scale hydrogen generation has yet to be created, unit costs of hydrogen generation vary widely in the literature [47].

Consequently, electricity expenses include both the power required to generate hydrogen and the electricity required for steel production if an electric arc furnace is being utilized [29]. The additional distinction between blast furnace/basic furnace oxygen furnace and direct reduction iron/electric arc furnace is the raw material input, as the latter method requires iron ore to be pre-processed into iron pellets. The remaining cost elements, such as service and primary factor costs, are not significantly different from blast furnace/basic oxygen furnace technology.

3. Steel Waste Management

3.1. Slags

In a steel mill, the transformation of iron ore into steel produces co-products, or solid wastes, which are classified as slags and sludges. Precipitation sludges, which are generated in the treatment of effluents from galvanizing operations, are among the most prevalent methods for managing such categories of solid waste.

The blast furnace and electric arc furnace are the principal sources of slag in the steelmaking industry. The separation of impurities present in a metal bath, which is constituted of silicates and silicon (Si), aluminium (Al), calcium (Ca), and magnesium oxides (MgO), forms this sort of waste. The chemical make-up of the sources material and the type of refractory employed in the furnace wall determine the concentration of each of these elements [48,49]. After being separated from metal bath, the slag goes through a solidification process. Different kinds of solidification give the material different qualities, resulting in different applications. The two most prevalent procedures are air cooling, which forms a crystalline slag, and granulation, which causes the slag to cool rapidly and become amorphous. Owing to its propensity to absorb water and its feature of hardening after adding water, granulated slag has greater reuse opportunities [28].

Since the volume of slags produced in a steel mill is huge, there are several management options for dealing with this steel waste. In reality, this substance is commonly regarded as a by-product of steel-making production. Table 4 shows a variety of options for managing these wastes.

Table 4. Management options for steelmaking slags.

Type of Steel Waste Management Options	Blast Furnace Slag	Electric Arc Furnace Slag	Ladle Slag	References
Reuse/recycling in steelmaking	-	Roughly 30% of slag is recycled in blast furnace in European countries; however, the phosphorus concentration should not exceed 0.5%. The elimination of phosphorus element is still a subject of research	-	[35,50]
Utilize as aggregates	The samples were maintained in sealed bag for 28 days in a curing environment at a temperature of 21 °C and relative humidity of 70%, providing superior mechanical properties to aggregate slag concrete.	Required the curing process (demoulded after 24 h, then cured at 20 °C of water tank) for because to the high expandability of the electric arc and ladle slag. It is not only cost effective, but it also has advantages in terms of material properties		[51–53]
Conventional cement manufacture	Owing to the hydraulicity of granulated slag, the residue used as a partial replacement for clinker material that leads to lower raw material and energy consumption, reduced pollution in cement manufacturing and enhanced finished material qualities. All the samples were cured in the range temperature of 20–35 °C	These residues obtain lesser hydraulic characteristics than blast furnace slag and can replace a portion of the clinker. Additionally, due to the expandability properties, such slags should go through the curing procedure for 28 days		[49,54,55]
Catalyst for the manufacture of biofuels	The effective catalyst for the synthesis of biodiesel was proven due to the slag crystallinity			[56–58]
Manufacturing of glass ceramic	The utilization of steel waste is widely known and commonly used. Glass ceramic structures are formed by the crystallization vitreous materials, such as slag under regulated conditions.			[59,60]
Absorbent materials	Higher reactivity and better specific surface area was obtained by slag materials when compared to the conventional absorbent			[61,62]

All of the slags mentioned in Table 4 can be employed in civil construction area, resulting in raw material savings and enhanced mechanical qualities of the finished product. Furthermore, replacing clinker with slag during the manufacturing of cements saves energy since slag does not require a calcination process. This management route also generates an abundance of CO_2, but using slag may minimize air pollution [28].

Meanwhile, a high expandability of slag related to reactive free oxides elements such as magnesium oxide (MgO) and calcium oxide (CaO) requires an extra care with the use of ladle slag and electric arc furnace slag for such a purpose. This issue can be remedied by ensuring that the material is free of these chemical elements or that any reactions that cause the substance to expand in volume have already occurred [48]. It was pointed out that ladle slag and electric arc furnace slag have a higher iron concentration and typically experience a magnetic separation process to separate the metallic portion (which is recycled in the steelmaking industry) from the non-metallic portion (which is usually allocated to other management routes). Yet, to avoid phosphorus accumulation in the steel, the metallic fraction should have a phosphorus level below than 0.5% [35].

3.2. Sludges

Sludges are formed in a wet gas cleaning system, which is used to treat the gases produced during the manufacturing process. Sludges containing zinc (Zn) and lead (Pb) are produced during the manufacturing of pig iron in a blast furnace. Substantial levels of iron (Fe) and carbon (C) are also present in the sludges' particles, which might be recycled in the furnace [63]. The blast furnace and ladle sludges are the most typical sludges produced in the steelmaking industry. These sludges are mainly composed of calcium, silicon, iron, manganese, and aluminium oxides. The amount of iron in these sludges is usually rather significant [64]. Table 5 shows the various options for dealing with these sludges.

Table 5. Management options for steelmaking sludges.

Type of Steel Waste Management Options	Blast Furnace Sludge	Ladle Sludge	References
Reuse/recycling in steelmaking	Lead and zinc must be eliminated from the dry sludges before they may be recycled directly. Since the majority of these elements are concentrated in the fine fraction, the coarser fraction of sludges could be recycled after classification during the steelmaking process. The reuse of the fine fraction still necessitates further research into removal of undesirable materials.		[37,39,65]
Utilize as adsorbent material	Preferably an effective adsorbents for copper, zinc, lead, chromium, and cadmium in various concentrations	-	[66,67]
Ceramic materials incorporation	As a result of the process, energy is saved, and waste disposal cost is reduced.		[68,69]

Considering the large proportion of iron oxide in blast furnace and ladle sludges (around 70%), the current trend in managing this type of steel waste is to reuse/recycle it in the steel industry through procedures such as sintering [70]. The inclusion of chemical components such as zinc, on the other hand, hinders direct recycling. Typically introduced as a raw material into blast furnace, zinc elements react with the refractory material, forming a crust around the reactor walls. Furthermore, zinc and lead have been reported to concentrate in the fine portion of the sludge [65]. The zinc-rich sludges from the overflow are stored or landfilled, while the sludge from the underflow is reused in the sinter plant. Conversely, pyrometallurgical or hydrometallurgical processes can be used to eliminate zinc from sludges in a very efficient and cost-effective manner [71].

Hydrometallurgical processes offer greater plant flexibility and correspond to be more cost-effective than pyrometallurgical technologies due to reduced capital and operational cost. By comparing to pyrometallurgy, hydrometallurgical process offer environmental benefits since no off-gases or dust nuisance are identified; nonetheless, effluents developed by these processes should be adequately handled [72].

To summarize, most of steel waste in the last ten years could be recycled/reused in various applications, especially in civil construction field. However, most of the management routes cause a detrimental impact on the environmental and mechanical qualities. Hence, another innovative route has been suggested. For example, steel waste could be utilized as an alkali-activated material in the production of greener construction materials [73–75].

3.3. Incorporating Alkali-Activated Cement Based Steel Waste

This section primarily focuses on a research project involving the incorporation of steel waste into alkali-activated materials, with the goal of examining the possibility of using an alkali activation system to enhance the mechanical, thermal, and chemical properties of alkali-activated based-steel waste.

One or more alumina-silicate sources and one or more alkaline activators make up an alkali-activated system. The activator solutions promote a pH environment that is in an acidic medium (e.g., silicate, sulphates, carbonated or hydroxides). A pre-mixed alumina-silicate source and alkaline activator in the form of dry powder can be used to produce a dry binder, which can then be mixed with water and aggregates to make mortar or concrete. Alternatively, the alkaline activator solution can be added to the alumina-silicate source separately, and then the wet binder can be mixed with extra water (if the concentration of the alkaline solution needs to be diluted), aggregates and additive materials to build a concrete or mortar. Instead, alkali-activated cement can be produced by mixing alumina-silicate source, alkaline activator, water, aggregates, and admixture without pre-producing the alkali-activated binder [76–78].

Any raw materials that consist of major elements of alumina (Al_2O_3) and silica (SiO_2) can be used for alkali activation production. These materials are copiously located in the Earth's crust and play a crucial role in providing fundamental sources of Al^{3+}- and Si^{4+}-free cations in the binding system. Commonly, the total compositions of Al_2O_3 and SiO_2 are more than 70% of clay and fly ash material, respectively. Meanwhile, for steel waste, the elements could decrease to the range of 30–50% and mostly appear in the reactive amorphous phase [79]. In this further section, the potential of steel waste minerals in alkali-activated slag cement, mortar, and composites are examined.

3.4. *Variability of Steel Waste*

Most steel waste is often chosen as the aluminosilicate material for alkali activation either as the main or blended binder. The use of steel waste by-products, such as ground granulated blast furnace slag [80,81], steel slag [82,83], nickel slag [84,85] and ladle slag [85,86], may influence these properties due to the variety chemical composition compositions apart from alkali-activated slag cement primary component (SiO_2 and Al_2O_3) as tabulated in Table 6. Steel waste composition varies significantly depending on the type of slag and the stage of steelmaking production. The use of steel waste as an alternative material in alkali activation technology has been extensively studied over previous decades. The studies mainly focused on the effects of different parameters such as Si/Al, Na/Al, SiO_2/Na_2O and Al_2O_3/Na_2O molar ratio, slag replacement (weight %), alkali concentration of liquid content, glass content, curing conditions, aggregate size, and slag particle size [87–91].

Table 6. Chemical composition of slag from different region and sources.

Type of Slag	Chemical Composition (wt %)				
	SiO_2	Al_2O_3	CaO	MgO	Fe_2O_3
Steel slag (Shandong Sheng, China) [92]	19.13	4.87	37.42	5.55	18.77
Steel slag (Wuhan, China) [89]	15.0	6.7	44.2	10.9	15.4
Steel slag (Wuhan, China) [83]	15.1	2.32	42.98	5.77	21.13
Steel slag (Jiangxi, China) [93]	18.48	3.76	45.18	4.83	19.45
High-magnesium slag (Nanjing, China) [84]	52.3	6.2	8.8	26.9	4.2
High-magnesium nickel slag (Jiangsu, China) [93]	50.37	4.65	1.72	32.22	7.94
Copper nickel slag (Murmansk, Russia) [94]	36.87	7.44	2.11	11.92	2.47
Copper nickel slag (Xinjiang, China) [95]	29.68	1.473	3.253	6.212	55.45
Copper slag (Aspropyrgos, Greece) [96]	39.95	3.30	4.08	1.77	44.41
Ferronickel slag (Larymna, Greece) [97]	32.74	8.32	3.73	2.76	0.76
Ferronickel slag (Marousi, Greece) [98]	40.29	10.11	3.65	5.43	37.69
Ferronickel slag (New Caledonia. France) [99]	52.52	2.33	0.27	33.16	10.80
Ferrochrome slag (Elazig, Turkey) [100]	33.8	25.48	1.1	35.88	-
Ferrochrome slag (Bhubaneswar, India) [101]	27.8	23.6	3.51	23.7	3.6
Ferrochrome slag (Malatya, Turkey) [102]	33.80	25.48	1.10	35.88	0.61
Ground granulated blast furnace slag (Chhattisgarh, India) [103]	32.97	17.97	35.08	10.31	0.72
Granulated blast furnace slag (Melbourne, Australia) [104]	33.8	13.68	42.56	5.34	0.4

Table 6. *Cont.*

Type of Slag	Chemical Composition (wt %)				
	SiO_2	Al_2O_3	CaO	MgO	Fe_2O_3
Ground granulated blast furnace slag (Paris, France) [105]	35.7	11.21	39.4	10.74	0.42
Granulated blast furnace slag (Dabrowa Goronicza, Poland) [106]	38.73	8.18	45.09	4.33	0.90
Granulated blast furnace slag (Cairo-Egypt) [107]	36.95	10.01	33.07	6.43	1.48
Blast furnace slag (Jiangsu, China) [108]	34.2	14.2	41.7	6.7	0.43
Ladle furnace slag (Taipei, Taiwan) [86]	23.7	4.2	48.6	8.1	-
Ladle slag (Lappohja, Finland) [109]	8.6	28.3	46.3	7.4	5.0
Blast oxygen furnace (Indiana, USA) [110]	8.35	60.8	5.21	8.89	2.35

A unique characteristic of steel waste, the production of alkali-activated materials, has attracted huge attention among researchers. In order to develop better properties of alkali-activated material, steel waste is mixed with other aluminosilicate source materials, such as the binder or filler in the matrix. For example, Samantasinghar et al. [111] incorporated ground granulated blast furnace slag (GGBFS) into class-F type fly ash to enhance compressive strength. A higher availability of leachable alumina-silicates and the presence of calcium oxide (CaO) as one of slag component resulted the strength development as proven by the significant compressive strength. Another study was performed by Gao et al. [112] with the utilization of slag in volcanic ash based alkali-activated. The effect of slag loading (50 and 100%) and activator modulus (0.8, 1.6, and 2.4) in preparing alkali-activated materials on reaction state and chemical environment of molecules were investigated. The higher activator modulus leads to a reduction of slag reactivity. However, it is worth noting that an alkali-activated system with a low CaO concentration can also achieve a high strength (>60 MPa), as stated by Li et al. [113]. A possible explanation for this might be the optimal and correct combination of CaO, Al_2O_3 and SiO_2.

A different study by Bouaissi et al. [114] reported that magnesium plays a role as an addition to calcium in a microstructure, which reflected in the development of mechanical properties. The incorporation of high-magnesium nickel slag (HNMS) in the ground granulated blast furnace slag/fly ash (GGBFS/FA) leads to a strength improvement, resulting in the formation of calcium beryllium praseodymium oxide ($CaBePr_2O_5$), which consists of an orthorhombic crystallography and space group Pnma. The crystalline phase transformation was believed to be attributed to the addition of ground granulated blast furnace slag (GGBFS) and high-magnesium nickel slag (HNMS).

The steel industry generates a vast volume and a diverse range of solid residues, all of which are characterized by a high percentage of metal in structural compositions. Instead of being utilized, these metals burdens are frequently discarded in industrial landfills. As potentially valuable alumina-silicate sources, steel waste from steel industry plants all over the world were characterized using an alkali activation technology. Apart from the use of steel waste in alkali-activated systems, alkali-activated composites, cement and mortar as matrices have also been investigated in the current literature. Since this paper reviews the alkali-activated system for steel waste management, the related works on the use of alkali activator solutions and reaction mechanisms are discussed further.

4. Alkaline Activator Solution

Alkali metal solution was used as a liquid component in the alkali activation process. The alkali solution is based on potassium or sodium, which includes hydroxides, alkali silicates, carbonates, and aluminates. The aluminosilicate sources dissolve quickly in a high alkaline environment, releasing AlO_4 and SiO_4 tetrahedral components and assisting the polycondensation process [115]. Frequently, the type of alkaline reactant applied in alkali-activated systems is a mixture of hydroxides (KOH or NaOH) and silicates solutions (K_2SiO_3 or $Na_2O{\cdot}nSiO_2{\cdot}mH_2O$) [116].

4.1. Hydroxide Alkali Solution

Sodium hydroxide (NaOH) and potassium hydroxide (KOH) solutions are widely used for various sources of aluminosilicates due to their leaching ability. Conventionally, a higher concentration of alkali solution accelerates the dissolution of aluminosilicates sources. The dissolution ability of a geopolymer is frequently the determining factor in its ultimate strength. Nevertheless, most geopolymer researchers claimed that a NaOH solution had a better leaching ability than the KOH solution.

Particularly, Sithole et al. [117] tested the unconfined compressive strength (UCS) of slag minerals activated by different activators (KOH and NaOH). As mentioned before, the alkali-activated slag in the NaOH solution possesses higher UCS than the KOH solution. NaOH showed a 45% increment in UCS at similar 15 M concentration compared to KOH. A contradicting result was reported by Altan et al. [118], where they claimed that the KOH activation yields a 10–15% higher compressive strength than NaOH at elevated temperatures. At similar concentrations, KOH contains a higher quantity of solid than the NaOH solution, thus contributing to a higher activating solid-to-slag ratio. Meanwhile, at ambient temperature conditions, the compressive strength of alkali-activated slag in the NaOH solution surpasses the KOH solution. Hence, NaOH activation is preferable over KOH as part of the activator component due to its economic value.

4.2. Chemistry of Alkali Hydroxide and Alkali Silicate Solution

As previously mentioned, alkali hydroxide is required at the early stage of alkali activation for the dissolution of aluminosilicates, while alkali silica functions are required as binders or plasticizers [119,120]. Alternatively, silica fumes are often used as supplementary materials to the sodium/potassium silicate. Occasionally, silica fumes are added to enhance the silica species in the design and boost the gelation and silicates precipitation.

Numerous studies on the production of alkali-activated material using only alkali solutions and a mixture of alkali hydroxide and metal silicate liquid were conducted. Most of the comparative studies determined that the chemistry of alkali hydroxide and alkali silicate solution is crucial and developed the preferable microstructure and mechanical properties. The alkali silicate solution induces a unit of soluble SiO_2 to produce alkali-activated main chain [121]. Based on Singh et al. [122], NaOH-activated slag/blend achieved the optimum compressive strength of 35 MPa at 14 M and had a decrement of up to 16 M. The excess of sodium cation produced sodium carbonate crystal which resulted in an unstable geopolymer edifice.

Meanwhile, Shariati et al. [123] concluded that the excess of OH^- ions during the alkali activation process caused a negative influence, which resulted in higher crack appearance and weak paste structure formation (Figure 2). In alkali-activated systems, increasing NaOH concentration increased the concentration of Na components, which could produce brittle samples once chemically bonded into the main structure of calcium silicate hydrate (C-S-H). Furthermore, according to Cihangir et al. [124], the pore refinements in alkali-activated slag concretes could occur in an acceptable level of Na_2O concentration. Moreover, increasing the silicate species in alkali activator solutions resulted in a denser microstructure by promoting the chemical reaction between Si and Ca elements in granulated blast furnace slag paste [125]. The denser and more homogenous microstructure is the result of the chemical reaction between silica and calcium components, which suggested the existence of calcium silicate ($CaSiO_3$ or Ca_2SiO_4).

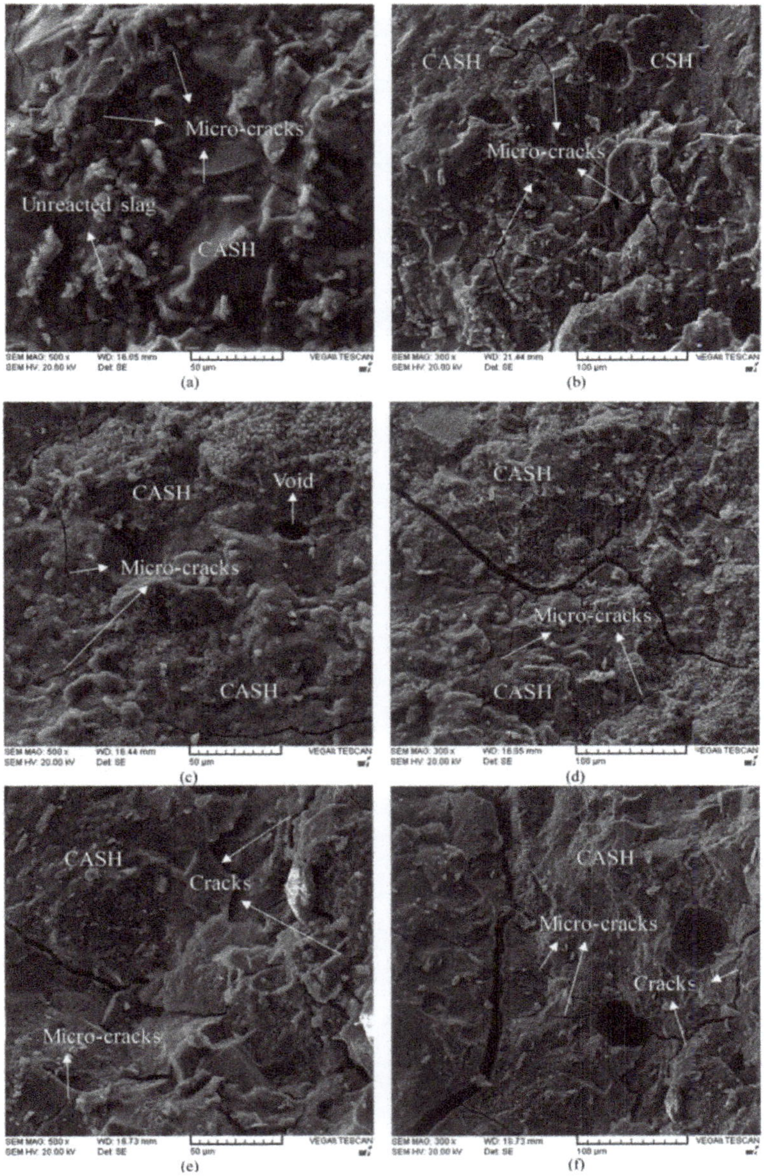

Figure 2. SEM images of AAS paste with various NaOH concentrations at different magnifications: (**a**) 8 M NaOH (500×), (**b**) 8 M NaOH (300×), (**c**) 12 M NaOH (500×), (**d**) 12 M NaOH (300×), (**e**) 16 M NaOH (500×) and (**f**) 16 M NaOH (300×) [123].

4.3. Reaction Mechanism of Slag Alkali Activation

Slag has a glassy phase that contains a large amount of calcium, which differs from metakaolin and fly ash in terms of the alkali activation reaction mechanism. Thus, the alkali hydration of a slag corresponds to a complex process that comprises several steps of chemical processes, including the initial dissolution of the slag and polymerization of the final product. As illustrated in Figure 3 [126], the dissolution mechanism of high calcium in the slag system accommodates both divalent and monovalent network-modifying metal cations. The major difference between the Na^+ and Ca^{2+} illustrates the greater extent of "destruction" caused by the shifting of both monovalent and divalent cations in the main alumina-silicate system.

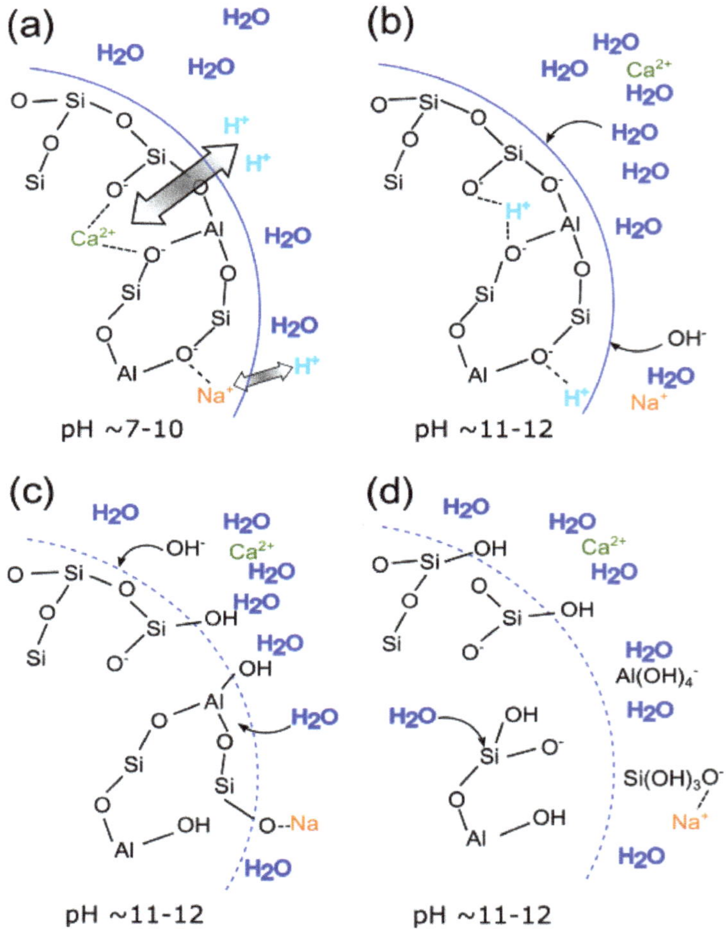

Figure 3. Dissolution mechanism of alkali activation of slag at early stage of (**a**) shifting of H^+ to Ca^{2+} and Na^+, (**b**) hydrolysis of Al-O-Si bonds, (**c**) breakdown of the depolymerized glass network, and (**d**) Si and Al detached from the network [126].

Slag alkali activation is an exothermic reaction that is similar to other aluminosilicate source materials. The chemical reaction predicted that the process is carried out through either dimers or trimers that allocate the existing component of the 3-D macromolecular structure. The slag alkali activation begins with a destruction of slag bonds Ca-O, Si-O, Al-O, Mg-O, and Fe-O, then produces a stronger Si-Al layer all over the surface of slag grains, ending with hydration products such as tobermorite or calcite.

In the study by Jamil et al. [127] composing oxides components (CaO, SiO_2, Al_2O_3 and MgO) were partly dissolved in the alkaline solution during the early stage of slag alkali activation, indicating that the Ca^{2+} was released from slag and bonded with OH- in alkali solutions to form calcium hydroxide ($Ca(OH)_2$), which then reacted with carbon dioxide (CO_2) in an open environment to form calcite.

Bouaissi et al. [114] discovered that the free cation Mg^{2+} leads to the formation of intermolecular bonding with Si^{4+} and Al^{3+} by the sharing of oxygen atoms. as depicted in Figure 4. A similar model was proposed by Zhang et al. [128], who also stated that the presence of Mg^{2+} provides chemical stability (interatomic bonding) in the geopolymer matrix as reflected in the formation of Si-O-Mg, Si-O-Al/Si and Ca-O-Si. The reaction mechanism of the alkali activation of slag is more complicated than fundamental geopolymers due to the significant amounts of calcium, magnesium and iron. Hence, it is essential to study the role of these elements in each slag waste material and hydration process.

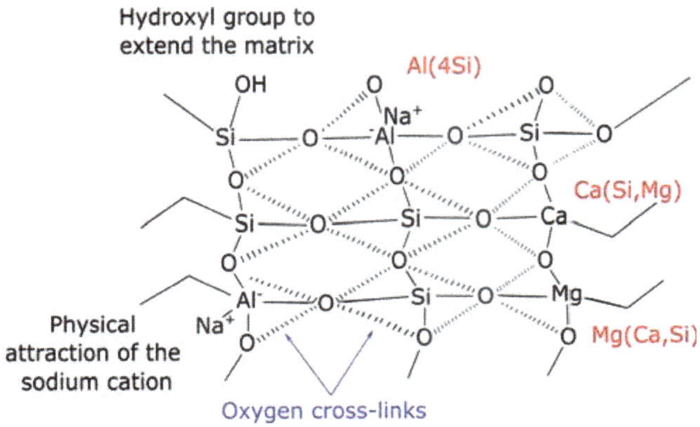

Figure 4. Proposed model of ternary C-A-S-H, and C-M-S-H gel phases [114].

4.4. Alkali-Activated Cement

Alkali-activated cement has been implemented as a replacement material to enhance mechanical properties and fire-resistant abilities, as tabulated in Table 7. Li et al. [129] pioneered the alkali-activated slag to develop high early strength by incorporating a proper mixing of Na_2CO_3 and $NaOH\text{-}Na_2O{\cdot}2SiO_2$. The ternary activators not only achieved a reasonable compressive strength, but also obtained the lowest porosity distribution. Another study was performed by Kim et el. [130] with the utilization of the cenosphere in alkali-activated slag cement. The effect of cenosphere replacement (50, 60, and 70%) in alkali-activated slag cement on compressive strength and thermal conductivity were investigated. The use of 70% of the cenosphere was clarified as a floating structural member for freshwater and marine applications.

Table 7. Research work utilizing alkali-activated slag cement.

No	Researcher	Materials	Findings
1	Kim et al. [130]	Blast furnace slag and cenosphere	• Increase water abscrption rate • Decrease density, compressive strength and thermal conductivity
2	Li et al. [129]	Ground blast furnace slag and river sand	• Shortened initial setting time • Increase compressive strength
3	He at al. [131]	Ground blast furnace slag, water glass and hydrated lime	• Increase compressive strength at early age • Decreased drying shrinkage
4	Hyeok-Jung et al. [132]	Ground granulated blast furnace slag and red mud	• Increase compressive strength • Increase efflorescence area
5	Nikolic et al. [133]	Electric arc furnace slag and electric arc furnace dust	• Deterioration of mechanical properties • Higher porous structure

Aside from its use in enhancing mechanical properties, alkali-activated slag cement has been used to magnify fire-resistance properties. For example, Shahari et al. [134] incorporated fibres into alkali-activated slag cement to enhance their fire-resistance properties. Different types of fibres were incorporated, including polypropylene fibres, basalt fibres, and glass fibres at 0.5%, 1.0% and 1.5%. With the appropriate fibres loading, the compressive strength was developed after been exposed up to 200 °C. Glass and basalt fibre show a better resistance compared to polypropylene fibre owing to superior fire resistance.

4.5. Alkali-Activated Mortar

Another reported alkali-activated materials, the implementation of steel waste as alkali-activated mortar (Table 8). For instance, Zhang et al. [135] initiated the utilization of alkali activate mortar with the addition seawater and coral sand. The effect of the modulus of sodium silicate, coral sand/sea sand replacement ratio and water/binder ratio on flexural and compressive strength was investigated. It was pointed out that alkali-activated material produced a hydration product that corresponded to the improvement of the interfacial microstructure between slurry and coral sand. Due to the self-curing of the coral aggregate, the drying shrinkage of the mortar was reduced.

Table 8. Research work utilizing alkali-activated slag mortar.

No	Researcher	Materials	Findings
1	Zhang et al. [135]	Ground granulated blast furnace slag, fly ash, silica fume, coral sand	• Reduction in drying shrinkage • Higher mechanical performance
2	Rovnanik et al. [136]	Slag, quartz, cement	• Large number of micropores • Remarkable self-sensing properties
3	Oh et al. [137]	Portland cement, superabsorbent polymers, and granulated blast furnace slag	• Significant reduction in shrinkage • Higher compressive strength
4	Kumarappa et al. [138]	Slag cement and shale lightweight aggregate	• The development of autogenous shrinkage can be controlled • Reduce surface tension

Rovnanik et al. [136] fabricated alkali-activated slag for electrical properties, such as ash resistance and capacitance, and the self-sensing functionality of mortar. The resistivity of alkali-activated slag mortar is nine times lower than that of cement mortar at low AC frequencies; nevertheless, as the AC frequency grows, the resistivity of both materials diminishes, and the values become similar at 500 Hz and above.

Oh et al. [137] found that the superabsorbent polymer is essential for reducing the shrinkage of alkali-activated slag mortar, which is a major limitation in related applications. The role of superabsorbent polymers of storing water inside the matrix initiated the hydration particle reaction.

Therefore, this review classifies the utilization of steel waste as an alkali-activated material with broad applications. The application as a cement replacement, aggregate, mortar and composite material proved that the steel waste could be implemented with any kind of material depending on the desired application.

4.6. Alkali-Activated Composites

Aside from alkali-activated cement, another alkali-activated material, such as alkali-activated composites, which consists of fibres, aggregate, and reinforcement materials, is incorporated with steel waste to impart exceptional mechanical properties on composite materials (Table 9). For an example, Nedeljkovic et al. [139] incorporated alkali-activated slag/fly ash and sand aggregates into composites at 2% of polyvinyl alcohol fibres. The contribution of fibres loading resulted in a sufficient bonding between the matrix and fibres. However, fibre failure was obtained for pullout and rupture observations due to fibre rupture. The small amount of fibres limits the capability of mechanical transfer load which believed the fibre rupture at higher strength.

Table 9. Research work utilizing on alkali-activated slag composite.

No	Researcher	Materials	Findings
1	Nedeljkovic et al. [139]	Slag/fly ash, sand aggregates, polyvinyl alcohol fibers	• Stronger bond formed between matrix and fibres • Limit the capability of mechanical transfer load
2	Cui et al. [140]	Ground granulated blast furnace slag, polycarboxylate, graphite-modified microencapsulated, and carbon fibre	• Higher compressive strength compared to conventional cement • Lower thermal conductivity • Achieve good thermal storage
3	Jiape et al. [141]	Ground granulated blast furnace slag, cement and epoxy resin	• Longer setting time required • Good bond and uncracking microstructure
4	Kan et al. [142]	Incineration fly ash, ground granulated blast furnace slag, polycarboxylate-based high-range water reducing mixture	• Better mechanical properties • Larger tensile strain capacity • Good for immobilizing toxic heavy metals
5	Cristelo et al. [143]	Steel slag, fly ash and silica sand	• Obtain superior mechanical performance • Well-graduated transition zone developed

By employing alkali-activated composites, Cui et al. [140] compared the thermal conductivity and mechanical properties of the composites that were incorporated with graphite-modified microencapsulated and carbon fibres. The inclusion of carbon fibres assist in retraining crack pulling and reflected the higher flexural strength of the composites. Meanwhile, the addition of both fibres enhances thermal conductivity.

The most significant importance factor is that the alkali-activated material (inorganic component) could be bonded with an organic component (fibre–polymer type). It is proven that the mechanical and thermal performance of the alkali-activated materials are developed with the acceptable proportion of fibre loadings into composite materials.

5. Conclusions and Future Works

In this review, the current development of the steelmaking industry, including cost analysis, energy consumption, and slag variability in the last ten years has been discussed. Additionally, the utilization of steel waste to develop an alkali-activated material has been reported. Their effect on the mechanical and thermal performance of alkali-activated materials is well explained. It was proved that the alkali activation process/technology could be one of the steel waste management in order to solve the landfill problem, environmental issue and economic growth.

Therefore, this review classifies the utilization of steel waste as alkali-activated materials in extensive application. The applications as a cement replacement, aggregate, mortar and composite material proved that the steel waste could be implemented with any kind of material, depending on the desired application. In the steel waste industry, alkali activation can be utilized to create green alternative materials for conventional cement replacement. Hence, the utilization of slag industrial waste should be carried out for the extensive future works, in order to mitigate the disposal nature of steel production and environmental issues. There is an abundance of works involving the manufacturing of alkali-activated slag with exceptional properties for extensive applications. Nevertheless, in order to produce exceptional properties of alkali-activated slag, the details of process chemistry, reaction mechanism, and material characteristic are elaborated. Alkali concentration and liquid/solid ratios are evaluated as the influencing parameters that affect the properties of alkali-activated slag.

Based on the identified gaps in this manuscript, future works on the alkali activation of varied steel waste material are listed below:

- Material and energy flow mechanisms in the steelmaking industry are still poorly understood, especially in the variable setting of steel production. Consequently, the quality of material and energy fluxes, as well as steel waste characteristics, necessitates greater consideration.
- The reaction mechanism and reaction products of alkali-activated cement are contributed to by prime materials and alkaline activators, hence the details of alkali-activation-based steel waste require more attention.
- It is also recommended that the number of steel waste management routes are increased, allowing the environmental impact to be reduced due to the introduction of more efficient technologies. As a result, organizations who embrace such approaches may save costs, add value to industrial waste, and develop the profitability and competitiveness of the manufacturing process.
- The evaluation of landfill cost avoidance benefits as part of production costs is important for the consideration of the impact on the steelmaking industry.

Author Contributions: Conceptualization, writing, review and editing, I.H.A.; supervision and resources, M.M.A.B.A. and M.A.A.M.S.; data curation, P.V.; validation, P.V., A.V.S. and O.N.; funding acquisition, A.V.S. and O.N.; methodology, L.Y.M., L.Y.L. and S.N.M. All authors have read and agreed to the published version of the manuscript.

Funding: This work was supported by Ministry of Higher Education, Malaysia and funded by the "Partnership for Research in Geopolymer Concrete" in the framework of Marie Sklodowska-Curie RISE Grant Agreement (689857 H2020-MSCA-RISE-2015).

Institutional Review Board Statement: Not applicable.

Informed Consent Statement: Not applicable.

Data Availability Statement: Not applicable.

Acknowledgments: The authors gratefully acknowledge the Centre of Excellence Geopolymer and Green Technology (CEGeoGTech) and the Faculty of Chemical Engineering Technology, UniMAP for their expertise and support. The authors wish to thank for the funding support obtained from the Fundamental Research Grant Scheme under the Malaysian Ministry of Education (MOE) and the support gained from "Partnership for Research in Geopolymer Concrete" in the framework of Marie Skłodowska-Curie RISE Grant Agreement (689857 H2020-MSCA-RISE-2015).

Conflicts of Interest: The authors declare no conflict of interest.

References

1. Medarac, H.; Moya, J.; Somers, J. Production costs from iron and steel industry in the EU and third countries. *JRC Tech. Rep.* **2020**, *10*, 705636.
2. Li, Y.; Zhu, L. Cost of energy saving and CO_2 emissions reduction in China's iron and steel sector. *Appl. Energy* **2014**, *130*, 603–616. [CrossRef]
3. Pinto, R.G.D.; Szklo, A.S.; Rathmann, R. CO_2 emissions mitigation strategy in the Brazilian iron and steel sector–From structural to intensity effects. *Energy Policy* **2018**, *114*, 380–393. [CrossRef]
4. Hasanbeigi, A.; Arens, M.; Price, L. Alternative emerging ironmaking technologies for energy-efficiency and carbon dioxide emissions reduction: A technical review. *Renew. Sustain. Energy Rev.* **2014**, *33*, 645–658. [CrossRef]

5. Morfeldt, J.; Nijs, W.; Silveira, S. The impact of climate targets on future steel production–an analysis based on a global energy system model. *J. Clean. Prod.* **2015**, *103*, 469–482. [CrossRef]
6. Burchart-Korol, D. Life cycle assessment of steel production in Poland: A case study. *J. Clean. Prod.* **2013**, *54*, 235–243. [CrossRef]
7. Suopajärvi, H.; Umeki, K.; Mousa, E.; Hedayati, A.; Romar, H.; Kemppainen, A.; Wang, C.; Phounglamcheik, A.; Tuomikoski, S.; Norberg, N. Use of biomass in integrated steelmaking–Status quo, future needs and comparison to other low-CO2 steel production technologies. *Appl. Energy* **2018**, *213*, 384–407. [CrossRef]
8. Pardo, N.; Moya, J.A. Prospective scenarios on energy efficiency and CO2 emissions in the European Iron & Steel industry. *Energy* **2013**, *54*, 113–128.
9. Skoczkowski, T.; Verdolini, E.; Bielecki, S.; Kochański, M.; Korczak, K.; Węglarz, A. Technology innovation system analysis of decarbonisation options in the EU steel industry. *Energy* **2020**, *212*, 118688. [CrossRef]
10. Sanjuán, M.Á.; Estévez, E.; Argiz, C.; del Barrio, D. Effect of curing time on granulated blast-furnace slag cement mortars carbonation. *Cem. Concr. Compos.* **2018**, *90*, 257–265. [CrossRef]
11. Shen, D.; Jiao, Y.; Kang, J.; Feng, Z.; Shen, Y. Influence of ground granulated blast furnace slag on early-age cracking potential of internally cured high performance concrete. *Constr. Build. Mater.* **2020**, *233*, 117083. [CrossRef]
12. Shen, D.; Liu, K.; Wen, C.; Shen, Y.; Jiang, G. Early-age cracking resistance of ground granulated blast furnace slag concrete. *Constr. Build. Mater.* **2019**, *222*, 278–287. [CrossRef]
13. Duraman, S.B.; Richardson, I.G. Microstructure & properties of steel-reinforced concrete incorporating Portland cement and ground granulated blast furnace slag hydrated at 20 °C. *Cem. Concr. Res.* **2020**, *137*, 106193.
14. Adolfsson, D.; Robinson, R.; Engström, F.; Björkman, B. Influence of mineralogy on the hydraulic properties of ladle slag. *Cem. Concr. Res.* **2011**, *41*, 865–871. [CrossRef]
15. Zhang, S.; Ghouleh, Z.; Liu, J.; Shao, Y. Converting ladle slag into high-strength cementing material by flue gas carbonation at different temperatures. *Resour. Conserv. Recycl.* **2021**, *174*, 105819. [CrossRef]
16. Henríquez, P.A.; Aponte, D.; Ibáñez-Insa, J.; Bizinotto, M.B. Ladle furnace slag as a partial replacement of Portland cement. *Constr. Build. Mater.* **2021**, *289*, 123106. [CrossRef]
17. Fang, K.; Wang, D.; Zhao, J.; Zhang, M. Utilization of ladle furnace slag as cement partial replacement: Influences on the hydration and hardening properties of cement. *Constr. Build. Mater.* **2021**, *299*, 124265. [CrossRef]
18. Cristelo, N.; Garcia-Lodeiro, I.; Rivera, J.F.; Miranda, T.; Palomo, Á.; Coelho, J.; Fernández-Jiménez, A. One-part hybrid cements from fly ash and electric arc furnace slag activated by sodium sulphate or sodium chloride. *J. Build. Eng.* **2021**, *44*, 103298. [CrossRef]
19. Ozturk, M.; Akgol, O.; Sevim, U.K.; Karaaslan, M.; Demirci, M.; Unal, E. Experimental work on mechanical, electromagnetic and microwave shielding effectiveness properties of mortar containing electric arc furnace slag. *Constr. Build. Mater.* **2018**, *165*, 58–63. [CrossRef]
20. Lam, M.N.-T.; Le, D.-H.; Jaritngam, S. Compressive strength and durability properties of roller-compacted concrete pavement containing electric arc furnace slag aggregate and fly ash. *Constr. Build. Mater.* **2018**, *191*, 912–922. [CrossRef]
21. Sosa, I.; Thomas, C.; Polanco, J.A.; Setien, J.; Tamayo, P. High performance self-compacting concrete with electric arc furnace slag aggregate and cupola slag powder. *Appl. Sci.* **2020**, *10*, 773. [CrossRef]
22. Aziz, I.H. The characterization of steel slag by alkali activation. *Open Access Libr. J.* **2017**, *4*, 1. [CrossRef]
23. Liao, Y.; Jiang, G.; Wang, K.; Al Qunaynah, S.; Yuan, W. Effect of steel slag on the hydration and strength development of calcium sulfoaluminate cement. *Constr. Build. Mater.* **2020**, *265*, 120301. [CrossRef]
24. Liu, Y.; Zhang, Z.; Hou, G.; Yan, P. Preparation of sustainable and green cement-based composite binders with high-volume steel slag powder and ultrafine blast furnace slag powder. *J. Clean. Prod.* **2021**, *289*, 125133. [CrossRef]
25. Zhang, N.; Wu, L.; Liu, X.; Zhang, Y. Structural characteristics and cementitious behavior of basic oxygen furnace slag mud and electric arc furnace slag. *Constr. Build. Mater.* **2019**, *219*, 11–18. [CrossRef]
26. Wen, T.; Yang, L.; Dang, C.; Miki, T.; Bai, H.; Nagasaka, T. Effect of basic oxygen furnace slag on succession of the bacterial community and immobilization of various metal ions in acidic contaminated mine soil. *J. Hazard. Mater.* **2020**, *388*, 121784. [CrossRef]
27. Lu, T.-H.; Chen, Y.-L.; Shih, P.-H.; Chang, J.-E. Use of basic oxygen furnace slag fines in the production of cementitious mortars and the effects on mortar expansion. *Constr. Build. Mater.* **2018**, *167*, 768–774. [CrossRef]
28. Lobato, N.C.C.; Villegas, E.A.; Mansur, M.B. Management of solid wastes from steelmaking and galvanizing processes: A brief review. *Resour. Conserv. Recycl.* **2015**, *102*, 49–57. [CrossRef]
29. Mayer, J.; Bachner, G.; Steininger, K.W. Macroeconomic implications of switching to process-emission-free iron and steel production in Europe. *J. Clean. Prod.* **2019**, *210*, 1517–1533. [CrossRef]
30. Sun, W.; Wang, Q.; Zhou, Y.; Wu, J. Material and energy flows of the iron and steel industry: Status quo, challenges and perspectives. *Appl. Energy* **2020**, *268*, 114946. [CrossRef]
31. Stewart, D.J.; Barron, A.R. Pyrometallurgical removal of zinc from basic oxygen steelmaking dust–A review of best available technology. *Resour. Conserv. Recycl.* **2020**, *157*, 104746. [CrossRef]
32. Mousa, E.; Wang, C.; Riesbeck, J.; Larsson, M. Biomass applications in iron and steel industry: An overview of challenges and opportunities. *Renew. Sustain. Energy Rev.* **2016**, *65*, 1247–1266. [CrossRef]

33. Sadek, D.M. Effect of cooling technique of blast furnace slag on the thermal behavior of solid cement bricks. *J. Clean. Prod.* **2014**, *79*, 134–141. [CrossRef]
34. Zainullin, L.; Epishin, A.Y.; Karelin, V.; Artov, D.; Spirin, N. Analysis of economic and energy efficiency of using electric-arc reduction for iron-containing materials. *Metallurgist* **2018**, *62*, 642–647. [CrossRef]
35. Marhual, N.; Pradhan, N.; Mohanta, N.; Sukla, L.; Mishra, B. Dephosphorization of LD slag by phosphorus solubilising bacteria. *Int. Biodeterior. Biodegrad.* **2011**, *65*, 404–409. [CrossRef]
36. Gurtubay, L.; Gallastegui, G.; Elias, A.; Rojo, N.; Barona, A. Accelerated ageing of an EAF black slag by carbonation and percolation for long-term behaviour assessment. *J. Environ. Manag.* **2014**, *140*, 45–50. [CrossRef]
37. Hamann, C.; Spanka, M.; Stolle, D.; Auer, G.; Weingart, E.; Al-Sabbagh, D.; Ostermann, M.; Adam, C. Recycling of blast-furnace sludge by thermochemical treatment with spent iron (II) chloride solution from steel pickling. *J. Hazard. Mater.* **2021**, *402*, 123511. [CrossRef] [PubMed]
38. Mombelli, D.; Gonçalves, D.L.; Mapelli, C.; Barella, S.; Gruttadauria, A. Processing and Characterization of Self-Reducing Briquettes Made of Jarosite and Blast Furnace Sludges. *J. Sustain. Metall.* **2021**, *7*, 1603–1626. [CrossRef]
39. Andersson, A.; Andersson, M.; Mousa, E.; Kullerstedt, A.; Ahmed, H.; Björkman, B.; Sundqvist-Ökvist, L. The potential of recycling the high-zinc fraction of upgraded BF sludge to the desulfurization plant and basic oxygen furnace. *Metals* **2018**, *8*, 1057. [CrossRef]
40. Rieger, J.; Colla, V.; Matino, I.; Branca, T.A.; Stubbe, G.; Panizza, A.; Brondi, C.; Falsafi, M.; Hage, J.; Wang, X. Residue Valorization in the Iron and Steel Industries: Sustainable Solutions for a Cleaner and More Competitive Future Europe. *Metals* **2021**, *11*, 1202. [CrossRef]
41. Ning, G.; Zhang, B.; Liu, C.; Li, S.; Ye, Y.; Jiang, M. Large-scale consumption and zero-waste recycling method of red mud in steel making process. *Minerals* **2018**, *8*, 102. [CrossRef]
42. Martín, M.; López, F.; Torralba, J.M. Production of sponge iron powder by reduction of rolling mill scale. *Ironmak. Steelmak.* **2012**, *39*, 155–162. [CrossRef]
43. Fischedick, M.; Marzinkowski, J.; Winzer, P.; Weigel, M. Techno-economic evaluation of innovative steel production technologies. *J. Clean. Prod.* **2014**, *84*, 563–580. [CrossRef]
44. Sabat, K.C.; Murphy, A.B. Hydrogen plasma processing of iron ore. *Metall. Mater. Trans. B* **2017**, *48*, 1561–1594. [CrossRef]
45. Anderl, M.; Freudenschuß, A.; Friedrich, A.; Haider, S.; Jobstmann, H.; Köther, T.; Kriech, M.; Kuschel, V.; Lampert, C.; Pazdernik, K. Austria's National Inventory Report 2011. In *Submission under the United Nations Framework Convention on Climate Change and under the Kyoto Protocol*; Umweltbundesamt: Berlin, Germany, 2020.
46. Renda, A.; Pelkmans, J.; Egenhofer, C.; Marcu, A. *Assessment of Cumulative Cost Impact for the Steel Industry*; RSC Publishing: London, UK, 2013.
47. Bhandari, R.; Trudewind, C.A.; Zapp, P. Life cycle assessment of hydrogen production via electrolysis–A review. *J. Clean. Prod.* **2014**, *85*, 151–163. [CrossRef]
48. Waligora, J.; Bulteel, D.; Degrugilliers, P.; Damidot, D.; Potdevin, J.; Measson, M. Chemical and mineralogical characterizations of LD converter steel slags: A multi-analytical techniques approach. *Mater. Charact.* **2010**, *61*, 39–48. [CrossRef]
49. Kim, H.-S.; Park, J.-W.; An, Y.-J.; Bae, J.-S.; Han, C. Activation of ground granulated blast furnace slag cement by calcined alunite. *Mater. Trans.* **2011**, *52*, 210–218. [CrossRef]
50. Kim, J.; Azimi, G. Valorization of electric arc furnace slag via carbothermic reduction followed by acid baking–water leaching. *Resour. Conserv. Recycl.* **2021**, *173*, 105710. [CrossRef]
51. Gesoğlu, M.; Güneyisi, E.; Mahmood, S.F.; Öz, H.Ö.; Mermerdaş, K. Recycling ground granulated blast furnace slag as cold bonded artificial aggregate partially used in self-compacting concrete. *J. Hazard. Mater.* **2012**, *235*, 352–358. [CrossRef]
52. Abu-Eishah, S.I.; El-Dieb, A.S.; Bedir, M.S. Performance of concrete mixtures made with electric arc furnace (EAF) steel slag aggregate produced in the Arabian Gulf region. *Constr. Build. Mater.* **2012**, *34*, 249–256. [CrossRef]
53. Choi, S.-J.; Bae, S.-H.; Lee, J.-I.; Kim, J.-H. Strength and Durability Characteristics of Cement Composites with Recycled Water and Blast Furnace Slag Aggregate. *Materials* **2021**, *14*, 2156. [CrossRef] [PubMed]
54. Yang, H.-M.; Kwon, S.-J.; Myung, N.V.; Singh, J.K.; Lee, H.-S.; Mandal, S. Evaluation of strength development in concrete with ground granulated blast furnace slag using apparent activation energy. *Materials* **2020**, *13*, 442. [CrossRef] [PubMed]
55. Sadowski, Ł.; Nikoo, M.; Shariq, M.; Joker, E.; Czarnecki, S. The nature-inspired metaheuristic method for predicting the creep strain of green concrete containing ground granulated blast furnace slag. *Materials* **2019**, *12*, 293. [CrossRef] [PubMed]
56. Santos, J.O.; Jahangiri, H.; Bashir, M.A.; Hornung, A.; Ouadi, M. The upgrading of bio-oil from the intermediate pyrolysis of waste biomass using steel slag as a catalyst. *ACS Sustain. Chem. Eng.* **2020**, *8*, 18420–18432. [CrossRef]
57. Kamil, F.H.; Ali, S.; Shahruzzaman, R.M.H.R.; Hussien, I.R.; Omer, R. Characterization and application of molten slag as catalyst in pyrolysis of waste cooking oil. *Bull. Chem. React. Eng. Catal.* **2020**, *15*, 119–127.
58. Liu, G.; Yang, J.; Xu, X. Synthesis of hydrotalcite-type mixed oxide catalysts from waste steel slag for transesterification of glycerol and dimethyl carbonate. *Sci. Rep.* **2020**, *10*, 1–14. [CrossRef]
59. Chen, L.; Long, Y.; Zhou, M.; Wang, H. Structure and Crystallization of High-Calcium, CMAS Glass Ceramics Synthesized with a High Content of Slag. *Materials* **2022**, *15*, 657. [CrossRef]
60. Jordanov, N.B.; Georgiev, I.; Karamanov, A. Sintered Glass-Ceramics, Self-Glazed Materials and Foams from Metallurgical Waste Slag. *Materials* **2021**, *14*, 2263. [CrossRef]

61. Nguyen, V.H.; Nguyen, V.C.; Nguyen, T.C.; Tran, D.M.T.; Nguyen, T.T.T.; Vu, Q.T.; Nguyen, D.T.; Thai, H. Treatment of yellow phosphorus slag and reuse of it as an absorbent of chromium (VI) ions and methylene blue. *J. Chem.* **2020**, *2020*, 1834829. [CrossRef]
62. Wang, P.; Chen, H.; Chen, P.; Pan, J.; Xu, Y.; Wang, H.; Shen, W.; Cao, K. Effect of internal curing by super absorbent polymer on the autogenous shrinkage of alkali-activated slag mortars. *Materials* **2020**, *13*, 4318. [CrossRef]
63. Yi, H.; Xu, G.; Cheng, H.; Wang, J.; Wan, Y.; Chen, H. An overview of utilization of steel slag. *Procedia Environ. Sci.* **2012**, *16*, 791–801. [CrossRef]
64. Földi, C.; Dohrmann, R.; Mansfeldt, T. Mercury in dumped blast furnace sludge. *Chemosphere* **2014**, *99*, 248–253. [CrossRef] [PubMed]
65. Trung, Z.H.; Kukurugya, F.; Takacova, Z.; Orac, D.; Laubertova, M.; Miskufova, A.; Havlik, T. Acidic leaching both of zinc and iron from basic oxygen furnace sludge. *J. Hazard. Mater.* **2011**, *192*, 1100–1107. [CrossRef] [PubMed]
66. Wang, Y.; Li, H.; Cui, S.; Wei, Q. Adsorption Behavior of Lead Ions from Wastewater on Pristine and Aminopropyl-Modified Blast Furnace Slag. *Water* **2021**, *13*, 2735. [CrossRef]
67. Doušová, B.; Bedrnová, E.; Reiterman, P.; Keppert, M.; Koloušek, D.; Lhotka, M.; Mastný, L. Adsorption Properties of Waste Building Sludge for Environmental Protection. *Minerals* **2021**, *11*, 309. [CrossRef]
68. Khater, G.A.; Nabawy, B.S.; El-Kheshen, A.A.; Abdel Latif, M.A.-B.; Farag, M.M. Use of Arc Furnace Slag and Ceramic Sludge for the Production of Lightweight and Highly Porous Ceramic Materials. *Materials* **2022**, *15*, 1112. [CrossRef]
69. Cechin, L.; Mymrine, V.; Avanci, M.A.; Povaluk, A.P. Ceramics composites from iron ore tailings and blast furnace slag. *Ceram. Int.* **2021**. [CrossRef]
70. Gao, D.; Wang, F.-P.; Wang, Y.-T.; Zeng, Y.-N. Sustainable utilization of steel slag from traditional industry and agriculture to catalysis. *Sustainability* **2020**, *12*, 9295. [CrossRef]
71. Cantarino, M.V.; de Carvalho Filho, C.; Mansur, M.B. Selective removal of zinc from basic oxygen furnace sludges. *Hydrometallurgy* **2012**, *111*, 124–128. [CrossRef]
72. Brožová, S.; Lisińska, M.; Saternus, M.; Gajda, B.; Simha Martynková, G.; Slíva, A. Hydrometallurgical Recycling Process for Mobile Phone Printed Circuit Boards Using Ozone. *Metals* **2021**, *11*, 820. [CrossRef]
73. Kou, Y.; Jiang, H.; Ren, L.; Yilmaz, E.; Li, Y. Rheological properties of cemented paste backfill with alkali-activated slag. *Minerals* **2020**, *10*, 288. [CrossRef]
74. De Hita, M.J.; Criado, M. Influence of the Fly Ash Content on the Fresh and Hardened Properties of Alkali-Activated Slag Pastes with Admixtures. *Materials* **2022**, *15*, 992. [CrossRef] [PubMed]
75. Jeon, I.K.; Ryou, J.S.; Jakhrani, S.H.; Kim, H.G. Effects of light-burnt dolomite incorporation on the setting, strength, and drying shrinkage of one-part alkali-activated slag cement. *Materials* **2019**, *12*, 2874. [CrossRef] [PubMed]
76. Amer, I.; Kohail, M.; El-Feky, M.; Rashad, A.; Khalaf, M.A. A review on alkali-activated slag concrete. *Ain Shams Eng. J.* **2021**, *12*, 1475–1499. [CrossRef]
77. Alharbi, Y.R.; Abadel, A.A.; Salah, A.A.; Mayhoub, O.A.; Kohail, M. Engineering properties of alkali activated materials reactive powder concrete. *Constr. Build. Mater.* **2021**, *271*, 121550. [CrossRef]
78. El-Feky, M.; Kohail, M.; El-Tair, A.M.; Serag, M. Effect of microwave curing as compared with conventional regimes on the performance of alkali activated slag pastes. *Constr. Build. Mater.* **2020**, *233*, 117268. [CrossRef]
79. Zawrah, M.; Gado, R.; Feltin, N.; Ducourtieux, S.; Devoille, L. Recycling and utilization assessment of waste fired clay bricks (Grog) with granulated blast-furnace slag for geopolymer production. *Process Saf. Environ. Prot.* **2016**, *103*, 237–251. [CrossRef]
80. Karthik, A.; Sudalaimani, K.; Kumar, C.V. Investigation on mechanical properties of fly ash-ground granulated blast furnace slag based self curing bio-geopolymer concrete. *Constr. Build. Mater.* **2017**, *149*, 338–349. [CrossRef]
81. Islam, A.; Alengaram, U.J.; Jumaat, M.Z.; Bashar, I.I.; Kabir, S.A. Engineering properties and carbon footprint of ground granulated blast-furnace slag-palm oil fuel ash-based structural geopolymer concrete. *Constr. Build. Mater.* **2015**, *101*, 503–521. [CrossRef]
82. Khan, M.; Castel, A.; Akbarnezhad, A.; Foster, S.J.; Smith, M. Utilisation of steel furnace slag coarse aggregate in a low calcium fly ash geopolymer concrete. *Cem. Concr. Res.* **2016**, *89*, 220–229. [CrossRef]
83. Bai, T.; Song, Z.-G.; Wu, Y.-G.; Hu, X.-D.; Bai, H. Influence of steel slag on the mechanical properties and curing time of metakaolin geopolymer. *Ceram. Int.* **2018**, *44*, 15706–15713. [CrossRef]
84. Zhang, Z.; Zhu, Y.; Yang, T.; Li, L.; Zhu, H.; Wang, H. Conversion of local industrial wastes into greener cement through geopolymer technology: A case study of high-magnesium nickel slag. *J. Clean. Prod.* **2017**, *141*, 463–471. [CrossRef]
85. Adesanya, E.; Ohenoja, K.; Kinnunen, P.; Illikainen, M. Properties and durability of alkali-activated ladle slag. *Mater. Struct.* **2017**, *50*, 255. [CrossRef]
86. Wang, W.-C.; Wang, H.-Y.; Tsai, H.-C. Study on engineering properties of alkali-activated ladle furnace slag geopolymer. *Constr. Build. Mater.* **2016**, *123*, 800–805. [CrossRef]
87. Sun, Z.; Lin, X.; Vollpracht, A. Pervious concrete made of alkali activated slag and geopolymers. *Constr. Build. Mater.* **2018**, *189*, 797–803. [CrossRef]
88. Lemougna, P.N.; Nzeukou, A.; Aziwo, B.; Tchamba, A.; Wang, K.-T.; Melo, U.C.; Cui, X.-M. Effect of slag on the improvement of setting time and compressive strength of low reactive volcanic ash geopolymers synthetized at room temperature. *Mater. Chem. Phys.* **2019**, *239*, 122077. [CrossRef]

89. Hu, S.; Wang, H.; Zhang, G.; Ding, Q. Bonding and abrasion resistance of geopolymeric repair material made with steel slag. *Cem. Concr. Compos.* **2008**, *30*, 239–244. [CrossRef]
90. Furlani, E.; Maschio, S.; Magnan, M.; Aneggi, E.; Andreatta, F.; Lekka, M.; Lanzutti, A.; Fedrizzi, L. Synthesis and characterization of geopolymers containing blends of unprocessed steel slag and metakaolin: The role of slag particle size. *Ceram. Int.* **2018**, *44*, 5226–5232. [CrossRef]
91. Alex, T.; Kalinkin, A.; Nath, S.; Gurevich, B.; Kalinkina, E.; Tyukavkina, V.; Kumar, S. Utilization of zinc slag through geopolymerization: Influence of milling atmosphere. *Int. J. Miner. Process.* **2013**, *123*, 102–107. [CrossRef]
92. Zhang, Y.J.; Liu, L.C.; Xu, Y.; Wang, Y.C. A new alkali-activated steel slag-based cementitious material for photocatalytic degradation of organic pollutant from waste water. *J. Hazard. Mater.* **2012**, *209*, 146–150. [CrossRef]
93. Mo, L.; Yang, S.; Huang, B.; Xu, L.; Feng, S.; Deng, M. Preparation, microstructure and property of carbonated artificial steel slag aggregate used in concrete. *Cem. Concr. Compos.* **2020**, *113*, 103715. [CrossRef]
94. Kalinkin, A.; Kumar, S.; Gurevich, B.; Alex, T.; Kalinkina, E.; Tyukavkina, V.; Kalinnikov, V.; Kumar, R. Geopolymerization behavior of Cu-Ni slag mechanically activated in air and in CO_2 atmosphere. *Int. J. Miner. Process.* **2012**, *112*, 101–106.
95. Zhang, T.; Jin, H.; Guo, L.; Li, W.; Han, J.; Pan, A.; Zhang, D. Mechanism of Alkali-Activated Copper-Nickel Slag Material. *Adv. Civ. Eng.* **2020**, *2020*, 7615848. [CrossRef]
96. Miltiadis, S.K.; Giannopoulou, I.; Tahir, M.F.M.; Hashim, M.F.A.; Panias, D. Upgrading Copper Slags to Added Value Fire Resistant Geopolymers. *Waste Biomass Valorization* **2019**, *11*, 3811–3820. [CrossRef]
97. Komnitsas, K.; Zaharaki, D.; Perdikatsis, V. Geopolymerisation of low calcium ferronickel slags. *J. Mater. Sci.* **2007**, *42*, 3073–3082. [CrossRef]
98. Maragkos, I.; Giannopoulou, I.P.; Panias, D. Synthesis of ferronickel slag-based geopolymers. *Miner. Eng.* **2009**, *22*, 196–203.
99. Thompson, A.; Saha, A.K.; Sarker, P.K. Comparison of the alkali-silica reactions of ferronickel slag aggregate in fly ash geopolymer and cement mortars. *Eur. J. Environ. Civ. Eng.* **2019**, *26*, 1–14. [CrossRef]
100. Karakoç, M.B.; Türkmen, İ.; Maraş, M.M.; Kantarci, F.; Demirboğa, R.; Toprak, M.U. Mechanical properties and setting time of ferrochrome slag based geopolymer paste and mortar. *Constr. Build. Mater.* **2014**, *72*, 283–292. [CrossRef]
101. Jena, S.; Panigrahi, R. Performance assessment of geopolymer concrete with partial replacement of ferrochrome slag as coarse aggregate. *Constr. Build. Mater.* **2019**, *220*, 525–537. [CrossRef]
102. Özcan, A.; Karakoç, M.B. Evaluation of sulfate and salt resistance of ferrochrome slag and blast furnace slag-based geopolymer concretes. *Struct. Concr.* **2019**, *20*, 1607–1621. [CrossRef]
103. Kumar, S.; Kumar, R.; Mehrotra, S. Influence of granulated blast furnace slag on the reaction, structure and properties of fly ash based geopolymer. *J. Mater. Sci.* **2010**, *45*, 607–615. [CrossRef]
104. Ismail, I.; Bernal, S.A.; Provis, J.L.; San Nicolas, R.; Hamdan, S.; van Deventer, J.S. Modification of phase evolution in alkali-activated blast furnace slag by the incorporation of fly ash. *Cem. Concr. Compos.* **2014**, *45*, 125–135. [CrossRef]
105. Puligilla, S.; Mondal, P. Role of slag in microstructural development and hardening of fly ash-slag geopolymer. *Cem. Concr. Res.* **2013**, *43*, 70–80. [CrossRef]
106. Mozgawa, W.; Deja, J. Spectroscopic studies of alkaline activated slag geopolymers. *J. Mol. Struct.* **2009**, *924*, 434–441. [CrossRef]
107. Rashad, A.M.; Essa, G.M. Effect of ceramic waste powder on alkali-activated slag pastes cured in hot weather after exposure to elevated temperature. *Cem. Concr. Compos.* **2020**, *111*, 103617. [CrossRef]
108. Yunsheng, Z.; Wei, S.; Qianli, C.; Lin, C. Synthesis and heavy metal immobilization behaviors of slag based geopolymer. *J. Hazard. Mater.* **2007**, *143*, 206–213. [CrossRef]
109. Nguyen, H.; Carvelli, V.; Adesanya, E.; Kinnunen, P.; Illikainen, M. High performance cementitious composite from alkali-activated ladle slag reinforced with polypropylene fibers. *Cem. Concr. Compos.* **2018**, *90*, 150–160. [CrossRef]
110. Gevaudan, J.P.; Caicedo-Ramirez, A.; Hernandez, M.T.; Srubar, W.V., III. Copper and cobalt improve the acid resistance of alkali-activated cements. *Cem. Concr. Res.* **2019**, *115*, 327–338. [CrossRef]
111. Samantasinghar, S.; Singh, S.P. Effect of synthesis parameters on compressive strength of fly ash-slag blended geopolymer. *Constr. Build. Mater.* **2018**, *170*, 225–234. [CrossRef]
112. Gao, X.; Yu, Q.; Brouwers, H. Apply ^{29}Si, ^{27}Al MAS NMR and selective dissolution in identifying the reaction degree of alkali activated slag-fly ash composites. *Ceram. Int.* **2017**, *43*, 12408–12419. [CrossRef]
113. Li, N.; Shi, C.; Wang, Q.; Zhang, Z.; Ou, Z. Composition design and performance of alkali-activated cements. *Mater. Struct.* **2017**, *50*, 1–11. [CrossRef]
114. Bouaissi, A.; Li, L.-Y.; Abdullah, M.M.A.B.; Bui, Q.-B. Mechanical properties and microstructure analysis of FA-GGBS-HMNS based geopolymer concrete. *Constr. Build. Mater.* **2019**, *210*, 198–209. [CrossRef]
115. Jia, D.; He, P.; Wang, M.; Yan, S. *Geopolymer and Geopolymer Matrix Composites*; Springer: Singapore, 2020; pp. 81–129.
116. Hu, X.; Shi, C.; Zhang, Z.; Hu, Z. Autogenous and drying shrinkage of alkali-activated slag mortars. *J. Am. Ceram. Soc.* **2019**, *102*, 4963–4975. [CrossRef]
117. Sithole, N.T.; Mashifana, T. Geosynthesis of building and construction materials through alkaline activation of granulated blast furnace slag. *Constr. Build. Mater.* **2020**, *264*, 120712. [CrossRef]
118. Altan, E.; Erdoğan, S.T. Alkali activation of a slag at ambient and elevated temperatures. *Cem. Concr. Compos.* **2012**, *34*, 131–139. [CrossRef]

119. Mohseni, E. Assessment of Na_2SiO_3 to NaOH ratio impact on the performance of polypropylene fiber-reinforced geopolymer composites. *Constr. Build. Mater.* **2018**, *186*, 904–911. [CrossRef]
120. Cihangir, F.; Ercikdi, B.; Kesimal, A.; Ocak, S.; Akyol, Y. Effect of sodium-silicate activated slag at different silicate modulus on the strength and microstructural properties of full and coarse sulphidic tailings paste backfill. *Constr. Build. Mater.* **2018**, *185*, 555–566. [CrossRef]
121. Tchadjie, L.; Ekolu, S. Enhancing the reactivity of aluminosilicate materials toward geopolymer synthesis. *J. Mater. Sci.* **2018**, *53*, 4709–4733. [CrossRef]
122. Singh, B.; Rahman, M.; Paswan, R.; Bhattacharyya, S. Effect of activator concentration on the strength, ITZ and drying shrinkage of fly ash/slag geopolymer concrete. *Constr. Build. Mater.* **2016**, *118*, 171–179. [CrossRef]
123. Shariati, M.; Shariati, A.; Trung, N.T.; Shoaei, P.; Ameri, F.; Bahrami, N.; Zamanabadi, S.N. Alkali-activated slag (AAS) paste: Correlation between durability and microstructural characteristics. *Constr. Build. Mater.* **2020**, *267*, 120886. [CrossRef]
124. Cihangir, F.; Ercikdi, B.; Kesimal, A.; Deveci, H.; Erdemir, F. Paste backfill of high-sulphide mill tailings using alkali-activated blast furnace slag: Effect of activator nature, concentration and slag properties. *Miner. Eng.* **2015**, *83*, 117–127. [CrossRef]
125. Phoo-ngernkham, T.; Maegawa, A.; Mishima, N.; Hatanaka, S.; Chindaprasirt, P. Effects of sodium hydroxide and sodium silicate solutions on compressive and shear bond strengths of FA–GBFS geopolymer. *Constr. Build. Mater.* **2015**, *91*, 1–8. [CrossRef]
126. Duxson, P.; Provis, J.L. Designing precursors for geopolymer cements. *J. Am. Ceram. Soc.* **2008**, *91*, 3864–3869. [CrossRef]
127. Jamil, N.H.; Abdullah, M.M.A.B.; Pa, F.C.; Mohamad, H.; Ibrahim, W.M.A.W.; Chaiprapa, J. Influences of SiO_2, Al_2O_3, CaO and MgO in phase transformation of sintered kaolin-ground granulated blast furnace slag geopolymer. *J. Mater. Res. Technol.* **2020**, *9*, 14922–14932. [CrossRef]
128. Zhang, Y.; Li, T.; Hou, D.; Zhang, J.; Jiang, J. Insights on magnesium and sulfate ions' adsorption on the surface of sodium alumino-silicate hydrate (NASH) gel: A molecular dynamics study. *Phys. Chem. Chem. Phys.* **2018**, *20*, 18297–18310. [CrossRef]
129. Li, N.; Shi, C.; Zhang, Z. Understanding the roles of activators towards setting and hardening control of alkali-activated slag cement. *Compos. Part B: Eng.* **2019**, *171*, 34–45. [CrossRef]
130. Kim, T. Characteristics of alkali-activated slag cement-based ultra-lightweight concrete with high-volume cenosphere. *Constr. Build. Mater.* **2021**, *302*, 124165. [CrossRef]
131. He, J.; Bai, W.; Zheng, W.; He, J.; Sang, G. Influence of hydrated lime on mechanical and shrinkage properties of alkali-activated slag cement. *Constr. Build. Mater.* **2021**, *289*, 123201. [CrossRef]
132. Hyeok-Jung, K.; Kang, S.-P.; Choe, G.-C. Effect of red mud content on strength and efflorescence in pavement using alkali-activated slag cement. *Int. J. Concr. Struct. Mater.* **2018**, *12*, 1–9. [CrossRef]
133. Nikolić, I.; Đurović, D.; Marković, S.; Veselinović, L.; Janković-Častvan, I.; Radmilović, V.V.; Radmilović, V.R. Alkali activated slag cement doped with Zn-rich electric arc furnace dust. *J. Mater. Res. Technol.* **2020**, *9*, 12783–12794. [CrossRef]
134. Shahari, S.; Fathullah, M.; Abdullah, M.M.A.B.; Shayfull, Z.; Mia, M.; Darmawan, V.E.B. Recent developments in fire retardant glass fibre reinforced epoxy composite and geopolymer as a potential fire-retardant material: A review. *Constr. Build. Mater.* **2021**, *277*, 122246. [CrossRef]
135. Zhang, B.; Zhu, H.; Shah, K.W.; Feng, P.; Dong, Z. Optimization of mix proportion of alkali-activated slag mortars prepared with seawater and coral sand. *Constr. Build. Mater.* **2021**, *284*, 122805. [CrossRef]
136. Rovnaník, P.; Kusák, I.; Bayer, P.; Schmid, P.; Fiala, L. Comparison of electrical and self-sensing properties of Portland cement and alkali-activated slag mortars. *Cem. Concr. Res.* **2019**, *118*, 84–91. [CrossRef]
137. Oh, S.; Choi, Y.C. Superabsorbent polymers as internal curing agents in alkali activated slag mortars. *Constr. Build. Mater.* **2018**, *159*, 1–8. [CrossRef]
138. Kumarappa, D.B.; Peethamparan, S.; Ngami, M. Autogenous shrinkage of alkali activated slag mortars: Basic mechanisms and mitigation methods. *Cem. Concr. Res.* **2018**, *109*, 1–9. [CrossRef]
139. Nedeljković, M.; Luković, M.; van Breugel, K.; Hordijk, D.; Ye, G. Development and application of an environmentally friendly ductile alkali-activated composite. *J. Clean. Prod.* **2018**, *180*, 524–538. [CrossRef]
140. Cui, H.; Feng, T.; Yang, H.; Bao, X.; Tang, W.; Fu, J. Experimental study of carbon fiber reinforced alkali-activated slag composites with micro-encapsulated PCM for energy storage. *Constr. Build. Mater.* **2018**, *161*, 442–451. [CrossRef]
141. Jiapei, D.; Yuhuan, B.; Xuechao, C.; Zhonghou, S.; Baojiang, S. Utilization of alkali-activated slag based composite in deepwater oil well cementing. *Constr. Build. Mater.* **2018**, *186*, 114–122. [CrossRef]
142. Kan, L.; Shi, R.; Zhao, Y.; Duan, X.; Wu, M. Feasibility study on using incineration fly ash from municipal solid waste to develop high ductile alkali-activated composites. *J. Clean. Prod.* **2020**, *254*, 120168. [CrossRef]
143. Cristelo, N.; Coelho, J.; Miranda, T.; Palomo, Á.; Fernández-Jiménez, A. Alkali activated composites–An innovative concept using iron and steel slag as both precursor and aggregate. *Cem. Concr. Compos.* **2019**, *103*, 11–21. [CrossRef]

materials MDPI

Review

A State-of-the-Art Review on Innovative Geopolymer Composites Designed for Water and Wastewater Treatment

Ismail Luhar [1], Salmabanu Luhar [2,3,4,*], Mohd Mustafa Al Bakri Abdullah [2,*], Rafiza Abdul Razak [2], Petrica Vizureanu [5,*], Andrei Victor Sandu [5,6,7,*] and Petre-Daniel Matasaru [8]

1 Department of Civil Engineering, Shri Jagdishprasad Jhabarmal Tibrewala University, Rajasthan 333001, India; jprraj2017@gmail.com
2 Center of Excellence Geopolymer and Green Technology (CEGeoGTech), Universiti Malaysia Perlis (UniMAP), Perlis 01000, Malaysia; rafizarazak@Unimap.edu.my
3 Frederick Research Center, P.O. Box 24729, Nicosia 1303, Cyprus
4 Department of Civil Engineering, Frederick University, Nicosia 1036, Cyprus
5 Faculty of Materials Science and Engineering, Gheorghe Asachi Technical University of Iasi, D. Mangeron 41, 700050 Iasi, Romania
6 Romanian Inventors Forum, St. P. Movila 3, 700089 Iasi, Romania
7 National Institute for Research and Development in Environmental Protection INCDPM, Splaiul Independentei 294, 060031 Bucuresti, Romania
8 Faculty of Electronics, Telecommunications and Information Technology, Technical University "Gheorghe Asachi", Carol I Bvd, nr. 11 A, 700506 Iasi, Romania; dmatasaru@etti.tuiasi.ro
* Correspondence: ersalmabanu.mnit@gmail.com (S.L.); mustafa_albakri@unimap.edu.my (M.M.A.B.A.); peviz@tuiasi.ro (P.V.); sav@tuiasi.ro (A.V.S.)

Citation: Luhar, I.; Luhar, S.; Abdullah, M.M.A.B.; Razak, R.A.; Vizureanu, P.; Sandu, A.V.; Matasaru, P.-D. A State-of-the-Art Review on Innovative Geopolymer Composites Designed for Water and Wastewater Treatment. *Materials* **2021**, *14*, 7456. https://doi.org/10.3390/ma14237456

Academic Editor: Sabino De Gisi

Received: 27 October 2021
Accepted: 29 November 2021
Published: 4 December 2021

Abstract: There is nothing more fundamental than clean potable water for living beings next to air. On the other hand, wastewater management is cropping up as a challenging task day-by-day due to lots of new additions of novel pollutants as well as the development of infrastructures and regulations that could not maintain its pace with the burgeoning escalation of populace and urbanizations. Therefore, momentous approaches must be sought-after to reclaim fresh water from wastewaters in order to address this great societal challenge. One of the routes is to clean wastewater through treatment processes using diverse adsorbents. However, most of them are unsustainable and quite costly e.g. activated carbon adsorbents, etc. Quite recently, innovative, sustainable, durable, affordable, user and eco-benevolent Geopolymer composites have been brought into play to serve the purpose as a pretty novel subject matter since they can be manufactured by a simple process of Geopolymerization at low temperature, lower energy with mitigated carbon footprints and marvellously, exhibit outstanding properties of physical and chemical stability, ion-exchange, dielectric characteristics, etc., with a porous structure and of course lucrative too because of the incorporation of wastes with them, which is in harmony with the goal to transit from linear to circular economy, i.e., "one's waste is the treasure for another". For these reasons, nowadays, this ground-breaking inorganic class of amorphous alumina-silicate materials are drawing the attention of the world researchers for designing them as adsorbents for water and wastewater treatment where the chemical nature and structure of the materials have a great impact on their adsorption competence. The aim of the current most recent state-of-the-art and scientometric review is to comprehend and assess thoroughly the advancements in geo-synthesis, properties and applications of geopolymer composites designed for the elimination of hazardous contaminants viz., heavy metal ions, dyes, etc. The adsorption mechanisms and effects of various environmental conditions on adsorption efficiency are also taken into account for review of the importance of Geopolymers as most recent adsorbents to get rid of the death-defying and toxic pollutants from wastewater with a view to obtaining reclaimed potable and sparkling water for reuse offering to trim down the massive crisis of scarcity of water promoting sustainable water and wastewater treatment for greener environments. The appraisal is made on the performance estimation of Geopolymers for water and wastewater treatment along with the three-dimensional printed components are characterized for mechanical, physical and chemical attributes, permeability and Ammonium (NH_4^+) ion removal competence of

Geopolymer composites as alternative adsorbents for sequestration of an assortment of contaminants during wastewater treatment.

Keywords: wastewater; geopolymers; geopolymer composite; water treatment; geopolymer adsorbent; heavy metal removal; nutrient recovery

1. Introduction

"Water is life" for all breathing organisms. Incredibly, our mother planet earth is 70% covered by water; however, we have merely 3% freshwater useable by lives breathing on it! Apprehensively, over recent decades, a signaling boost in water demand has cropped up as a major predicament of "water scarcity", which is directly linked to "water stress" or "water crisis", i.e., the lack of fresh, clean and potable water resources in order to meet the standard and rising exigency. Staggeringly, the emerging clean water crisis is turning out to be an international concern since it is impacting 785 million persons and 1.1 billion are those who lack access to water, while 2.7 billion are experiencing water shortages. In accordance with the estimation, by 2025, almost 1800 million people will be living in the areas having absolute scarcity of water, while about 66% of the world population will pass through the stressful conditions [1]. Shockingly, the other estimation assesses that more than 40% of the global population will have to live in severe water scarcity areas by 2050 [2]. This happening is associated with the mushrooming international population and the eco-unfriendly industrial growth that has accelerated the incredible demand for water and also the pollution of water to unprecedented echelons. Of which, the latter mounds the pressure on the accessible sources to deal with the demand. For this reason, water and wastewater treatment are of prime significance as they can enhance the supply of water. Recently, the World Economic Forum report has also regarded water scarcity as the gravest concern for our society. Even in nations with enough water resources, water dearth is not uncommon. Consequently, the EU has publicized the Water Framework Directives for the adoption of specific legislation to handle the situation [1,2].

On the other hand, the pollution of water is one of the most severe environmental concerns as it carries heavy metals and lethal pollutants with it, which are difficult to biodegrade. The terrifying organic pollutants encompass plasticizers, phenols, poly-nuclear aromatic hydrocarbons (PAHs), polychlorinated biphenyls (PCBs), poly-brominated diphenyl ethers (PBDEs), pesticides and drug residues, while the chief inorganic contaminants are toxic metals together with nutrients, such as nitrate and phosphate [3]. Heavy metal cations are regarded as one of the most perilous contaminants in water and wastewater because of their very high toxicity and ability to accumulate in living tissues [4,5]. These contaminants, when entering the human body via the food chain, accumulate and can cause chronic poisoning [6].

Cadmium (Cd) and Lead (Pb) are the most noxious heavy metals responsible for numerous diseases [7]. Under these terrible situations, an approach to get clean water from wastewater by removing its enclosed polluting agents seems to be a momentous viable solution, helpful for both predicaments of a scarcity of water, as well as wastewater management. Simply speaking, "wastewater" is the water that has been used at least once or more and gets polluted. It is generated from the release of an assortment of pollutants from domestic, municipal, industrial, businesses, etc., sectors into freshwater bodies directly or indirectly. It encloses human wastes, oils, grease, soaps, food scraps, diverse chemicals, toxic materials, storm runoff (such as debris, woods, rocks, plastics, at times hazardous heavy metals), dissolved gases (such as hydrogen sulfide (H_2S)) bacteria, viruses, disease-causing pathogens, a few discarded pharmaceuticals, dyes, surfactants, pesticides, personal hygiene care items and even dead animals, etc. However, nature owns an astonishing aptitude to deal with smaller quantities of contaminants in wastewater, but it would be overwhelmed if the daily generation of billions of gallons of wastewater and sewage are not subjected to systematic treatment prior to release back to the environment. It is a universal fact that water is a widespread solvent; hence, it is considered a noteworthy source of diverse pollutions and infections. The world health organization (WHO) in its alarming account warned that around 80 % of diseases are assigned to pollution of water since the stated WHO standards are not being followed appropriately and stringently [8].

Clean water is highly needed for life and is also a great playground for us all and, therefore, wastewater should be treated methodically in order to have potable water and advantages for the environment, the health of living beings, ecology, water bodies, etc. If wastewater management is not made systematic, it will negatively impact land populations, aquatic lives, fauna and flora, wildlife; it will contaminate drinking and surface water; lead to oxygen (O_2) depletion, beach closures, limitations on recreational water use, restrictions on fish and shellfish harvesting, etc. The decaying of organic matter and debris consume the dissolved O_2 in water bodies, so aquatic biota cannot survive. Analogously, the chlorine compounds and inorganic chloramines are toxic to algae, fish and invertebrates living in water bodies. An excessive presence of nutrients of phosphorus (P) and nitrogen (N), including ammonia (NH_3), carried along by wastewater can cause "excessive fertilization or eutrophication" of receiving waters, which may be noxious to organisms existing in the water. Not only that, they encourage unwarranted extreme plant growth, trim down O_2, damage the spawning grounds, modify the habitat and lead to a turndown of specific species. With a view to deal with water scarcity, wastewater can be converted to potable through different treatments with the use of absorbents to remove pollutants, such as heavy metals in the form of lead (Pb), cadmium (Cd), chromium (Cr), cobalt (Co), arsenic (As), mercury (Hg), radionuclide [9], etc., which can have chronic and acute toxic influences, and can also bio-accumulate [10]. Thus, the subsequent reuse of wastewater can offer a significant rise in clean water supply, addressing the setback of water paucity. Wastewater treatment includes a number of methods, e.g., heavy metal removal can be made possible through chemical precipitation, coagulation-flocculation, floatation and reverse osmosis [11]. Ion-exchange, adsorption, supercritical fluid extraction, advanced oxidation course and membrane bio-reactors, filtration, electro-dialysis, microbial system, electro-chemical procedures, Fenton oxidation, photo-degradation, etc. [12–14], are utilized for the subtraction of heavy metals, dyes and other contaminants from wastewater, each has their own pros and cons. The process's efficiency, eco-friendliness and financial viability restrict the use of the conventional techniques mentioned. However, adsorption has proved to be most useful and most expansively employed on account of not only its simplicity and efficiency but also for its low cost amongst all the above-listed routes for the exclusion of heavy metals, dyes and other contaminants (viz., zeolites, activated carbon, resins, fly ash, chitosan, alumina and silica, as well as diverse materials, such as ZnO nanoparticles, CuO nanoparticles, active graphene oxide, boron nitride, etc. [15–18]) represent higher competence levels, but their higher production cost is a hindrance to their across-the-board application.

Therefore, in order to address all of the aforementioned predicaments, an orderly treatment of wastewater is highly essential to follow for sustainable eco-system, human health, the breathability of living beings, financial productivity in the course of recovery of valuable materials, etc. The key objective of the wastewater treatment process is to get rid of as much of the suspended solid contaminants as possible from wastewater and convert it into the remaining water, known as "effluent", which is returned to the water cycle in the environment. Once back in the water cycle, the effluent creates the standard impact on the eco-system or is reused for an assortment of objectives, known as "water reclamation". The treatment process performed in a wastewater treatment plant depends upon the types of wastewater, such as domestic or municipal wastewater or sewage treatment plant, or industrial wastewater treatment plant by and large subsequent to a few forms of prior treatments; along with the agricultural wastewater treatment plants and leachate treatment plants. On the whole, the procedures in practice include phase separation, such as sedimentation, the biological and chemical course of actions, e.g., oxidation or polishing. The primary by-product generated from wastewater treatment plants is a kind of sludge that is normally treated in the same or a different wastewater treatment plant. If an anaerobic treatment progression is utilized, then biogas (Methane—CH_4) may be an additional by-product. The primary processes of wastewater treatment include screening, pumping, aerating, and elimination of organic matters, i.e., sludge, removal of scum consisting of grease, oils, plastics and soap; filtration, usually through sand by the action of gravity to remove all the bacteria, which decreases the turbidity and color, takes away odors, mitigates iron quantity and removes most other solid particles that remained. Sometimes, in order to remove organic particles, filtering through carbon particles is followed. Ultimately, with a view to kill the bacteria, the wastewater flows into a "chlorine contact" tank, where the chemical chlorine is added. Thus, the effluent, i.e., the treated water, is then discharged into a local river or the ocean. "Wastewater Residuals" is one more portion of wastewater treatment that deals with the solid waste materials, which are kept for 20 to 30 days in larger enclosed and heated tanks known as "digesters", whereby the bacteria digest the material, resulting in a reduction in its volume, odors and removal of disease-causing organisms. The end-product is, for the most part, dumped into landfills; however, at times, it is found useful as fertilizer.

Accordingly, wastewater can be regarded as a resource that is so valuable that no throwing away is acceptable, particularly on our globe where there exists an increasing water scarcity. For this reason, nowadays, the reuse of treated wastewater and water conservation are increasingly becoming more significant. Appreciably, the reclamation of wastewater frees up freshwater, which can be utilized at another place where it is needed, for example, for drinking purposes, thus conserving drinking water and offers water for irrigational and industrial applications. Incredibly, even astronauts of the International Space Station drink urine following reclamation because, in outer space, water is at a premium, and therefore, not even a single drop is to be squandered.

On the other hand, quite recently, an innovative geopolymer technology has attracted world concrete researchers mostly due to its sustainable, durable and affordable approaches. The hitch of lack of sustainability in several other common practices of treatment technologies, viz., adsorption, ion exchange, chemical treatment, coagulation, etc. [11,19]. Among the list, adsorption gives the impression as the most promising, efficient and within one's means for wastewater treatment, particularly using geopolymer adsorbents [20]. The "geo-synthesis" of geopolymers involves the activation of rich alumina-silicates with strong alkali activators at low temperatures and atmospheric pressure only, popularly known as "geopolymerization" in an alkaline medium. Geopolymers are amorphous to semi-crystalline in structures, which are produced by an exothermal chemical reaction between silica and alumina-rich precursors with alkaline activators in an alkaline medium, keeping the temperature low at atmospheric pressure with low energy consumption, representing a lower carbon footprint. Since the chemical structure of geopolymers is comprised of a negatively-charged alumina-silicate framework, the charge-balancing can be exchanged with the cations present in the activator solution, making them suitable for employing as valuable adsorbents for wastewater treatment.

Geopolymers are versatile materials that have created a center of attention in scores of environmental applications, for instance, they have been investigated as "Sorbents" for wastewater treatment which is assigned to the key drivers in form of excellent mechanical and chemical stability together with a relatively simple, low operational energy, and lower carbon footprint bearing production process for this emerging interest [20]. Practically, Geopolymer sorbents can conveniently be installed in pipes or columns, whereby the water is pumped by means of the lattice structure to interact with the active surface sites enclosing cations like Na^+, etc. which are exchangeable [20]. Accordingly, the geopolymers are competent enough to offer viable alternatives for competing materials for application for wastewater treatment in form of conventional ceramics, synthetic zeolites, or polymeric components in virtue of not only cost but also in performance and significantly in eco-system impacts. In the interest to prevent eutrophication, the legislative need for the elimination of nitrogen present in the wastewaters generated from municipal and industrial operations is turning out to be increasingly widespread [19,20].

The objective of the present review manuscript is not only to lend a hand to comprehend the significance of geopolymers as most modern adsorbents to eliminate perilous and noxious pollutants from wastewater in order to get potable and clean reclaimed water for reuse, extending relief to the gigantic impasse of a scarcity of water but also encourage sustainable water and wastewater treatment for green eco-systems. The assessment is to be made for the performance evaluation of geopolymers for water and wastewater treatment.

Web of Science, ScienceDirect, ResearchGate, SpringerLink, and other databases were used to find out more about novel geopolymer composites for water and waste water treatment. The keywords "geopolymer, waste water treatment, water treatment, and geopolymerization" were identified in the Scopus database. The goal of this study is to highlight current research and application status in the domains of geopolymer -water and wastewater treatment technology. For reference, the related references provided in the literature were also used.

2. Introduction of Geopolymers

Geopolymers are a class of ceramic-like inorganic polymers analogous to the Felspathoid family of minerals in their mineralogical attributes, which can be developed at a considerably low temperature of below 100 °C and merely at atmospheric pressure by means of an exothermal process of geopolymerization, i.e., through a polycondensation reaction obtained by alkali activation of an alumina-silicate source as a precursor, such as alumina-silicate minerals [21] (metakaolin) or industrial wastes, e.g., fly ash, ground granulated blast furnace slag (GGBFS), coal gangue, red mud, etc., and alkali-activating solutions as activators, viz., sodium hydroxide (NaOH), sodium water glass (Na_2SiO_3), etc., in this process, which is also known as "geo-synthesis", formulating amorphous polymers possessing the chains or networks of mineral molecules linked to covalent bonds [22]. Besides, acid-based activators, such as phosphoric acid, and aluminum phosphate-based activators have been discovered to be a viable alternative to alkali-based activators. According to research, geopolymer produced by phosphate-based activators has better mechanical and microstructural properties than geopolymer produced by alkali-based activators, and the absence of alkali ions and an increase in bridging oxygen in the former geopolymer than the latter are the reasons for superior performance [23].

Geopolymerization occurs in three separate but interrelated stages—firstly, the dissolution, i.e., alumina-silicate precursors, get dissolved in an alkali activator solution in order to form free AlO_4 and SiO_4 tetrahedral units; secondly, the condensation, whereby two tetrahedral units form a long chain because of the condensation reaction kinetics; and thirdly, the poly-condensation reactions stage wherein the reactions among the long chains to produce gel-like materials, which are mainly amorphous N–A–S–H (sodium–alumino–silicate–hydrate) and C–S–H (calcium-silicate-hydrate) or the Al-substituted C–A–S–H (Calcium–Aluminium–Silicate–Hydrate) gel in their form [24]. Following the reaction kinetics, the rock-hardened rigid material developed which contains an amorphous three dimensional (3D) structure comprising of AlO_4— and SiO_4-tetrahedra linked alternatively by sharing oxygen (O) atoms, coined as poly-sialates by French scientist Joseph Davidovits [24]. The motive behind the use of alkali silicate is three-fold, i.e., the silicate molecules are taking part in the formulation of the poly-sialate, the alkali portion existing in the solution causes the cleavage of the alumina-silicate precursor; and this activator solution is also the source of the metal cations for charge balancing which is an indispensable precondition. Thus, the negative charge of tetrahedral-coordinated Al-atoms inside the network is balanced by cations sourced from the activating solution. The vital phase is of hydrated sodium alumina-silicates (N–A–S–H) gel, which is a flowable paste, and it is a network built of silicon (Si) and aluminum (Al) tetrahedra, linked together through oxygen bridges. The oxygen tetrahedra holding Al^{3+} ions hold a negative charge that is neutralized by an alkali metal cation. Consequently, geopolymers are the result of a mineral polycondensation reaction, which involves dissolution and condensation. Cations of Na^{+}, K^{+}, Li^{+}, Ca^{2+}, etc., must be present in the voids of the poly-sialate with a view to balancing the negative charge of the tetravalent aluminum (Al).

The empirical formula as per Davidovits for a poly-sialate can be portrayed as follows [24]:

$$Mn\ [(\text{-}SiO_2)z\text{-}AlO_2]n.wH_2O, \quad (1)$$

whereby M—an alkali cation (Na^{+}, K^{+}, Li^{+}, etc.); n—degree of polycondensation; and z—an atomic ratio of Si:Al, which may be equivalent to 1 to 32.

Poly-sialates will form an amorphous or semi-crystalline matrix at ambient temperatures and low water contents relying upon the reaction conditions. The atomic ratios of Si:Al in the poly-sialates determine the physical characteristics and uses of the end-product. The atomic ratio of Si:Al plays an important role in the diverse final products. The condensation can take place, even at room temperature, and for this reason, geopolymers are frequently considered cementitious materials. Even so, at elevated temperatures, crystalline phases develop and sintering reactions result in the development of ceramic products. Evidentially, the geopolymer system exhibited enhanced strength, durability, resistance against acid attack, thermal and fire in comparison to most cementitious matrices and Portland cement/concrete. On top of that, they are user and eco-friendly with low operational energy. That is why their numerous potential applications are proposed. Fly ash-based geopolymers are identified as an eco-friendly, purer and chemically homogeneous replacement of ordinary Portland cement (OPC) while those metakaolin based are being further employed in more significant technical applications viz., encapsulation or immobilization of radioactive nuclear wastes. Recently, a novel potential application of geopolymers for wastewater treatment was found to be eye-catching. Admirably, this approach is advantageous because of its higher strength, low permeability, user and eco-benevolent nature, together with fire-resistance and improved dimensional stability [22,25]. Several studies are directed in searching for low-cost raw materials appropriate for zeolite synthesis [25].

The chemistry of geopolymers and the reaction kinetics of geopolymerization have ended the days of high temperatures and intense energy processes to obtain the materials of ceramic-like structure, demonstrating equivalent properties as a conventional system. Consequently, no more energy-intensive and elevated temperature reactions are necessary anymore, as found in that of the cement system. In fact, the investigations have proven that the geopolymers are zeolite-like structures comprising $[SiO_4]^{-4}$ and $[AlO_4]^{-5}$ tetrahedra linked by oxygen atoms, wherein the negative charges of the $[AlO_4]^{-5}$ tetrahedra are balanced by alkali cations accessible from either the activator or precursors. Nevertheless, the geopolymers are amorphous or semi-crystalline, with heterogeneous inner micro-pores, such as zeolites [22–25].

Even so, the application of geopolymers as adsorbents is quite recent, and the breathing literature is, therefore, also limited [26,27]. The study reveals that both Metakaolin-based and coal fly ash-based geopolymers [27] have displayed an uptake aptitude for lead (Pb) more or less 100 and 80 mg/g, in that order. That simply means that the latter exhibits the luminous practicability of utilizing geopolymers as heavy metals adsorbents. The middle-of-the-road investigations have thrown lights upon the incorporation of diverse wastes used as precursors in the geopolymer's production, such as fly ash, metakaolin, blast furnace slag, etc. Extensively, the industrial by-product of Coal fly ash has been regarded as an alumina-silicate source. Geopolymer composites were produced with a view to enhance the strength of the original material by supplementing chamotte by Musil et al. [28], whereas Mohseni [29] and Noushini et al. [30] used a mix of polypropylene along with other synthetic fibers.

Additionally, Saafi et al. [31] and Yan et al. [32] have employed a mixture of carbon nanotubes and graphene. As the promising adsorbents for wastewater treatment, porous geopolymers, were formulated, their performance is confined to their inherent definite surface area and pore structure. It might be enhanced by using pore-forming additions that can optimize the porous structure of geopolymers from the view of a composite. The coal gangue—a by-product generated during the coal mining process, which is not only a significant resource for the manufacturing of ceramic, cement, and construction materials [33,34] but also portrayed by a layered structure made up of the silicate, promising the admirable adsorption attributes subsequently to correct modification. Yan et al. [35–37] made obvious the good-quality adsorption characteristics of microspheres of coal gangue that makes them competent environmental adsorbents: the adsorption potential of methylene blue on microspheres of coal gangue can reach ~30 mg/g. The microspheres of coal gangue might be employed to manufacture porous geopolymer composites, which can be employed for wastewater treatment. The viability of the incorporation of wastes not only extends to reduce the environmental impact but also lowers the cost of production of geopolymers. A modest work is performed on the adsorption attributes of heavy metal ions. The structure of geopolymers makes their applications valuable for water remediation, i.e., photo-degradation of risky organic compounds, heavy metals adsorption, energy evolution/storage, antibacterial uses, etc. Their good adsorption performance, simple, reduced carbon footprint, low-temperature and lower operational energy process of geopolymerization and the cost-effectiveness and accessibility of wastes as raw materials have altogether made them popular and received momentous attention, from both researchers and industries, for water and wastewaters treatments [38–41]

3. Hazardous and Toxic Ions of Heavy Metals as Pollutants in Industrial Wastewater

The huge quantity of industrial wastewater is a solemn and enduring environmental setback on account of it carrying heavy metals with it [40]. Mostly the industrial products, such as heavy metals, pesticides, colorants, etc., can cause water and wastewater pollution resulting in eco-degradation. The non-biodegradable and extremely venomous heavy metals, such as cobalt (Co), lead (Pb), manganese (Mn), copper (Cu), cadmium (Cd), etc., are frequently found present as contaminants in a range of industrial wastewater, surface and subsurface waters. That is why it is highly essential to remove these metal ions from the industrial wastewater and water utilized for human use [41]. They are found to occur in wastewater generated from industries of the major source of heavy metal cations release, namely, electroplating and chemical-processing plants, as well as the coal washing, end-products of battery production, etc., posing progressively more grave pollution of ecology. The water pollution due to the inclusion of heavy metal ions is regarded as very perilous since these ions have a tendency to bio-accumulate in living organisms over time. The heavy metal ions cross the threshold of freshwater supplies through mining and industrial activities or from acid rain, which breaks down soils and liberates heavy metal ions into surface water bodies, viz., streams, lakes, rivers, and of course, sometimes sub-surface water too. For these reasons, it is extremely urgent and highly essential to remove them with a view of making the water clean through the treatment of enclosing wastewater [42,43], especially by means of the alternative affordable user and eco-friendly geopolymer adsorbents. By and large, the following heavy metals are found associated with wastewater, making the job of its management very challenging.

3.1. Lead (Pb)

Lead is one of the heavy metals, and it causes industrial pollution. Having a chemical symbol (Pb) and valences of ions as Pb^{+2} and Pb^{+4}, it is one of the most widespread heavy metals found enclosed in industrial wastewaters, which is identified to be lethal for living beings [44]; hence, it is essential that its elimination from wastewaters is of prime significance since it is the most toxic of heavy metals answerable for so many diseases. The heavy metal contamination occurring from the industries has piloted to plentiful ecological and security crises. For instance, the accumulation of Pb^{+2} ions can harm different organs of human beings, viz., the kidneys, central nervous system, heart and the immune system; it can cause cancer and even death, besides causing physical developmental disorders in innocent children [45–50]. Hence, stringent standards are made for allowed Pb^{+2} concentrations in drinking and surface waters. For illustration, the World Health Organization (WHO), the European Union (EU) and India have narrowed the highest level of Pb^{+2} in drinking water down to 0.01 mg/L [51]. Consequently, significant notice is paid to the taking away of Pb^{+2} from the wastewater urgently through methods, namely, adsorption, ion exchange, membrane separation and precipitation [52–54]. However, adsorption is found with plenty of competitive advantages amongst all listed techniques, such as its simplicity, low price, and effectiveness, along with ease to design, regenerate and operate [55]. A variety of adsorbents, such as geopolymers, were investigated for the exclusion of Pb^{+2} [56]. For the most part, the studies on removing Pb^{+2} ions [57–59] are concentrated on the static removal mechanism. Nevertheless, the dynamic removal is found more in agreement with the practical usefulness [60]. Praiseworthily, geopolymers extend affordable, renewable and eco-benevolent widely-accepted adsorbents.

3.2. *Cobalt (Co)*

Cobalt (Co) is a chemical element having its ions as Co^{+2} and Co^{+3}. It is an imperative trace element, and it is being extensively employed in plenty of industries as a part of diverse anthropogenic actions, such as mining, colorants in glass, paints, ceramics, discarded batteries, alloys manufacturing, petro-chemical, dye, metal industries, etc. [61]. By and large, it enters the environment from the eruptions of volcanoes, wildfires in forests, and overflow or leaching of the rainwater through natural rocks enclosing cobalt. It is a key metal ingredient of vitamin B12, which is essential for maintaining human health and in the treatment of anemia by helping to produce red blood cells, but it can be hazardous if consumed in extreme amounts. The permissible limits for cobalt to be present in drinking water, internal surface water, public sewers, irrigation water and industry are recommended as 0.01, 0.05, 0.5, 1.0 and 0.2 mg/L, respectively [62,63]. Beyond these limits, cobalt can cause a few discomforts, such as diarrhea, skin degeneration, vomiting, pneumonia, skin rashes, allergies, bone defects and impacts on organs, such as lungs resulting in asthma, etc. [64], and weight loss, as well as passing it on from mother to fetus during pregnancy and through milk during the period of breastfeeding. Therefore, the recommended standard level for cobalt proportion in drinking water, which is 2 μg/L, must be maintained [65]. In recent years, owing to these negative impacts of cobalt, a few endeavors to get rid of cobalt contamination from water and industrial wastewater have amplified with the help of geopolymer adsorbents in order to make them free from this hazardous pollutant. Therefore, it is necessary to remove Co^{+2} ions from the water and wastewater using geopolymers, kaolinite, activated carbon, palygorskite and sepiolite [63,66,67].

3.3. *Chromium (Cr)*

Chromium (Cr^{+2}, Cr^{+3}, Cr^{+6}) ions are considered as heavy metal ions discharged from natural or industrial sources, such as tanneries, electroplating and other related chromium-generating industries [68]. In the environment, the chromium occurs primarily in two valence states, i.e., trivalent chromium (Cr^{+3}) and hexavalent chromium (Cr^{+6}), which have been spotted in wastewaters. However, Cr^{+6} is considered to be as much more risky and noxious than Cr^{+3} since it can damage human skin and internal organs directly. For that reason, the research studies have been concentrated on the mitigation of Cr^{+6} levels through cutback to the lesser venomous Cr^{+3}. Nevertheless, Cr^{+3} is effortlessly re-oxidized to the perilous Cr^{+6} on its exposure to the oxidative environment [69], and so, the Cr^{+3} should also be eliminated.

3.4. *Copper (Cu)*

Copper is a chemical element with a symbol (Cu). It is an extensively utilized metal in copious industries. To name a few, leather tanning, mining, electroplating, brass manufacturing, petroleum refining (i.e., copper sweetening), coating, smelting [70]. Copper (Cu^{+2}) ion is the chief heavy metal cation found in waste effluents generated from the electroplating industry [71]. Among the ions of Cu, the cupric (Cu^{+2}) ion is commonly found in water as a more widespread oxidation state, forming complexes with carbonate ($CO_3{}^{-2}$) and hydroxide (OH^{-1}) ions. The development of insoluble mineral Malachite with the chemical composition [$Cu_2(OH)_2$ CO_3] in water is a key domineering feature of the level of free cupric (Cu^{+2}) ions in an aqueous solution. Accordingly, drinking water gets contaminated with salts of Cu^{+2}, negatively influencing living organisms with short and long-term impacts [72–74]. An excessive intake of Cu^{+2} ions pilots to accumulation in the human liver and can cause other solemn human health crises, such as mucosal irritation, damage to the renal or central nervous system [75], damage to the liver [76], since it is highly toxic to living beings and organisms even at a very low concentration. Therefore, taking into account the incessant eco-damage and the necessary need to remove these hazardous heavy metal ions from the water, as well as wastewater, the water and wastewater must be made clear by freeing them from these pollutants before their release into the environment and/or water bodies. It is recommended that the upper limit for the concentration of Cu^{+2} must not go beyond 2.0 mg/L in drinking water [77]. For this reason, different methods are being used for its removal from water and wastewaters; however, adsorption by means of geopolymers is found most suitable and effective. Alshaaer et al. [78] investigated the adsorption behavior of zeolitic tuff-metakaolin geopolymers for Cu^{+2} ions elimination and monitored the highest adsorption competence of 7.8 mg/g of adsorbent at an early zeolitic tuff:metakaolin ratio as 0.5.

3.5. Zinc (Zn)

Zinc (Zn) is also a metal, and its ions (Zn^{+2}) are generated by industries of galvanizing, pigments, paints, etc., pollute water streams. Even if zinc (Zn) is valuable for a lot of constructive biological functions, the exposure to a higher level of Zn has been confirmed accountable for causing detrimental influences on human health in the form of muscular stiffness, gastro-intestinal distress, irritation, growth retardation, cancer, lung disorders, etc. [79]. Consequently, the treatment of both water and polluted waters has turned out to be a pressing need that must be addressed. The methods, such as absorption through cost-effective geopolymers, are becoming more and more popular because of their competence, ease and eco-friendliness.

3.6. Manganese (Mn)

Manganese (Mn) is a vital nutrient for both—humans and animals. however, the chronic exposure of this heavy metal to higher doses is injurious to health. It may be generated from an assortment of sources and decreasing quality. Non-biodegradable manganese can settle in living organisms, utilizing contaminated water enclosing Mn, which causes diverse diseases to take place [80]. In accordance with the World Health Organization (WHO, Geneva, Switzerland), the highest admissible Mn concentration in drinking water is considered to be 0.05 mg/L. The surplus Mn found in water is causing issues, such as Parkinson's disease and bronchitis, as well as damaging respiratory and nervous systems. To address these predicaments, different adsorbents for Mn^{+2} ions have been examined, including volcanic ashes, fly ash, zeolite fruit nutshell, kaolinite, glycine modified chitosan, vermiculite [66,80–84].

3.7. Nickel (Ni)

Nickel is a metal with a chemical symbol of (Ni). Heavy metal nickel ions (Ni^{+2}) are responsible for polluting water and wastewater and hence should be removed from them. Ge et al. [85] accounted that the metakaolin-geopolymer membrane can powerfully remove the Ni^{+2} ions from wastewater and predicted that such geopolymeric membranes can potentially find their way to be utilized for the removal of Ni^{+2}, as well as other heavy metal ions from industrial wastewater.

4. Applications of Geopolymers for Water and Wastewater Treatment

Nowadays, geopolymers are regarded as versatile materials, which have drawn attention in scores of environmental applications, and hence, are being employed nearly in all fields of technology, for instance, in water and wastewater treatment as adsorbents or ion-exchangers, high-pressure membranes and filtration media, photo-catalysts, anti-microbial materials, pH buffers, carrier media in bioreactors and for the solidification or stabilization of water and wastewater treatment residues. The applications are not merely those; they can be employed as anti-bacterial materials following their incorporation with copper (Cu) or silver (Ag) in the course of ion exchange or supplementation of nanoparticles, namely, silver-silica nano-composites into the geopolymeric alumina-silicate matrix [86–88]. Thus, they are useful as anti-microbial binders with analogously modified plentiful zeolites to designate their potential usage for the disinfection of wastewaters or sub-surface water [89–91].

Since the discovery of geopolymers four decades ago, they have come quite a long way successfully [92] and gained attention, mostly due to the effortlessness for geo-synthesis with lesser emission of greenhouse gases (GHG) [93]. Their admirable properties, such as toughness, heat and fire resistance, refractory nature and radiation hardness, altogether make them promising for applications for radioactive waste containment, as well as for pozzolanic action, making them multi-functional. In realism, lots of industrial materials are derived using geopolymers. For example, fiber-based geopolymer composites are regarded as fire-resistant. Quite a lot of geopolymeric composites are positioned in metal tool coatings and the construction of cabinets of airplanes and buildings, with a view to trim down the intensity of inferno incidents [94]. The polymeric chain-like structure of geopolymers contributes higher chemical resistance, lower shrinkage and enhanced resistance to abrasion with high early mechanical strength [95].

At present, geopolymers are an emerging class of materials for the eco-benevolent refurbishment of decrepit infrastructures, recovery of marsh environments and sustainable reinforcement of structural amenities. The highly workable properties of geopolymers, such as water retention aptitude, when owning shear stress of ~80 Pa at a shear rate of 110 s1 and compressive strengths of ~40 MPa at 7 days curing, facilitate the tailoring of them. Furthermore, they possess attributes that make geopolymers a significant class of promising materials for the containment and capping of nuclear wastes perils [96]. Moreover, the development of geopolymer sorbents necessitates considerations for how to employ them in general practice. It is feasible to dose powdered sorbent constantly; however, then an added process phase is needed to separate the utilized sorbent, and operational costs may go higher. As a result, a choice of alternative methods is accessible to manufacture highly porous and permeable geopolymers, e.g., direct foaming, sacrificial template method, freeze casting, granulation or additive manufacturing (AM), i.e., 3D printing.

4.1. Geopolymers as Adsorbents or Ion-Exchangers

The key drivers for this emerging approach for the application of geopolymers as adsorbents or ion exchangers are their outstanding mechanical and chemical stability coupled with a fairly uncomplicated, lower carbon, low-energy, user-friendly and cost-effective manufacturing process of geopolymerization [97]. There exist numerous efficient and familiar wastewater treatments to get rid of heavy metal cations present in it, including adsorption [98–100], ion-exchange, photo-catalytic degradation, i.e., as photo-catalysts, membrane filtration (or separation) materials, chemical precipitation, bio-remediation sedimentation, pH adjustment agents, the solidification or stabilization of water treatment residues reverse and forward osmosis [101–115]. Consequently, geopolymers offer viable alternatives to competing materials for water and wastewater treatment, such as conventional ceramics or synthetic zeolites, polymeric components in terms of cost, an assortment of choice, eco-impacts, and of course, performance.

In recent times, on account of the significant attributes, such as a simple method, low-cost, environ-friendly nature, higher competence and mild micro-spheres synthesis method, geopolymers are successfully developed and commonly used as adsorbents, inorganic membranes, catalysts and for the immobilization of dangerous metal ions. Adsorption is regarded as the most efficient, uncomplicated and widespread technique for water decontamination [116–120]. Undeniably, activated carbon is the standard adsorbent material and is well-known as the most utilized adsorbent for water and wastewater treatment with an excellent demonstration of higher adsorption competencies for the removal of heavy metal cations, but its use gets hindered due to its comparatively high cost [121]. Hence, in spite of the exceptional adsorption ability of this material, its higher production costing is the root cause of wide-ranging research for low-cost yet effective options in recent times. Significantly, the unique porous structure of geopolymers and the existence of negative charges on the aluminum-tetrahedra of the resulting gels, they can efficiently and successfully absorb metal ions, namely, Pb^{+2}, Cu^{+2}, Cd^{+2}, Mg^{+2}, Hg^{+2}, Cr^{+3}, etc. [122–127].

The performance of heavy metal adsorption by geopolymers relies mostly on the structures formulated by the geopolymer gels; nevertheless, owing to the presence of numerous inert crystalline phases in them, the clear-cut correlations among the geopolymeric gels and performance of adsorption by them cannot be verified all the time. The earlier research works deal with the hydrochloric acid (HCl) dissolution to verify the degree of reaction of geopolymerization, and as inert crystalline phases are insoluble in acid, the gels and other dissolved matters can be measured based on the mass alterations. The adsorbents in powder form may necessitate the support materials use, such as porous ceramics and polymer foams, [127] to permit their industrial utilization or a separation step, such as pressure filtration after wastewater treatment course of action, both being harmful to the wastewater treatment costing besides the rising complexity of the procedures. The geopolymers can fruitfully be used as green adsorbents or ion exchangers by manufacturing them with wastes as precursors extending a systematic solution for their disposal too. Thus, the adsorption using these environ-benign adsorbents is a superb and very much progressive technique for the elimination of not merely inorganic but also organic contaminants from aqueous mediums, particularly through geopolymers, which can be produced employing solid wastes as precursors [128].

The ever-increasing diverse wastes from various fields, such as fly ash, metakaolin or low-calcium alumina-silicate solid wastes, accumulating as landfills are not only hazardous to the environmental sustainability but are also resource-wasting and polluting water, soil, etc., with lots of negative impacts on the health of living beings, and can advantageously be used as precursor resources. For instance, fly ash (FA) is being used as higher value-added applications, viz., the synthesis of FA-ZnO enclosing nanofibers for adsorption and photo-catalytic degradation of organic dyes, etc. [129]. Furthermore, the production of geopolymers enables benefiting industrial, agricultural, etc., wastes containing alumina and silica to be utilized as a resource for precursors, such as fly ash, clay, slag, raw kaolin, metakaolin and rice husks. Lots of research works used rice husk as a precursor, considering it as a silicon source useful for preparing a range of geopolymers making the costing comparatively low. Moreover, their uses for disinfection made ground, gaining the geopolymer technology substantial attention from not only researchers but also industries for concentrating on cleaner manufacturing. This is an additional advantage to offer systematic disposal of all such kinds of wastes that can be used to design a geopolymer useful for water and wastewater treatment and reclamation as well [130]. These compounds raise quite remarkable questions about their surface and structure, e.g., the almost total exclusion of diffusion of sequestered metal ions and avoidance of issues linked to leaching of geopolymer adsorbents [131].

As referred, geopolymers are famous for occupying inter-crossed linked bonds with cationic ends on the surface, structurally permitting the entrapment of toxic and radioactive metals through charge balancing. On the other hand, the micro-structure of geopolymer in the nano-metric scale of 5 to 10 nm frequently consists of numerous pores inside a highly porous –Al–O–Si– repetitive unit, and these voids are probable to provide room for ionic integration, replacement and balances too. Extraordinarily, geopolymers own a key structural unit analogous to zeolites; Provis et al. [132] unveiled that they possess an amorphous gel phase and nano-crystalline zeolites agglomerates. According to them, a proposal has been put forward that the geopolymers and zeolites could be inter-converted under suitable conditions. Thus, the conversion of geopolymers into zeolites is regarded as an exceptional strategy since the well-organized recycling of solid wastes from industries and the economical production of costly zeolites can be recognized concurrently. Beforehand, Cui et al. [133] authenticated using a high-resolution transmission electron microscope and particular area electron diffraction that numerous nanometer-ordered structures were detected in the amorphous geopolymer ($Al_2O_32SiO_2Na_2O7H_2O$), successfully acquiring the Na-A type molecular sieve from this geopolymer through a hydrothermal reaction. Afterward, a series of sodium (Na)-zeolites, including sodium (Na)-A, faujasite, sodium (Na)–P and sodium (Na)-X were attained via the hydrothermal alteration of geopolymers when sodium cations (Na^+) were present, which displayed huge potential for membrane separation and bulky adsorbent utilizations. It is eminent that the development and nucleation of zeolites rely upon the concentration of alkali metal cations and their kind [23,134–137].

The zeolites, which are derived from geopolymers, are acquired from a solution of Na^+ cations resulting in restraining their diversity. Hence, it is imperative to develop an extensive variety of zeolites from geopolymers for utilizations linking bulky adsorption and membrane separation. On the other hand, in a research study, Rossi blended a biomass fly ash as 75% with 25% geological natural metakaolin and used the mixture as a precursor for geo-synthesis through geopolymerization and achieved an end-product with a huge specific surface area of 56.35 m^2/g together with a brilliant pressure resistance of up to ~10 MPa. Furthermore, a noteworthy improvement in adsorption competence and compressive strength was monitored in a study whereby the supplement of cork waste residue was made into metakaolin-based geopolymers [138]. Moreover, the existence of dyes in wastewaters is considered a somber crisis for ecology [139]. Cation-enclosing dyes, such as methylene blue or methylthioninium chloride with a formula $C_{16}H_{18}ClN_3S$, is a salt that is cationic and used as a dye, mostly in the textile industry for cotton, wool and leather dyeing, paper and plastics, as well as furniture coloring. Methylene blue is very injurious to the health of living beings since it can cause blindness, allergy, abdominal disorders, asthma, etc. [110,139]. As a result, such dyes must be removed from aqueous solutions by means of ultra-filtration, photo-degradation or ion-exchange methods. Still, adsorption is one of the most efficient and affordable routes for water and wastewater treatments [2,140]. For the reason that it is a cationic dye, hence, positively charged; the application of geopolymers with negatively charged networks as adsorbents is the most promising solution in this context. Furthermore, the likelihood of the development of zeolites in the course of the geopolymerization of metakaolin can optimistically influence the procedures of adsorption.

The zeolite is portrayed by a higher selectivity headed for methylene blue [141]. The few modern research works are throwing light on the application of zeolite-type geopolymeric green nano-adsorbents considering their higher effectiveness, steadiness, trouble-free production system, exceptional binding characteristic and attractive lower costing [99,141–144]. They are found highly porous and can be manufactured at as low as room temperature through a simple process of geopolymerization among aluminum and silicon sources in an alkali medium [99,143]. Zeolites resembling geopolymers are appreciably useful for a variety of eco-friendly applications, viz., the taking away of heavy metals and photo-catalytic degradation of a few organic compounds, such as benzophenone, metronidazole, p-nitrophenol and eriochrome black T [145–150], along with some significant industrial uses, namely, diversion of methanol (CH_3OH) to olefins and phenol hydroxylation for water and wastewater treatment.

4.2. Thermo-Dynamics of Adsorption

The necessary data with regard to the nature and thermo-dynamic viability of the adsorption progression can be obtained with the help of the thermo-dynamic parameters, standard change of entropy (DS0), change of standard free energy (DG0) and change of standard enthalpy (DH0). In the context of geopolymer adsorbents, the accounted DHO values of adsorption are commonly found as positive [27,102,113], designating with respect to the adsorption competence that it escalates at elevated temperatures. Nevertheless, one differing upshot as DH0 10 kJ/mol was found reported by Li et al. [151] in the case of elimination of methylene by fly ash-based geopolymer. This is additionally supported by the fact that the values of DS0 are constantly positive, signifying that the system entropy enhances are subsequent to the adsorption course of action [113,151,152]. This can be comprehended with the separation of hydrated water molecules from the metal ion, i.e., inner-sphere complexes are formulated, and specific adsorption takes place previous to the attachment to the surface sites of geopolymers, which is regarded as an energy-requiring procedure [27,113,152].

4.3. Adsorption of Radioisotopes

Fantastically, geopolymer composites are employed to adsorb a number of elements that have radioisotopes, viz., caesium (Cs), strontium (Sr), radium (Ra) and cobalt (Co). The elimination of radioisotopes from water has turned out to be progressively more significant in the aftermath of the incident of the accident at Fukushima nuclear site, Japan, in 2011 [153]. In the aqueous environment, caesium, which is highly soluble and stable, occurs as Cs. The radioactive isotope 137Cs is of great concern because it has a long half-life of 30.2 years. On the other hand, strontium is found present chiefly as Sr^{+2} in the aqueous environment and possesses quite a lot of radioactive isotopes. Of which, 90Sr with a half-life of 28.9 years is the most noteworthy [154]. Usually, all radium (Ra) isotopes are found present as Ra^{+2} in low salinity conditions with half-lives ranging between a few days and 1600 years. The metakaolin-based geopolymer is potentially effective for the adsorption of the caesium (Cs), and therefore, preferable to lead (Pb^{+2}), copper (Cu^{+2}), cadmium (Cd^{+2}), nickel (Ni^{+2}) and zinc (Zn^{+2}). The adsorption competence is not influenced by the presence of sodium chloride (NaCl) up to 10 weight-% concentration. Chuang and Liao [155] adsorbed caesium (Cs^{+}) ions by phosphate-based geopolymer enclosing potassium (K^{+}), zinc (Zn^{+2}) cations and Ferrocyanide anion $[(Fe(CN)_6)]^{-4}$. Lee et al. [156] utilized porous geopolymeric blocks manufactured with fly ash and blast furnace slag with a view for eliminating Cs^{+}. Geopolymer composites are compared satisfactorily with so many other types of adsorbents for the removal of Cs^{+}, which show the highest adsorption competence [153]. Chen et al. [157] investigated the elimination of Cs^{+}, Sr^{+2} and Co^{+2} ions by metakaolin and fly ash-based geopolymer composites and achieved very high adsorption competencies. Moreover, it is possible to adsorb these ions with 0.1 M HCl, which suggests that the adsorbent could be regenerated [157]. Geopolymer composites using fly ash are reported to be potentially efficient for taking away cobalt (Co^{+2}). At this juncture, it is interesting to note that both the radioisotopes and non-radioactive isotopes exhibit analogous aqueous chemistry. In the end, radium isotopes can be significantly removed by foamed geopolymers where there is an incorporation of barium sulfate ($BaSO_4$), and the material is suggested to be employed as a passive filtration material. Thus, geopolymer composites can brilliantly remove different polluting ions efficiently.

4.4. Adsorption of Heavy Metal Ions

Regrettably, the perilous and lethal wastewater carrying with it a range of heavy metal ions has turned out to be a universal environmental apprehension attributing to not merely so many human and animal diseases through contaminated water but also fueling the issue of the freshwater scarcity on our planet [158–161]. The degradation of an organic molecule and heavy metal confiscation from wastewater is a chief topic of public attention because of its relevance to the environment and human health. Thus, the heavy metal contamination in wastewater is exceedingly creating a negative impact on the environment due to their toxicity and its direct drain in natural surface water bodies [39]. For these reasons, plenty of types of materials are being used as adsorbents to solve the purpose of removing heavy metals from water and wastewater; however, the complexity of their synthesis and the higher costing of their precursors has been compelled to limit their widespread use. Therefore, geopolymers are being used as an adsorbent for the removal of ions of different heavy metals, namely, chromium (Cr), calcium (Ca), cesium or caesium (Cs), magnesium (Mg), arsenic (As), nickel (Ni^{+2}), zinc (Zn), copper (Cu^{+2}), cadmium (Cd^{+2}), lead (Pb^{+2}), NH^{+4}, manganese (Mn^{+2}) and Co^{+2}, as well as methyl violet and methylene blue along with anions—phosphate, fluoride, and radionuclide of 137Cs and 90Sr, plus dyes, etc. [162–171]—from water and industrial wastewaters, which were found superbly suitable and valuable for adsorption, proving to be the most extensive and useful method.

Just recently, some endeavors to utilize both fly ash and metakaolin-based geopolymers as adsorbents for heavy metal ions and dye removal from wastewaters were undertaken. The fly ash-based geopolymers are employed to eliminate Cu^{+2} ions and Pb^{+2} with significant success [26,171]. In contrast, Cheng et al. [172] investigated a metakaolin-based geopolymer playing the role of an adsorbent for removing Pb^{+2}, Cu^{+2}, Cr^{+2} and Cd^{+2} from water and monitored the efficacy for Pb^{+2} subtraction in particular. López et al. [170] optimized the chemistry of geopolymers keeping the molar ratio of Si:Al as 2.0 for the choosy adsorption of Cs^{+} and Pb^{+2} from the solution mix of other heavy metal ions. Tang et al. [173] studied the fabricated porous metakaolin-geopolymer spheres having 54 m^2/g surface area, 15 nm pore size and 60% porosity to take out Cu^{+2}, Pb^{+2}, Ca^{+2} and other polluting ions from wastewater and projected that the porous geopolymer spheres are cost-effective, expedient and environ-benevolent adsorbent for metal ion removal.

On the other hand, Andrejkovicova et al. [174] examined the high performance of metal removal for the metakaolin-based geopolymer mix and clinoptilolite, which was used as a filler and accounted for the adsorption of metal cations on geopolymers that is found to be well fitted by the Langmuir model and 25% clinoptilolite adding on provided the optimum upshots for the subtraction of Pb^{+2}, Cd^{+2} and Zn^{+2}. Furthermore, Barbosa [163] made a comparison of methyl violet 10B dye adsorption behavior of the metakaolin-geopolymer and its permeable counterpart and found that the adsorption competence of the geopolymer was enhanced in porosity, with the greatest adsorption capability of mesoporous geopolymer reported as 276.9 mg/g.

Geopolymeric adsorbents can be employed as powders, granules and monoliths too; however, in the case of packed beds, the powdered adsorbents cannot be used in contrast to monoliths [167]. The remediation of contaminated water is of prime significance because of the escalating freshwater consumption of water, resulting in a grave shortage in most large parts of the world. To address this challenge, water and wastewater treatments, by means of geopolymers, have proven to be unique owing to their three-dimensional (3D) network structure, which marvelously makes geopolymers the most promising adsorbents [175,176]. Previous studies have revealed the immobilization of different heavy metals in geopolymers, for instance, Pb^{+2}, and shown that geopolymers are competent enough to absorb the heavy metals Pb^{+2}, Cu^{+2}, Cr^{+3}, Cd^{+2}, etc. [172,177]. The synthesis of a kaolin-based geopolymer was performed by Naghsh and Shams [126] in interest to review adsorption competence for Ca^{+2} and Mg^{+2}, and utilized "cetyl trimethylammonium bromide" to modify geopolymers and to enhance the adsorption ability for Cu^{+2}.

Noteworthily, Bentonite—a naturally accessible low-cost clay—possesses one-of-a-kind attributes, namely, high surface area, higher alumina/silica ratio, high amorphous content, chemical stability, elevated porosity and far above the ground capability of exchange of cations, making it valuable for dissimilar utilizations for geopolymers [178,179]. Both the perilous organic molecules and heavy metals are identified as noxious originators for the sickness of the common people [180]. First and foremost, they are discharged into the atmosphere through the actions of humans and/or industry [181]. For this reason, research on multi-functional materials that deal with concurrent energy manufacture and environmental remediation is related to modern-day technologies associated with cleaner production.

Presently, apart from geopolymer composite-based adsorbents, a range of other adsorbents are rising; however, most of them are found in powder form [182,183]. It is tricky to recover and regenerate these adsorbents, which may cause secondary heavy metal sludge contamination and is not conducive to constant function. When the powder is much finer, then it floats on the wastewater and is uncomfortable to recycle, and the pressure passed through the column is found to be much higher. The outsized particles are simple to post-process; however, the pressure passed through the column is much smaller, and hence, it may diminish the adsorption competence. The micro-spheric adsorbents are regarded as one of the most widespread adsorbents employed in the column for dynamic removal since they possess a fairly huge diffusion resistance and are trouble-free to disassemble and separate, as well as the pressure through the column being opposite [59].

4.5. Dyes

The subtraction of dyes from industrial wastewaters is of paramount significance, not merely because most of them are toxic and demonstrate carcinogenic attributes, but also for the reason that treated industrial wastewaters can play a role of an imperative source of clean water to provide relief to the most pressing need of freshwater demand from the society. The existence of trace quantities of dyes for <1 ppm in industrial wastewaters is tremendously evident, which is redundant [184]. One of the most universally employed dyes is methylene blue, a cationic dye, which is proven to cause blindness, abdominal disorders and respiratory distress [110,184]. The dyes and their breakdown yields are also venomous, mutagenic and carcinogenic and cause irritation of eyes, sore throat, skin, asthma and allergic contact dermatitis on exposure to organic dyes [185,186]. That is why its subtraction from wastewaters is highly obligatory. Quite a lot of techniques, such as photo-degradation, ion exchange and ultrafiltration, are utilized for methylene blue elimination; however, adsorption is still regarded as one of the most successful, simple and low-priced methods. Even though activated carbon represents elevated adsorption competence, its extensive application is confined due to the higher manufacturing cost [110,184–187]. Consequently, new-fangled and cost-effective options are practiced. One such alternative of geopolymers is emerging swiftly since they are inherently porous, having a negatively-charged alumina-silicate network balanced by cations of Na^+, K^+, etc., symptomatic of the practicability of being applied as adsorbents [188–192]. There are reports on their application as powders for dye removal and as granules, as well as monoliths for heavy metal pulling out from wastewaters [26,188]. The synthesis of fly ash-ZnO containing nanofibers is used not only for the adsorption but also for the photo-catalytic degradation of organic dyes. On the other hand, one significant cationic dye of methylene blue present in wastewaters can be absorbed by a number of mechanisms, such as electrostatic or ionic interactions, active binding by oxygen groups, photo-degradation and π–π conjugation [189]. One more imperative occurrence is the "adsorbate transport" from the bulk solution to the sorbent surface, which realizes in quite a lot of steps. Generally, the course of action of adsorption can be controlled by only one step, e.g., film or external diffusion, surface diffusion, pore diffusion and adsorption on the surface of the pore, or the grouping of more than one step [2,190]. However, the application of porous geopolymer monoliths for methylene blue (MB) extraction from wastewaters was also accounted for [2,190]. The MB elimination capability of the monolithic bodies figured 15.4 mg/g, while the adsorbent can be reused up to five times. Thus, the hopeful outcomes exhibited the practicability of the use of porous bodies, and not powders, to pull out MB from contaminated wastewater. Nonetheless, the exclusive competence of adsorbents is trimmed down considerably, to about 65%, when the MB early concentration touched 50 ppm. Consequently, more investigations for addressing the employment of bulk porous geopolymers should throw light on the most dominant parameters influencing the adsorption of dyes by the geopolymers. One option to improve the MB adsorption capacity of geopolymers, compared to the utilization of cylindrical discs with a thickness of 1/4 3 mm and d 1/4 22 mm, could be the application of porous geopolymer spheres (GS) with a size of 2 to 3 mm [2].

The adsorption of dyes using geopolymer composites as adsorbents is a significant topic, and hence, must be understood thoroughly. Quite a lot of geopolymers have been applied so far for the subtraction of the basic cationic dyes, such as methylene blue and crystal violet. Yousef et al. [191] prepared an effectual methylene blue adsorbent with a Langmuir capacity of 4.75 mg/g using sodium hydroxide (NaOH) for activating kaolin and employing zeolite as a filler. Furthermore, Li et al. [151] manufactured a fly ash-based geopolymer by means of a solid-state fusion technique with sodium hydroxide (NaOH) and obtained higher capacities, to some extent, of 18.3 and 17.2 mg/g for methylene blue and crystal violet, correspondingly. With a view of absorbing methylene blue, a competent geopolymer was manufactured, and it obtained an adsorption capacity of 50.7 mg/g experimentally. Furthermore, their adsorbent demonstrated enhanced porosity, and it is found floatable because of hydrogen peroxide (H_2O_2) supplementation. Geopolymers are exhibiting promising characteristics for the elimination of dye, namely, the aptitude to be regenerated [184]. Barbosa et al. [163] manufactured a geopolymer based on metakaolin, rice husk ash and soybean oil as a mesostructured directing agent and employed it for the adsorption of crystal violet and achieved a comparatively higher maximum capacity of 276.9 mg/g.

4.6. Adsorption of Ammonium, Sulfate, Fecal Coliforms and Phosphorous

Geopolymer composites are valuable for the adsorption of ammonium ions (NH_4^+), sulfate, fecal coliforms and phosphorous (P). Attractively, one precise, unique and interesting application of geopolymers is for the subtraction of ammonium (NH_4^+) ions from wastewater [162,192]. Ammonium (NH_4^+) ions are generally found in nitrogen species, such as untreated wastewater. The elimination of NH_4^+ ions from wastewater is investigated expansively by means of ion exchange on natural and synthetic zeolites as an option to the customarily employed microbial nitrification–denitrification [193–196]. With a view to prevent "eutrophication", the legislative requisites for removing nitrogen (N) from municipal and industrial wastewaters turn out to be progressively more common in the present time. Indeed, ammonium (NH^{+4}) is the nitrogen species that contributes the highest to the "eutrophication" of water bodies, whereby nitrogen (N) is the nutrient in poor supply [197].

The anaerobic digestion followed by natural zeolite-based ion exchange is potentially applicable for the recovery of nitrogen with a low operational cost and superior performance for nitrogen removal when compared to the traditional Anammox procedures or nitrification–denitrification [198]. The conventional methods to get rid of the nitrogen (N) are found based on the microbial nitrification–denitrification processes and technologies, which may become ineffectual at lower temperatures with high operational costing when compared to the process based on ion-exchanging [199]. Marvelously, the geopolymers have a very high affinity for NH_4^+ for the charge-balancing Na^+ cations in the alumina-silicate network of geopolymers, which may be exchanged to NH_4^+ at nearly 100% competence [200], provided that the charge-bearing sites are within reach. The ion exchange aptitudes of powdered metakaolin geopolymer for NH_4^+ is accounted as 18 to 32 mg/g [162], relying upon the examined water matrix, i.e., raw wastewater subsequent to the screening or following the advanced primary treatment, which is analogous to a lot of synthetic or natural zeolites. Furthermore, the granules of the geopolymer are investigated for ammonium (NH_4^+) exclusion on a bench-scale column experiment occurring in a plant for wastewater treatment: < 4 mg/L NH_4^+ is constantly reached at a water temperature of 10 °C; generally, 12 °C is the maximum limit for a well-organized operation of nitrification–denitrification processes, and the material can be reproduced with chloride of sodium (NaCl)/sodium hydroxide (NaOH) [192]. Appealingly, the elimination of ammonium (NH_4^+) ions from wastewater by means of exchange with sodium (Na^+) ions is one precise application of geopolymer sorbents prepared through three-dimensional (3D) printing, i.e., additive manufacturing [162]. Moreover, they have been examined as "adsorbent" or "ion exchanger" [27,102,103,105,110,164] materials for membrane agents for pH adjustment photo-catalyst, for solidification or stabilization of water treatment residues in water and wastewater treatment.

Geopolymers have a higher affinity for ammonium (NH_4^+) ions, the charge-balancing Na^+ cations in the geopolymer alumina-silicate network and may be exchanged with NH_4^+ ions up to nearly 100% competence provided the charge-bearing sites are available. Luukkonen et al. [162,165,192] utilized a metakaolin-based geopolymer composite and achieved 21.07 mg/g ammonium (NH_4^+) adsorption capacity in model solutions, which is greater than using typical natural zeolites. Furthermore, they reported that the elimination was based on ion exchange, and the material can be regenerated with a solution having 0.2 M sodium chloride (NaCl) and 0.1 M sodium hydroxide (NaOH), and the adsorbent was found effective for site-treatment when landfills initially leachate NH_4^+ at 55 mg/L at lower temperatures [165].

In one more research study by Luukkonen et al. [162], where they employed a central composite design method to optimize the metakaolin-based geopolymer composite manufacturing in order to maximize ammonium (NH_4^+) adsorption capacity. When higher quantities of silicate and hydroxide are employed, a low content of metakaolin and sodium (Na^+) ions as a charge-balancing cation in spite of potassium during the production resulted in the highest capacity. This type of optimization has likely augmented the highest adsorption capacity from 21.07 to 31.70 mg/g. Luukkonen et al. [192] also investigated a metakaolin geopolymer produced with the granulation-geopolymerization technique for employing it as a filter media. It is probable to achieve constantly less than 4 mg/L ammonium ion (NH_4^+) concentration in wastewater from municipal drainage with initial NH_4^+ and 23 mg/L at a lower temperature of more or less 10 °C and regenerate material multiple times with sodium chloride (NaCl)/sodium hydroxide (NaOH). Bai and Colombo [201] investigated metakaolin-based geopolymeric foams as monolithic porous filters manufactured by means of 3D printing, utilizing the direct ink writing approach. The filter is found capable of taking away up to 95.3% of the ammonium (NH_4^+) from the initial concentration of 3 mg/L of NH_4^+.

On the other hand, sulfate (SO_4^{-2}) is a naturally-occurring ever-present anion that is not regarded as toxic, even though it can be responsible for the salinization of surface water bodies. For that reason, the elimination of sulfate (SO_4^{-2}) should be managed in numerous industries of mining, such as drainage from acid mine and plants of desalination, wherein reverse osmosis reject water. Runtti et al. [113] tailored a blast furnace slag-based geopolymer through ion-exchanging charge-balancing sodium (Na^+) cations into barium and utilized the achieved material to get rid of sulfate from synthetic wastewater with (SO_4^{-2}) at 1000 mg/L, as well as mine effluents with (SO_4^{-2}) at 900 mg/L. The monitored adsorption capacity was comparatively higher, at around 119 mg/g, and it was likely to obtain very low sulfate concentrations of 2 mg/L. The projected removal mechanism was the precipitation of the tremendously lower solubility of barium sulfate ($BaSO_4$) or surface complexation [113]. Zhang et al. [202] displayed, by molecular dynamics simulation, that cations, viz., sodium (Na^+) or magnesium (Mg^{+2}), are adsorbed on the surface hydroxyls of the N–A–S–H gel and attract the ions of (SO_4^{-2}).

4.7. Surfactants

The name "surfactant" is derived from the mixture of words surface-active agent, meaning "active agents on the surface". Such compounds trim down the surface or interfacial tension among two fluids, or one fluid and gas, or a liquid and a solid active as wetting and foaming agents, detergents, dispersants and emulsifiers. They are one of the hottest up-and-coming contaminants contributing to boosting the pollution of global water resources. Basically, they are organic surface-active agents employed for a range of applications in the field of polymers, fibers, metals, paints, pharmaceuticals, paper and pulp, micro-electronics, cosmetics, foods, etc. [203]. The category of surfactants is being confirmed on the data of their charge on the head group. A total of three classes of surfactants, viz., non-ionic, ionic(anionic and cationic) and amphoteric (enclosing both the types of charges), are found. More or less 60% of the surfactants developed are from the anionic class, and roughly 12 M tons of surfactants are utilized internationally [204–206].

Globally, the consumption of an elevated quantity of anionic surfactants is assigned to their simple synthesis and good-quality cleaning, as well as the foaming attributes [207]. Alas, the negatively-charged surfactants are accountable for detrimental impacts to not only humans but also fish, vegetation, etc. For instance, they can be the reason for the eutrophication of effluent treatment plants and produce foam in freshwater-bodies on the earth. Notably, the concentration of surfactants was found in ranges from 1 to 10 mg/L in domestic wastewater, while in industrial wastewater, it reaches the level of 300 mg/L [208]. It is quite noteworthy at this stage that the standards of the World DOE permit for merely 1.0 mg/L of anionic surfactants in surface water bodies. Consequently, it is very much essential to get rid of these most modern pollutants in the form of anionic surfactants from industrial effluents before their discharge into freshwater resources on account of their higher level of usage, concentration limits, as well as negative impacts on environments and health of living beings. With a view to remove anionic surfactants from the wastewater, loads of techniques for its treatment are being used presently, such as chemical and electrochemical oxidation, ion-exchange and membrane separation, microbial treatment, foam separation, coagulation and some other special adsorption techniques [209–211]. However, a few of the mentioned techniques are restricted on the basis of efficacy, costing and eco-benevolence. Appreciably, adsorption is the leading and fitting process for eliminating surfactants owing to its practicability, cost and course convenience. Scores of adsorbents, namely, multi-walled carbon nanotubes (MWCNTs), geopolymers, activated carbon, chitosan, graphene, coal, zeolites and fly ash, are being used for the purpose of surfactants subtraction from wastewater [212–216]. Additionally, MWCNTs have a very interesting evolution of the dielectric characteristics in frequency and temperature variation due to the polarization mechanisms present. At frequencies in the MHz order, it appears that the polarization of orientation increases with the rise of temperature. The electrical conduction of this material increases with increasing frequency. If at low frequency this material has an insulating character, with increasing the frequency, it has a semi-conductor character [217].

Among all these, the fly ash-based geopolymer exhibited first-rate performance to address the removal predicament of an anionic surfactant. The surface area can be improved by changing the geopolymer chemistry. When surfactants are being employed, they stabilize the formulated foam, and the resultant materials contain extremely interconnected pores and permeability akin to granular filters [218]. Furthermore, they can be prepared on-site by supplementing vegetable oils, i.e., triacylglycerols, to the alkaline geopolymer suspension resulting in the saponification reaction, meaning decomposition into glycerol and soaps due to higher pH [219]. Significantly, the albumin powder obtained from the chicken egg is examined for use as a surfactant for foaming geopolymers with hydrogen peroxide (H_2O_2) [201].

5. Geopolymers as Membranes and Filters

The application of monolithic bodies might permit the direct use of geopolymers in packed beds as membranes that would appreciably make the procedure simpler. The production of fly ash-based geopolymeric micro-filtration membranes has drawn increasing interest [220]. Furthermore, zeolites are being classed as inorganic alumina-silicates, which typically enclose a number of channels smaller than 2 nm in size and zeolite membranes. These materials are receiving escalating interest for an assortment of applications, such as catalyst reactors and flow batteries [221]. Among the present accessible methods of eliminating heavy metals, membrane separation offers a distinctive benefit by treating the ions irrespective of their concentrations [222]. Specifically, the membrane can improve trace ions in a solution to a comparatively higher concentration for analytical chemistry utilizations [223,224], whereas for the ions with higher concentrations, the membrane can additionally enhance them for recovery purposes [225]. In reality, before the flame atomic absorption spectrometric determinations, several investigations employed membranes for the pre-concentration of trace ions in environmental specimens. A lot of industrial areas brought them into play for practical utilization for their accurateness and operability. Although studies on the enhancement of ions with comparatively higher concentrations for advanced higher value-added application are somewhat inadequate, as such investigations have a key focus on immediately sought-after ions, such as lithium (Li) and magnesium (Mg) [226]. Even though a few reports emphasize the recovery of membrane-separated heavy metal ions; they are focusing on ion concentration effectiveness. Conversely, some of the examinations have concentrated on the on-site application of heavy metal ions attached to the membrane [227–230]. Even so, loads of investigations have authenticated that geopolymers with concentrated mesoporous or macroporous structures can productively be produced, and a few researchers tested their membrane separation performance [85]. What is more, geopolymers show evidence of superior thermo-stability and resistance to corrosion that is expedient for advanced alteration subsequent to the separation when compared to conventional organic membranes [39].

5.1. *Pressure-Driven Membranes*

The pressure-driven membrane course of action is employed to either refine or concentrate the weak solutions or dispersals where the size of molecules or separated particles, as well as chemical attributes of the solvent, are crucial in making the choice for a suitable membrane. On the basis of the separation capacity, the pressure-driven membranes are categorized as micro-filtration, ultra-filtration, nano-filtration and reverse osmosis. Among these types, the geopolymeric membranes are characteristically found in the range of micro or ultra-filtration with pore sizes between 20 and 100 nm [85]. The intrinsic porosity of the geopolymer composites is valuable for membrane utilization. It means that its making procedures do not fundamentally necessitate any precise additives or foaming procedures.

Nevertheless, one of the early geopolymer membranes was made by employing the sacrificial filler technique, wherein the self-assembled strata from nano-size polystyrene spheres are prepared and then covered with a slurry made up of a mix of metakaolin/alkali silicate, after that, 200 MPa pressure is applied for 1 h and curing is executed at 50 °C for one full day, i.e., 24 h, and the polystyrene template is dissolved into a solvent [231]. Furthermore, Ge et al. [85] manufactured a geopolymeric membrane straightforwardly, blending the metakaolin with sodium silicate (Na_2SiO_3) and curing at 60 °C, whereby the total porosity membrane possessed was 62.64%, and water flux values at steady-state were roughly ranging between 21 and 237 kg/m^2 (9 h) at 0.1 MPa, relying upon the ratio of H_2O/Na_2O. What is more, the membrane can absorb Ni^{+2} ions with a capacity of 43.36 mg/g [85]. Xu et al. [111] prepared a similar membrane using metakaolin with an optimum ratio of H_2O/Na_2O as 18 with a water flux of 185 kg/m^2 (9 h) at 0.3 MPa when the membrane thickness was 5.0 mm, and there was a 100% rejection rate for nano-sized Al_2O_3-particles. Additionally, the geopolymer membranes having the potential to eliminate Ca^{+2} and Mg^{+2} are also found reported [232].

Geopolymer membranes based on blast furnace slag were manufactured by the hydraulic pressing technique wherein the optimum conditions for manufacturing were 400 bar pressure, 6 h drying time and 5 min pressing time, as well as 225 bar pressure, 8 h drying time and 30 min pressing time for chemical oxygen demand (COD) removal as 100% and permeation flux optimization at 1960 kg/m^2 (9 h), correspondingly [233,234]. One more technique for the application of geopolymer technology for membrane manufacturing is to first prepare an amorphous geopolymer material and alter it into crystalline zeolite membrane by means of a hydrothermal aging procedure, e.g., self-supporting membranes of Na-A zeolite manufactured with geopolymeric material by immersing them into water or a dilute solution of sodium hydroxide (NaOH) at 90 °C for 6 to 48 h [234]. Moreover, if the ratio of sodium (Na)/silica (Si) is boosted up to 1.25, the conversion from amorphous geopolymer to zeolite can take place through a plain heat curing at 60 °C devoid of hydrothermal treatment [234]. Furthermore, geopolymers can be cured as the faujasite—a type of zeolite—by alternatively supplementing oleic acid ($C_{18}H_{34}O_2$) and powder of aluminum (Al) to endorse the permeability [235]. Though the referred membranes are utilized for the water/ethanol separation, it is found displayed that the zeolite membranes also demonstrated the carry potential in water treatment utilizations [236].

5.2. Geopolymeric View on Filtration Media

Magnificently, geopolymer composites are not only valuable for the high pressure-driven membrane application but also useful for filtration media in sand filters, point-of-use water, permeable reactive barriers, treatment filters, etc. This simply means that besides simple physical filtration, geopolymer materials are competent enough to serve the purpose as catalytically-active media and/or as adsorptive. Their use as filtration media essentially necessitates pores in the range of micro or millimeters besides micro and mesoporosity with a view to possess adequately higher permeability. Their production techniques of exceedingly porous geopolymers categorized as direct foaming, replica method, additive manufacturing or 3D printing, sacrificial filler method and a few others, such as granulation, are also considered [97]. Among these, direct foaming is the most extensively employed method, which includes mechanical blending or blowing of agents to induce gas bubbles into the fresh geopolymeric paste [97].

Estimably, geopolymer foam is found competent enough to absorb 95% and 87% of ammonium (NH^+) and copper (Cu^{+2}) ions, in that order, from synthetic wastewater at the time when the initial concentration was kept as 3 mg/L for each of them [201]. The total and open porosities of these geopolymer foams are reported ranging from 67 to 88 and 60 to 85 vol%, correspondingly [201]. Landi et al. [237] manufactured geopolymeric filter material using metakaolin, potassium silicate ($K_2\ SiO_3$) and fumed silica with exceedingly interconnected macro-pores by means of elemental silicon as a blowing agent. The filter possessed $NH_4{}^+$ ion exchange capability, though it is concluded that merely a smaller segment of the internal volume pores was found accessible because of the cavity dimensions. Pervious geopolymer concrete made up of fly ash, nano-silica and ordinary Portland cement (OPC) could be employed for the elimination of fecal coliform and phosphorus from wastewater on account of a boost in pH and leaching of calcium, correspondingly [238]. The cautiously controlled size, shape and amounts of pores of geopolymeric materials can be developed through a novel technique of 3D printing or additive manufacturing [97].

The geopolymers are investigated as catalysts or catalyst supports for the degradation of a contaminant in both liquid and gaseous phase reactions [239]. Thus far, geopolymers are applied as photo-catalysts for the degradation of an assortment of dyes in the treatment process of water and wastewater. In this case, the reactive radicals are produced when the catalyst undergoes irradiation with the equivalent or greater energy than the bandgap. Furthermore, geopolymer catalysts can be utilized in some other sorts of advanced oxidation course of actions and for other recalcitrant contaminants of organic origin; however, such investigations are still undersupplied. Captivatingly, geopolymers are attention-grabbing materials as catalysts since they can be designed to have higher permeability, durable chemistry, surface area and, of course, accepted mechanical strength along with an uncomplicated, economical and reduced operational energy synthesis with a low carbon footprint. For the most part, the method of production of adsorbents or porous materials is being followed for the manufacturing of geopolymeric catalysts and catalyst supports, for instance, the mesoporous microstructure of geopolymer catalysts.

The metals that are active catalytically can be brought into the structure through blending them into the early-state geopolymer paste as salts or nanoparticles, or to the cured material using direct ion exchange or through firstly converting the geopolymer into ammonium ($NH_4{}^+$). The benefit of carrying out the Na^+ to $NH_4{}^+$ ion exchange stage first is that the succeeding exchange of ions by means of catalytically-active metal is more competent [240]. What is more, the modus operandi of wet and incipient wetness impregnation has been brought into play for the supplement of catalytically-active components to geopolymers. Additionally, a few raw materials from wastes already enclose catalytically-active metals with them, as found reported in a study by Zhang and Liu [241], whereby fly ash was a precursor for the production of geopolymers. Here, fly ash is found to contain an adequate quantity of oxide of metal semi-conductors, Fe_2O_3 as 4.78 weight% and TiO_2 as 0.94 weight%, to contribute to photo-catalytic activity to degrade the dye. The cadmium (Cd^{+2}) suspension-geopolymer spheres (CdS-GS) are used for photo-catalytic degradation of methyl orange by the suspension–solidification technique along with the photochemical synthesis practice wherein the CdS crystals development is encouraged by UV-irradiation [115]. The solutions with 5 mg/L concentration of methyl orange are constantly treated with 0.4 mL/minute at 25 °C in a quartz tube reactor packed with catalysts of CdS-GS under ultra-violet light. The methyl orange could be eliminated up to a total of 93%, i.e., 39% adsorption plus 54% photo-degradation, by means of a bed of 5 g of CdS-GS and a 2 h contact time. Likewise, the resistance against catalyst poisoning is examined via the repetition of five cycles: the methyl orange abatement was still beyond 90%, signifying that the CdS geopolymer-spheres had a comparatively stable photo-catalytic recital. The methylene blue, another dye, is photo-catalytically degraded by making the use of geopolymers based on fly ash or metakaolin [115,241].

Incorporation of the active component, TiO_2, with the metakaolin-based geopolymers by way of ion exchange is performed by Gasca-Tirado et al. [240]. The geopolymers treated through the ion exchange technique have exhibited larger surface areas than those with no active component. The greatest Ti-content is obtained with the geopolymer cured at 90 °C. Following 90 min of reaction, the color of the methylene blue was found to have vanished absolutely. Zhang and Liu [241] brought into play the fly ash-based geopolymeric photo-catalyst with no adding up of active metals using a contact time of six hours and successfully achieved a 93% abatement of methylene blue. Nevertheless, the adsorption was found accountable for 89% of the methylene blue subtraction. Fallah et al. [242] reported from their study that the majority of the methylene blue degradation is also on the grounds of adsorption of the dye onto the Cu_2O nanoparticle-blended metakaolin-based geopolymer with an inconsequential mechanism of UV-induced degradation.

Zhang et al. [243] manufactured a geopolymer composite employing bottom ash and graphene enclosing manganese (Mn^{+2}) ions and copper oxide (CuO) as active metals. The metals are induced through conducting the first ion exchange of Na^+ to $NH_4{}^+$ and immersing the material into the solutions of $Cu(NO_3)_2.3H_2O$ and $Mn(NO_3)_2$. The calcination at 400 °C for four hours is carried out to alter copper into oxide. The catalyst is examined for hydrogen (H_2) gas production, as well as the photo-catalytic degradation of aqueous direct sky blue 5B dye fruitfully.

5.3. Geopolymers as a pH Adjustment Agent

In the course of wastewater sludge bio-gasification, a significant plunge in pH can take place owing to the development of the acid through bacteria metabolism, posing a challenge for the upholding of a constant pH. It is known that the geopolymers are potential pH-buffering materials since they enclose free leachable alkalis in the pore solution [1,106,109,167,244]. Moreover, the fixed-film wastewater treatment courses of actions using bio-film reactors employ diverse kinds of floating carrier media to facilitate the adhesion and biofilm development, e.g., the carrier materials comprising polystyrene and light-weight expanded clay aggregate are already utilized. Silva et al. [245] manufactured geopolymeric bio-film carrier media of size 2–3 cm from thermally-treated tungsten-containing mine waste mud, sodium silicate (Na_2SiO_3) and sodium hydroxide (NaOH) at 800 °C, for 2 h. One of the mixtures exhibited good-quality potential in the context of pH, which should be inferior to 8 and should have stability when immersed in water. For this reason, the geopolymer-based bio-film carrier media is an economical option for wastewater treatment work.

6. Geopolymer Interaction Mechanism with Heavy Metals

Geopolymer removes heavy metals via an adsorption process in which heavy metal ions, or adsorbates, cling to the accessible binding surfaces of geopolymer. The adsorption isotherm is required to explore the nature of the interaction between geopolymer and heavy metal ions and to improve the application. In this scenario, the Langmuir and Freundlich models are typically used. Table 1 shows the adsorption of metals through a geopolymer and Table 2 represents adsorption of dyes in different types of geopolymers. Ge et al. [164] also showed that when MK-based porous geopolymer spheres were employed to remove Cu^{2+} the Langmuir model fit better than the Freundlich model. These data imply that Cu adsorption by FA-based standard geopolymers is monolayer. In the instance of Pb^{2+} removal, FA-based traditional geopolymers and found that the Langmuir model is more suited to represent the adsorption process than the Freundlich model [26]. As a result of the superior match of the Langmuir model seen in most studies, it is possible to conclude that heavy metal ion adsorption onto the geopolymer adsorbent is classified as monolayer. The interaction between Cu^{2+} and FA-based conventional geopolymer within a temperature range of 25–45°C using both the Langmuir and Freundlich models and found that both models had a good correlation coefficient, however the Langmuir model fit better [164]. The removal effectiveness of Ni^{2+} using LD-slag based traditional geopolymer and discovered that the Langmuir model suited better than the Freundlich model, implying that the adsorption is monolayer adsorption [152]. On the other hand, classified the adsorption process based on the kind of bond produced, which are physisorption and chemisorption [215]. The former is preferable because the weak Van der Waal force allows for easy renewal of geopolymer as an adsorbent by simple or steam washing, chemical or thermal treatment. It has been observed that after 6–10 consecutive times of use, the adsorption capacity of geopolymer was lowered by roughly 1–10%. The latter, on the other hand, was observed to occur in the majority of the investigations, which is related to the fitting of the Langmuir isotherm. Fourier-transform infrared spectroscopy (FTIR) was used to understand the adsorption mechanism at the microscopic level, despite the fact that very little literature was available. Typically, wavenumbers ranging from 400 or 450 cm^{-1} to 4000 cm^{-1} were used. Furthermore, no significant change in the absorption band of Al–O–Si bending vibrations and Si–O–Si bending vibrations was seen following heavy metal adsorption [169]. However, following Mn^{2+} and Cu^{2+} adsorption, the adsorption band of H–O–H bending vibration at 1647 cm^{-1} was relocated to 1637 cm^{-1}, and the peak at 1450 cm^{-1} was definitely shifted to 1431 cm^{-1} after Cu^{2+} adsorption. These spectral changes suggested that when Mn^{2+} and Cu^{2+} were attached to MK-based conventional geopolymer, a complexion with –OH groups occurred. When Ni^{2+} was adherent to LD-slag based conventional geopolymer, the peak of the spectra at 963 cm^{-1} got broader, and postulated that this was due to the formation of a metal ion layer surrounding the LD-slag based conventional geopolymer matrix. Another tiny signal was found at 1447 cm^{-1}, showing that heavy metal ions may be chemically bound to the LD-slag-based conventional geopolymer matrix [152].

Table 1. Adsorption of metals through a geopolymer.

Source Material for Geopolymer	Adsorbate	Alkaline Activator	Adsorption Capacity (mg/g)	References
Metakaolin, Rice Husk Ash	Crystal Violet	KOH	276.9	[163]
Fly Ash	Cd^{2+}	NaOH, Na_2SiO_3	9.02	[246]
Pyrophyllite	Cd^{2+}	NaOH	7.82	[247]
Pyrophyllite	Co^{2+}	NaOH	7.1	[247]
Metakaolin	Co^{2+}	NaOH, Na_2SiO_3	69.23	[169]
Metakaolin	Mn^{2+}	NaOH, Na_2SiO_3	72.34	[169]
Pyrophyllite	Ni^{2+}	NaOH	7.28	[247]

Table 1. *Cont.*

Source Material for Geopolymer	Adsorbate	Alkaline Activator	Adsorption Capacity (mg/g)	References
Pyrophyllite	Pb^{2+}	NaOH	7.54	[247]
Ld Slag	Zn^{2+}	NaOH, Na_2SiO_3	86	[248]
Fly Ash, Blast Furnace Slag	Cs^+	NaOH,	15.24	[156]
Metakaolin	Cu^{2+}	KOH, silica fume	40	[249]
Metakaolin	Cu^{2+}	NaOH, Na_2SiO_3	62.5	[250]
Metakaolin	Ni^{2+}	NaOH, Na_2SiO_3	42.61	[169]
Blast Furnace Slag	$SO_4{}^{2-}$	NaOH, Na_2SiO_3	119	[113]
Metakaolin	Cd^{2+}	NaOH, Na_2SiO_3	98.10	[174]
Metakaolin, Clinoptilolite	Cr^{3+}	NaOH, Na_2SiO_3	21.84	[174]
Fly Ash, Iron Ore Tailings	Cu^{2+}	NaOH, Na_2SiO_3	113.41	[103]
Metakaolin	Cu^{2+}	NaOH, Na_2SiO_3	44.73	[174]
Metakaolin, Clinoptilolite	Pb^{2+}	NaOH, Na_2SiO_3	261.22	[174]
Volcanic Tuff	Zn^{2+}	NaOH	14.83	[251]
Metakaolin, Clinoptilolite	Zn^{2+}	NaOH, Na_2SiO_3	35.88	[174]
Metakaolin, Al_2O_3	Methylene Blue	H_3PO_4	4.26	[110]
Metakaolin	Ca^{2+}	NaOH	24	[173]
Fly Ash	Cu^{2+}	NaOH	152.3	[27]
Metakaolin	Cu^{2+}	NaOH	34.5	[173]
Metakaolin	Pb^{2+}	NaOH	45.1	[173]
Metakaolin	Cu^{2+}	NaOH, Na_2SiO_3	52.63	[164]
Fly Ash	Methylene Blue	NaOH, Na_2SiO_3	50.7	[188]
Metakaolin	$NH_4{}^+$	NaOH, Na_2SiO_3	21.07	[165]
Blast Furnace Slag	As(III)	NaOH, Na_2SiO_3	0.52	[107]
Fly Ash	Co^{2+}	NaOH, Na_2SiO_3	66	[105]
Fly Ash	Co^{2+}	NaOH, Na_2SiO_3	59	[252]
Fly Ash	Co^{2+}	NaOH, Na_2SiO_3	52	[252]
Fly Ash	Cu^{2+}	NaOH, Na_2SiO_3	77	[105]
Blast Furnace Slag	Ni^{2+}	NaOH, Na_2SiO_3	4.42	[107]
Fly Ash	Pb^{2+}	NaOH, Na_2SiO_3	118.6	[102]
Fly Ash	Pb^{2+}	NaOH, Na_2SiO_3	6.34	[1]
Blast Furnace Slag	Sb(III)	NaOH, Na_2SiO_3	0.34	[165]
Metakaolin	$NH_4{}^+$	NaOH, Na_2SiO_3	32	[192]

Table 2. Adsorption of dyes in different types of geopolymers.

Geopolymer as a Adsorbent	Dye	Adsorption Capacity (mg/g)	Efficiency Degradation (%)	References
TiO_2 geopolymer composite	MB	20.11	97	[253]
Phosphoric acid-based geopolymer	MB	3.01		[110]
Fly ash geopolymer	MB	37.04	–	[254]
Metakaolin-based geopolymer	MB	43.48	-	[255]
Magnetic geopolymer	AR97	1814.27		[256]
Geopolymer	CR	–	100	[257]
Fly ash-based geopolymer	BY	36.364		[258]

Table 2. *Cont.*

Geopolymer as a Adsorbent	Dye	Adsorption Capacity (mg/g)	Efficiency Degradation (%)	References
Magnetic geopolymer	AG	183.17		[259]
Metakaolin-based geopolymer	MV10B	276.9		[163]
Metakaolin geopolymer	MO	0.333	–	[260]
Magnetic geopolymer	PR	39.21		[259]
Alkali-activated phosphorous slag	BV	46.58		[261]
Fly ash geopolymer	MB	–	92.79	[8]
Alkali-activated phosphorous slag	MGO	46.36		[261]
Magnetic geopolymer	AG16	400		[262]
Geopolymer	MV	–	91.16	[263]

Methylene blue—MB, Basic yellow 2—BY, Acid green—AG, Procion red—PR, Basic violet—BV, Malachite green oxalate—MGO, Acid green 16—AG16, Methyl violet 10B—MV10B, Acid red 97—AR97, Methyl orange—MO, Crystal violet—CV, Congo red—CR, Methyl violet—MV.

7. Pros and Cons of the Main Conventional Methods Used for the Treatment of Polluted Industrial Wastewater

7.1. Electrochemical Coagulation

Pros

Electrocoagulation is a simple procedure. Because it has few moving components, it may be remotely monitored while requiring less monitoring and maintenance. If necessary, the method may usually be altered to accept varying numbers of particles with little effort. In addition, the EC procedure may target several pollutants with a single system and, in certain situations, a single treatment pass. Because it does not include usual chemical additives, it creates lower amounts of sludge that are normally non-hazardous, quickly dewatered, and less expensive to process and dispose of.

Cons

An EC system may require the addition of acids or bases to change the pH, hence it is not totally additive-free. Furthermore, because of the nature of the process, the electrodes are sacrificial and will corrode with time, necessitating their replacement. It can clean plates using a clean-in-place (CIP) procedure that includes acid in the cleaning cycle. The nature of the procedure necessitates the use of electricity as well. While it may not need much at once, power may be more expensive in various parts of the world, which might boost running costs.

7.2. Chemical Coagulation

Pros

The fundamental reason for using chemical coagulation is that it shortens the time it would take for the solids to settle on their own. As a result, the overall detention time of the wastewater treatment process is reduced. Chemical coagulation can also help finer colloidal particles and mineral impurities settle. These particles may not settle during the sedimentation process and instead travel through a filtering system.

Cons

Chemical coagulation is fundamentally an additive process. It can lower the number of solids in a solution, but it still requires the addition of chemicals to do so. Adding these ingredients can be difficult and time-consuming, necessitating considerable jar testing. The doses must be quite precise in order to process the influent optimally. Dosage may need to be adjusted on a constant basis because to the fluctuating composition of the wastewater source. The addition of chemicals also leads in the development of a considerable amount of sludge, which must be treated and disposed of after treatment. Because of the nature of the elements included, this sludge is also dangerous. Because sludge cannot be easily dewatered, its bulk and toxicity can push up disposal prices.

7.3. UV Disinfection

Pros

Because UV disinfection is a completely physical process, there are no toxic chemicals to deal with. There are no potentially dangerous leftover byproducts in the treated water. It is extremely efficient against the majority of viruses, bacteria, spores, and cysts and requires less contact time than other tertiary wastewater treatment procedures. Furthermore, it has a small footprint for its disinfecting capabilities.

Cons

High quantities of total suspended solids (TSS) can render light inefficient for decontaminating a solution. If the preceding treatment technique is effective in removing TSS, this is a non-issue. Low UV light dosages may be ineffective against certain viruses, spores, and cysts, necessitating longer contact durations or higher-intensity exposure. There is also the possibility of photoreactivation in microorganisms, in which the organisms repair themselves after treatment if the UV dosage is insufficient.

7.4. Chlorine Disinfection

Pros

Chlorine is quite cheap and widely available. Furthermore, because it is such a strong oxidizing agent, it may be highly successful at turning huge numbers of hazardous germs inert with a sufficient reaction time.

Cons

Chlorine is very flammable and can produce disinfection byproducts (DBPs) that are hazardous to humans, animals, and aquatic life. It must be handled with care in order to be delivered, stored, and utilized securely. Chlorine disinfection has no effect on viruses, Giardia lamblia, or Cryptosporidium.

8. Challenges and Future Perspectives

Despite the fact that several studies have shown that geopolymer can be a possible alternative adsorbent to the existing commercial absorbent, there are certain limitations that need to be addressed further. The first source of concern is a scarcity of raw resources [264–267]. A number of aluminosilicate materials and industrial wastes can be utilized to make geopolymers [268]. In this regard, the physiochemistry features of those materials often vary, which makes ensuring the quality of geopolymer generated problematic [269–274]. Aside from that, the strength of the geopolymer is a source of worry. It is critical for geopolymer adsorbents, particularly those in spherical and monolith forms, to be strong enough to keep their shape when used. The compressive strength of geopolymer changes with the addition of heavy metals, and these variations are difficult to anticipate. Use of chromium oxide or an element of chromium as a filler resulted in a more compact matrix, which has a significant impact on the compressive strength of geopolymer. Furthermore, adding PbO boosted compressive strength whereas adding $PbSO_4$ and PbS compounds lowered compressive strength. The following are some drawbacks of the geopolymer production method—Alkali solution readily rusted vessels; there is no single standard for the beginning material for geopolymer production, limiting the larger scale application of geopolymer in industry; and; the complexity of alkaline activator preparation in real conditions. Furthermore, the majority of the current research was conducted on manufactured wastewater that only included a single type of heavy metal ion, which could not accurately represent actual wastewater containing many types of contaminants. Utilizing actual wastewater and discovered a decrease in removal efficiency when compared to the experiment using generated wastewater. This might be attributable to competition between heavy metal ions and other cationic species present in the solution. As a result, more research on real-world wastewater should be conducted in the future to gain a better understanding of the competition between different types of contaminants on the removal effectiveness of geopolymer. As a result, more research into the ability of geopolymers to remove additional types of heavy metal ions and pollutants should be conducted. It is also advised that significant research be conducted on the interaction mechanism of geopolymer with heavy metal ions, particularly at the molecular level, to offer a better understanding and help in the construction of a higher performance geopolymer adsorbent. These are some of the major problems that are limiting larger-scale study and implementation of geopolymer adsorbents. As a result, these characteristics should be thoroughly investigated in order to expand the body of knowledge of geopolymer adsorbents for more stable and promising applications.

Geopolymers have a high potential utility in environmental remediation technology. Its high flexural strength, low carbon footprint, quick curing time and solidification, and strong fire and toxin resistance open up new avenues for researchers. When developed properly, it has the potential to outperform activated carbon and zeolites in the adsorption of hazardous chemicals in both air and water contaminated environments. This is owing to the high porosity of the material. The adsorption effectiveness of geopolymer is determined by the raw ingredients employed in its manufacture. Waste water may be readily transformed to clean water by using geopolymer as a purifier with the right precursors and synthetic techniques. Furthermore, geopolymers can serve as effective photocatalyst supports owing to their unique composition, which yields a variety of metal oxides. However, further research is needed to develop scalable systems that use geopolymers as adsorbents and photocatalyst supports. Surface nano structuring, coating, and composites synthesis are projected to be important undertakings in the near future toward this objective. In all of these scenarios, synthesis and process controls will be important to optimize. Geopolymers are typically found to be highly successful in the removal of cationic chemicals, however it is critical to assess the removal efficiency towards anionic contaminants prevalent in waste water systems. Different raw materials-based geopolymers with different activators should also be tested on nuclear wastes with multicomponent systems in continuous operation to assess the sequestration mechanism of low and intermediate level toxic nuclear materials, as well as their selectivity based on electrostatic/ionic bonding with geopolymer charged sites. Geopolymeric materials can be recommended to develop a socio-economic assessment framework including socio-environment, climate smart, and resilience aspects of global sustainability and innovative engineering for the progressive development of civil infrastructure and society based on the concept of sustainable geotechnical engineering. Previous authors substantiated the sustainable concept of geopolymers by calculating the global carbon budget, oligomer synthesis life cycle assessment, and total energy need for the synthesis cycle. Several research have stated that the geopolymer's low thermal conductivity and greater durability may be adjusted based on the low operational cost/total energy budget and environmental impact evaluation. The chemical composition of the starting materials and other synthesis factors/designs such as temperature and/or humidity determine the reaction pathways and formation of cross-linked microstructure monomers for the rapid and sustainable polymerization of geopolymers, which are commonly used in civil and geotechnical engineering. This will assist various industries including the environment and ecology in operating under economically feasible and viable operational circumstances in order to maintain with little economic pressure in the new era of sustainability.

9. Discussion

In this day and age, on the one hand, water pollution has emerged as the most critical titanic global challenge, and on the other hand, the global paradigm is moving from a linear to circular economy is proving to be quite significant. The valorization of diverse and profound wastes of dissimilar origin is one of the pillars of this novel concept. The innovative geopolymer technology facilitates the application of numerous waste streams as either a precursor or as add-on materials, such as fly ash (FA) and glass waste [272–274], which reduces the consumption of natural mineral resources, such as metakaolin (MK), besides evident eco-pros. These assortments of wastes have raised great concerns vis-à-vis their management strategies. Most commonly, these solid wastes are dumped in open spaces, which may otherwise be valuable for other industrial or agricultural activities, converting them into landfills that are accountable for pollution of the environment, surface and subsurface waters, soils and more dangerously for health hazards. This is highly unsustainable in following the concept of the circular economy. That is why the incorporation of the geopolymeric-adsorbent production for water and wastewater treatment will surely reduce not only landfilling practices but also extend systematic waste management, helping to shift from a linear to a circular world economy. Table 3 display the geopolymers application in water and wastewater treatment.

Table 3. Geopolymers application in water and wastewater treatment.

Geopolymers Application in Water and Wastewater Treatment	Geopolymer Source Materials	References
Removal of Adsorption and organic pollutants	FA	[275]
Removal of Air particulate matter	MK	[276]
Antimicrobial and membrane filtration	MK	[277]
Desalinization by pervaporation	MK + HZ	[236]
	AS	[278]
Pervaporation	MK	[279]
	GGBFS	[135]
	AS	[276]
	L	[280]
Removal of Heavy metals	MK	[85]
	FA	[281]
	MK + F	[282]
Household wastewater treatment	FA	[283]
Method of Ion exchange	MK + F	[237]
	MK	[284]
Method of Oil separation	FA	[285]
	FA	[286]
	FA + B	[287]
	FA + Q + C	[288]
	GGBFS	[233]
Removal of Organic pollutants	MK + FS	[289]
Textile wastewater treatment	FA	[290]
Method of Green liquor treatment	MK	[291]
Method of Water desalination	MK	[232]
	MK + FS	[292]
	FA	[293]
	MK	[294]
Removal of turbidity	MK	[130]
Method of Oil separation from Removal of organic pollutants	MK	[295]

FA—Fly ash; MK- Metakaolin; HZ—Hydroxysodalite zeolite; AS—Aluminium-silica powder; GGBFS—Grand Granulated Blast furnace slag; L—Laterite; F—Fumed silica; B-Bauxite; Q—quartz; C—calcium carbonate.

As a fact, water is crucial for life, yet humans trash it anyway, which should be avoided and should purify it to make it reclaimed for reuse. Internationally, around 80% of wastewater is dumped back into water bodies without treatment. The burgeoning universal population urbanization, industrialization and laxity in the correct use of water resources have led to the challenging pollution of freshwater resources. In accordance with the WHO and the United Nations Children Fund, roughly 2.5 billion citizens do not have access to safe potable water to drink, which is a basic necessity to keep one's self alive next to oxygen to breathe. More or less, 245,000 km^2 of marine eco-systems are destroyed on account of the discharge of untreated wastewater into seas and oceans. A variety of organic and inorganic contaminants in nature are found mixed with water, and some of them are deeply persistent, noxious and causing loads of deadly diseases, viz., cancer, due to the presence of heavy metal ions. Therefore, getting access to clean water from the purification of wastewater is one of the best strategies to deal with the water crisis.

Until now, quite a lot of scientific investigations and approaches, such as conventional coagulation, ion exchange, adsorption, chemical precipitation, electrolysis, reverse osmosis, electro-dialysis, are being brought to play for their eliminations from wastewater in order to conserve it as a freshwater resource. However, it seems that the conventional methods are in short supply to cope with the existing gigantic dilemma of wastewater treatment. The novel geopolymer composites are found interesting, which are emerging as the adsorbent material of choice, not only because of their surface area criterion but also for their functioning, as well as chemical accessibility. Furthermore, geopolymer composites exhibit versatile and enhanced properties, such as corrosion resistance, high specific stiffness, thermal insulation, low density, specific strength and great absorbance quality for heavy metal ions. Furthermore, adding-on various waste materials for obtaining brand-new GP materials is possible.

10. Conclusions

The present review drives to conclude that "Clean water is a human right" and there must be alternative water resources for a step forward towards "Zero water crisis". With a view to having pure potable drinking water for all the human beings and lives on the planet, the approach of water and wastewater treatment is highly essential to hit the goal of the reuse of reclaimed water. The diverse methods to clean water by removing impurities of extremely perilous and noxious heavy metal ions are widespread for wastewater treatment; however, geopolymer technology seems to be highly appreciated and promising through its adsorbent composites. For that reason, significant enhancement and interest have been taken during the last decade in this regard. Revolutionarily, the geopolymer composites came forward and became known, gaining tremendous popularity as heavy-ion adsorbents on account of copious valuable attributes, especially of mechanical strength, porosity, durability, ion-exchange competence, etc., proving them superior to other competing materials.

Geopolymerization, the process of production of geopolymer composite adsorbents, is quite simple with low-energy and low-temperature requirements and significantly mitigated carbon footprints. Commercially, the shift of a linear economy into a circular one is possible with geopolymer composites due to their readiness to incorporate diverse wastes of dissimilar origin with them as precursors or industrial wastes bringing not only their systematic waste management into play but also heading lucratively towards cost-effectiveness. This cutting-edge technology has evidently revealed a promising potential for heavy-metal removal from water and wastewater with simultaneous recovery of different metals, rare earth elements, ammonium, sulfate, organic dyes, etc., boosting its profitability.

The research studies throwing light on their successful application as catalysts or catalyst supports related to the use of geopolymer composites point towards their great potential for photo-catalytic degradation of recalcitrant organic compounds as micro-pollutants. From the standpoint of techno-applications; the key magnetism of geopolymers arrives from the actuality that their production is readily scalable, and the process of production is rendered as "Green" which brings about enormous mitigation of wastes enabling cleaner production systems in the manufacturing industries. Nevertheless, it is noteworthy that the surface redox reactions, in general, and origins of photo-activity mostly stay open issues with respect to geopolymers. While geopolymer composites have already made it to numerous commercial applications, their fundamental surface chemistry and typical chemical structure will persistently offer opportunities to explore them further. Geopolymeric membranes exploit the built-in mesoporosity of the materials, making their application fit for micro or ultra-filtration. Their extensive use for water and wastewater treatment include lower pressure filtration media, water treatment residues, such as various sludge and spent adsorbents, pH buffering materials, anti-microbial materials, carrier media for biological treatment processes. Although a lot of successful work is found on the adsorption properties and performances of geopolymers in the reviewed utilizations that exhibits absolute potential, still more, advanced studies are essential on a large scale and for a long duration to establish the techno-scientific readiness of them fully in order to prove them totally promising materials to extend confidence to industries, engineers and field workers. Financially, the advantages need to recover valuable metals, nutrients, higher energy products from side streams, incorporation of wastes, etc., and will surely bring about a circular economy towards the next echelon. Interestingly, it is also possible to cover the costs of the wastewater treatment course of action from the revenue income of eliminated materials, instituting a prominent position for them.

The cheapest and newest method of geopolymer technology will promote the reuse of the treated effluent for a variety of purposes, such as agriculture. Geopolymers are found far above the ground affinity for NH_4^+ ions as the charge-balancing Na^+ cations in their alumina-silicate network. Feasibly, the porous geopolymer components produced by means of direct ink writing technique of 3D printing for the exclusion of NH_4^+ ions from wastewater are found to be very significant. The printed geopolymer adsorbents demonstrate excellent cation exchange capacity with NH_4^+ subtraction efficiency. The admirable capacity of 3D-printed geopolymers for brilliant ion exchange is assigned to their higher mechanical properties and elevated permeability. That means printed geopolymer composites are more efficient than synthetic zeolites and conventional ceramics because of their lesser energy-intensive production at a low temperature using locally profound wastes. The synergy between material and geopolymerization contributes to the effectual customization and tailoring of adsorbents with optimal porosity, mostly favoring permeability and optimum mechanical properties.

Author Contributions: Writing—original draft preparation and conceptualization, I.L. and S.L.; supervision and project administration, S.L. methodology and data curation, M.M.A.B.A.; methodology and formal analysis, R.A.R.; funding acquisition and visualization, P.V.; data curation and writing—review and editing, A.V.S.; data curation, visualization and resources, P.-D.M. All authors have read and agreed to the published version of the manuscript.

Funding: This research was supported by CNFIS Romania, Grant no CNFIS-FDI-2021-0354.

Institutional Review Board Statement: Not applicable.

Informed Consent Statement: Not applicable.

Data Availability Statement: Data is contained within the article.

Conflicts of Interest: The authors declare no conflict of interest.

References

1. Novais, R.M.; Buruberri, L.H.; Seabra, M.P.; Labrincha, J.A. Novel porous fly-ash containing geopolymer monoliths for lead adsorption from wastewaters. *J. Hazard. Mater.* **2016**, *318*, 631–640. [CrossRef] [PubMed]
2. Novais, R.M.; Ascensao, G.; Tobaldi, D.M.; Seabra, M.P.; Labrincha, J.A. Biomass fly ash geopolymer monoliths for effective methylene blue removal from wastewaters. *J. Clean. Prod.* **2018**, *171*, 783–794. [CrossRef]
3. Unuabonah, E.I.; Taubert, A. Clay–polymer nanocomposites (CPNs): Adsorbents of the future for water treatment. *Appl. Clay Sci.* **2014**, *99*, 83–92. [CrossRef]
4. Malandrinos, G.; Hadjiliadis, N. Cu(II)–histones interaction related to toxicity-carcinogenesis. *Coord. Chem. Rev.* **2014**, *262*, 55–71. [CrossRef]
5. Panhwar, A.H.; Kazi, T.G.; Afridi, H.I.; Arain, S.A.; Arain, M.S.; Brahaman, K.D.; Arain, S.S. Correlation of cadmium and aluminum in blood samples of kidney disorder patients with drinking water and tobacco smoking: Related health risk. *Environ. Geochem. Health* **2016**, *38*, 265–274. [CrossRef] [PubMed]
6. Abdulla, S.M.; Jamil, D.M.; Aziz, K.H.H. Investigation in heavy metal contents of drinking water and fish from darbandikhan and dokan lakes in sulaimaniyah Province—Iraqi kurdistan region. *IOP Conf. Ser. Earth Environ. Sci.* **2020**, *612*, 012023. [CrossRef]
7. Lü, L.; Chen, L.; Shao, W.; Luo, F. Equilibrium and kinetic modeling of Pb(II) biosorption by a chemically modified orange peel containing cyanex 272. *J. Chem. Eng. Data* **2010**, *55*, 4147–4153. [CrossRef]

8. Van Den Berg, H.H.J.L.; Friederichs, L.; Versteegh, J.F.M.; Smeets, P.W.M.H.; Husman, A.M.D.R. How current risk assessment and risk management methods for drinking water in the Netherlands cover the WHO water safety plan approach. *Int. J. Hyg. Environ. Health* **2019**, *222*, 1030–1037. [CrossRef]
9. Gu, P.; Zhang, S.; Li, X.; Wang, X.; Wen, T.; Jehan, R.; Alsaedi, A.; Hayat, T.; Wang, X. Recent advances in layered double hydroxide-based nanomaterials for the removal of radionuclides from aqueous solution. *Environ. Pollut.* **2018**, *240*, 493–505. [CrossRef]
10. Da'na, E. Adsorption of heavy metals on functionalized-mesoporous silica: A review. *Microporous Mesoporous Mater.* **2017**, *247*, 145–157. [CrossRef]
11. Fu, F.; Wang, Q. Removal of heavy metal ions from wastewaters: A review. *J. Environ. Manag.* **2011**, *92*, 407–418. [CrossRef]
12. Lesmana, S.O.; Febriana, N.; Soetaredjo, F.E.; Sunarso, J.; Ismadji, S. Studies on potential applications of biomass for the separation of heavy metals from water and wastewater. *Biochem. Eng. J.* **2009**, *44*, 19–41. [CrossRef]
13. Guieysse, B.; Norvill, Z.N. Sequential chemical–biological processes for the treatment of industrial wastewaters: Review of recent progresses and critical assessment. *J. Hazard. Mater.* **2014**, *267*, 142–152. [CrossRef]
14. Boamah, P.O.; Huang, Y.; Hua, M.; Zhang, Q.; Wu, J.; Onumah, J.; Sam-Amoah, L.; Boamah, P.O. Sorption of heavy metal ions onto carboxylate chitosan derivatives—A mini-review. *Ecotoxicol. Environ. Saf.* **2015**, *116*, 113–120. [CrossRef]
15. Jafari, M.; Rahimi, M.R.; Ghaedi, M.; Dashtian, K. ZnO nanoparticles loaded different mesh size of porous activated carbon prepared from Pinus eldarica and its effects on simultaneous removal of dyes: Multivariate optimization. *Chem. Eng. Res. Des.* **2017**, *125*, 408–421. [CrossRef]
16. Li, J.; Wang, X.; Zhao, G.; Chen, C.; Chai, Z.; Alsaedi, A.; Hayat, T.; Wang, X. Metaleorganic framework-based materials: Superior adsorbents for the capture of toxic and radioactive metal ions. *Chem. Soc. Rev.* **2018**, *47*, 2322–2356. [CrossRef] [PubMed]
17. Dashamiri, S.; Ghaedi, M.; Dashtian, K.; Rahimi, M.R.; Goudarzi, A.; Jannesar, R. Ultrasonic enhancement of the simultaneous removal of quaternary toxic organic dyes by CuO nanoparticles loaded on activated carbon: Central composite design, kinetic and isotherm study. *Ultrason. Sonochem.* **2016**, *31*, 546–557. [CrossRef] [PubMed]
18. Yu, S.; Wang, X.; Pang, H.; Zhang, R.; Song, W.; Fu, D.; Hayat, T.; Wang, X. Boron nitride-based materials for the removal of pollutants from aqueous solutions: A review. *Chem. Eng. J.* **2018**, *333*, 343–360. [CrossRef]
19. Gupta, V.K.; Ali, I.; Saleh, T.A.; Nayak, A.; Agarwal, S. Chemical treatment technologies for waste-water recycling—An overview. *RSC Adv.* **2012**, *2*, 6380–6388. [CrossRef]
20. Guo, S.; Dan, Z.; Duan, N.; Chen, G.; Gao, W.; Zhao, W. Zn (II), Pb (II), and Cd (II) adsorption from aqueous solution by magnetic silica gel: Preparation, characterization, and adsorption. *Environ. Sci. Pollut. Res.* **2018**, *25*, 30938–30948. [CrossRef]
21. Luhar, I.; Luhar, S. Rubberized Geopolymer Composites: Value-Added Applications. *J. Compos. Sci.* **2021**, *5*, 312. [CrossRef]
22. Luhar, S.; Chaudhary, S.; Luhar, I. Thermal resistance of fly ash based rubberized geopolymer concrete. *J. Build. Eng.* **2018**, *19*, 420–428. [CrossRef]
23. Luhar, S.; Chaudhary, S.; Luhar, I. Development of rubberized geopolymer concrete: Strength and durability studies. *Constr. Build. Mater.* **2019**, *204*, 740–753. [CrossRef]
24. Davidovits, J. *Geopolymer Chemistry and Applications*; Geopolymer Institute: Saint-Quentin, France, 2008.
25. Grela, A.; Łach, M.; Bajda, T.; Mikuła, J.; Hebda, M. Characterization of the products obtained from alkaline conversion of tuff and metakaolin. *J. Therm. Anal. Calorim.* **2018**, *133*, 217–226. [CrossRef]
26. Al-Zboon, K.; Al-Harahsheh, M.S.; Hani, F.B. Fly ash-based geopolymer for Pb removal from aqueous solution. *J. Hazard. Mater.* **2011**, *188*, 414–421. [CrossRef] [PubMed]
27. Al-Harahsheh, M.S.; Al Zboon, K.; Al-Makhadmeh, L.; Hararah, M.; Mahasneh, M. Fly ash based geopolymer for heavy metal removal: A case study on copper removal. *J. Environ. Chem. Eng.* **2015**, *3*, 1669–1677. [CrossRef]
28. Musil, S.S.; Kriven, W.M. In situ mechanical properties of chamotte particulate reinforced, potassium geopolymer. *J. Am. Ceram. Soc.* **2014**, *97*, 907–915. [CrossRef]
29. Mohseni, E. Assessment of Na2SiO3 to NaOH ratio impact on the performance of polypropylene fiber-reinforced geopolymer composites. *Constr. Build. Mater.* **2018**, *186*, 904–911. [CrossRef]
30. Noushini, A.; Hastings, M.; Castel, A.; Aslani, F. Mechanical and flexural performance of synthetic fibre reinforced geopolymer concrete. *Constr. Build. Mater.* **2018**, *186*, 454–475. [CrossRef]
31. Saafi, M.; Andrew, K.; Tang, P.L.; McGhon, D.; Taylor, S.; Rahman, M.; Yang, S.; Zhou, X. Multifunctional properties of carbon nanotube/fly ash geopolymeric nanocomposites. *Constr. Build. Mater.* **2013**, *49*, 46–55. [CrossRef]
32. Yan, S.; He, P.; Jia, D.; Duan, X.; Yang, Z.; Wang, S.; Zhou, Y. In-situ preparation of fully stabilized graphene/cubic-leucite composite through graphene oxide/geopolymer. *Mater. Des.* **2016**, *101*, 301–308. [CrossRef]
33. Yang, M.; Guo, Z.; Deng, Y.; Xing, X.; Qiu, K.; Long, J.; Li, J. Preparation of CaOAl2O3-SiO2 glass ceramics from coal gangue. *Int. J. Miner. Process.* **2012**, *102*, 112–115. [CrossRef]
34. Liu, X.; Zhang, N.; Yao, Y.; Sun, H.; Feng, H. Micro-structural characterization of the hydration products of bauxite-calcination-method red mud-coal gangue based cementitious materials. *J. Hazard. Mater.* **2013**, *262*, 428–438. [CrossRef]
35. Yan, S.; Huo, W.; Yang, J.; Zhang, X.; Wang, Q.; Wang, L.; Pan, Y.; Huang, Y. Green synthesis and influence of calcined temperature on the formation of novel porous diatomite microspheres for efficient adsorption of dyes. *Powder Technol.* **2018**, *329*, 260–269. [CrossRef]

36. Yan, S.; Pan, Y.; Wang, L.; Liu, J.; Zhang, Z.; Huo, W.; Yang, J.L.; Huang, Y. Synthesis of low-cost porous ceramic microspheres from waste gangue for dye adsorption. *J. Adv. Ceram.* **2018**, *7*, 30–40. [CrossRef]
37. Yan, S.; He, P.; Jia, D.; Wang, Q.; Liu, J.; Yang, J.; Huang, Y. Synthesis of novel low-cost porous gangue microsphere/geopolymer composites and their adsorption properties for dyes. *Int. J. Appl. Ceram. Technol.* **2018**, *15*, 1602–1614. [CrossRef]
38. Onutai, S.; Kobayashi, T.; Thavorniti, P.; Jiemsirilers, S. Porous fly ash-based geopolymer composite fiber as an adsorbent for removal of heavy metal ions from wastewater. *Mater. Lett.* **2019**, *236*, 30–33. [CrossRef]
39. Rasaki, S.A.; Zhang, B.; Guarecuco, R.; Thomas, T.; Yang, M. Geopolymer for use in heavy metals adsorption, and advanced oxidative processes: A critical review. *J. Clean. Prod.* **2019**, *213*, 42–58. [CrossRef]
40. Wang, F.; Lu, X.; Li, X.-Y. Selective removals of heavy metals (Pb^{2+}, Cu^{2+}, and Cd^{2+}) from wastewaters by gelation with alginate for effective metal recovery. *J. Hazard. Mater.* **2016**, *38*, 75–83. [CrossRef] [PubMed]
41. Anguile, J.J.; Mbega, M.G.O.; Makani, T.; Mbadcam, J.K. Adsorption of manganese (II) ions from aqueous solution on to volcanic ash and geopolymer based volcanic ashes. *Int. J. Basic Appl. Chem.* **2013**, *3*, 7–18.
42. Sharma, Y.C.; Srivastava, V.; Singh, V.K.; Kaul, S.N.; Weng, C.H. Nano-adsorbents for the removal of metallic pollutants from water and wastewater. *Environ. Technol.* **2009**, *30*, 583–609. [CrossRef]
43. Mishra, V.K.; Upadhyaya, A.R.; Pandey, S.K.; Tripathi, B.D. Heavy metal pollution induced due to coal mining effluent on surrounding aquatic ecosystem and its management through naturally occurring aquatic macrophytes. *Bioresour. Technol.* **2008**, *99*, 930–936. [CrossRef]
44. Sone, H.; Fugetsu, B.; Tanaka, S. Selective elimination of lead (II) ions by alginate/polyurethane composite foams. *J. Hazard. Mater.* **2009**, *162*, 423–429. [CrossRef]
45. Zhang, S.S.; Du, Q.; Sun, Y.Q.; Song, J.P.; Yang, F.; Tsang, D.C.W. Fabrication of L-cysteine stabilized α-FeOOH nanocomposite on porous hydrophilic biochar as an effective adsorbent for Pb^{2+} removal. *Sci. Total Environ.* **2020**, *720*, 137415. [CrossRef] [PubMed]
46. Yogeshwaran, V.; Priya, A.K. Adsorption of lead ion concentration from the aqueous solution using tobacco leaves. *Mater. Today Proc.* **2021**, *37*, 489–496. [CrossRef]
47. Zhao, H.; Ouyang, X.K.; Yang, L.Y. Adsorption of lead ions from aqueous solutions by porous cellulose nanofiber-sodium alginate hydrogel beads. *J. Mol. Liq.* **2021**, *324*, 115122. [CrossRef]
48. Feng, T.; Xu, J.; Yu, C.; Cheng, K.; Wu, Y.; Wang, Y.; Li, F. Graphene oxide wrapped melamine sponge as an efficient and recoverable adsorbent for Pb (II) removal from fly ash leachate. *J. Hazard. Mater.* **2019**, *367*, 26–34. [CrossRef]
49. Wheeler, D.C.; Boyle, J.; Raman, S.; Nelson, E.J. Modeling elevated blood lead level risk across the United States. *Sci. Total Environ.* **2021**, *769*, 145237. [CrossRef] [PubMed]
50. Li, S.S.; Yang, F.; Zhang, Y.Y.; Lan, Y.B.; Cheng, K. Performance of lead ion removal by the three-dimensional carbon foam supported nanoscale zero-valent iron composite. *J. Clean. Prod.* **2021**, *294*, 125350. [CrossRef]
51. Bhattacharjee, S.; Chakrabarty, S.; Maity, S.; Kar, S.; Thakur, P.; Bhattacharyya, G. Removal of lead from contaminated water bodies using sea nodule as an adsorbent. *Water Res.* **2003**, *37*, 3954–3966. [CrossRef]
52. Yang, F.; Zhang, S.S.; Cho, D.W.; Du, Q.; Song, J.P. Porous biochar composite assembled with ternary needle-like iron-manganese-sulphur hybrids for high efficiency lead removal. *Bioresour. Technol.* **2018**, *272*, 415–420. [CrossRef]
53. Zhang, P.; Ouyang, S.; Li, P.; Sun, Z.; Ding, N.; Huang, Y. Ultrahigh removal performance of lead from wastewater by tricalcium aluminate via precipitation combining flocculation with amorphous aluminum. *J. Clean. Prod.* **2020**, *246*, 118728. [CrossRef]
54. Arbabi, M.; Hemati, S.; Amiri, M. Removal of lead ions from industrial wastewater: A review of removal methods. *Int. J. Epidemiol. Res.* **2015**, *2*, 105–109.
55. Zheng, H.; Meng, X.; Chen, J.; Que, M.; Wang, W.; Liu, X.; Yang, L.; Zhao, Y. In situ phase evolution of TiO2/Ti3C2Tx heterojunction for enhancing adsorption and photocatalytic degradation. *Appl. Surf. Sci.* **2021**, *545*, 149031. [CrossRef]
56. Ghorbani, M.; Seyedin, O.; Aghamohammadhassan, M. Adsorptive removal of lead (II) ion from water and wastewater media using carbon-based nanomaterials as unique sorbents: A review. *J. Environ. Manag.* **2020**, *254*, 109814. [CrossRef] [PubMed]
57. Gao, R.; Fu, Q.; Hu, H.; Wang, Q.; Liu, Y.; Zhu, J. Highly-effective removal of Pb by co-pyrolysis biochar derived from rape straw and orthophosphate. *J. Hazard. Mater.* **2019**, *371*, 191–197. [CrossRef] [PubMed]
58. Xiong, C.; Wang, W.; Tan, F.; Luo, F.; Chen, J.; Qiao, X. Investigation on the efficiency and mechanism of Cd (II) and Pb (II) removal from aqueous solutions using MgO nanoparticles. *J. Hazard. Mater.* **2015**, *299*, 664–674. [CrossRef]
59. Tang, Q.; Wang, K.; Yaseen, M.; Tong, Z.; Cui, X. Synthesis of highly efficient porous inorganic polymer microspheres for the adsorptive removal of Pb^{2+} from wastewater. *J. Clean. Prod.* **2018**, *193*, 351–362. [CrossRef]
60. Yusuf, M.; Song, K.; Li, L. Fixed bed column and artificial neural network model to predict heavy metals adsorption dynamic on surfactant decorated graphene. *Colloids Surf. A Physicochem. Eng. Asp.* **2020**, *585*, 124076. [CrossRef]
61. Swelam, A.A.; Awad, M.B.; Salem, A.M.A.; El-Feky, A.S. An economically viable method for the removal of cobalt ions from aqueous solution using raw and modified rice straw. *HBRC J.* **2019**, *14*, 255–263. [CrossRef]
62. Javadian, H. Application of kinetic, isotherm and thermodynamic models for the adsorption of Co(II) ions on polyaniline/polypyrrole copolymer nanofibers from aqueous solution. *J. Ind. Eng. Chem.* **2014**, *20*, 4233–4241. [CrossRef]
63. He, M.Y.; Zhu, Y.; Yang, Y.; Han, B.P.; Zhang, Y.M. Adsorption of cobalt(II) ions from aqueous solutions by palygorskite. *Appl. Clay Sci.* **2011**, *54*, 292–296. [CrossRef]
64. Manohar, D.M.; Noeline, B.F.; Anirudhan, T.S. Adsorption performance of Alpillared bentonite clay for the removal of cobalt(II) from aqueous phase. *Appl. Clay Sci.* **2006**, *31*, 194–206. [CrossRef]

65. Gil, R.A.; G´asquez, J.A.; Olsina, R.; Martinez, L.D.; Cerutti, S. Cloud point extraction for cobalt preconcentration with on-line phase separation in a knotted reactor followed by ETAAS determination in drinking waters. *Talanta* **2008**, *76*, 669–673. [CrossRef]
66. Mihaly-Cozmuta, L.; Mihaly-Cozmuta, A.; Peter, A.; Nicula, C.; Tutu, H.; Silipas, D.; Indrea, E. Adsorption of heavy metal cations by Na-clinoptilolite: Equilibrium and selectivity studies. *J. Environ. Manag.* **2014**, *137*, 69–80. [CrossRef]
67. Kara, M.; Yuzer, H.; Sabah, E.; Celik, M.S. Adsorption of cobalt from aqueous solutions onto sepiolite. *Water Res.* **2003**, *37*, 224–232. [CrossRef]
68. Wu, Z.J.; Chen, R.; Gan, Q.; Li, J.; Zhang, T.Q.; Ye, M.M. Mesoporous Na^+-SiO_2 spheres for efficient removal of Cr^{3+} from aqueous solution. *J. Environ. Chem. Eng.* **2018**, *6*, 1774–1782. [CrossRef]
69. Ma, H.R.; Hei, Y.N.; Hua, L.; Guo, Y.Y.; Yang, Y.Z.; Yu, C.L.; Qiao, X.R. Fabrication of zirconium-pillared montmorillonite porous ceramic as adsorbents for Cr^{3+} removal and recycling. *Ceram. Int.* **2016**, *42*, 14903–14909. [CrossRef]
70. Anush, S.M.; Vishalakshi, B. Modified chitosan gel incorporated with magnetic nanoparticle for removal of Cu(II) and Cr(VI) from aqueous solution. *Int. J. Biol. Macromol.* **2019**, *133*, 1051–1062. [CrossRef] [PubMed]
71. Scarazzato, T.; Panossian, Z.; Tenório, J.A.S.; Pérez-Herranz, V.; Espinosa, D.C.R. A review of cleaner production in electroplating industries using electrodialysis. *J. Clean. Prod.* **2017**, *168*, 1590–1602. [CrossRef]
72. Bulcke, F.; Santofimia-Castaño, P.; Gonzalez-Mateos, A.; Dringen, R. Modulation of copper accumulation and copper-induced toxicity by antioxidants and copper chelators in cultured primary brain astrocytes. *J. Trace Elem. Med. Biol.* **2015**, *32*, 168–176. [CrossRef]
73. Benhangi, H.M.; Ahmadi, S.; Hakimi, M.; Molafilabi, A.; Faraji, H.; Mashkani, B. Protective effects of isatin and its synthetic derivatives against iron, copper and lead toxicity. *Toxicol. Vitr.* **2019**, *54*, 232–236. [CrossRef] [PubMed]
74. Cambrollé, J.; García, J.L.; Ocete, R.; Figueroa, M.E.; Cantos, M. Growth and photosynthetic responses to copper in wild grapevine. *Chemosphere* **2013**, *93*, 294–301. [CrossRef] [PubMed]
75. Nagarajan, D.; Venkatanarasimhan, S. Kinetics and mechanism of efficient removal of Cu(II) ions from aqueous solutions using ethylenediamine functionalized cellulose sponge. *Int. J. Biol. Macromol.* **2020**, *148*, 988–998. [CrossRef] [PubMed]
76. Fanni, D.; Fanos, V.; Gerosa, C.; Piras, M.; Dessi, A.; Atzei, A.; Van, E.P.; Gibo, Y.; Faa, G. Effects of iron and copper overload on the human liver: An ultrastructural study. *Curr. Med. Chem.* **2014**, *21*, 3768–3774. [CrossRef] [PubMed]
77. Pu, X.; Yao, L.; Yang, L.; Jiang, W.; Jiang, X. Utilization of industrial waste lithium-silicon-powder for the fabrication of novel nap zeolite for aqueous Cu(II) removal. *J. Clean. Prod.* **2020**, *265*, 121822. [CrossRef]
78. Alshaaer, M.; Zaharaki, D.; Komnitsas, K. Microstructural characteristics and adsorption potential of a zeolitic tuff-metakaolin geopolymer. *Desalin. Water Treat.* **2015**, *56*, 338–345. [CrossRef]
79. Wang, T.; Zhang, P.; Wu, D.; Sun, M.; Deng, Y.; Frost, R.L. Effective removal of zinc(II) from aqueous solutions by tricalcium aluminate (C3A). *J. Colloid Interface Sci.* **2015**, *443*, 65–71. [CrossRef]
80. Al-Wakeel, K.Z.; Abd El Monem, H.; Khalil, M.M. Removal of divalent manganese from aqueous solution using glycine modified chitosan resin. *J. Environ. Chem. Eng.* **2015**, *3*, 179–186. [CrossRef]
81. Suguna, M.; Kumar, N.S.; Subbaiah, M.V.; Krishnaiah, A. Removal of divalent manganese from aqueous solution using *Tamarindus indica* fruit nut shell. *J. Chem. Pharm. Res.* **2010**, *2*, 7–20.
82. Belviso, C.; Cavalcante, F.; Di Gennaro, S.; Lettino, A.; Palma, A.; Ragone, P.; Fiore, S. Removal of Mn from aqueous solution using fly ash and its hydrothermal synthetic zeolite. *J. Environ. Manag.* **2014**, *137*, 16–22. [CrossRef]
83. da Fonseca, M.G.; de Oliveira, M.M.; Arakaki, L.N.H. Removal of cadmium, zinc, manganese and chromium cations from aqueous solution by a clay mineral. *J. Hazard. Mater.* **2006**, *137*, 288–292. [CrossRef] [PubMed]
84. Yavuz, O.; Altunkaynak, Y.; Guzel, F. Removal of copper, nickel, cobalt and manganese from aqueous solution by kaolinite. *Water Res.* **2003**, *37*, 948–952. [CrossRef]
85. Ge, Y.Y.; Yuan, Y.; Wang, K.T.; He, Y.; Cui, X.M. Preparation of geopolymer-based inorganic membrane for removing Ni^{2+} from wastewater. *J. Hazard. Mater.* **2015**, *299*, 711–718. [CrossRef]
86. Hashimoto, S.; MacHino, T.; Takeda, H.; Daiko, Y.; Honda, S.; Iwamoto, Y. Antimicrobial activity of geopolymers ion exchanged with copper ions. *Ceram. Int.* **2015**, *41*, 13788–13792. [CrossRef]
87. O'Connor, S.J.; MacKenzie, K.J.D.; Smith, M.E.; Hanna, J.V. Ion exchange in the charge-balancing sites of aluminosilicate inorganic polymers. *J. Mater. Chem.* **2010**, *20*, 10234–10240. [CrossRef]
88. Adak, D.; Sarkar, M.; Maiti, M.; Tamang, A.; Mandal, S.; Chattopadhyay, B. Anti-microbial efficiency of nano silver-silica modified geopolymer mortar for eco-friendly green construction technology. *RSC Adv.* **2015**, *5*, 64037–64045. [CrossRef]
89. Akhigbe, L.; Ouki, S.; Saroj, D.; Lim, X.M. Silver-modified clinoptilolite for the removal of Escherichia coli and heavy metals from aqueous solutions. *Environ. Sci. Pollut. Res.* **2014**, *21*, 10940–10948. [CrossRef]
90. Dolic, M.B.; Rajakovic-Ognjanovic, V.N.; Strbac, S.B.; Rakocevic, Z.L.; Veljovic, D.N.; Dimitrijevic, S.I.; Rajakovic, L.V. The antimicrobial efficiency of silver activated sorbents. *Appl. Surf. Sci.* **2015**, *357*, 819–831. [CrossRef]
91. Johari, S.A.; Kalbassi, M.R.; Soltani, M.; Yu, I.J. Application of nanosilver-coated zeolite as water filter media for fungal disinfection of rainbow trout (*Oncorhynchus mykiss*) eggs. *Aquac. Int.* **2016**, *24*, 23–38. [CrossRef]
92. Davidovits, J. Solid Phase Synthesis of a Mineral Blockpolymer by Low Temperature Polycondensation of Aluminosilicate Polymers. In Proceedings of the International Symposium on Macromolecules, Stockholm, Sweden, 1976.
93. Noor ul, A.; Faisal, M.; Muhammad, K.; Gul, S. Synthesis and characterization of geopolymer from bagasse bottom ash, waste of sugar industries and naturally available China clay. *J. Clean. Prod.* **2016**, *129*, 491–495. [CrossRef]

94. Salwa, S.S.M.; Mustafa, M.A.; Abdullah, M.M.A.B.; Kamarudin, H.; Ruzaidi, C.; Binhussain, M.; Syed Zuber, S.Z. Review on current geopolymer as a coating material. *Aust. J. Basic Appl. Sci.* **2013**, *7*, 246–257.
95. Aizat, A.E.; Abdullah, M.M.A.B.; Ming, L.; Yong, H.; Kamarudin, H.; Aziz, A. Review of Geopolymer Materials for Thermal Insulating Applications. *Key Eng. Mater.* **2015**, *660*, 17–22.
96. Perera, D.; Blackford, M.; Vance, R.; Hanna, V.E.; Finnie, K.J.; Nicholson, L. Geopolymers for the immobilization of radioactive waste. *Mater. Res. Soc. Symp. Proc.* **2004**, *824*, CC8.35. [CrossRef]
97. Bai, C.; Colombo, P. Processing, properties and applications of highly porous geopolymers: A review. *Ceram. Int.* **2018**, *44*, 16103–16118. [CrossRef]
98. Abdelrahman, E.A.; Hegazey, R.M. Exploitation of Egyptian insecticide cans in the fabrication of Si/Fe nanostructures and their chitosan polymer composites for the removal of Ni(II), Cu(II), and Zn(II) ions from aqueous solutions. *Compos. Part B Eng.* **2019**, *166*, 382–400. [CrossRef]
99. Abdelrahman, E.A.; Hegazey, R.M.; Alharbi, A. Facile synthesis of mordenite nanoparticles for efficient removal of Pb(II) ions from aqueous media. *J. Inorg. Organomet. Polym. Mater.* **2020**, *30*, 1369–1383. [CrossRef]
100. Khalifa, M.E.; Abdelrahman, E.A.; Hassanien, M.M.; Ibrahim, W.A. Application of mesoporous silica nanoparticles modified with Dibenzoylmethane as a novel composite for efficient removal of Cd(II), Hg(II), and Cu(II) ions from aqueous media. *J. Inorg. Organomet. Polym. Mater.* **2020**, *30*, 2182–2196. [CrossRef]
101. Horpibulsuk, S.; Suksiripattanapong, C.; Samingthong, W.; Rachan, R.; Arulrajah, A. Durability against wetting-drying cycles of water treatment sludge-fly ash geopolymer and water treatment sludge-cement and silty clay-cement systems. *J. Mater. Civ. Eng.* **2016**, *28*, 1–9. [CrossRef]
102. Liu, Y.; Yan, C.; Zhang, Z.; Wang, H.; Zhou, S.; Zhou, W. A comparative study on fly ash, geopolymer and faujasite block for Pb removal from aqueous solution. *Fuel* **2016**, *185*, 181–189. [CrossRef]
103. Duan, P.; Yan, C.; Zhou, W.; Ren, D. Development of fly ash and iron ore tailing based porous geopolymer for removal of Cu(II) from wastewater. *Ceram. Int.* **2016**, *42*, 13507–13518. [CrossRef]
104. Novais, R.M.; Gameiro, T.; Carvalheiras, J.; Seabra, M.P.; Tarelho, L.A.C.; Labrincha, J.A.; Capela, I. High pH buffer capacity biomass fly ash-based geopolymer spheres to boost methane yield in anaerobic digestion. *J. Clean. Prod.* **2018**, *178*, 258–267. [CrossRef]
105. Mužek, M.N.; Svilović, S.; Ugrina, M.; Zelić, J. Removal of copper and cobalt ions by fly ash-based geopolymer from solutions-equilibrium study, Desalin. *Water Treat.* **2016**, *57*, 10689–10699. [CrossRef]
106. Bumanis, G.; Rugele, K.; Bajare, D. The effect of alkaline material particle size on adjustment ability of buffer capacity. *Medziagotyra* **2015**, *21*, 405–409. [CrossRef]
107. Luukkonen, T.; Runtti, H.; Niskanen, M.; Tolonen, E.T.; Sarkkinen, M.; Kemppainen, K.; Rämö, J.; Lassi, U. Simultaneous removal of Ni(II), as(III), and Sb(III) from spiked mine effluent with metakaolin and blast-furnace-slag geopolymers. *J. Environ. Manag.* **2016**, *166*, 579–588. [CrossRef]
108. Gasca-Tirado, J.R.; Manzano-Ramírez, A.; Vazquez-Landaverde, P.A.; Herrera-Díaz, E.I.; Rodríguez-Ugarte, M.E.; Rubio-Ávalos, J.C.; Amigó-Borrás, V.; Chávez-Páez, M. Ion exchanged geopolymer for photocatalytic degradation of a volatile organic compound. *Mater. Lett.* **2014**, *134*, 222–224. [CrossRef]
109. Novais, R.M.; Seabra, M.P.; Labrincha, J.A. Porous geopolymer spheres as novel pH buffering materials. *J. Clean. Prod.* **2017**, *143*, 1114–1122. [CrossRef]
110. Khan, M.I.; Min, T.K.; Azizli, K.; Sufian, S.; Ullah, H.; Man, Z. Effective removal of methylene blue from water using phosphoric acid based geopolymers: Synthesis, characterizations and adsorption studies. *RSC Adv.* **2015**, *5*, 61410–61420. [CrossRef]
111. Xu, M.; He, Y.; Wang, C.; He, X.; He, X.; Liu, J.; Cui, X. Preparation and characterization of a self-supporting inorganic membrane based on metakaolin based geopolymers. *Appl. Clay Sci.* **2015**, *115*, 254–259. [CrossRef]
112. Ascensão, G.; Seabra, M.P.; Aguiar, J.B.; Labrincha, J.A. Red mud-based geopolymers with tailored alkali diffusion properties and pH buffering ability. *J. Clean. Prod.* **2017**, *148*, 23–30. [CrossRef]
113. Runtti, H.; Luukkonen, T.; Niskanen, M.; Tuomikoski, S.; Kangas, T.; Tynjälä, P.; Tolonen, E.T.; Sarkkinen, M.; Kemppainen, K.; Rämö, J.; et al. Sulphate removal over barium-modified blast-furnace-slag geopolymer. *J. Hazard. Mater.* **2016**, *317*, 373–384. [CrossRef] [PubMed]
114. Bumanis, G.; Bajare, D. The effect of porous alkali activated material composition on buffer capacity in bioreactors. *Int. J. Chem. Nucl. Metall. Mater. Eng.* **2014**, *8*, 1040–1046.
115. Min Li, C.; He, Y.; Tang, Q.; Tuo Wang, K.; Min Cui, X.; Min Li, C.; He, Y.; Tang, Q.; Tuo Wang, K.; Min Cui, X. Study of the preparation of CdS on the surface of geopolymer spheres and photocatalyst performance. *Mater. Chem. Phys.* **2016**, *178*, 204–210.
116. Li, Z.; Wang, L.; Meng, J.; Liu, X.; Xu, J.; Wang, F.; Brookes, P. Zeolite-supported nanoscale zero-valent iron: New findings on simultaneous adsorption of Cd(II), Pb (II), and as(III) in aqueous solution and soil. *J. Hazard. Mater.* **2018**, *344*, 1–11. [CrossRef]
117. Ji, Z.; Pei, Y. Immobilization efficiency and mechanism of metal cations (Cd^{2+}, Pb^{2+} and Zn^{2+}) and anions (${AsO_4}^{3-}$ and ${Cr_2O_7}^{2-}$) in wastes-based geopolymer. *J. Hazard. Mater.* **2020**, *384*, 121290. [CrossRef]
118. Xie, W.M.; Zhou, F.P.; Bi, X.L.; Chen, D.D.; Li, J.; Sun, S.Y.; Liu, J.Y.; Chen, X.Q. Accelerated crystallization of magnetic 4A-zeolite synthesized from red mud for application in removal of mixed heavy metal ions. *J. Hazard. Mater.* **2018**, *358*, 441–449. [CrossRef]
119. Zhang, Y.J.; Han, Z.C.; He, P.Y.; Chen, H. Geopolymer-based catalysts for cost effective environmental governance: A review based on source control and end-of pipe treatment. *J. Clean. Prod.* **2020**, *263*, 121556. [CrossRef]

120. Huang, J.; Li, Z.; Zhang, J.; Zhang, Y.; Cui, X. In-situ synchronous carbonation and self-activation of biochar/geopolymer composite membrane: Enhanced catalyst for oxidative degradation of tetracycline in water. *Chem. Eng. J.* **2020**, *397*, 125528. [CrossRef]
121. Mestre, A.S.; Pires, R.A.; Aroso, I.; Fernandes, E.M.; Pinto, M.L.; Reis, R.L.; Andrade, M.A.; Pires, J.; Silva, S.P.; Carvalho, A.P. Activated carbons prepared from industrial pre-treated cork: Sustainable adsorbents for pharmaceutical compounds removal. *Chem. Eng. J.* **2014**, *253*, 408–417. [CrossRef]
122. Sanguanpak, S.; Wannagon, A.; Saengam, C.; Chiemchaisri, W.; Chiemchaisri, C. Porous metakaolin-based geopolymer granules for removal of ammonium in aqueous solution and anaerobically pretreated piggery wastewater. *J. Clean. Prod.* **2021**, *297*, 126643. [CrossRef]
123. Maleki, A.; Hajizadeh, Z.; Sharifi, V.; Emdadi, Z. A green, porous and eco-friendly magnetic geopolymer adsorbent for heavy metals removal from aqueous solutions. *J. Clean. Prod.* **2019**, *215*, 1233–1245. [CrossRef]
124. Lan, T.; Guo, S.; Li, X.; Guo, J.; Bai, T.; Zhao, Q.; Yang, W.; Li, P. Mixed precursor geopolymer synthesis for removal of Pb(II) and Cd(II). *Mater. Lett.* **2020**, *274*, 127977. [CrossRef]
125. Yu, Z.; Song, W.; Li, J.; Li, Q. Improved simultaneous adsorption of Cu(II) and Cr (VI) of organic modified metakaolin-based geopolymer. *Arab. J. Chem.* **2020**, *13*, 4811–4823. [CrossRef]
126. Naghsh, M.; Shams, K. Synthesis of a kaolin-based geopolymer using a novel fusion method and its application in effective water softening. *Appl. Clay Sci.* **2017**, *146*, 238–245. [CrossRef]
127. Zhang, Z.; Li, L.; He, D.; Ma, X.; Yan, C.; Wang, H. Novel self-supporting zeolitic block with tunable porosity and crystallinity for water treatment. *Mater. Lett.* **2016**, *178*, 151–154. [CrossRef]
128. Abdelrahman, E.A. Synthesis of zeolite nanostructures from waste aluminum cans for efficient removal of malachite green dye from aqueous media. *J. Mol. Liq.* **2018**, *253*, 72–82. [CrossRef]
129. Pant, B.; Ojha, G.P.; Kim, H.-Y.; Park, M.; Park, S.-J. Fly-ash-incorporated electrospun zinc oxide nanofibers: Potential material for environmental remediation. *Environ. Pollut.* **2019**, *245*, 163–172. [CrossRef] [PubMed]
130. Prachasaree, W.; Limkatanyu, S.; Hawa, A.; Samakrattakit, A. Development of equivalent stress block parameters for fly-ash-based geopolymer concrete. *Arab. J. Sci. Eng.* **2014**, *39*, 8549–8558. [CrossRef]
131. Cioffi, R.; Maffucci, L.; Santoro, L. Optimization of geopolymer synthesis bycalcination and polycondensation of a kaolinitic residue. *Resour. Conserv. Recycl.* **2003**, *40*, 27–38. [CrossRef]
132. Provis, J.L.; Lukey, G.C.; Deventer, J.S.J.V. Do geopolymers actually contain nanocrystalline zeolites? A reexamination of existing results. *Chem. Mater.* **2005**, *17*, 3075–3085. [CrossRef]
133. Cui, X.; He, Y.; Liu, L.; Chen, J. NaA zeolite synthesis from geopolymer precursor. *MRS Commun.* **2011**, *1*, 49–51.
134. Van, T.L.; Gobechiya, E.; Breynaert, E.; Martens, J.A.; Kirschhock, C.E. Alkaline cations directing the transformation of FAU zeolites into five different framework types. *Chem. Commun.* **2013**, *49*, 11737–11739.
135. Azarshab, M.; Mohammadi, F.; Maghsoodloorad, H.; Mohammadi, T. Ceramic membrane synthesis based on alkali activated blast furnace slag for separation of water from ethanol. *Ceram. Int.* **2016**, *42*, 15568–15574. [CrossRef]
136. Wang, H.; Yan, C.; Li, D.; Zhou, F.; Liu, Y.; Zhou, C.; Komarneni, S. In situ transformation of geopolymer gels to self-supporting NaX zeolite monoliths with excellent compressive strength. *Microporous Mesoporous Mater.* **2017**, *261*, 164–169. [CrossRef]
137. Tang, Q.; Ge, Y.Y.; Wang, K.T.; He, Y.; Cui, X.M. Preparation of porous P-type zeolite spheres with suspension solidification method. *Mater. Lett.* **2015**, *161*, 558–560. [CrossRef]
138. Sudagar, A.; Andrejkovicová, S.; Patinha, C.; Velosa, A.; Mcadam, A.; Silva, E.F.; Rocha, F. A novel study on the influence of cork waste residue on metakaolin zeolite based geopolymers. *Appl. Clay Sci.* **2018**, *152*, 196–210. [CrossRef]
139. Sivalingam, S.; Kella, T.; Maharana, M.; Sen, S. Efficient sono-sorptive elimination of methylene blue by fly ash-derived nano-zeolite X: Process optimization, isotherm and kinetic studies. *J. Clean. Prod.* **2019**, *208*, 1241–1254. [CrossRef]
140. Mouni, L.; Belkhiri, L.; Bollinger, J.C.; Bouzaza, A.; Assadi, A.; Tirri, A.; Dahmoune, F.; Madani, K.; Remini, H. Removal of Methylene Blue from aqueous solutions by adsorption on Kaolin: Kinetic and equilibrium studies. *Appl. Clay Sci.* **2018**, *153*, 38–45. [CrossRef]
141. Sapawe, N.; Jalil, A.A.; Triwahyono, S.; Shah, M.I.A.; Jusoh, R.; Salleh, N.F.M.; Hameed, B.H.; Karim, A.H. Cost-effective microwave rapid synthesis of zeolite NaA for removal of methylene blue. *Chem. Eng. J.* **2013**, *229*, 388–398. [CrossRef]
142. Subaihi, A.; Morad, M.; Hameed, A.M.; Alharbi, A.; Abou, Y.G.; Algethami, F.K.; Hegazey, R.M.; Abdelrahman, E.A. Studying some analytical parameters affecting the removal of Mn(II) ions from aqueous media using facilely synthesised analcime. *Int. J. Environ. Anal. Chem.* **2020**, 1–12. [CrossRef]
143. Nasiri-ardali, M.; Nezamzadeh-ejhieh, A. A Comprehensive Study on the Kinetics and Thermodynamic Aspects of Batch and Column Removal of Pb(II) by the Clinoptilolite—Glycine Adsorbent. *Mater. Chem. Phys.* **2020**, *240*, 122142. [CrossRef]
144. al Sadat Shafiof, M.; Nezamzadeh-Ejhieh, A. A comprehensive study on the removal of Cd(II) from aqueous solution on a novel pentetic acid-clinoptilolite nanoparticles adsorbent: Experimental design, kinetic and thermodynamic aspects. *Solid State Sci.* **2020**, *99*, 106071.
145. Derikvandi, H.; Nezamzadeh-Ejhieh, A. Increased photocatalytic activity of NiO and ZnO in photodegradation of a model drug aqueous solution: Effect of coupling, supporting, particles size and calcination temperature. *J. Hazard. Mater.* **2017**, *321*, 629–638. [CrossRef] [PubMed]

146. Derikvandi, H.; Nezamzadeh-Ejhieh, A. Comprehensive study on enhanced photocatalytic activity of heterojunction ZnS-NiS/zeolite nanoparticles: Experimental design based on response surface methodology (RSM), impedance spectroscopy and GC-MASS studies. *J. Colloid Interface Sci.* **2017**, *490*, 652–664. [CrossRef] [PubMed]
147. Tamiji, T.; Nezamzadeh-Ejhieh, A. Sensitive voltammetric determination of bromate by using ion-exchange property of a Sn(II)-clinoptilolite-modified carbon paste electrode. *J. Solid State Electrochem.* **2019**, *23*, 143–157. [CrossRef]
148. Nezamzadeh-Ejhieh, A.; Khorsandi, M. Photodecolorization of Eriochrome Black T using NiS-P zeolite as a heterogeneous catalyst. *J. Hazard. Mater.* **2010**, *176*, 629–637. [CrossRef]
149. Esmaili-Hafshejani, J.; Nezamzadeh-Ejhieh, A. Increased photocatalytic activity of Zn(II)/Cu(II) oxides and sulfides by coupling and supporting them onto clinoptilolite nanoparticles in the degradation of benzophenone aqueous solution. *J. Hazard. Mater.* **2016**, *316*, 194–203. [CrossRef] [PubMed]
150. Babaahamdi-Milani, M.; Nezamzadeh-Ejhieh, A. A comprehensive study on photocatalytic activity of supported Ni/Pb sulfide and oxide systems onto natural zeolite nanoparticles. *J. Hazard. Mater.* **2016**, *318*, 291–301. [CrossRef]
151. Li, L.; Wang, S.; Zhu, Z. Geopolymeric adsorbents from fly ash for dye removal from aqueous solution. *J. Colloid Interface Sci.* **2006**, *300*, 52–59. [CrossRef]
152. Sarkar, C.; Basu, J.K.; Samanta, A.N. Removal of Ni^{2+} Ion from waste water by geopolymeric adsorbent derived from LD Slag. *J Water Proc. Eng.* **2017**, *17*, 237–244. [CrossRef]
153. Liu, X.; Chen, G.R.; Lee, D.J.; Kawamoto, T.; Tanaka, H.; Chen, M.L.; Luo, Y.K. Adsorption removal of cesium from drinking waters: A mini review on use of biosorbents and other adsorbents. *Bioresour. Technol.* **2014**, *160*, 142–149. [CrossRef]
154. Guille'n, F.J.; Baeza, A.; Salas, A. Strontium. In *Radionuclides in the Environment*; Atwood, D.A., Ed.; Wiley: Chichester, UK, 2010; pp. 79–96.
155. Chuang, L.; Liao, C. Sorption of cesium using KZnFc on phosphoric acid-based geopolymer. *Ceram. Silik.* **2015**, *59*, 125–134.
156. Lee, N.K.; Khalid, H.R.; Lee, H.K. Adsorption characteristics of cesium onto mesoporous geopolymers containing nanocrystalline zeolites. *Microporous Mesoporous Mater.* **2017**, *242*, 238–244. [CrossRef]
157. Chen, Y.L.; Tong, Y.Y.; Pan, R.W.; Tang, J. The research on adsorption behaviors and mechanisms of geopolymers on Sr^{2+}, Co^{2+} and $Cs^{+1,2}$. *Adv. Mater. Res.* **2013**, *704*, 313–318. [CrossRef]
158. Cao, C.Y.; Cui, Z.M.; Chen, C.Q.; Song, W.-G.; Cai, W. Ceria hollow nanospheres produced by a template-free microwave-assisted hydrothermal method for heavy metal ion removal and catalysis. *J. Phys. Chem. C* **2010**, *114*, 9865–9870. [CrossRef]
159. Kaşgöz, H.; Özgümüş, S.; Orbay, M. Modified polyacrylamide hydrogels and their application in removal of heavy metal ions. *Polymer* **2003**, *44*, 1785–1793. [CrossRef]
160. Kiani, G.; Sheikhloie, H.; Arsalani, N. Heavy metal ion removal from aqueous solutions by functionalized polyacrylonitrile. *Desalination* **2011**, *269*, 266–270. [CrossRef]
161. Hu, J.S.; Zhong, L.S.; Song, W.G.; Wan, L.J. Synthesis of hierarchically structured metal oxides and their application in heavy metal ion removal. *Adv. Mater.* **2008**, *20*, 2977–2982. [CrossRef]
162. Luukkonen, T.; Tolonen, E.T.; Runtti, H.; Kemppainen, K.; Perämäki, P.; Rämö, J.; Lassi, U. Optimization of the metakaolin geopolymer preparation for maximized ammonium adsorption capacity. *J. Mater. Sci.* **2017**, *52*, 9363–9376. [CrossRef]
163. Barbosa, T.R.; Foletto, E.L.; Dotto, G.L.; Jahn, S.L. Preparation of mesoporous geopolymer using metakaolin and rice husk ash as synthesis precursors and its use as potential adsorbent to remove organic dye from aqueous solutions. *Ceram. Int.* **2018**, *44*, 416–423. [CrossRef]
164. Ge, Y.; Cui, X.; Kong, Y.; Li, Z.; He, Y.; Zhou, Q. Porous geopolymeric spheres for removal of Cu(II) from aqueous solution: Synthesis and evaluation. *J. Hazard. Mater.* **2015**, *283*, 244–251. [CrossRef]
165. Luukkonen, T.; Sarkkinen, M.; Kemppainen, K.; Rämö, J.; Lassi, U. Metakaolin geopolymer characterization and application for ammonium removal from model solutions and landfill leachate. *Appl. Clay Sci.* **2016**, *119*, 266–276. [CrossRef]
166. Javadian, H.; Ghorbani, F.; Allah Tayebi, H.; Asl, S.M.H. Study of the adsorption of Cd (II) from aqueous solution using zeolite-based geopolymer, synthesized from coal fly ash; kinetic, isotherm and thermodynamic studies. *Arab. J. Chem.* **2015**, *8*, 837–849. [CrossRef]
167. Novais, R.M.; Carvalheiras, J.; Tobaldi, D.M.; Seabra, M.P.; Pullar, R.C.; Labrincha, J.A. Synthesis of porous biomass fly ash-based geopolymer spheres for efficient removal of methylene blue from wastewaters. *J. Clean. Prod.* **2019**, *207*, 350–362. [CrossRef]
168. Minelli, M.; Medri, V.; Papa, E.; Miccio, F.; Landi, E.; Doghieri, F. Geopolymers as solid adsorbent for CO_2 capture. *Chem. Eng. Sci.* **2016**, *148*, 267–274. [CrossRef]
169. Kara, I.; Tunc, D.; Sayin, F.; Akar, S.T. Study on the performance of metakaolin based geopolymer for Mn(II) and Co(II) removal. *Appl. Clay Sci.* **2018**, *161*, 184–193. [CrossRef]
170. Lopez, F.J.; Sugita, S.; Kobayashi, T. Cesium-adsorbent geopolymer foams based on silica from rice husk and metakaolin. *Chem. Lett.* **2013**, *43*, 128–130. [CrossRef]
171. Muzek, M.N.; Svilovic, S.; Zelic, J. Fly ash-based geopolymeric adsorbent for copper ion removal from wastewater. *Desalin. Water Treat.* **2014**, *52*, 2519–2526. [CrossRef]
172. Cheng, T.W.; Lee, M.L.; Ko, M.S.; Ueng, T.H.; Yang, S.F. The heavy metal adsorption characteristics on metakaolin-based geopolymer. *Appl. Clay Sci.* **2012**, *56*, 90–96. [CrossRef]
173. Tang, Q.; Ge, Y.Y.; Wang, K.T.; He, Y.; Cui, X.M. Preparation and characterization of porous metakaolin-based inorganic polymer spheres as an adsorbent. *Mater. Des.* **2015**, *88*, 1244–1249. [CrossRef]

174. Andrejkovicova, S.; Sudagar, A.; Rocha, J.; Patinha, C.; Hajjaji, W.; da Silva, E.F.; Velosa, A.; Rocha, F. The effect of natural zeolite on microstructure, mechanical and heavy metals adsorption properties of metakaolin based geopolymers. *Appl. Clay Sci.* **2016**, *126*, 141–152. [CrossRef]
175. Vu, T.H.; Gowripalan, N. Mechanisms of heavy metal immobilisation using geopolymerisation techniques—A review. *J. Adv. Concr. Technol.* **2018**, *16*, 124–135. [CrossRef]
176. Siyal, A.A.; Shamsuddin, M.R.; Khan, M.I.; Rabat, N.E.; Zulfiqar, M.; Man, Z.; Siame, J.; Azizli, K.A. A review on geopolymers as emerging materials for the adsorption of heavy metals and dyes. *J. Environ. Manag.* **2018**, *224*, 327–339. [CrossRef] [PubMed]
177. Guo, B.; Pan, D.A.; Liu, B.; Volinsky, A.A.; Fincan, M.; Du, J.; Zhang, S. Immobilization mechanism of Pb in fly ash-based geopolymer. *Constr. Build. Mater.* **2017**, *134*, 123–130. [CrossRef]
178. Bhattacharyya, K.G.; Gupta, S.S. Adsorption of a few heavy metals on natural and modified kaolinite and montmorillonite: A review. *Adv. Colloid Interface Sci.* **2008**, *140*, 114–131. [CrossRef]
179. Bhatt, A.S.; Sakaria, P.L.; Vasudevan, M.; Pawar, R.R.; Sudheesh, N.; Bajaj, H.C.; Mody, H.M. Adsorption of an anionic dye from aqueous medium by organoclays: Equilibrium modeling, kinetic and thermodynamic exploration. *RSC Adv.* **2012**, *2*, 8663–8671. [CrossRef]
180. Jaishankar, M.; Tseten, T.; Anbalagan, N.; Mathew, B.B.; Beeregowda, K.N. Toxicity, mechanism and health effects of some heavy metals. *Interdiscipl. Toxicol.* **2014**, *7*, 60–72. [CrossRef]
181. Azimi, A.; Azari, A.; Rezakazemi, M.; Ansarpour, M. Removal of heavy metals from industrial wastewaters: A review. *ChemBioEng Rev.* **2017**, *4*, 37–59. [CrossRef]
182. Khodadadi, M.; Malekpour, A.; Ansaritabar, M. Removal of Pb(II) and Cu(II) from aqueous solutions by NaA zeolite coated magnetic nanoparticles and optimization of method using experimental design. *Microporous Mesoporous Mater.* **2017**, *248*, 256–265. [CrossRef]
183. Uogint e, I.; Lujaniene, G.; Mazeika, K. Study of Cu (II), Co (II), Ni (II) and Pb (II) removal from aqueous solutions using magnetic Prussian blue nano-sorbent. *J. Hazard. Mater.* **2019**, *369*, 226–235. [CrossRef]
184. Rafatullah, M.; Sulaiman, O.; Hashim, R.; Ahmad, A. Adsorption of methylene blue on low-cost adsorbents: A review. *J. Hazard. Mater.* **2010**, *177*, 70–80. [CrossRef] [PubMed]
185. Cheng, S.; Zhang, L.; Ma, A.; Xia, H.; Peng, J.; Li, C.; Shu, J. Comparison of activated carbon and iron/cerium modified activated carbon to remove methylene blue from wastewater. *J. Environ. Sci.* **2018**, *65*, 92–102. [CrossRef]
186. Merouani, S.; Hamdaoui, O.; Saoudi, F.; Chiha, M.; Pétrier, C. Influence of bicarbonate and carbonate ions on sonochemical degradation of Rhodamine B in aqueous phase. *J. Hazard. Mater.* **2010**, *175*, 593–599. [CrossRef] [PubMed]
187. Kannan, N.; Sundaram, M.M. Kinetics and mechanism of removal of methylene blue by adsorption on various carbons e a comparative study. *Dye. Pigment.* **2001**, *51*, 25–40. [CrossRef]
188. Liu, Y.; Yan, C.; Zhang, Z.; Gong, Y.; Wang, H.; Qiu, X. A facile method for preparation of floatable and permeable fly ash-based geopolymer block. *Mater. Lett.* **2016**, *185*, 370–373. [CrossRef]
189. Lan Huong, P.T.; Tu, N.; Lan, H.; Thang, L.H.; Van Quy, N.; Tuan, P.A.; Dinh, N.X.; Phan, V.N.; Le, A.T. Functional manganese ferrite/graphene oxide nanocomposites: Effects of graphene oxide on the adsorption mechanisms of organic MB dye and inorganic As(v) ions from aqueous solution. *RSC Adv.* **2018**, *8*, 12376–12389. [CrossRef]
190. Ioannou, Z.; Karasavvidis, C.; Dimirkou, A.; Antoniadis, V. Adsorption of methylene blue and methyl red dyes from aqueous solutions onto modified zeolites. *Water Sci. Technol.* **2013**, *67*, 1129–1136. [CrossRef]
191. Yousef, R.I.; El-Eswed, B.; Alshaaer, M.; Khalili, F.; Khoury, H. The influence of using Jordanian natural zeolite on the adsorption, physical, and mechanical properties of geopolymers products. *J. Hazard. Mater.* **2009**, *165*, 379–387. [CrossRef]
192. Luukkonen, T.; Věžníková, K.; Tolonen, E.T.; Runtti, H.; Yliniemi, J.; Hu, T.; Kemppainen, K.; Lassi, U. Removal of ammonium from municipal wastewater with powdered and granulated metakaolin geopolymer. *Environ. Technol.* **2018**, *39*, 414–423. [CrossRef]
193. Couto, R.S.D.P.; Oliveira, A.F.; Guarino, A.W.S.; Perez, D.V.; Marques, M.R.D.C. Removal of ammonia nitrogen from distilled old landfill leachate by adsorption on raw and modified aluminosilicate. *Environ. Technol.* **2016**, *38*, 816–826. [CrossRef]
194. He, H.; Xu, S.; Han, R.; Wang, Q. Nutrient sequestration from wastewater by using zeolite Na-P1 synthesized from coal fly ash. *Environ. Technol.* **2016**, *38*, 1022–1029. [CrossRef]
195. Leyva-Ramos, R.; Monsivais-Rocha, J.E.; Aragon-Pina, A.; Berber-Mendoza, M.S.; Guerrero-Coronado, R.M.; Alonso-Davila, P.; Mendoza-Barron, J. Removal of ammonium from aqueous solution by ion exchange on natural and modified chabazite. *J. Environ. Manag.* **2010**, *91*, 2662–2668. [CrossRef] [PubMed]
196. Ye, Z.; Wang, J.; Sun, L.; Zhang, D.; Zhang, H. Removal of ammonium from municipal landfill leachate using natural zeolites. *Environ. Technol.* **2015**, *36*, 2919–2923. [CrossRef]
197. Wetzel, R.G. *Limnology: Lake and River Ecosystems*, 3rd ed.; Elsevier: San Diego, CA, USA, 2001.
198. Lin, Y.; Guo, M.; Shah, N.; Stuckey, D.C. Economic and environmental evaluation of nitrogen removal and recovery methods from wastewater. *Bioresour. Technol.* **2016**, *215*, 227–238. [CrossRef]
199. Hwang, J.H.; Oleszkiewicz, J.A. Effect of cold-temperature shock on nitrification. *Water Environ. Res.* **2007**, *79*, 964–968. [CrossRef]
200. Gasca-Tirado, J.R.; Manzano-Ramírez, A.; RiveraMuñoz, E.M.; Velázquez-Castillo, R.; Apátiga-Castro, M.; Nava, R.; Rodríguez-López, A. Ion exchange in Geopolymers. In *New Trends Ion Exchange Studies*; Karakus, S., Ed.; IntechOpen: London, UK, 2018; pp. 71–82.

201. Bai, C.; Colombo, P. High-porosity geopolymer membrane supports by peroxide route with the addition of egg white as surfactant. *Ceram. Int.* **2017**, *43*, 2267–2273. [CrossRef]
202. Zhang, Y.; Li, T.; Hou, D.; Zhang, J.; Jiang, J. Insights on magnesium and sulfate ions' adsorption on the surface of sodium alumino-silicate hydrate (NASH) gel: A molecular dynamics study. *Phys. Chem. Chem. Phys.* **2018**, *20*, 18297–18310. [CrossRef] [PubMed]
203. Meesters, R.; Schroder, H.F. Perfluorooctane sulfonate-a quite mobile anionic anthropogenic surfactant, ubiquitously found in the environment. *Water Sci. Technol.* **2004**, *50*, 235–242. [CrossRef] [PubMed]
204. Karray, F.; Mezghani, M.; Mhiri, N.; Djelassi, B.; Sayadi, S. Scale-down studies of membrane bioreactor degrading anionic surfactants wastewater: Isolation of new anionic-surfactant degrading bacteria. *Int. Biodeterior. Biodegrad.* **2016**, *114*, 14–23. [CrossRef]
205. Edser, C. Status of global surfactant markets. *Focus Surfactants* **2008**, *11*, 1–2. [CrossRef]
206. Beltran-Heredia, J.; Sanchez-Martín, J.; Barrado-Moreno, M. Long-chain anionic surfactants in aqueous solution. Removal by moringa oleifera coagulant. *Chem. Eng. J.* **2012**, *180*, 128–136. [CrossRef]
207. Zanoletti, A.; Federici, S.; Borgese, L.; Bergese, P.; Ferroni, M.; Depero, L.E.; Bontempi, E. Embodied energy as key parameter for sustainable materials selection: The case of reusing coal fly ash for removing anionic surfactants. *J. Clean. Prod.* **2017**, *141*, 230–236. [CrossRef]
208. Zhang, C.; Valsaraj, K.T.; Constant, W.D.; Roy, D. Aerobic biodegradation kinetics of four anionic and nonionic surfactants at sub-and supra-critical micelle concentrations (CMCs). *Water Res.* **1999**, *33*, 115–124. [CrossRef]
209. Kundu, S.; Gupta, A.K. Arsenic adsorption onto iron oxide-coated cement (IOCC): Regression analysis of equilibrium data with several isotherm models and their optimization. *Chem. Eng. J.* **2006**, *122*, 93–106. [CrossRef]
210. Santamaría, L.; Vicente, M.A.; Korili, S.A.; Gil, A. Progress in the removal of pharmaceutical compounds from aqueous solution using layered doublehydroxides as adsorbents: A review. *J. Environ. Chem. Eng.* **2020**, *8*, 104577. [CrossRef]
211. Siyal, A.A.; Shamsuddin, M.R.; Rabat, N.E.; Zulfiqar, M.; Man, Z.; Low, A. Fly ash based geopolymer for the adsorption of anionic surfactant from aqueous solution. *J. Clean. Prod.* **2019**, *229*, 232–243. [CrossRef]
212. Valizadeh, S.; Younesi, H.; Bahramifar, N. Highly mesoporous K_2CO_3 and KOH/ activated carbon for SDBS removal from water samples: Batch and fixed-bed column adsorption process. *Environ. Nanotechnol. Monit. Manag.* **2016**, *6*, 1–13. [CrossRef]
213. Luhar, I.; Luhar, S.; Abdullah, M.M.A.; Nabialek, M.; Sandu, A.V.; Szmidla, J.; Jurczynska, A.; Razak, R.A.; Aziz, I.H.A.; Jamil, N.H.; et al. Assessment of the Suitability of Ceramic Waste in Geopolymer Composites: An Appraisal. *Materials* **2021**, *14*, 3279. [CrossRef]
214. Sandu, A.V.; Vasilache, V.; Sandu, I.G.; Sieliechi, J.M.; Kouame, I.K.; Matasaru, P.D.; Sandu, I. Characterization of the Acid-Base Character of Burned Clay Ceramics Used for Water Decontamination. *Materials* **2019**, *12*, 3836. [CrossRef] [PubMed]
215. Nergis, D.D.B.; Vizureanu, P.; Ardelean, I.; Sandu, A.V.; Corbu, O.C.; Matei, E. Revealing the Influence of Microparticles on Geopolymers' Synthesis and Porosity. *Materials* **2020**, *13*, 3211. [CrossRef]
216. Burduhos Nergis, D.D.; Abdullah, M.M.A.B.; Sandu, A.V.; Vizureanu, P. XRD and TG-DTA Study of New Alkali Activated Materials Based on Fly Ash with Sand and Glass Powder. *Materials* **2020**, *13*, 343. [CrossRef]
217. Matasaru, D.; Scarlatache, V.A.; Pruteanu, A.; Florean, B.; Viziteu, G. Analysis of dielectric loss at high frequency, for a nanocomposite polymer matrix, based on Polypropylene with insertion of multi-walled carbon nanotubes (MWCNTS), 2012. In Proceedings of the International Conference and Exposition on Electrical and Power Engineering (EPE 2012), Iasi, Romania, 25–27 October 2012; pp. 116–119, ISBN 978-1-4673-1171-7.
218. Strozi Cilla, M.; Raymundo Morelli, M.; Colombo, P. Effect of process parameters on the physical properties of porous geopolymers obtained by gelcasting. *Ceram. Int.* **2014**, *40*, 13585–13590. [CrossRef]
219. Cilla, M.S.; de Mello Innocentini, M.D.; Morelli, M.R.; Colombo, P. Geopolymer foams obtained by the saponification/ peroxide/gelcasting combined route using different soap foam precursors. *J. Am. Ceram. Soc.* **2017**, *100*, 3440–3450. [CrossRef]
220. Zou, D.; Qiu, M.; Chen, X.; Drioli, E.; Fan, Y. One step co-sintering process for lowcost fly ash based ceramic microfiltration membrane in oil-in-water emulsion treatment. *Sep. Purif. Technol.* **2019**, *210*, 511–520. [CrossRef]
221. Zhou, C.; Wang, N.; Qian, Y.; Liu, X.; Caro, J.; Huang, A. Efficient synthesis of dimethyl ether from methanol in a bifunctional zeolite membrane reactor. *Angew. Chem. Int. Ed.* **2016**, *55*, 12678–12682. [CrossRef]
222. Duran, A.; Tuzen, M.; Soylak, M. Preconcentration of some trace elements via using multiwalled carbon nanotubes as solid phase extraction adsorbent. *J. Hazard. Mater.* **2009**, *169*, 466–471. [CrossRef]
223. Chitpong, N.; Husson, S.M. Polyacid functionalized cellulose nanofiber membranes for removal of heavy metals from impaired waters. *J. Membrane. Sci.* **2017**, *523*, 418–429. [CrossRef]
224. Güell, R.; Anticó, E.; Salvadó, V.; Fontàs, C. Efficient hollow fiber supported liquid membrane system for the removal and preconcentration of Cr (VI) at trace levels. *Sep. Purif. Technol.* **2008**, *62*, 389–393. [CrossRef]
225. Guha, A.K.; Yun, C.H.; Basu, R.; Sirkar, K.K. Heavy metal removal and recovery by contained liquid membrane permeator. *AIChE J.* **1994**, *40*, 1223–1237. [CrossRef]
226. Sakamoto, H.; Kimura, K.; Shono, T. Lithium separation and enrichment by protondriven cation transport through liquid membranes of lipophilic crown nitrophenols. *Anal. Chem.* **1987**, *59*, 1513–1517. [CrossRef]
227. Zhang, Q.; Wang, N.; Zhao, L.; Xu, T.; Cheng, Y. Polyamidoamine dendronized hollow fiber membranes in the recovery of heavy metal ions. *ACS. Appl. Mater. Inter.* **2013**, *5*, 1907–1912. [CrossRef] [PubMed]

228. Roviello, G.; Chianese, E.; Ferone, C.; Ricciotti, L.; Roviello, V.; Cioffi, R.; Tarallo, O. Hybrid geopolymeric foams for the removal of metallic ions from aqueous waste solutions. *Materials* **2019**, *12*, 4091. [CrossRef]
229. Panda, B.; Paul, S.C.; Mohamed, N.A.N.; Tay, Y.W.D.; Tan, M.J. Measurement of tensile bond strength of 3D printed geopolymer mortar. *Measurement* **2018**, *113*, 108–116. [CrossRef]
230. Li, Z.; Zhang, S.; Zuo, Y.; Chen, W.; Ye, G. Chemical deformation of metakaolin based geopolymer. *Cement. Concrete. Res.* **2019**, *120*, 108–118. [CrossRef]
231. Mallicoat, S.; Sarin, P.; Kriven, W. Novel, alkali-bonded, ceramic filtration membranes. *Ceram. Eng. Sci. Proc.* **2005**, *26*, 37–44.
232. Li, Q.; He, Y.; Xu, M.; Liu, J.; He, M.; Huang, L.; Cui, X. Study on the removal of Ca^{2+} and Mg^{2+} in water by the geopolymer-based inorganic membrane. *Gongneng. Cailiao.* **2017**, *48*, 01215–01220.
233. Mohammadi, F.; Mohammadi, T. Optimal conditions of porous ceramic membrane synthesis based on alkali activated blast furnace slag using Taguchi method. *Ceram. Int.* **2017**, *43*, 14369–14379. [CrossRef]
234. Xu, M.; He, Y.; Wang, Y.; Cui, X. Preparation of a nonhydrothermal NaA zeolite membrane and defect elimination by vacuum-inhalation repair method. *Chem. Eng. Sci.* **2017**, *158*, 117–123. [CrossRef]
235. Liu, Y.; Yan, C.; Zhang, Z.; Li, L.; Wang, H.; Pu, S. One-step fabrication of novel porous and permeable self-supporting zeolite block from fly ash. *Mater. Lett.* **2017**, *196*, 328–331. [CrossRef]
236. Kazemimoghadam, M. New nanopore zeolite membranes for water treatment. *Desalination* **2010**, *251*, 176–180. [CrossRef]
237. Landi, E.; Medri, V.; Papa, E.; Dedecek, J.; Klein, P.; Benito, P.; Vaccari, A. Alkali-bonded ceramics with hierarchical tailored porosity. *Appl. Clay Sci.* **2013**, *73*, 56–64. [CrossRef]
238. Jo, M.; Soto, L.; Arocho, M.; St John, J.; Hwang, S. Optimum mix design of fly ash geopolymer paste and its use in pervious concrete for removal of fecal coliforms and phosphorus in water. *Constr. Build Mater.* **2015**, *93*, 1097–1104. [CrossRef]
239. Asim, N.; Alghoul, M.; Mohammad, M.; Amin, M.H.; Akhtaruzzaman, M.; Amin, N.; Sopian, K. Emerging sustainable solutions for depollution: Geopolymers. *Constr. Build Mater.* **2019**, *199*, 540–548. [CrossRef]
240. Gasca-Tirado, J.R.; Manzano-Ramırez, A.; Villasenor-Mora, C.; Muniz-Villarreal, M.S.; Zaldivar-Cadena, A.A.; Rubio-Avalos, J.C.; Borras, V.A.; Mendoza, R.N. Incorporation of photoactive TiO_2 in an aluminosilicate inorganic polymer by ion exchange. *Microporous Mesoporous Mater.* **2012**, *153*, 282–287. [CrossRef]
241. Zhang, Y.; Liu, L. Fly ash-based geopolymer as a novel photocatalyst for degradation of dye from wastewater. *Particuology* **2013**, *11*, 353–358. [CrossRef]
242. Fallah, M.; MacKenzie, K.J.D.; Hanna, J.V.; Page, S.J. Novel photoactive inorganic polymer composites of inorganic polymers with copper(I) oxide nanoparticles. *J. Mater. Sci.* **2015**, *50*, 7374–7383. [CrossRef]
243. Zhang, Y.J.; He, P.Y.; Yang, M.Y.; Kang, L. A new graphene bottom ash geopolymeric composite for photocatalytic H2 production and degradation of dyeing wastewater. *Int. J. Hydr. Energy* **2017**, *42*, 20589–20598. [CrossRef]
244. Lloyd, R.R.; Provis, J.L.; van Deventer, J.S. Pore solution composition and alkali diffusion in inorganic polymer cement. *Cem. Concr. Res.* **2010**, *40*, 1386–1392. [CrossRef]
245. Silva, I.; Castro-Gomes, J.; Albuquerque, A. Mineral waste geopolymeric artificial aggregates as alternative materials for wastewater-treatment processes: Study of structural stability and pH variation in water. *J. Mater. Civ. Eng.* **2012**, *24*, 623–628. [CrossRef]
246. Onutai, S.; Kobayashi, T.; Thavorniti, P.; Jiemsirilers, S. The adsorption of cadmium ions on fly ash based geopolymer particles. In *Key Engineering Materials (KEM)*; Trans Tech Publications Ltd.: Kapellweg, Switzerland, 2018; Volume 766, pp. 65–70.
247. Panda, L.; Rath, S.S.; Rao, D.S.; Nayak, B.B.; Das, B.; Misra, P.K. Thorough understanding of the kinetics and mechanism of heavy metal adsorption onto a pyrophyllite mine waste based geopolymer. *J. Mol. Liq.* **2018**, *263*, 428–441. [CrossRef]
248. Sarkar, C.; Basu, J.K.; Samanta, A.N. Synthesis of mesoporous geopolymeric powder from LD slag as superior adsorbent for Zinc (II) removal. *Adv. Powder Technol.* **2018**, *29*, 1142–1152. [CrossRef]
249. Singhal, A.; Gangwar, B.P.; Gayathry, J.M. CTAB modified large surface area nanoporous geopolymer with high adsorption capacity for copper ion removal. *Appl. Clay. Sci.* **2017**, *150*, 106–114. [CrossRef]
250. Ge, Y.; Cui, X.; Liao, C.; Li, Z. Facile fabrication of green geopolymer/alginate hybrid spheres for efficient removal of Cu (II) in water: Batch and column studies. *Chem. Eng. J.* **2017**, *311*, 126–134. [CrossRef]
251. Al-Zboon, K.; Al-smadi, B.M.; Al-Khawaldh, S. Natural volcanic tuff-based geopolymer for Zn removal: Adsorption isotherm, kinetic, and thermodynamic study. *Water Air Soil Pollut.* **2016**, *227*, 1–22. [CrossRef]
252. Muzek, M.N.; Svilovic, S.; Zelic, J. Kinetic studies of cobalt ion removal from aqueous solutions using fly ash-based geopolymer and zeolite NaX as sorbents. *Sep. Sci. Technol.* **2016**, *51*, 2868–2875. [CrossRef]
253. Falah, M.; MacKenzie, K.J.; Knibbe, R.; Page, S.J.; Hanna, J.V. New composites of nanoparticle Cu (I) oxide and titania in a novel inorganic polymer (geopolymer) matrix for destruction of dyes and hazardous organic pollutants. *J. Hazard. Mater.* **2016**, *318*, 772–782. [CrossRef]
254. El Alouani, M.; Alehyen, S.; El Achouri, M.; Taibi, M. Removal of Cationic Dye—Methylene Blue—From Aque Ous Solution by Adsorption on Fly Ash—Based Geopolymer. *J. Mater. Environ. Sci.* **2018**, *9*, 32–46.
255. El Alouani, M.; Alehyen, s.; El Achouri, M.; Taibi, M. Preparation, characterization, and application of metakaolin-based geopolymer for removal of methylene blue from aqueous solution. *J. Chem.* **2019**, *2019*, 4212901. [CrossRef]

256. Netto, M.S.; Rossatto, D.L.; Jahn, S.L.; Mallmann, E.S.; Dotto, G.L.; Foletto, E.L. Preparation of a novel magnetic geopolymer/zero-valent iron composite with remarkable adsorption performance towards aqueous Acid Red 97. *Chem. Eng. Commun.* **2020**, *207*, 1048–1061. [CrossRef]
257. Zhang, Y.J.; Liu, L.C.; Ni, L.L.; Wang, B.L. A facile and low-cost synthesis of granulated blast furnace slag-based cementitious material coupled with Fe_2O_3 catalyst for treatment of dye wastewater. *Appl. Catal. B Environ.* **2013**, *138*, 9–16. [CrossRef]
258. Acisli, O.; Acar, I.; Khataee, A. Preparation of a fly ash-based geopolymer for removal of a cationic dye: Isothermal, kinetic and thermodynamic studies. *J. Ind. Eng. Chem.* **2020**, *83*, 53–63. [CrossRef]
259. Hua, P.; Sellaoui, L.; Franco, D.; Netto, M.S.; Dotto, G.L.; Bajahzar, A.; Belmabrouk, H.; Bonilla-Petriciolet, A.; Li, Z. Adsorption of acid green and procion red on a magnetic geopolymer based adsorbent: Experiments, characterization and theoretical treatment. *Chem. Eng. J.* **2020**, *383*, 123113. [CrossRef]
260. Fumba, G.; Essomba, J.S.; Tagne, G.M.; Nsami, J.N.; élibi, P.D.B.B.; Mbadcam, J.K. Equilibrium and kinetic adsorption studies of methyl orange from aqueous solutions using kaolinite, metakaolinite and activated geopolymer as low cost adsorbents. *J. Acad. Ind. Res.* **2014**, *3*, 156–163.
261. Salehi, A.; Najafi Kani, E. Green cylindrical mesoporous adsorbent based on alkal-activated phosphorous slag: Synthesis, dye removal, and RSM modeling. *Adsorption* **2018**, *24*, 647–666. [CrossRef]
262. Rossatto, D.L.; Netto, M.S.; Jahn, S.L.; Mallmann, E.S.; Dotto, G.L.; Foletto, E.L. Highly efficient adsorption performance of a novel magnetic geopolymer/Fe_3O_4 composite towards removal of aqueous acid green 16 dye. *J. Environ. Chem. Eng.* **2020**, *8*, 103804. [CrossRef]
263. Zhang, Y.J.; Yang, M.Y.; Zhang, L.; Zhang, K.; Kang, L. A new graphene/geopolymer nanocomposite for degradation of dye wastewater. *Integr. Ferroelectr. Int. J.* **2016**, *171*, 38–45. [CrossRef]
264. Luhar, S.; Cheng, T.W.; Nicolaides, D.; Luhar, I.; Panias, D.; Sakkas, K. Valorization of glass wastes for the development of geopolymer composites—Durability, thermal and microstructural properties: A review. *Constr. Build. Mater.* **2019**, *222*, 673–687. [CrossRef]
265. Luhar, S.; Nicolaides, D.; Luhar, I. Fire Resistance Behaviour of Geopolymer Concrete: An Overview. *Buildings* **2021**, *11*, 82. [CrossRef]
266. Luhar, S.; Luhar, I.; Gupta, R. Durability performance evaluation of green geopolymer concrete. *Eur. J. Environ Civ. Eng.* **2020**, *1*, 1–49. [CrossRef]
267. Luhar, S.; Luhar, I.; Nicolaides, D.; Gupta, R. Durability Performance Evaluation of Rubberized Geopolymer Concrete. *Sustainability* **2021**, *13*, 5969. [CrossRef]
268. Luhar, S.; Dave, U.V.; Chaudhary, S.; Khandelwal, U. A brief review on geopolymer concrete. In Proceedings of the 5th Nirma University International Conference on Engineering, Ahmedabad, India, 26–28 November 2015.
269. Luhar, S.; Rajamane, N.P.; Corbu, O.; Luhar, I. Impact of incorporation of volcanic ash on geopolymerization of eco-friendly geopolymer composites: A review. *IOP Conf. Ser. Mater. Sci. Eng.* **2019**, *572*, 012001. [CrossRef]
270. Luhar, S.; Dave, U. Investigations on mechanical properties of fly ash and slag based geopolymer concrete. *Ind. Concr. J.* **2016**, 34–41.
271. Luhar, S.; Luhar, I.; Abdullah, M.M.A.B.; Hussin, K. Challenges and prospective trends of various industrial and solid wastes incorporated with sustainable green concrete. In *Advances in Organic Farming*; Woodhead Publishing: Sawston, UK, 2021; pp. 223–240.
272. Luhar, S.; Chaudhary, P.; Luhar, I. Influence of steel crystal powder on performance of aggregate concrete. *IOP Conf. Ser. Mater. Sci. Eng.* **2018**, *431*, 102003. [CrossRef]
273. Luhar, S.; Luhar, I.; Shaikh, F.U.A. Review on Performance Evaluation of Autonomous Healing of Geopolymer Composites. *Infrastructures* **2021**, *6*, 94. [CrossRef]
274. Luhar, I.; Luhar, S.; Savva, P.; Theodosiou, A.; Petrou, M.F.; Nicolaides, D. Light Transmitting Concrete: A Review. *Buildings* **2021**, *11*, 480. [CrossRef]
275. Shao, N.; Tang, S.; Li, S.; Chen, H.; Zhang, Z. Defective analcime/geopolymer composite membrane derived from fly ash for ultrafast and highly efficient filtration of organic pollutants. *J Hazard Mater.* **2020**, *388*, 121736. [CrossRef]
276. Wang, J.T.; Ge, Y.Y.; He, Y.; Xu, M.X.; Cui, X.M. A porous gradient geopolymer-based tube membrane with high PM removal rate for air pollution. *J Clean Prod.* **2019**, *217*, 335–343. [CrossRef]
277. Luukkonen, T.; Yliniemi, J.; Sreenivasan, H.; Ohenoja, K.; Finnilä, M.; Franchin, G.; Colombo, P. Ag- or Cu-modified geopolymer filters for water treatment manufactured by 3D printing, direct foaming, or granulation. *Sci. Rep.* **2020**, *10*, 1–14.
278. He, Y.; Cui, X.M.; Liu, X.D.; Wang, Y.P.; Zhang, J.; Liu, K. Preparation of self-supporting NaA zeolite membranes using geopolymers. *J Membr. Sci.* **2013**, *447*, 66–72. [CrossRef]
279. Zhang, J.; He, Y.; Wang, Y.P.; Mao, J.; Cui, X.M. Synthesis of a selfsupporting faujasite zeolite membrane using geopolymer gel for separation of alcohol/water mixture. *Mater. Lett.* **2014**, *116*, 167–170. [CrossRef]
280. Haris, A.; Irhamsyah, A.; Permatasari, A.D.; Desa, S.S.; Irfanita, R.; Wahyuni, S. Pervaporation membrane based on laterite zeolite-geopolymer for ethanol water separation. *J Clean Prod.* **2020**, *249*, 119413.
281. He, P.Y.; Zhang, Y.J.; Chen, H.; Han, Z.C.; Liu, L.C. Low-cost and facile synthesis of geopolymer-zeolite composite membrane for chromium(VI) separation from aqueous solution. *J Hazard Mater.* **2020**, *392*, 122359. [CrossRef] [PubMed]

282. Li, C.J.; Zhang, Y.J.; Chen, H.; He, P.Y. High value-added utilization of silica fume to synthesize ZSM-35 zeolite membrane for Cd2 removal. *Mater. Lett.* **2020**, *260*, 126940. [CrossRef]
283. Amir, N.; Fazli, S.; Khraisheh, M.; Al-Bakri, M.; Saeed, G. Porosity control of self-supported geopolymeric membrane through hydrogen peroxide and starch additives. *Desalin. Water Treat.* **2019**, *152*, 11–15.
284. Anu Karthi Swaghatha, A.I.; Cindrella, L. Self-humidifying novel chitosan-geopolymer hybrid membrane for fuel cell applications. *Carbohydr. Polym.* **2019**, *223*, 115073.
285. Fang, J.; Qin, G.; Wei, W.; Zhao, X.; Jiang, L. Elaboration of new ceramic membrane from spherical fly ash for micro filtration of rigid particle suspension and oil-in-water emulsion. *Desalination* **2013**, *311*, 113–126. [CrossRef]
286. Singh, G.; Bulasara, V.K. Preparation of low-cost microfiltration membranes from fly ash. *Desalin. Water Treat.* **2015**, *3994*, 1–9. [CrossRef]
287. Chen, M.; Zhu, L.; Dong, Y.; Li, L.; Liu, J. Waste-to-resource strategy to fabricate highly porous whisker-structured mullite ceramic membrane for simulated oil-in-water emulsion wastewater treatment. *ACS Sustain. Chem. Eng.* **2016**, *4*, 2098–2106. [CrossRef]
288. Suresh, K.; Pugazhenthi, G.; Uppaluri, R. Fly ash based ceramic microfiltration membranes for oil–water emulsion treatment: Parametric optimization using response surface methodology. *J. Water Process Eng.* **2016**, *13*, 27–43. [CrossRef]
289. Chen, H.; Zhang, Y.J.; He, P.Y.; Li, C.J.; Li, H. Coupling of selfsupporting geopolymer membrane with intercepted Cr(III) for dye wastewater treatment by hybrid photocatalysis and membrane separation. *Appl. Surf. Sci.* **2020**, *515*, 146024. [CrossRef]
290. Jedidi, I.; Saïdi, S.; Khmakem, S.; Larbot, A.; Elloumi-Ammar, N.; Fourati, A.; Charfi, A.; Amar, R.B. New ceramic microfiltration membranes from mineral coal fly ash. *Arab. J. Chem.* **2009**, *2*, 31–39. [CrossRef]
291. Xu, M.; He, Y.; Liu, Z.; Tong, Z.; Cui, X. Preparation of geopolymer inorganic membrane and purification of pulp-papermaking green liquor. *Appl. Clay Sci.* **2019**, *168*, 269–275. [CrossRef]
292. Zhang, Y.J.; Chen, H.; He, P.Y.; Li, C.J. Developing silica fume-based self-supported ECR-1 zeolite membrane for seawater desalination. *Mater. Lett.* **2019**, *236*, 538–541. [CrossRef]
293. Fang, J.; Qin, G.; Wei, W.; Zhao, X. Preparation and characterization of tubular supported ceramic microfiltration membranes from fly ash. *Sep. Purif. Technol.* **2011**, *80*, 585–591. [CrossRef]
294. Liu, J.; He, Y.; Yuan, Y.; Huang, J.L.; Cui, X.M. The preparation and characterization of geopolymer based inorganic membranes. *Key Eng. Mater.* **2014**, *602–603*, 80–83. [CrossRef]
295. Song, Y.; Li, Z.; Zhang, J.; Tang, Y.; Ge, Y.; Cui, X. A low-cost biomimetic heterostructured multilayer membrane with geopolymer microparticles for broad-spectrum water purification. *ACS Appl. Mater. Interfaces* **2020**, *12*, 12133–12142. [CrossRef] [PubMed]

materials

MDPI

Article

Development of High-Tech Self-Compacting Concrete Mixtures Based on Nano-Modifiers of Various Types

Sergey A. Stel'makh [1], Evgenii M. Shcherban' [1], Alexey Beskopylny [2,*], Levon R. Mailyan [3], Besarion Meskhi [4], Nikita Beskopylny [5] and Yuriy Zherebtsov [6]

1 Department of Engineering Geology, Bases, and Foundations, Don State Technical University, 344003 Rostov-on-Don, Russia; sergej.stelmax@mail.ru (S.A.S.); au-geen@mail.ru (E.M.S.)

2 Department of Transport Systems, Faculty of Roads and Transport Systems, Don State Technical University, 344003 Rostov-on-Don, Russia

3 Department of Roads, Don State Technical University, 344003 Rostov-on-Don, Russia; lrm@aaanet.ru

4 Department of Life Safety and Environmental Protection, Faculty of Life Safety and Environmental Engineering, Don State Technical University, 344003 Rostov-on-Don, Russia; spu-02@donstu.ru

5 Department Hardware and Software Engineering, Don State Technical University, 344003 Rostov-on-Don, Russia; beskna@yandex.ru

6 Department of Technological Engineering and Expertise in the Construction Industry, Don State Technical University, 344003 Rostov-on-Don, Russia; yuri.zherebtscv@gmail.com

* Correspondence: besk-an@yandex.ru; Tel.: +7-863-273-8454

Abstract: Promising areas of concrete material science are maximum greening, reducing the carbon footprint, and, at the same time, solving the problems of increasing the cost of raw materials using industrial waste as modifiers for self-compacting concrete mixtures. This study aimed to review, investigate and test from the point of view of theory and practice the possibility of using various industrial types as a nano-modifier in self-compacting concrete with improved performance. The possibility of nano-modification of self-compacting concrete with a complex modifier based on industrial waste has been proved and substantiated theoretically and experimentally. The possibility of improving the technological properties of concrete mixtures using such nanomodifiers was confirmed. The recipe and technological parameters of the process were revealed and their influence on the characteristics of concrete mixes and concretes were expressed and determined. Experimental technological and mathematical dependencies between the characteristics of the technological process and raw materials and the characteristics of concrete mixtures and concretes were determined. The optimization of these parameters was carried out, a theoretical substantiation of the obtained results was proposed, and a quantitative picture was presented, expressed in the increment of the properties of self-compacting concrete mixtures using nano-modifiers from industrial waste concretes based on them. The mobility of the concrete mixture increased by 12%, and the fluidity of the mixture increased by 83%. In relation to the control composition, the concrete strength increased by 19%, and the water resistance of concrete increased by 22%. The ultimate strains decreased by 14%, and elastic modulus increased by 11%.

Keywords: self-compacting concrete; self-compacting concrete mixture; nano-modifier; nano-modified concrete; technological characteristics

Citation: Stel'makh, S.A.; Shcherban', E.M.; Beskopylny, A.; Mailyan, L.R.; Meskhi, B.; Beskopylny, N.; Zherebtsov, Y. Development of High-Tech Self-Compacting Concrete Mixtures Based on Nano-Modifiers of Various Types. *Materials* **2022**, *15*, 2739. https://doi.org/10.3390/ma15082739

Academic Editors: Petrica Vizureanu and Wenzhong Zhu

Received: 27 February 2022
Accepted: 6 April 2022
Published: 8 April 2022

1. Introduction

The growing requirements for the quality of building materials, products, and structures [1–3], the emergence of unique buildings and structures, high and long-span, set the primary areas of scientific tasks in building materials, innovative technologies, production methods, compositions, technological, and highly functional concretes of a new generation. In this regard, it is conditionally possible to divide the promising areas of concrete science into two aspects. The first is the achievement of high-performance characteristics of the resulting concrete for the construction of unique buildings and structures of increased responsibility from them. The second is the maximum manufacturability and ergonomics of the construction processes themselves, which are primarily due to the possibility of obtaining these reinforced concrete products, structures, buildings, and structures with acceptable quality under challenging conditions and within a reasonable time.

Thus, this article focuses attention on high-tech processes for constructing reinforced concrete and concrete buildings and structures of a new type. As is known, in current conditions all over the world, the main requirements are not only high quality and speed of construction of buildings and structures, but also the maximum environmental friendliness and efficiency of such construction in the face of rising prices for energy, raw materials, machinery and equipment, and increased requirements for the wages of workers. Thus, one of the main directions is also greening and achieving maximum efficiency of construction processes in the construction of reinforced concrete buildings and structures. In this regard, one of the promising areas is the development of the most technologically advanced, highly functional concrete mixtures of a new generation, which allows achieving savings using a minimum amount of means for transportation, laying, compaction, care of hardening concrete, the possibility of the fastest construction of buildings and structures with the most attentive ensuring proper quality. Self-compacting concrete mixtures are mixtures of a new generation. They allow bypassing such problems of current construction as the high-density of reinforcement and the impossibility of manual or machine compaction of the concrete mixture in hard-to-reach concrete structures, minimize costs for all stages of the life cycle of the construction process, and ensure the excellent quality of the construction process as soon as possible in the construction of reinforced concrete buildings and structures. From this point of view, concerning improving the quality and manufacturability of self-compacting concrete, many works are known.

At the same time, as mentioned above, a critical condition is maximum greening, reducing the carbon footprint, using the possibility of maximum greening of the environment, reducing the environmental burden, and solving the problems of increasing the cost of raw materials. In this regard, the search for technological, constructive, recipe approaches, the use of several types of waste, industrial and agricultural and other wastes of economic entities and humans as modifiers of self-compacting concrete mixtures, are also seen as promising areas.

Superplasticizers can produce self-compacting concretes (SCC) containing various particulate materials such as Portland cement, mineral additives, fillers, and fine and exceptionally fine sand. The addition of a superplasticizer makes it possible to preserve the rheological properties of the concrete mixture for a time sufficient for the transportation and molding of products and structures [4]. The influence of various dosages and types of superplasticizers on the workability of self-compacting concrete mixtures and the parameters of the hardened SCC were considered in [5–7]. For example, various types of superplasticizers are based on polycarboxylates, lignosulfonic acid, melamine-formaldehyde sulfonic acid naphthalene-formaldehyde sulfate acid, had different effects on the flow and loss of fluidity of concrete mixtures [6]. At the same time, the mechanism of action of the additive is also essential, as well as its exact dosage for each specific case of the use of self-compacting concrete [7].

The use of finely dispersed mineral additives makes it possible to improve the technological characteristics of self-compacting concrete mixtures and the physical and mechanical properties of concrete. At the same time, the most significant effect in the mentioned improvement can be achieved when using rationally selected combinations and dosages of mineral additives together with a superplasticizer, optimizing the granulometric composition of aggregates, which makes it possible to obtain high-strength concrete as a result [8–10]. The works [11–28] describe the mechanisms of action of various mineral additives; for example, fly ash from thermal power plants, ground blast-furnace granulated slag, metakaolin, micro-silica, and present various dosages of compositions of self-compacting concrete mixtures both with one type of mineral additive and with several. Improvements in the characteristics of self-compacting concrete mixes and concrete obtained in [11–28] from finely dispersed mineral additives confirm the effectiveness of their use and the need for accurate selection of the concrete composition based on them. Furthermore, an essential role in forming the structure of concrete and interparticle interactions is played using finely dispersed mineral material-micro- (nano-) silica. Due to its properties and granulometry, it creates a dense three-dimensional framework of chains and aggregates with numerous coagulation contacts with other larger particles. Therefore, the use of silica fume and other mineral additives listed above enhances the effect and further improves the technological characteristics of self-compacting concrete mixtures and the physical, mechanical properties [29–33].

The structure-forming role of mineral additives is to reduce the induction period of structure formation due to the adsorption of hydrolysis products and to increase the time to reach supersaturation of the liquid phase. Thus, the mineral additive activates the processes of hydration of the binder and promotes an increase in the volume and degree of crystallinity of the formed hydrates, which contributes to the compaction of the structure at the point of contact with the additive with a significant increase in the microhardness of hydrate aggregates [34].

The joint inclusion of a superplasticizer and a mineral additive into a self-compacting concrete mix increases the effect compared with the use of only one of these types of additives. In the example of cement stone, the differences in the relative value of the modulus of elasticity are presented when using only a superplasticizer (an increase of up to 10%) and when using a superplasticizer with a mineral additive (an increase of up to 20%) [27,28].

Thus, the use of complex additives (chemical and mineral) makes it possible to reduce the consumption of cement, maintain the workability of concrete mixtures for a long time, increase the density of the structure, and, as a result, the strength, durability, and resistance of concrete under aggressive operating conditions [10].

In [35], the authors studied the "effect of granulated blast-furnace slag, and two types of superplasticizers on the properties of self-compacting concrete. The results showed that concrete mixtures with a polycarboxylate-based superplasticizer provided more excellent workability and higher compressive strength for all ages than those with a naphthalenesulfonate-based superplasticizer." In [29], the authors conducted experimental studies to assess the effect of micro-silica and a viscosity modifier on the mechanical and rheological properties of self-compacting concrete.

One of the most "focused on the maximum use of industrial waste as modifying additives for self-compacting concrete" are works [10,36–38]. The authors of these works carried out theoretical and experimental studies, which are of particular interest to us. In [10], the influence of a polyfunctional modifier on the properties of concrete mixtures and concrete was studied. For this, the rheological properties of cement pastes used in the compositions of SCC were determined. It was established that a polyfunctional modifier in the form of a concentrated suspension of microsilica in a solution of sodium hydroxide and a superplasticizer based on polymethylene naphthalenesulfonate reduces plastic viscosity and slightly reduces the yield strength. The regularities of influence of rheological properties of cement paste with organomineral modifier on the properties of self-compacting concrete mix was established. As a replacement for cement in the amount of 35–45%, ground blast furnace granulated slag was used. It was also established that the multicomponent modifier improves the properties (increase in slump flow and slump retention) and hardened SCC [36]. This study, in many ways, sets itself the task of developing the existing ideas and achievements of the authors [10,36–38].

Having reviewed and analyzed the previously achieved results and technological methods obtained by other authors [1–45], it can be noted that the use of various industrial wastes as modifier additives have already been announced as promising areas. However, it should be emphasized that the use of such wastes in a processed form, crushed to the state of nanoparticles, thereby molds them into the form of nano-modifiers. Nanomodification of building materials, particularly concretes and self-compacting concretes, is one of the main fundamental areas of building materials science. Summing up the results of the literature review and the conclusion made, it can be formulated a working hypothesis of the study, define the study's purpose and objectives, and formulate the proposed scientific novelty and practical significance of the study.

2. Materials and Methods

2.1. Hypothesis, Aim, and Objectives of the Study

The working hypothesis of the study is the possibility of improving the manufacturability of self-compacting concrete mixtures using nano-modifying additives from several types of waste, thereby making it possible to achieve an increase in the manufacturability, efficiency, and environmental friendliness of the construction process.

The study is aimed to review, investigate, and evaluate from the point of view of theory and practice the possibility of using various types of industrial and agricultural waste as nano-modifiers in self-compacting concrete with improved characteristics.

The research objectives comprise:

- formulation of the problem based on data on the modification of concrete mixtures with a reorientation of the study in the direction of nano-modification with the waste of this type;
- conducting approbation studies with a detailed study of the initial components selected based on the most common wastes that accumulate in large quantities and pose the greatest potential threat to the environment;
- after selecting, determining, and establishing the initial characteristics of the base raw material-determining the working compositions for optimizing and testing the hypotheses put forward, conducting large-scale experimental studies with fixing the technological characteristics of self-compacting concrete mixtures as criteria and checking them to verify the resulting characteristics of highly functional concrete from such mixtures;
- summarizing the results of the study, determining the technological properties of concrete mixes and the performance characteristics of concrete, putting forward subsequent scientific hypotheses, and developing proposals for applied use in building materials science and technology of building processes, determining the goals and objectives of subsequent research, as well as proposals for industrial approbation of the results obtained after their detailed comparison with the results of other authors and understanding of the scientific novelty and place of the research in construction science and practice.

The program study is presented in Figure 1.

2.2. Materials

Considering the complexity of the study and its division into the study of the technological parameters of the concrete mixture, and the verification of the results achieved by the characteristics of the obtained concrete, a complete list of the materials and research methods used are presented.

When conducting experiments for the preparation of cement pastes and self-compacting concrete mixtures, Portland cement PC 500-D0-N produced by Holcim (Rus) LLC (Volsk, Russia) was used as a binder. The chemical and mineralogical composition of Portland cement clinker is presented in Tables 1 and 2, and the physical and mechanical properties of cement are presented in Table 3.

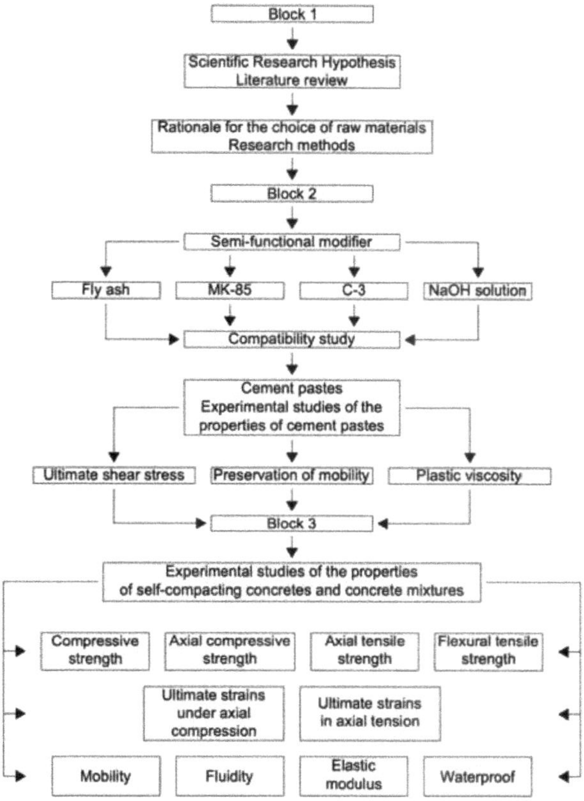

Figure 1. The research program.

Table 1. Chemical composition of Portland cement clinker.

Cement Type	Oxide Content, %											L.O.I.	Cl
	SiO_2	Al_2O_3	CaO	Fe2O3	MgO	TiO_2	P_2O_5	SO_3	Na_2O	K_2O	$Na_2O_{equiv.}$		
PC 500-D0-N	20.5	4.1	64.9	3.5	1.1	0.2	0.2	3.1	0.2	0.8	0.7	0.9	0.01

Table 2. Mineralogical composition of clinker.

Cement Type	Mineral Content, %				
	C_3S	C_2S	C_3A	C_4AF	$CaO_{fr.}$
PC 500-D0-N	74.0	8.6	4.0	11.8	0.7

Table 3. Physical and mechanical properties of Portland cement PC 500-D0-N.

Density, kg/m^3	Blaine Specific Surface Area, cm^2/g	Normal Density, %	Setting Time, min		Compressive Strength 28 Days, MPa	Bending Strength 28 Days, MPa
			Start	End		
3100	3512	26.4	160	212	59.2	6.9

As mineral additives, the following are accepted: fly ash from the Novocherkassk State District Power Plant (Novocherkassk, Russia) and micro-silica grade MS-85 produced by ZIPo LLC (Lipetsk, Russia). The chemical composition and physical and mechanical characteristics of fly ash are given in Table 4; the chemical composition of micro-silica is presented in Table 5.

Table 4. Chemical composition and physical and mechanical characteristics of fly ash from Novocherkassk GRES.

Indicators	Value
SiO_2, %	45.92
TiO_2, %	0.87
Al_2O_3, %	25.9
Fe_2O_3, %	9.38
CaO, %	0.81
MgO, %	1.35
MnO, %	0.36
K_2O, %	5.29
Na_2O, %	0.93
SO_3, %	1.28
P_2O_5, %	0.15
L.O.I., %	7.37
Bulk density, kg/m^3	480
True density, kg/m^3	2600

Table 5. Chemical composition of micro-silica MS-85.

Material	Oxide Content, %							
	SiO_2	Al_2O_3	Fe_2O_3	CaO	MgO	R_2O	SO_3	L.O.I.
MS-85	82.3	1.7	3.0	1.1	0.2	0.8	3.5	7.4

Granite crushed stone produced by Pavlovsknerud JSC (Pavlovsk, Russia) was used as a coarse aggregate, and quartz sand (M_f = 2.2) produced by Arkhipovsky Quarry JSC (Arkhipovskoe village, Russia) and quartz sand (M_f = 1.2) produced by Quartz Sands LLC (Semenov, Russia). Physical and mechanical properties of aggregates are presented in Table 6.

Table 6. Physical and mechanical properties of aggregates.

Material Title	Fineness Modulus M_f or Grain Sizes	Bulk Density, kg/m^3	True Density, kg/m^3	Void, %	Crushability, %
Medium quartz sand	2.2	1426	2649	46.8	-
Very fine quartz sand	1.2	1413	2653	44.2	-
Granite crushed stone	fraction 5–10 mm	1368	2670	45.6	11.8

The following are accepted as chemical additives: sodium hydroxide produced by OOO "KHIMEKS" (Moscow, Russia); superplasticizer C-3 manufactured by Component LLC (Vladimir, Russia). The qualitative characteristics of the additives used are presented in Tables 7 and 8.

Table 7. Qualitative characteristics of superplasticizer C-3.

Indicator Title	Value
Visual form	Homogeneous dark brown liquid
Density at 20 °C, no less, g/cm^3	1.17
Water, wt. % no more	68.0
Hydrogen ion activity index (pH), 2.5% aqueous solution	8.0 ± 1.0

Table 8. Qualitative characteristics of NaOH.

Indicator Title	Value
Main substance, %	99.487
Sodium carbonate (Na_2CO_3), %	0.5
Sodium chloride (NaCl), %	0.005
Sodium sulfate (Na_2SO_4), %	0.005
Iron in terms of Fe_2O_3, %	0.002
Mercury content (Hg), %	0.0001

2.3. Methods

The research methods can be conditionally divided into groups:

- analytical methods and literature data study;
- phenomenological methods for the selection and determination of raw materials, compositions, and the setting of allowable limits and ranges;
- experimental studies to verify the selected recipe and technological solutions;
- mathematical methods of substantiation and optimization of the developed proposals.

Experimental studies were carried out using standardized and unique methods, including:

- determination of the characteristics of the initial raw materials by standardized methods for binder and aggregates and research methods of laser granulometry for nano-modifying additives;
- determination of the technological characteristics of concrete mixes by standardized methods with verification of the results by the author's method;
- determination of the characteristics of the resulting concrete by standardized methods [46–56].

The determination of the granulometric composition of powdered raw materials (fly ash and micro-silica) was carried out using the method of laser granulometry. A Microsizer 201C laser particle analyzer (OOO VA Install, St. Petersburg, Russia) was used for granulometric analysis [52].

X-ray phase analysis (XPA) of mineral additives was carried out on an X-ray diffractometer HZG-4C (Freiberger Prazisionmechanik, Germany).

The rheological properties of cement pastes (plastic viscosity, ultimate shear stress) were studied using a B085-21 rotational viscometer (OOO RNPO RusPribor, St. Petersburg, Russia). The results were processed according to the Bingham rheological model [10]:

$$\tau = \tau_0 + \mu\gamma \quad (1)$$

where τ_0 is ultimate shear stress, (Pa); μ is plastic viscosity (Pa$\times$s); γ is shear rate gradient (s^{-1}).

The workability of a self-compacting concrete mixture is characterized by its ability to slump-flow and fill a given shape under the action of its mass or external forces while maintaining uniformity and solidity. Indicators of the technological properties of self-compacting concrete mixtures are such characteristics as mobility, expressed by the diameter of the flow of a standard cone and fluidity, expressed by the flow time of the concrete mixture from a V-shaped funnel. The mobility of the concrete mixture was estimated by the diameter of the flow in millimeters using a standard cone (RNPO RusPribor LLC, St. Petersburg, Russia), the appearance of the cone is shown in Figure 2a. The flow time of the self-compacting concrete mixture was determined using a V-shaped funnel Matest C171 (Treviolo, Italy). This funnel is made of stainless steel and mounted on a tripod. The inside has a smooth surface, the edges of the upper hole are reinforced with stiffeners, and the bottom hole is sealed (Figure 2b).

To evaluate and verify the results of standard methods, taking into account the specifics of the resulting mixtures, additional tests were carried out using a non-standard method, according to [57].

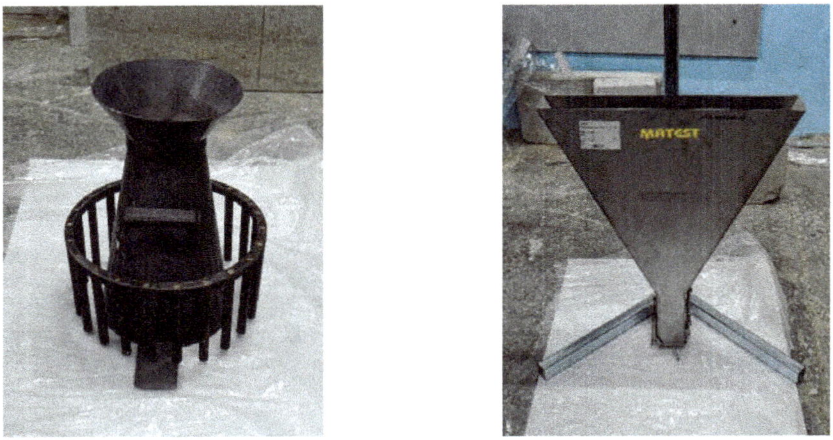

(**a**) (**b**)

Figure 2. Devices for monitoring the rheological characteristics of SCC mixtures: (**a**) a cone for determining mobility; (**b**) V-shaped flow funnel.

Compression tests were carried out on 100 mm × 100 mm × 100 mm sample cubes (Figure 3a) of six pieces for each SCC composition. In addition, axial and bending tensile strength tests were carried out on 100 mm × 100 mm × 400 mm prism specimens of three pieces for each composition and type of SCC strength according to the requirements GOST 10180 "Concretes. Methods for strength determination using reference specimens" [58].

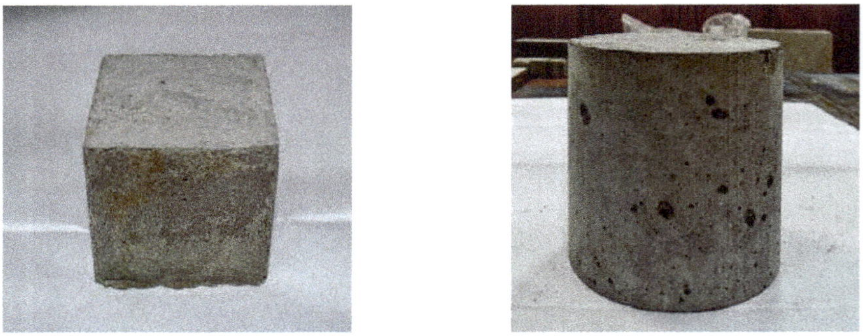

(**a**) (**b**)

Figure 3. Concrete samples: (**a**) cubes; (**b**) cylinders.

Determination of the elastic modulus and axial compressive strength (three prism samples 100 mm × 100 mm × 400 mm for each SCC composition) was carried out following the requirements of GOST 24452-80 "Concretes. Methods of prismatic, compressive strength, modulus of elasticity and Poisson's ratio determination" [59].

Measurements of the strains of the concrete samples were carried out with the chain of strain gauges with a base of 50 mm and dial indicators with a division value of 0.001 mm on the experimental prisms. In addition, tests of experimental prisms for axial compression and axial tension were carried out at a constant strain rate to obtain the strength and strain characteristics of concrete and its complete strain diagrams "σ-ε" with descending branches.

The water-resistance of SUB samples was determined by the "wet spot" method following the requirements of GOST 12730.5 "Concretes. Methods for determination of water tightness" [60] on cylindrical specimens with a diameter of 150 mm and a height of

150 mm (Figure 3b). Six prepared samples of each composition were stored in a standard hardening chamber at a temperature of 20 ± 2 °C and relative air humidity of at least 95%.

The water pressure was increased stepwise with 0.2 MPa for 1–5 min and maintained at each step for 16 h (Figure 4). The water-resistance of each sample was evaluated by the maximum pressure of water at which no leakage through the sample was observed in the form of a wet spot on the end surface of the sample opposite to its surface through which water was supplied under pressure. The water-resistance of a series of concrete samples was evaluated by the maximum water pressure at which no seepage was observed on four of the six samples.

(a) (b)

Figure 4. Concrete water resistance test: (**a**) UVF-6 installation; (**b**) concrete specimens in test cases.

All manufactured prototypes hardened under normal conditions at a temperature of 20 ± 2 °C for 28 days.

The study used mathematical planning of the experiment, which aims to increase the efficiency of experiments.

In particular, the Box–Wilson method was applied-orthogonal planning of the optimal experiment. The essence of the chosen method is as follows:

In the first stage, for some local area of existence of the object under study, the regularities of its behavior are determined by the method of planning the experiment, and as a result of their analysis, the direction to the optimum is determined, in which the parameters should be changed. Next, another experiment is carried out in a new area of the object's existence, and so on, until optimal conditions are reached.

Mathematical planning of an experiment is a formalized procedure for choosing the number and conditions for conducting experiments that are necessary and sufficient to solve a problem with the required accuracy. This procedure allows for the following:

(a) minimizing the number of experiments;
(b) changing the parameters that affect the state of the object according to a certain law;
(c) using a mathematical method that formalizes the actions of the experimenter in data processing and obtain a mathematical model of the research object;
(d) use the logical apparatus when making decisions based on the analysis of the resulting model.

Following the recommendations [10,36–38], it is possible to correctly assess the effectiveness of nano-modification of self-compacting concrete mixtures by analyzing the following experimental operations:

- revealing the influence of the type of dispersion medium on the efficiency of dispersion of the nano-modifier;
- to determine and evaluate the rheological properties of cement pastes with a complex nano-modifier;
- based on the data obtained, optimize the compositions according to the criteria of strength and deformability of concrete and the mobility of concrete mixtures.

3. Results

3.1. Determination of the Characteristics of Raw Components

The granulometric characteristics of the mineral additives are shown below in Figure 5.

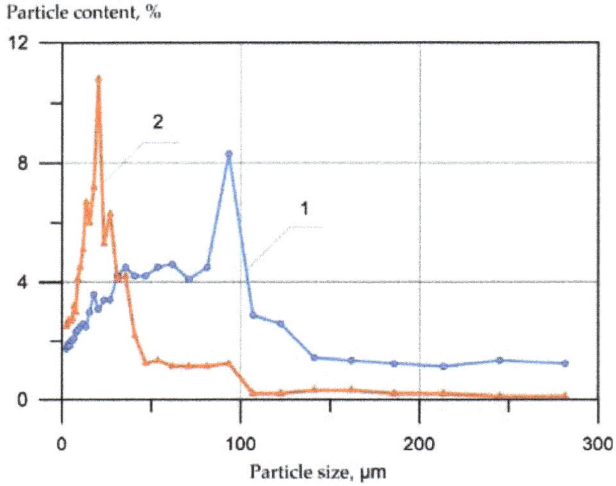

Figure 5. Particle size distribution determined on the laser analyzer "Microsizer 201C": 1—fly-ash; 2—micro-silica MS-85.

Following the data shown in Figure 5, it can be seen that the main size distribution range of the studied fly ash particles is in the range of 2–120 μm, and the main peak falls at 90 μm. On the other hand, as for micro-silica particles, the predominant part of the grains (more than 80%) is located in the range from 2 to 40 μm, and the main peak falls at 20 μm.

X-ray phase analysis of fly ash and micro-silica particles is shown in Figure 6.

X-ray diffraction analysis shows that the fly ash microspheres (Figure 6a) are a mixture of two phases: $Al_6Si_2O_{13}$ mullite and an X-ray amorphous phase referred to as a glass, while the main phase is the mullite phase. Micro-silica (Figure 6b) is represented by amorphous silica, minor impurities of iron, carbonaceous substances, and crystalline α-quartz.

3.2. Influence of the Type of Dispersion Medium on the Efficiency of Micro-Silica Dispersion

Following the recommendations [10,36–38], a comparative analysis of three variants of dispersion media was carried out which included:

- tap water;
- sodium hydroxide solution (concentration 1.5% in terms of Na_2O);
- solution of sodium hydroxide and superplasticizer (concentration 2.5%).

The preparation of an aqueous suspension of silica fume was carried out in a laboratory planetary ball mill "Activator-4M" for 8–12 min.

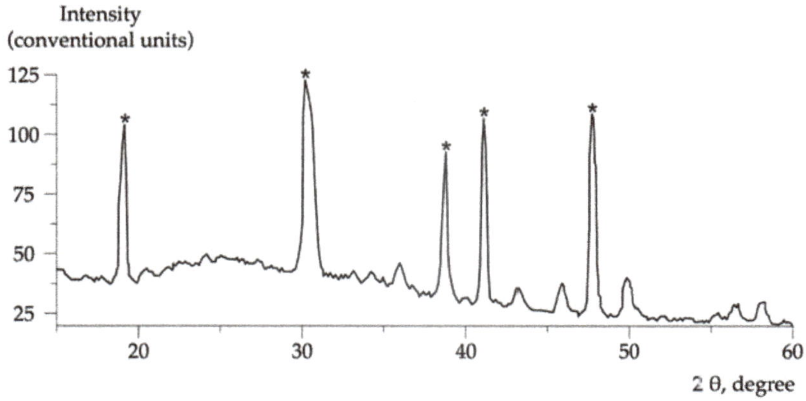

(**a**)

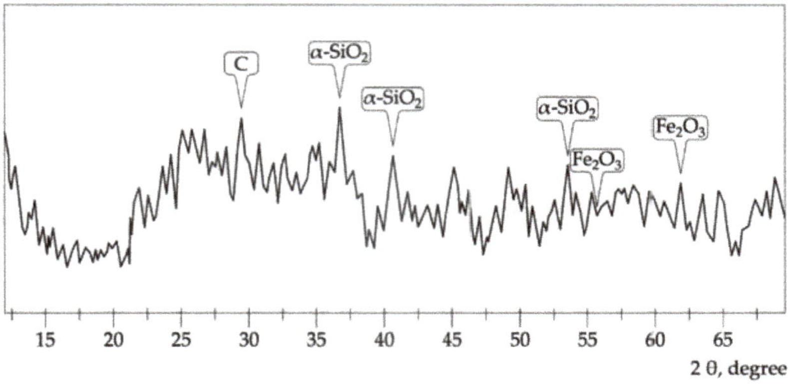

(**b**)

Figure 6. Diffractogram of samples of (**a**) fly ash (*—mullite phase); (**b**) micro-silica.

After the dispersion process, the predominant part of the micro-silica particles (more than 80%) is in the range of 2 to 35 μm, and the main peak falls at 10 μm.

The dispersing ability of liquids was assessed by the kinetics of sedimentation of aqueous suspensions of silica fume (Figure 7), the concentration of the MS-85 additive in the aqueous suspension was 35%.

The authors of works [10,36–38] provide a theoretical justification that the highest particle settling rate is observed when tap water is used as a liquid medium. In a sodium hydroxide solution, the precipitation rate decreases, indicating the resulting dispersion's finer granulometric composition. This is due to the fact that in a highly alkaline medium, along with the dispersion of micro-silica aggregates, it also dissolves.

In the presence of alkali, silicon dioxide goes into solution in the form of a silicate ion, followed by interaction with water to form soluble mono silicic acid. When the amount of alkali is small, mono silicic acid can polymerize to form stable colloidal particles. In the soluble sodium silicate formation process, an additional dispersion of amorphous silica occurs with a transition to a highly concentrated state-a lyosol nano dispersed system. In addition, the degree of micro-silica grinding is exclusively influenced by the self-heating temperature of the suspension during grinding, which is achieved mainly due to the work of friction.

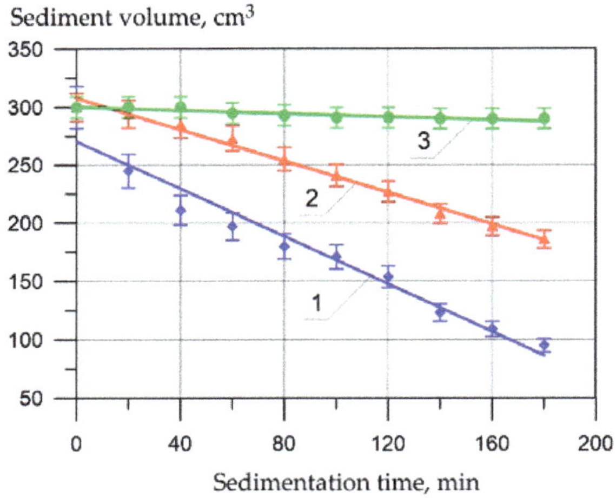

Figure 7. The kinetics of sedimentation of micro-silica suspension dispersed in the medium: 1—tap water; 2—sodium hydroxide solution; 3—sodium hydroxide solution + superplasticizer.

When micro-silica is dispersed in the medium of sodium hydroxide and superplasticizer solutions, a colloid system stabilized by a surfactant is formed, which retains sedimentation stability for more than 10 days [10,36–38].

Thus, a solution of sodium hydroxide and a superplasticizer (concentration 2.5%) was chosen as the dispersion medium.

3.3. *Rheological Properties of Cement Pastes with a Complex Nano-Modifier*

Further, following the developed program, it was necessary to evaluate the rheological properties of cement pastes with different content of complex nano-modifiers (Table 9 and Figure 8). The rheological characteristics of cement pastes were determined as an average of three samples of each composition and are presented in Table 9.

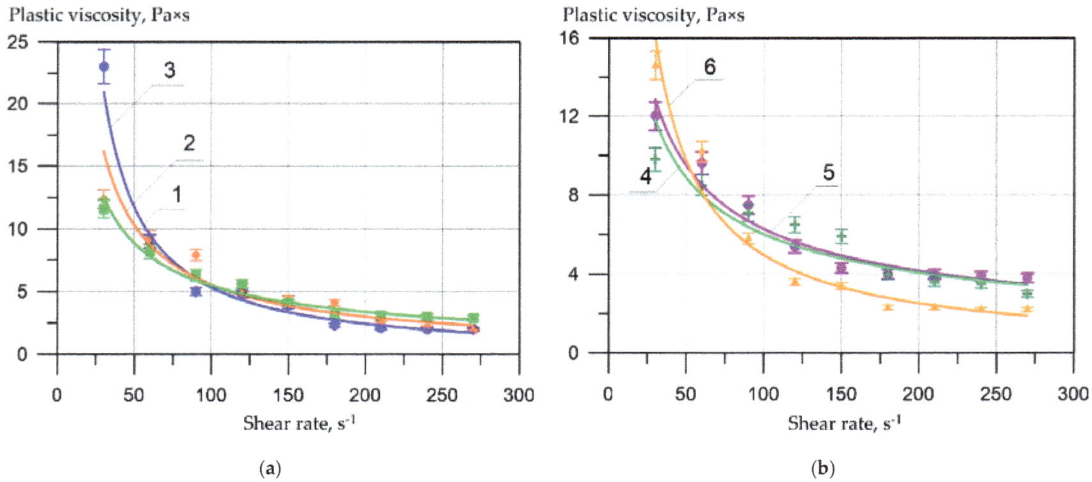

Figure 8. The plastic viscosity of cement pastes depends on shear rate: (**a**) 1—C1; 2—P1; 3—P2; (**b**) 4—P3; 5—P4; 6—P5.

Table 9. Composition and properties of cement paste with a complex nano-modifier.

Denotation	Component Content, g					W/W	Ultimate Shear Stress, τ_0, Pa
	PC	Fly-ash	MS-85	C-3	NaOH		
C1	500	-	-	-	-	0.25	59.7 ± 3.4
P1	500	-	-	5	-	0.20	56.7 ± 3.0
P2	-	500	-	-	1.5	0.29	56.0 ± 2.9
P3	300	200	-	-	1.5	0.26	57.6 ± 3.1
P4	300	200	-	5	1.5	0.22	58.1 ± 3.3
P5	300	200	24	5	1.5	0.24	58.0 ± 3.2

During experiments on the selected initial formulation and technological parameters, the data [10,36–38] confirmed that the maximum value of structural viscosity at a minimum shear rate gradient of 30 s the value of normal density. When a superplasticizer is added, the water demand of the cement paste (P1) decreases to W/W = 0.20. In this case, the value of structural viscosity decreases by almost two times, and the equilibrium plastic viscosity at a shear rate of 240–270 s-1 is the smallest among all the studied compositions of cement pastes. At the same time, the addition of sodium hydroxide has a weak plasticizing effect, reducing the water requirement of the fly ash paste (P2) to W/W = 0.29.

To summarize, it can be noted that the partial replacement of Portland cement with fly ash in 40% slightly increases the water demand and plastic viscosity of the cement paste (P3) as established [10,36–38]. At the same time, a positive effect of a complex additive in the form of a superplasticizer C-3 + NaOH (P4) was noted. On the other hand, the relatively low content of micro-silica increases both the water demand and the plastic viscosity of the cement paste (P5). This effect should have a positive impact on the SCC resistance to delamination. Also, as in [10,36–38], we have not established a significant effect of the complex nano-modifier on the indicators of the ultimate shear stress of the cement paste.

3.4. Optimization of the Composition of Complex Nano-Modifiers According to the Criterion of Concrete Strength and Mobility of Concrete Mixtures

To optimize the composition of concrete (modifier) in terms of the consumption of mineral and chemical additives, studies were carried out using the method of orthogonal compositional planning of the 2nd order of Box–Wilson, which, under specific optimal parameters, allows achieving the maximum strength of concrete at the age of 28 days while ensuring the required mobility of concrete mixtures. Thus, the compressive strength ($R_{b,cub}$), mobility (d_{sl}), and flow (T_{fl}) parameters served as optimization parameters. The basic composition of concrete was adopted for testing, determined according to the method [61]. As a result of calculations, the parameters of the composition of the concrete mix are shown in Table 10.

Table 10. Parameters of the composition of the concrete mixture.

Indicator Title	W/C	C, kg/m^3	W, l/m^3	CS, kg/m^3	S, kg/m^3	C-3, kg
Value	0.64	447	286	812	864	6.7

The values of the variation factors, their physical meaning, and variation levels are given in Table 11, the planning matrix-in Table 12. The compressive strength of concrete was determined on sample cubes with an edge of 100 mm at 28 days of normal hardening. The results of the experiments are shown in Table 13.

Table 11. Levels of input factors and intervals of their variation for 2nd order planning.

Input Factors		Level of Variation					Variation Interval δ
Factor Code	The Physical Meaning of the Factor	−2	−1	0	+1	+2	
X_1	NaOH concentration by weight of cement (N), %	0.2	0.4	0.8	1.2	1.4	0.4 and 0.2
X_2	Fly ash content (instead of part of cement) (F), %	25	30	40	50	55	10 and 5
X_3	C-3 content by weight of cement (C), %	1.1	1.2	1.5	1.7	1.8	0.2 and 0.1
X_4	The content of MS-85 by weight of cement (M), %	7	8	10	12	13	2 and 1

Table 12. Experiment design matrix.

Num	Variables				Natural			
	X_1	X_2	X_3	X_4	N	F	C	M
1	−1	−1	−1	−1	0.4	30	1.2	8
2	1	−1	−1	−1	1.2	30	1.2	8
3	−1	1	−1	−1	0.4	50	1.2	8
4	1	1	−1	−1	1.2	50	1.2	8
5	−1	−1	1	−1	0.4	30	1.7	8
6	1	−1	1	−1	1.2	30	1.7	8
7	−1	1	1	−1	0.4	50	1.7	8
8	1	1	1	−1	1.2	50	1.7	8
9	−1	−1	−1	1	0.4	30	1.2	12
10	1	−1	−1	1	1.2	30	1.2	12
11	−1	1	−1	1	0.4	50	1.2	12
12	1	1	−1	1	1.2	50	1.2	12
13	−1	−1	1	1	0.4	30	1.7	12
14	1	−1	1	1	1.2	30	1.7	12
15	−1	1	1	1	0.4	50	1.7	12
16	1	1	1	1	1.2	50	1.7	12
17	−2	0	0	0	0.2	40	1.5	10
18	2	0	0	0	1.4	40	1.5	10
19	0	−2	0	0	0.8	25	1.5	10
20	0	2	0	0	0.8	55	1.5	10
21	0	0	−2	0	0.8	40	1.1	10
22	0	0	2	0	0.8	40	1.8	10
23	0	0	0	−2	0.8	40	1.5	7
24	0	0	0	2	0.8	40	1.5	13
25	0	0	0	0	0.8	40	1.5	10
26	0	0	0	0	0.8	40	1.5	10
27	0	0	0	0	0.8	40	1.5	10
28	0	0	0	0	0.8	40	1.5	10
29	0	0	0	0	0.8	40	1.5	10
30	0	0	0	0	0.8	40	1.5	10
31	0	0	0	0	0.8	40	1.5	10

Having calculated the coefficients of orthogonal central planning of the 2nd order, the following regression Equations (2)–(4) were obtained:

$$\begin{aligned} R_{(b,cub)} = {} & 50.618 - 1.698 \cdot N - 6.941 \cdot F + 0.974 \cdot C - 0.283 \cdot M + 0.785 \cdot N \cdot F + 1.284 \cdot N \cdot C \\ & -0.586 \cdot N \cdot M + 0.150 \cdot F \cdot M - 0.349 \cdot C \cdot M - 3.828 \cdot N^2 - 3.370 \cdot F^2 - 1.746 \cdot C^2 - 2.229 \cdot M^2 \end{aligned} \tag{2}$$

$$\begin{aligned} d_{spr} = {} & 546.592 - 17.061 \cdot N - 1.248 \cdot F + 64.500 \cdot C - 7.074 \cdot M - 1.870 \cdot N \cdot F + 16.831 \cdot N \cdot C \\ & +6.857 \cdot N \cdot M + 3.117 \cdot F \cdot M - 5.610 \cdot C \cdot M + 9.713 \cdot N^2 - 18.395 \cdot F^2 - 1.040 \cdot C^2 - 18.395 \cdot M^2 \end{aligned} \tag{3}$$

$$\begin{aligned} T_{fl} = {} & 22.653 + 0.375 \cdot N + 0.458 \cdot F + 3.038 \cdot C + 1.290 \cdot M + 0.436 \cdot N \cdot F + 0.187 \cdot N \cdot C + 0.062 \cdot N \cdot M \\ & -0.187 \cdot F \cdot M + 2.307 \cdot C \cdot M - 0.326 \cdot N^2 + 0.015 \cdot F^2 - 1.348 \cdot C^2 - 2.588 \cdot M^2 \end{aligned} \tag{4}$$

The significance of the regression equation coefficients was checked using the Student's test, and the adequacy of the regression equations to the experimental data was checked using the Fisher criterion.

Table 13. Properties of concrete mixture and concretes.

Num	Yield, s	Slump-Flow, mm	Compressive Strength, MPa
1	14	500	45.9
2	15	440	39.2
3	15	480	32.5
4	16	450	29.7
5	15	610	49.7
6	16	630	48.5
7	17	600	30.8
8	17	620	31.7
9	13	490	47.9
10	12	460	37.5
11	15	450	33.6
12	16	440	28.7
13	26	520	48.4
14	25	610	45.9
15	21	600	31.3
16	26	610	28.9
17	23	670	40.7
18	24	460	35.8
19	24	460	52.8
20	26	450	27.8
21	13	410	45.8
22	26	640	47.9
23	14	460	44.9
24	15	450	44.9
25	23	540	46.3
26	21	550	45.8
27	22	550	46.8
28	20	540	45.5
29	24	540	44.9
30	23	540	47.9
31	25	550	46.9
Control	11	500	40.4

After verification, insignificant coefficients were discarded, resulting in Equations (5)–(7):

$$R_{(b,cub)} = 50.618 - 6.941 \cdot F - 3.828 \cdot N^2 - 3.370 \cdot F^2 \quad (5)$$

$$d_{spr} = 546.59 + 64.50 \cdot C \quad (6)$$

$$T_{fl} = 22.653 + 3.038 \cdot C + 2.307 \cdot C \cdot M - 2.588 \cdot M^2 \quad (7)$$

Graphical interpretation of the obtained polynomial dependencies is shown in Figures 9–11.

An analysis of the regression equations shows that the factors X_1 and X_2 have the greatest influence on the compressive strength of the SCC. The optimal dosage of fly ash introduced to replace part of the cement is a dosage of 40% by weight of cement. As for the dosage of NaOH, the optimal dosage is 0.8% by weight of cement.

It should be noted that for compositions No. 3, 4, 7, 8, 11, 12, 15, 16, where the content of fly ash in the SCC composition is 50%, lower values of compressive strength are observed in comparison with other compositions and vary from 28.9 MPa and up to 33.6 MPa; the minimum strength value was recorded for composition No. 22 with a fly ash dosage of 55% and is 27.8 MPa.

The X3 factor has a significant impact on mobility, increasing the dosage of the C-3 plasticizer from 1.1% to 1.8% increases the mobility of the self-compacting concrete mix. The maximum value of the mixture flow diameter was recorded for composition No. 22, where the C-3 dosage was 1.8% of the cement weight.

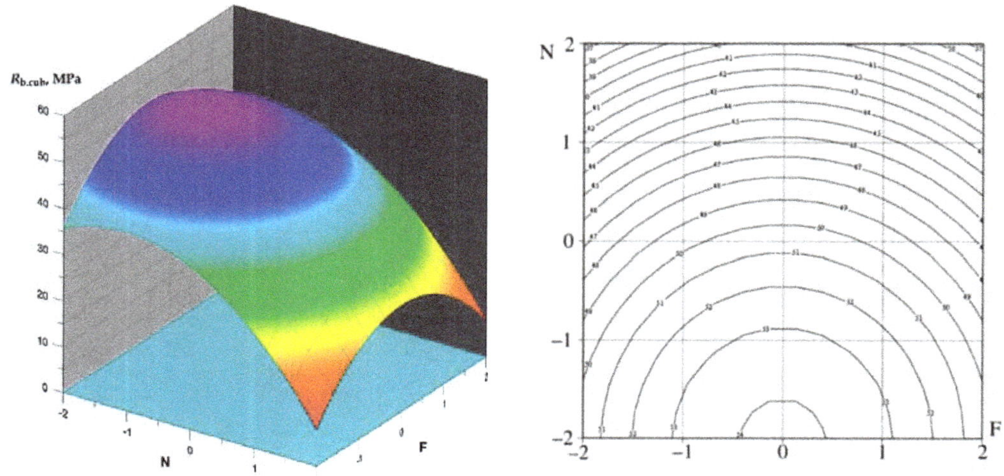

(a) (b)

Figure 9. Dependence of compressive strength on NaOH concentration and fly ash content: (**a**) plot surface; (**b**) surface contours.

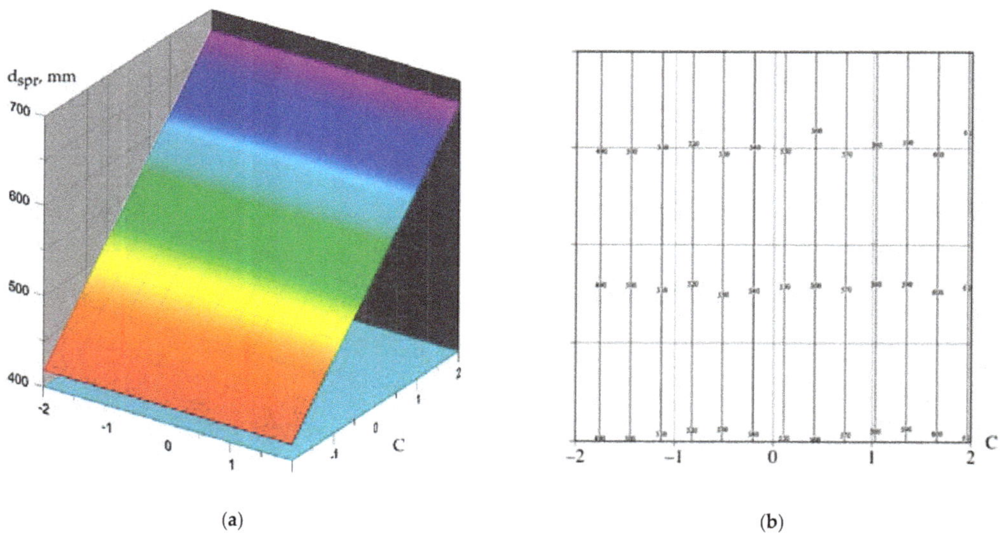

(a) (b)

Figure 10. The dependence of the diameter of the flow of the concrete mixture on the content of C-3: (**a**) dependence surface; (**b**) surface contours.

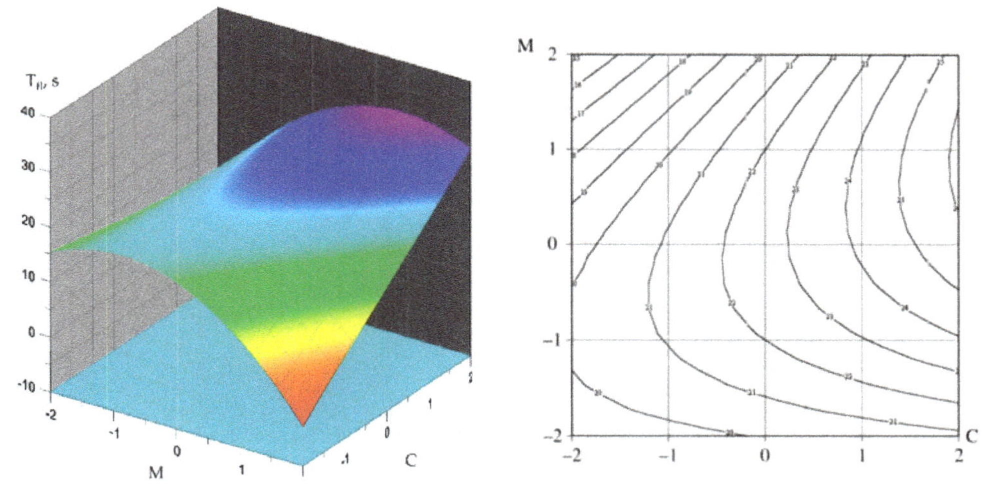

(**a**) (**b**)

Figure 11. Dependence of the fluidity of the concrete mix on the content of C-3 and silica fume: (**a**) the dependence surface; (**b**) surface contours.

As for such a parameter as the flowability of a self-compacting concrete mix, the factors X_3 (C-3 content by cement mass, %) and X_4 (MS-85 content by cement mass, %) have the greatest influence on this indicator. The highest values of fluidity were recorded for compositions No. 13, 14, 16, 20, and 22 and amounted to 25–26 s. The lowest values of fluidity are observed in compositions No. 1, 2, 3, 4, 5, 6, 7, 8, 9, 10, 11, 23 and vary from 13 to 17 s. An increase in the dosages of superplasticizer and micro-silica additives contributes to an increase in the flow time of the concrete mixture, which can be directly related to an increase in the viscosity of the mixture itself.

3.5. Rheological and Physical-Mechanical Properties of Self-Compacting Concrete with a Complex Nano-Modifier

Table 14 shows the compositions of the self-compacting concrete mixture of the base composition with a complex nano-modifier.

Table 14. Composition of self-compacting concrete mixes.

Mix	W/B	C, kg/m^3	CS, kg/m^3	S, kg/m^3	C-3, kg	NaOH, kg/m^3	MS-85, kg/m^3	Fly-Ash, kg/m^3
CC	0.48	447	812	864	6.7	-	-	-
CNM	0.48	268	812	864	4.8	2.14	26.8	179

Comparative tests of concretes and concrete mixtures of the control composition (CC) and with a complex nano-modifier (CNM) are presented in Table 15.

Comparative tests of the control composition of concrete and nano-modified one show that the alkaline activator in the composition of the complex nano-modifier allows partial replacement (40%) of Portland cement with fly ash while increasing the compressive strength of concrete by 17%, the axial compressive strength by 10%, the strength for axial tension by 19%, and the tensile strength in bending by 11%. Ultimate strains under axial compression and tension decreased by 14% and 12%, respectively. The modulus of elasticity increased by 11%. As for water resistance, its value increased by 22%.

Table 15. Rheological and physical-mechanical properties of self-compacting concrete with a complex nano-modifier.

Indicator Title	Value	
	CC	CNM
Slump-flow, mm	500	560
Fluidity, s	12	22
Compressive strength at the age of 2 days, MPa	21.9	25.9
Compressive strength at the age of 7 days, MPa	33.8	38.7
Compressive strength at the age of 28 days, MPa	46.9	51.9
Axial compressive strength at the age of 28 days, MPa	35.7	38.9
Axial tensile strength at the age of 28 days, MPa	3.2	3.8
Tensile strength in bending at the age of 28 days, MPa	5.6	6.2
Ultimate strains under axial compression, mm/m $\times 10^{-3}$	2.17	1.86
Ultimate strains in axial tension, mm/m $\times 10^{-4}$	1.26	1.09
Modulus of elasticity, GPa	31.4	34.8
Water resistance, MPa	1.25	1.53

As can be seen in Table 13, CNM concrete samples are characterized by higher strength in the early stages of hardening (by 18% on day 2 and by 14% on day 7) compared to CC samples.

Higher indications of the early strength of the CNM composition can be theoretically justified in accordance with [10,36–38] because primary aluminate phases begin to form more intensively, providing concrete strength indicators at an early age of hardening.

As noted earlier, organomineral additives introduced into the cement system in the form of compositions (mixtures) with a superplasticizer with an optimized particle size distribution of aggregates make it possible to obtain high-tech self-compacting concretes. At the same time, such concrete mixtures are characterized by sufficient viscosity to prevent mixture separation and the sedimentation of large aggregates [10,36–38].

Also, according to the test results, "stress-strain" diagrams of compression "ε_b-σ_b" and tension "ε_{bt}-σ_{bt}" were constructed. Graphic dependencies are shown in Figures 12 and 13.

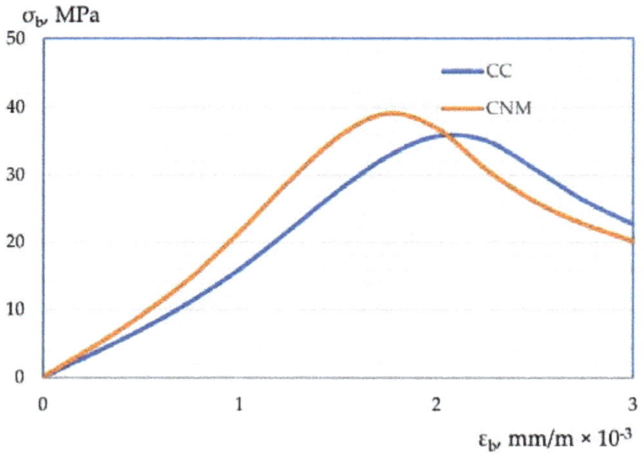

Figure 12. Stress-strain diagram for compression of self-compacting concrete.

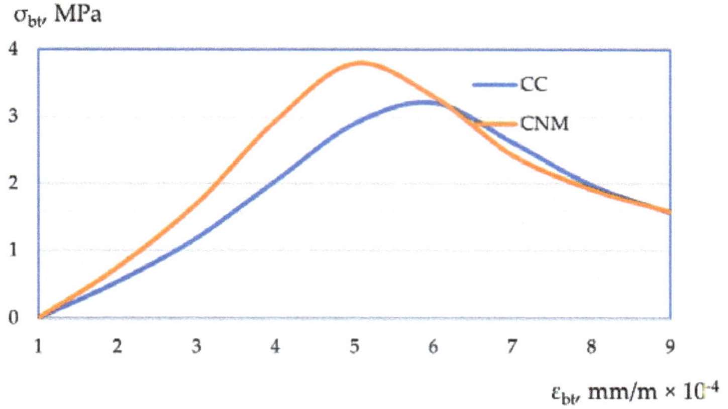

Figure 13. Tensile stress-strain diagram for self-compacting concrete.

After experimentally obtaining "stress-strain" diagrams for cases of axial compression and cases of axial tension and their changes depending on the composition, their more detailed interpretation should be given with an explanation of the results obtained. So, the resulting stress-strain diagrams, both in compression and tension are presented in Figures 12 and 13. Thus, the peak of the "stress-strain" diagrams with the most effective compositions, previously established according to the strength characteristics of concrete, shifts up and to the left relative to control samples made by a more traditional method. This confirms the theoretical and practical results of improving the mechanical and reducing the strain properties of nano-modified concrete obtained according to optimal parameters, both prescription and technological. The shift of the peak of both diagrams up and to the left refers to an increase in strength characteristics and better resistance of such concrete to strain, thereby confirming the hypothesis on the compaction of the concrete package with the help of nano-modifiers, a more structured arrangement of both macroparticles and aggregates of concrete, and microparticles, including dispersed nano-modifiers. Thus, we obtain concrete with an improved structure, which in turn is reflected in their stress-strain diagrams.

4. Discussion

A thorough evaluation of the data got in this research with the results of other authors should be made to assess the authors' contribution to the science and practice of production and construction. To better assess, it is necessary to set the criteria for a comparative assessment, which can be expressed in three aspects. First, this is a research aspect. The second is the technological aspect. And the third aspect is the result got from the point of view of operation of the final product in real construction that are special structures with hard-to-reach concrete. Their evaluation is carried out according to various characteristics and requirements that apply to these concretes.

The structure of large-scale studies conducted by Professor Zaichenko N.M. [10,36–38] earlier, and associate professor Egorova E.V. [10] was taken as a template and pattern concerning a full-fledged comprehensive study of self-compacting concrete. In addition, the works of the authors were considered using various modifiers, which are raw finely dispersed components used as an additive to self-compacting concrete mixtures with their verification by the rheology of the mixtures and the performance properties of the concrete obtained from them [4–9,11–35,39–45].

Summarizing the results of the identified research differences, it can be noted that a large-scale study was conducted in this paper on the nano-modification of self-compacting concrete mixtures with additives of various types. The goal was to improve the technological characteristics of the mixture and the performance characteristics of the resulting concrete. The nano-modifiers were fly ash formed during the operation of the existing state district power station, and micro-silica, which is a waste of industrial production. As a result of the study, high technological and operational indicators of concrete mix and concrete were revealed. The ecological and economic efficiency of the proposed research method, with the proposed technology formulation, has been proven. Thus, the prospects of this study and further vectors of the development of this study have been determined.

From the point of view of the technological properties of the obtained self-compacting concrete mixtures, a quantitative assessment of our results is presented. The quantitative increase was up to 83% in the technological and rheological characteristics of the self-compacting concrete obtained.

The third aspect of comparison is the result demonstrated by the concrete obtained on the technological mixtures proposed in terms of operational characteristics. The quantitative increase in the strength characteristics was up to 19% and up to 22% for water resistance. The reduction in ultimate strains was up to 14% and the elastic modulus increased by 11%.

The increase in strength characteristics is in good agreement with the results of the authors, who previously investigated the use of complex modifiers in self-compacting concrete [10,13,20,21,23,24,31–33]. However, the use of a complex modifier earlier in the form of blast-furnace granulated slag and fly ash [13] or limestone [20] improved the rheological properties but reduced the strength characteristics [13] or slightly increased them [20]. Also, the addition of only fly ash does not allow obtaining high increases in the characteristics of self-compacting concrete [16,17]. At the same time, the effective dosage of fly ash obtained in this research is comparable with the dosages of ash previously used by other authors, which provided the best results (30–50%) [16,17,23,24,33,40].

Summarizing all three aspects, several elements of novelty can be noted. First, this research is based on large-scale studies of our colleagues around the world, considering the experience of previous studies. Second, the theory of nano-modification of self-compacting concrete mixtures was continued and developed, and specific practical results were obtained, which are proposed for development. The vector of this development is bringing nano-modifiers to an even more dispersed state. In this case, an assessment is made in terms of raw materials and the possibility of establishing the effect of even greater grinding of these components. This makes it possible to identify opportunities for further improving the manufacturability of mixtures and improving the performance characteristics of concrete. The engineering aspect of our study is proposed for implementation in those construction organizations and enterprises to produce commercial self-compacting concrete. These enterprises have the highest ecological burden on the environment and access a large array of accumulated industrial waste. Thus, these enterprises are ready to provide initial raw materials that have already been verified in studies as suitable for use.

The observed results can be explained by the fact that the precipitation rate decreases in sodium hydroxide solution, which indicates a finer granulometric composition of the resulting dispersion. In a highly alkaline medium, along with the dispersion of micro-silica aggregates, its dissolution also occurs. In the presence of alkali, silicon dioxide goes into solution in the form of a silicate ion, followed by interaction with water to form soluble monosilicic acid. When the amount of alkali is small, monosilicic acid can polymerize to form stable colloidal particles. In the process of formation of soluble sodium silicate, an additional dispersion of amorphous silica occurs with a transition to a highly concentrated state-a lyosol nanodispersed system. When micro-silica is dispersed in the medium of sodium hydroxide and superplasticizer solutions, a colloidal system stabilized by a surfactant is formed, which retains sedimentation stability for more than 10 days. When replacing 40% of Portland cement with fly ash, a positive effect of a complex additive in the form of a superplasticizer C-3 + NaOH (P4) is observed in terms of SCC resistance to delamination [10,36–38].

Discussing the obtained results and comparing our conclusions with previous literature studies [10,13,20,21,23,24,31–33,36–38], the high degree of efficiency of our proposals can be noted. The complex nano-modifier, including such wastes as micro-silica and fly ash, presented positive results and demonstrated a synergistic effect in terms of saving raw materials and environmentally friendly disposal. This allows for recommending the proposed method for practical implementation in the construction industry and for monolithic concreting of complex densely reinforced structures.

For a visualization of the analysis of the proposed solutions, Figure 14 shows a cause-and-effect diagram of the problem of obtaining effective nano-modified self-compacting concrete mixed with improved technological characteristics.

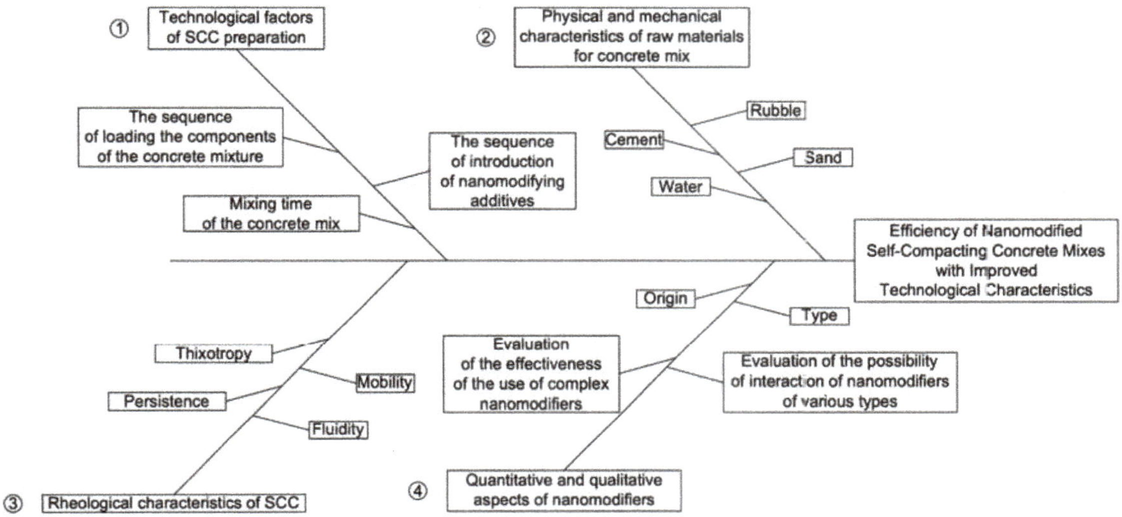

Figure 14. Causal diagram of obtaining effective nano-modified self-compacting concrete mixes with improved technological characteristics.

Thus, the most significant factors affecting the effectiveness of nano-modified self-compacting concrete mixtures with improved technological characteristics were grouped and analyzed.

Branch 1 (Figure 14) of the upper part of the diagram reflects the technological factors in preparing SCC mixtures that affect the quality of SCC mixtures as a final product. Branch 2 of the upper part of the diagram shows the physical and mechanical characteristics of the raw materials for the preparation of the concrete mixture, which also significantly affect the values of the final physical, mechanical and technological characteristics of the SCC mixture and the hardened SCC.

Branch 3 of the lower part of the diagram shows the rheological characteristics of SCC mixtures, the values of which directly affect the quality, scope and manufacturability of nano-modified SCC mixtures and composites based on them. On branch 4 of the scheme, the factor of influence of nano-modifiers is presented in quantitative and qualitative aspects, considering their type, origin, and joint interaction. The presented four branches play a decisive role in obtaining effective self-compacting concrete mixtures with improved technological characteristics and concrete based on them.

5. Conclusions

In conclusion, we theoretically and experimentally proved and substantiated the possibility of nano-modification using a complex modifier based on the industrial waste of self-compacting concrete and the possibility of improving the technological properties of concrete mixes using such nano-modifiers.

Experimental technological and mathematical dependencies between input factors—the characteristics of the technological process and raw materials and output parameters of the characteristics of concrete mixes and concretes—are determined. The optimization of these parameters was carried out, a theoretical substantiation of the obtained results has been proposed, and a quantitative picture has been presented, expressed in the increment of the properties of self-compacting concrete mixtures using nano-modifiers from industrial waste and concretes based on them. In terms of the mixture, these increases were 12% for slump-flow and 83% for fluidity concerning the control composition, while in terms of concrete, these increases were up to 19% for strength indicators and 22% for water tightness to the control composition. The ultimate strain was decreased by 14%, and elastic modulus increased by 11%.

A complex nano-modifier, including such waste as micro-silica and fly ash, gave positive results and demonstrated a synergistic effect in terms of saving raw materials and environmentally friendly disposal. This allows for recommending the proposed method for practical construction industry enterprises, as well as for monolithic concreting of complex densely reinforced structures.

An assessment of the results obtained is given and proposals for implementation in the practical industry are developed. According to preliminary estimates, the use of this nano-modifier leads to improving in the environmental situation because of the use of a nano-modifier based on industrial waste in concrete and results in a reduction of the process cost. The cost of the final product was decreased by reducing the cost of raw materials by up to 15%.

Author Contributions: Conceptualization, S.A.S., E.M.S., A.B.; methodology, S.A.S., E.M.S. and Y.Z.; software, S.A.S., E.M.S., A.B. and N.B.; validation, S.A.S., E.M.S. and A.B.; formal analysis, S.A.S., E.M.S.; investigation, L.R.M., S.A.S., E.M.S., A.B. and B.M.; resources, B.M.; data curation, S.A.S., E.M.S. and Y.Z.; writing—original draft preparation, S.A.S., E.M.S. and A.B.; writing—review and editing, S.A.S., E.M.S. and A.B.; visualization, S.A.S., E.M.S., A.B. and N.B.; supervision, L.R.M., B.M.; project administration, L.R.M., B.M.; funding acquisition, A.B., B.M. All authors have read and agreed to the published version of the manuscript.

Funding: This research received no external funding.

Institutional Review Board Statement: Not applicable.

Informed Consent Statement: Not applicable.

Data Availability Statement: The study did not report any data.

Acknowledgments: The authors would like to acknowledge the administration of Don State Technical University for their resources and financial support.

Conflicts of Interest: The authors declare no conflict of interest.

References

1. Smirnova, O.; Kazanskaya, L.; Koplík, J.; Tan, H.; Gu, X. Concrete Based on Clinker-Free Cement: Selecting the Functional Unit for Environmental Assessment. *Sustainability* **2021**, *13*, 135. [CrossRef]
2. Smirnova, O.M.; Menéndez Pidal de Navascués, I.; Mikhailevskii, V.R.; Kolosov, O.I.; Skolota, N.S. Sound-Absorbing Composites with Rubber Crumb from Used Tires. *Appl. Sci.* **2021**, *11*, 7347. [CrossRef]
3. Smirnova, O.M. Low-Clinker cements with low water demand. *J. Mater. Civ. Eng.* **2020**, *32*, 06020008. [CrossRef]
4. Sebök, T.; Krejčí, J.; Musil, A.; Šimoník, J. A contribution to the explanation of the action principles of organic plasticizers. *Cem. Concr. Res.* **2005**, *35*, 1551–1554. [CrossRef]
5. Felekoğlu, B.; Sarıkahya, H. Effect of chemical structure of polycarboxylate-based superplasticizers on workability retention of self-compacting concrete. *Constr. Build. Mater.* **2008**, *22*, 1972–1980. [CrossRef]
6. Chandra, S.; Björnström, J. Influence of superplasticizer type and dosage on the slump loss of Portland cement mortars—Part II. *Cem. Concr. Res.* **2002**, *32*, 1613–1619. [CrossRef]
7. Łazniewska-Piekarczyk, B. The influence of selected new generation admixtures on the workability, air-voids parameters and frost-resistance of self-compacting concrete. *Constr. Build. Mater.* **2012**, *31*, 310–319. [CrossRef]
8. Farooq, F.; Czarnecki, S.; Niewiadomski, P.; Aslam, F.; Alabduljabbar, H.; Ostrowski, K.A.; Śliwa-Wieczorek, K.; Nowobilski, T.; Malazdrewicz, S. A Comparative Study for the Prediction of the Compressive Strength of Self-Compacting Concrete Modified with Fly Ash. *Materials* **2021**, *14*, 4934. [CrossRef]
9. Kostrzanowska-Siedlarz, A.; Gołaszewski, J. Statistical Models Supporting the High-Performance Self-Compacting Concrete (HPSCC) Design Process for High Strength. *Materials* **2022**, *15*, 690. [CrossRef]

10. Zaichenko, M.; Yegorova, O. The rheological properties of cement pastes formulated for self-compacting concretes with multifunctional modifier. *Mod. Ind. Civ. Constr.* **2015**, *11*, 189–197.
11. Pang, L.; Liu, Z.; Wang, D.; An, M. Review on the Application of Supplementary Cementitious Materials in Self-Compacting Concrete. *Crystals* **2022**, *12*, 180. [CrossRef]
12. Xie, T.; Mohamad Ali, M.S.; Elchalakani, M.; Visintin, P. Modelling fresh and hardened properties of self-compacting concrete containing supplementary cementitious materials using reactive moduli. *Constr. Build. Mater.* **2021**, *272*, 121954. [CrossRef]
13. Zhao, H.; Sun, W.; Wu, X.; Gao, B. The properties of the self-compacting concrete with fly ash and ground granulated blast furnace slag mineral admixtures. *J. Clean. Prod.* **2015**, *95*, 66–74. [CrossRef]
14. Nepomuceno, M.; Oliveira, L.; Lopes, S.M.R. Methodology for mix design of the mortar phase of self-compacting concrete using different mineral additions in binary blends of powders. *Constr. Build. Mater.* **2012**, *26*, 317–326. [CrossRef]
15. Derabla, R.; Benmalek, M.L. Characterization of heat-treated self-compacting concrete containing mineral admixtures at early age and in the long term. *Constr. Build. Mater.* **2014**, *66*, 787–794. [CrossRef]
16. Bouzoubaâ, N.; Lachemi, M. Self-compacting concrete incorporating high volumes of class F fly ash: Preliminary results. *Cem. Concr. Res.* **2001**, *31*, 413–420. [CrossRef]
17. Sukumar, B.; Nagamani, K.; Raghavan, R.S. Evaluation of strength at early ages of self-compacting concrete with high volume fly ash. *Constr. Build. Mater.* **2008**, *22*, 1394–1401. [CrossRef]
18. Yazıcı, H. The effect of silica fume and high-volume Class C fly ash on mechanical properties, chloride penetration and freeze–thaw resistance of self-compacting concrete. *Constr. Build. Mater.* **2008**, *22*, 456–462. [CrossRef]
19. Güneyisi, E.; Gesoğlu, M.; Özbay, E. Strength and drying shrinkage properties of self-compacting concretes incorporating multi-system blended mineral admixtures. *Constr. Build. Mater.* **2010**, *24*, 1878–1887. [CrossRef]
20. Güneyisi, E.; Gesoğlu, M. Properties of self-compacting portland pozzolana and limestone blended cement concretes containing different replacement levels of slag. *Mater. Struct.* **2011**, *44*, 1399–1410. [CrossRef]
21. Güneyisi, E.; Gesoglu, M.; Özbay, E. Permeation properties of self-consolidating concretes with mineral admixtures. *ACI Mater. J.* **2011**, *108*, 150–158. Available online: https://www.researchgate.net/publication/282764863_Permeation_Properties_of_Self-Consolidating_Concretes_with_Mineral_Admixtures (accessed on 7 April 2022).
22. Uysal, M.; Yilmaz, K.; Ipek, M. The effect of mineral admixtures on mechanical properties, chloride ion permeability and impermeability of self-compacting concrete. *Constr. Build. Mater.* **2012**, *27*, 263–270. [CrossRef]
23. Dinakar, P.; Reddy, M.K.; Sharma, M. Behaviour of self compacting concrete using Portland pozzolana cement with different levels of fly ash. *Mater. Des.* **2013**, *46*, 609–616. [CrossRef]
24. Turk, K.; Karatas, M.; Gonen, T. Effect of Fly Ash and Silica Fume on compressive strength, sorptivity and carbonation of SCC. *KSCE J. Civ. Eng.* **2013**, *17*, 202–209. [CrossRef]
25. Siad, H.; Mesbah, H.A.; Mouli, M.; Escadeillas, G.; Khelafi, H. Influence of Mineral Admixtures on the Permeation Properties of Self-Compacting Concrete at Different Ages. *Arab. J. Sci. Eng.* **2014**, *39*, 3641–3649. [CrossRef]
26. Raisi, E.M.; Amiri, J.V.; Davoodi, M.R. Mechanical performance of self-compacting concrete incorporating rice husk ash. *Constr. Build. Mater.* **2018**, *177*, 148–157. [CrossRef]
27. Nesvetaev, G.; Koryanova, Y.; Zhilnikova, T.; Kolleganov, A. To the problem of assessing the level of self-stresses during the formation of the structure of self-compacting concrete. *Mater. Sci. Forum* **2019**, *974*, 293–298. [CrossRef]
28. Nesvetaev, G.; Koryanova, Y.; Korchagin, I. To the problem of the methodology for evaluating the effectiveness of the use of superplasticizers in concretes. *IOP Conf. Ser. Mater. Sci. Eng.* **2020**, *709*, 044056. [CrossRef]
29. Benaicha, M.; Roguiez, X.; Jalbaud, O.; Burtschell, Y.; Alaoui, A.H. Influence of silica fume and viscosity modifying agent on the mechanical and rheological behavior of self compacting concrete. *Constr. Build. Mater.* **2015**, *84*, 103–110. [CrossRef]
30. Singh, R.B.; Singh, B. Rheological behaviour of different grades of self-compacting concrete containing recycled aggregates. *Constr. Build. Mater.* **2018**, *161*, 354–364. [CrossRef]
31. Güneyisi, E.; Gesoglu, M.; Al-Goody, A.; İpek, S. Fresh and rheological behavior of nano-silica and fly ash blended self-compacting concrete. *Constr. Build. Mater.* **2015**, *95*, 29–44. [CrossRef]
32. Ardalan, R.B.; Joshaghani, A.; Hooton, R.D. Workability retention and compressive strength of self-compacting concrete incorporating pumice powder and silica fume. *Constr. Build. Mater.* **2017**, *134*, 116–122. [CrossRef]
33. Rahim, N.I.; Mohammed, B.S.; Abdulkadir, I.; Dahim, M. Effect of Crumb Rubber, Fly Ash, and Nanosilica on the Properties of Self-Compacting Concrete Using Response Surface Methodology. *Materials* **2022**, *15*, 1501. [CrossRef] [PubMed]
34. Olginskii, A.G. Peculiarities of Contact Formation in Cement Concrete with Mineral Microfiller. *Proc. DonNACEA* **2004**, *1*, 134–140.
35. Boukendakdji, O.; El-Hadj, K.; Kenai, S. Effects of granulated blast furnace slag and superplasticizer type on the fresh properties and compressive strength of self-compacting concrete. *Cem. Concr. Compos.* **2012**, *34*, 583–590. [CrossRef]
36. Marshdi, Q.S.R.; Al-Sallami, Z.H.A.; Zaichenko, N.M. Effect of multicomponent modifier on the properties of cement pastes formulated from self-compacting concrete. *Mag. Civ. Eng.* **2020**, *98*, 9805. [CrossRef]
37. Zaichenko, N.M.; Nefedov, V.V. Poly(ethylene terephthalate) composite material with modified fly ash filler. *Mag. Civ. Eng.* **2021**, *101*, 10103. [CrossRef]
38. Zaichenko, M.; Lakhtaryna, S.; Korsun, A. The Influence of Extra Mixing Water on the Properties of Structural Lightweight Aggregate Concrete. *Procedia Eng.* **2015**, *117*, 1036–1042. [CrossRef]

39. Ding, X.; Zhao, M.; Qiu, X.; Wang, Y.; Ru, Y. The Optimization of Mix Proportion Design for SCC: Experimental Study and Grey Relational Analysis. *Materials* **2022**, *15*, 1305. [CrossRef]
40. Kelechi, S.E.; Adamu, M.; Mohammed, A.; Ibrahim, Y.E.; Obianyo, I.I. Durability Performance of Self-Compacting Concrete Containing Crumb Rubber, Fly Ash and Calcium Carbide Waste. *Materials* **2022**, *15*, 488. [CrossRef]
41. Kelechi, S.E.; Adamu, M.; Mohammed, A.; Obianyo, I.I.; Ibrahim, Y.E.; Alanazi, H. Equivalent CO_2 Emission and Cost Analysis of Green Self-Compacting Rubberized Concrete. *Sustainability* **2022**, *14*, 137. [CrossRef]
42. Kathirvel, P.; Murali, G.; Vatin, N.I.; Abid, S.R. Experimental Study on Self Compacting Fibrous Concrete Comprising Magnesium Sulphate Solution Treated Recycled Aggregates. *Materials* **2022**, *15*, 340. [CrossRef] [PubMed]
43. Ding, X.; Geng, H.; Zhao, M.; Chen, Z.; Li, J. Synergistic Bond Properties of Different Deformed Steel Fibers Embedded in Mortars Wet-Sieved from Self-Compacting SFRC. *Appl. Sci.* **2021**, *11*, 10144. [CrossRef]
44. Muhammad, A.; Thienel, K.-C.; Sposito, R. Suitability of Blending Rice Husk Ash and Calcined Clay for the Production of Self-Compacting Concrete: A Review. *Materials* **2021**, *14*, 6252. [CrossRef] [PubMed]
45. Cabrera, M.; Martinez-Echevarria, M.J.; López-Alonso, M.; Agrela, F.; Rosales, J. Self-Compacting Recycled Concrete Using Biomass Bottom Ash. *Materials* **2021**, *14*, 6084. [CrossRef]
46. Khalyushev, A.; Stelmakh, S.; Shcherban, E.; Chernilnik, A. Structural Formation of the Cement Pastes Based on the Concrete Modified Dispersed Mineral Components. *Mater. Sci. Forum* **2020**, *1011*, 14–22. [CrossRef]
47. Khalyushev, A.; Stel'makh, S.; Shcherban', E.; Smolyanichenko, A.; Korobkin, A. Microfiller Influence on the Modified Cement Granulometric Composition. *Mater. Sci. Forum* **2021**, *1043*, 9–14. [CrossRef]
48. Mailyan, L.R.; Beskopylny, A.N.; Meskhi, B.; Stel'makh, S.A.; Shcherban, E.M.; Ananova, O. Optimization of Composition and Technological Factors for the Lightweight Fiber-Reinforced Concrete Production on a Combined Aggregate with an Increased Coefficient of Structural Quality. *Appl. Sci.* **2021**, *11*, 7284. [CrossRef]
49. Stel'makh, S.A.; Shcherban', E.M.; Beskopylny, A.N.; Mailyan, L.R.; Meskhi, B.; Butko, D.; Smolyanichenko, A.S. Influence of Composition and Technological Factors on Variatropic Efficiency and Constructive Quality Coefficients of Lightweight Vibro-Centrifuged Concrete with Alkalized Mixing Water. *Appl. Sci.* **2021**, *11*, 9293. [CrossRef]
50. Shcherban', E.M.; Stel'makh, S.A.; Beskopylny, A.; Mailyan, L.R.; Meskhi, B.; Varavka, V. Nanomodification of Lightweight Fiber Reinforced Concrete with Micro-silica and Its Influence on the Constructive Quality Coefficient. *Materials* **2021**, *14*, 7347. [CrossRef]
51. Beskopylny, A.N.; Shcherban', E.M.; Stel'makh, S.A.; Mailyan, L.R.; Meskhi, B.; El'shaeva, D. The Influence of Composition and Recipe Dosage on the Strength Characteristics of New Geopolymer Concrete with the Use of Stone Flour. *Appl. Sci.* **2022**, *12*, 613. [CrossRef]
52. Beskopylny, A.N.; Stel'makh, S.A.; Shcherban', E.M.; Mailyan, L.R.; Meskhi, B. Nano modifying additive micro-silica influence on integral and differential characteristics of vibrocentrifuged concrete. *J. Build. Eng.* **2022**, *51*, 104235. [CrossRef]
53. Stel'makh, S.A.; Shcherban', E.M.; Beskopylny, A.N.; Mailyan, L.R.; Meskhi, B.; Varavka, V. Quantitative and Qualitative Aspects of Composite Action of Concrete and Dispersion-Reinforcing Fiber. *Polymers* **2022**, *14*, 682. [CrossRef]
54. Stel'makh, S.A.; Shcherban', E.M.; Beskopylny, A.N.; Mailyan, L.R.; Meskhi, B.; Dotsenko, N. Enchainment of the Coefficient of Structural Quality of Elements in Compression and Bending by Combined Reinforcement of Concrete with Polymer Composite Bars and Dispersed Fiber. *Polymers* **2021**, *13*, 4347. [CrossRef]
55. Beskopylny, A.N.; Stel'makh, S.A.; Shcherban', E.M.; Mailyan, L.R.; Meskhi, B.; El'shaeva, D.; Varavka, V. Developing Environmentally Sustainable and Cost-Effective Geopolymer Concrete with Improved Characteristics. *Sustainability* **2021**, *13*, 13607. [CrossRef]
56. Shcherban', E.M.; Stel'makh, S.A.; Beskopylny, A.N.; Mailyan, L.R.; Meskhi, B. Influence of Mechanochemical Activation of Concrete Components on the Properties of Vibro-Centrifugated Heavy Concrete. *Appl. Sci.* **2021**, *11*, 10647. [CrossRef]
57. Zajchenko, N.M.; Khalyushev, A.K.; Stelmakh, S.A.; Shcherban, E.M.; Kholodnyak, M.G.; Chernilnik, A.A. Pat. 2716285 Russian Federation, MPK G01N 11/10 Rostov-on-Don. DSTU. N 2019129135; Declared 16.09.19; Publ. 11.03.20, Bull. N 8. 8p. Available online: https://patentimages.storage.googleapis.com/d2/1b/a3/fda70493adbec9/RU2716285C1.pdf (accessed on 22 February 2022).
58. GOST 10180 Concretes. Methods for Strength Determination Using Reference Specimens. Available online: http://docs.cntd.ru/document/1200100908 (accessed on 22 February 2022).
59. GOST 24452 Concretes. Methods of Prismatic, Compressive Strength, Modulus of Elasticity and Poisson's Ratio Determination. Available online: https://docs.cntd.ru/document/9056198 (accessed on 22 February 2022).
60. GOST 12730.5 Concretes. Methods for Determination of Water Tightness. Available online: https://docs.cntd.ru/document/1200163874 (accessed on 22 February 2022).
61. Su, N.; Hsu, K.-C.; Chai, H.-W. A simple mix design method for self-compacting concrete. *Cem. Concr. Res.* **2001**, *31*, 1799–1807. [CrossRef]

materials

MDPI

Article

A Study to Derive Equivalent Mechanical Properties of Porous Materials with Orthotropic Elasticity

Changmin Pyo [1], Younghyun Kim [1], Jaewoong Kim [1,*] and Sungwook Kang [2]

[1] Smart Mobility Materials and Components R&D Group, Korea Institute of Industrial Technology, Gwangju 61012, Korea; changmin@kitech.re.kr (C.P.); kyh1927@kitech.re.kr (Y.K.)

[2] Precision Mechanical Process and Control R&D Group, Korea Institute of Industrial Technology, Jinju 52845, Korea; swkang@kitech.re.kr

* Correspondence: kjw0607@kitech.re.kr; Tel.: +82-62-600-6480

Abstract: The need for diverse materials has emerged as industry becomes more developed, and there is a need for materials with pores in various industries, including the energy storage field. However, there is difficulty in product design and development using the finite element method because the mechanical properties of a porous material are different from those of a base material due to the pores. Therefore, in this study, a Python program that can estimate the equivalent property of a material with pores was developed and its matching was verified through comparison with the measurement results. For high-efficiency calculation, the pores were assumed to be circular or elliptical, and they were also assumed to be equally distributed in each direction. The material with pores was assumed to be an orthotropic material, and its equivalent mechanical properties were calculated using the equivalent strain and equivalent stress by using the appropriate material property matrix. The material properties of a specimen with the simulated pores were measured using UTM, and the results were compared with the simulation results to confirm that the degree of matching achieved 6.4%. It is expected that this study will contribute to the design and development of a product in the industrial field.

Keywords: equivalent mechanical properties; porosity; finite element method; orthotropic elasticity

Citation: Pyo, C.; Kim, Y.; Kim, J.; Kang, S. A Study to Derive Equivalent Mechanical Properties of Porous Materials with Orthotropic Elasticity. *Materials* **2021**, *14*, 5132. https://doi.org/10.3390/ma14185132

Academic Editor: Petrica Vizureanu

Received: 29 July 2021
Accepted: 1 September 2021
Published: 7 September 2021

1. Introduction

When developing a new product, the materials constituting the product must have target mechanical and electrical properties under its operating conditions. In the past, a product suitable for a specific situation was manufactured by using a single material, but it was difficult to satisfy complex design conditions. Recently, various materials including composite materials have been used to satisfy complex design conditions, contributing to the development of products and technologies.

In addition, the demand for porous materials is steadily increasing; specifically, the need for porous materials is rising in fields of chemical engineering such as pharmaceuticals and catalysts, as well as in the field of energy storage. Related research has also been ongoing [1–4].

Moreover, many materials such as bones, teeth, and other biomaterials in medicine and dentistry industries are porous materials [5], and porous tantalum for dental implantology has porosities [6].

Concept design–basic design–detail design–mockup production and verification procedures are essential to develop products and technologies. To minimize trial and error, it is advantageous to verify design feasibility through the finite element method at the design stage, and this is an approach widely used in academia as well as in industry. The finite element method has the advantage of being able to verify design feasibility at a much lower cost compared with experiments or deriving the optimal design through case studies under various conditions [7,8].

However, when the finite element method is used for a material with pores, the pores show properties that differ from the mechanical properties of an existing base material. This may distort the results of the finite element method, and is a limitation when using the finite element method for the materials. To solve this problem, a method is used to estimate and utilize equivalent mechanical properties while simplifying modeling by applying the homogenization method for composite materials [9–12].

However, there has been little research on the application of the homogenization method to materials with pores. Therefore, this study addressed how to calculate equivalent mechanical properties that reflect the pores, for a material that has pores. In prior studies, a simulation technique that can derive isotropic material properties under the plane stress condition, which is a material property that can be used in a 2D shell, was developed and was verified under the two-dimensional orthotropic elasticity in plane stress that is symmetrical in the in-plane direction [13,14]. In this study, the elastic modulus and Poisson's ratio, which are the main equivalent properties of porous materials, were derived under the condition of three-dimensional orthotropic elasticity. Assuming that the shape of pores is symmetrical in the in-plane direction, and that the pores are uniformly distributed in the in-plane direction, a unit cell was defined and equivalent properties were calculated. Thus, the elastic modulus and Poisson's ratio were measured through a universal test machine (UTM) by preparing a material specimen with the same simulation conditions. As a result, the similarity between the value obtained through the simulation and the value measured through the experiment was observed, so the validity of this simulation was verified.

2. Prior Research and Background Theory

2.1. *Study on Equivalent Properties Derivation*

The homogenization method was mainly used to derive the equivalent physical properties. Homogenization methods include the representative unit cell method [15] and the representative equivalent continuum method [16], but the most dominant homogenization method is the representative volume element (RVE) method [9–12]. The representative volume element method is applied to estimate the stiffness of fiber composites [9] or to estimate the equivalent properties of reinforced plastics mixed with multidirectional fibers [10]. It has also been used in studies to estimate the equivalent properties of an anisotropic material [11,12].

The representative volume element assumes that the strain energy before and after homogenization should be the same. When various materials or pores are distributed within a unit volume, it is assumed that all material properties within a unit volume are the same (Figure 1).

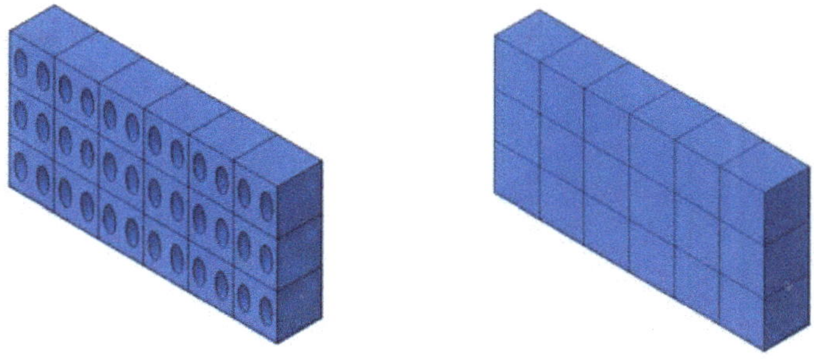

(a) (b)

Figure 1. (**a**) Shape of unit cell before homogenization. (**b**) Shape of unit cell after homogenization.

The related governing equations are shown as Equations (1)–(5) below [9–12]. Equation (1) means that the strain energy before and after homogenization is the same. Equation (2) means that the strain energy before homogenization has a relationship with the sum of the strain energies for each element in the entire unit volume. Equation (3) indicates that the energy after homogenization is related to the average strain energy, which is obtained by multiplying the average value of the stress for each element and the average value of the strain for each element, multiplied by the volume. Equation (4) is a method to calculate the average stress, and Equation (5) is a method to calculate the average strain.

$$U^* = U \tag{1}$$

$$U^* = \frac{1}{2}\int_{V_{RVE}} \sigma_{ij}\epsilon_{ij} dV \tag{2}$$

$$U = \frac{1}{2}\overline{\sigma_{ij}\epsilon_{ij}} V_{RVE} \tag{3}$$

$$\overline{\sigma_{ij}} = \frac{1}{V_{RVE}}\int_{V^*} \sigma_{ij}(x,y,z) dV \tag{4}$$

$$\overline{\varepsilon_{ij}} = \frac{1}{V_{RVE}}\int_{V^*} \varepsilon_{ij}(x,y,z) dV \tag{5}$$

U^*: strain energy before homogenization; U: strain energy after homogenization; V^*: total volume before homogenization; V_{RVE}: volume after homogenization; $\sigma_{ij}(x,y,z)$: stress per element; $\varepsilon_{ij}(x,y,z)$: strain per element; $\overline{\sigma_{ij}}$: stress weighted average for each element; $\overline{\varepsilon_{ij}}$: strain weighted average for each element.

2.2. *Orthotropic Elasticity*

In this study, it was assumed that the pores are asymmetric in the in-plane direction, and that the distribution of the pores is uniform in each direction (Figure 2). The size of pores is also irregular, and the distribution is not uniform, so it is correct to think that a material with pores has anisotropic properties. In this case, all materials should be verified through experiments and measurements. If a material is too thin or is toxic, it is difficult to measure it using UTM, etc., and it is not suitable for the purpose of performing simulation using FEM. Therefore, in this study, a material with pores was assumed to be an orthotropic material, which is thought to be a more suitable property for use in the industrial field.

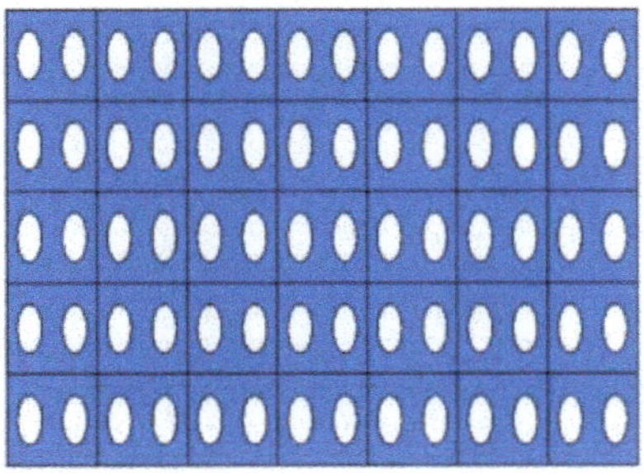

Figure 2. Shape of orthotropic elasticity.

The elastic modulus determinant of a perfectly anisotropic material is shown in Equation (6) [17]. In addition, the elastic modulus determinant of an orthotropic material is shown in Equation (7) [17]. Because the shape of pores is difficult to standardize and their distribution is not uniform, it is more realistic to assume that the material is a perfectly anisotropic material. As this study estimates the equivalent properties of a material by approximating the shape and distribution of pores, the material was assumed to be an orthotropic material, and the elastic modulus determinant was applied accordingly.

$$\begin{bmatrix} \overline{\sigma_x} \\ \overline{\sigma_y} \\ \overline{\sigma_x} \\ \overline{\sigma_{xy}} \\ \overline{\sigma_{xz}} \\ \overline{\sigma_{yz}} \end{bmatrix} = \begin{bmatrix} D_{1111} & D_{1122} & D_{1133} & D_{1112} & D_{1113} & D_{1123} \\ & D_{2222} & D_{2233} & D_{2212} & D_{2213} & D_{2223} \\ & & D_{3333} & D_{3312} & D_{3313} & D_{3323} \\ & Sym. & & D_{1212} & D_{1213} & D_{1223} \\ & & & & D_{1313} & D_{1323} \\ & & & & & D_{2323} \end{bmatrix} \begin{bmatrix} \overline{\epsilon_x} \\ \overline{\epsilon_y} \\ \overline{\epsilon_z} \\ \overline{\epsilon_{xy}} \\ \overline{\epsilon_{xz}} \\ \overline{\epsilon_{yz}} \end{bmatrix} \quad (6)$$

$$\begin{bmatrix} \varepsilon_{11} \\ \varepsilon_{22} \\ \varepsilon_{33} \\ \gamma_{12} \\ \gamma_{13} \\ \tau_{23} \end{bmatrix} = \begin{bmatrix} 1/E_1 & -\nu_{21}/E_2 & -\nu_{31}/E_3 & 0 & 0 & 0 \\ -\nu_{12}/E_1 & 1/E_2 & -\nu_{32}/E_3 & 0 & 0 & 0 \\ -\nu_{13}/E_1 & -\sigma_{23}/E_2 & 1/E_3 & 0 & 0 & 0 \\ 0 & 0 & 0 & 1/G_{12} & 0 & 0 \\ 0 & 0 & 0 & 0 & 1/G_{13} & 0 \\ 0 & 0 & 0 & 0 & 0 & 1/G_{23} \end{bmatrix} \begin{bmatrix} \sigma_{11} \\ \sigma_{22} \\ \sigma_{33} \\ \tau_{12} \\ \tau_{13} \\ \tau_{23} \end{bmatrix} \quad (7)$$

For reference, the elastic modulus determinant of an isotropic material under a plane stress condition is shown in Equation (8) [16], and the elastic modulus determinant of an orthotropic material under a plane stress condition is shown in Equation (9).

$$\begin{bmatrix} \overline{\epsilon_x} \\ \overline{\epsilon_y} \\ \overline{\gamma_{xy}} \end{bmatrix} = \begin{bmatrix} \frac{1}{E_x} & -\frac{\overline{\nu_{yx}}}{E_y} & 0 \\ -\frac{\overline{\nu_{xy}}}{E_x} & \frac{1}{E_y} & 0 \\ 0 & 0 & \frac{1}{G_{xy}} \end{bmatrix} \begin{bmatrix} \overline{\sigma_x} \\ \overline{\sigma_y} \\ \overline{\tau_{xy}} \end{bmatrix} \quad (8)$$

$$\begin{bmatrix} \overline{\epsilon_x} \\ \overline{\epsilon_y} \\ \overline{\gamma_{xy}} \end{bmatrix} = \frac{1}{\overline{E}} \begin{bmatrix} 1 & -\overline{\nu} & 0 \\ -\overline{\nu} & 1 & 0 \\ 0 & 0 & 2(1+\overline{\nu}) \end{bmatrix} \begin{bmatrix} \overline{\sigma_x} \\ \overline{\sigma_y} \\ \overline{\tau_{xy}} \end{bmatrix} \quad (9)$$

3. Deriving Equivalent Properties Using Simulation

3.1. Computer Software

In this study, Abaqus 2020 by Dassault Systèmes was used as a finite element method program. Abaqus has been widely used in academia and industry due to its technical feasibility [18,19]. Using the result file derived from the finite element method program, the elastic modulus matrix was derived, and Python (Version 3.8.5, Python Software Foundation Beaverton, OR, USA) was applied to derive the equivalent properties of a material with pores from the elastic modulus matrix. Python has been widely used in research for data processing and matrix calculation because its validity has been verified [20].

3.2. Equivalence Derivation Process

In this study, it was assumed that the size and spacing of pores were uniform. The material properties were derived by using pores of two shapes, and they were compared through experiments. The shape of circular pores is shown in Figure 3, while the shape of elliptical pores is shown in Figure 4. Modeling was performed for the shape of a specimen to be used in the measurement, and verification was carried out as in Section 4. Circular and elliptical shapes were selected for ease of processing of a material specimen.

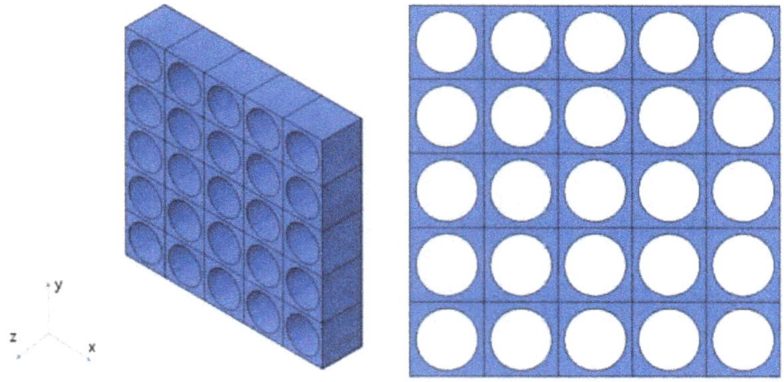

(**a**) (**b**)

Figure 3. The shape of circular pores. (**a**) Isotropic view. (**b**) Top view.

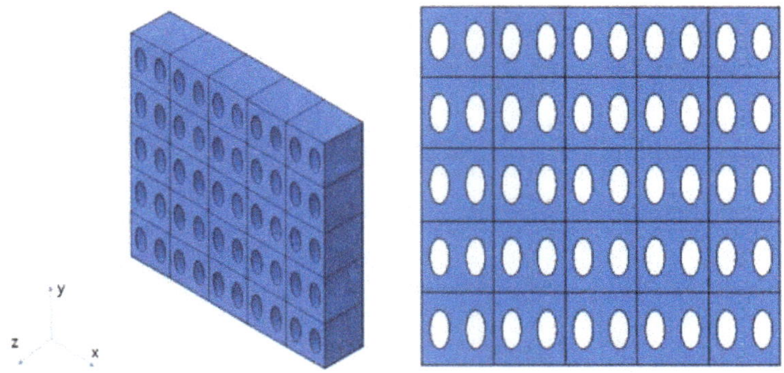

(**a**) (**b**)

Figure 4. The shape of elliptical pores. (**a**) Isotropic view. (**b**) Top view.

As shown in Figures 3 and 4, unit cells were modeled and a constant displacement was applied as a load condition for each direction, and stress values were derived for each direction when the condition was applied.

To derive the equivalent stress, a weighted average was applied to the stress value of a mesh according to the volume size. In addition to the tensile stress in the in-plane X and Y directions and the shear stress in the XY direction, the tensile stress in the Z direction and the shear stress in the XZ and YZ directions were derived, respectively.

The elastic modulus and Poisson's ratio of a representative volume element were calculated using the determinant obtained in Section 2.2 based on the equivalent strain in each direction, reflected as the load condition and the calculated equivalent stress in each direction.

3.3. Determination of Unit Cell Model and Its Shape

To verify the process in Section 3.2, a unit cell model that can be actually manufactured was selected. Through discussion with the specimen manufacturer, a model with a large number of pores in the smallest size was selected. The diameter of the circular pores was 120 μm, and the spacing between pores was 220 μm. The specimen was manufactured in a size of 25 mm in the X direction and 200 mm in the Y direction. For the comparison with the measurement results from UTM, they were manufactured in a size applicable to UTM. As a result, 113 pores in the X direction and 910 pores in the Y direction were obtained, and the shape of the designed specimen and that of the unit cell are shown in Figure 5. Actually, those tests were performed former studies [13,14], we used the test results from those studies.

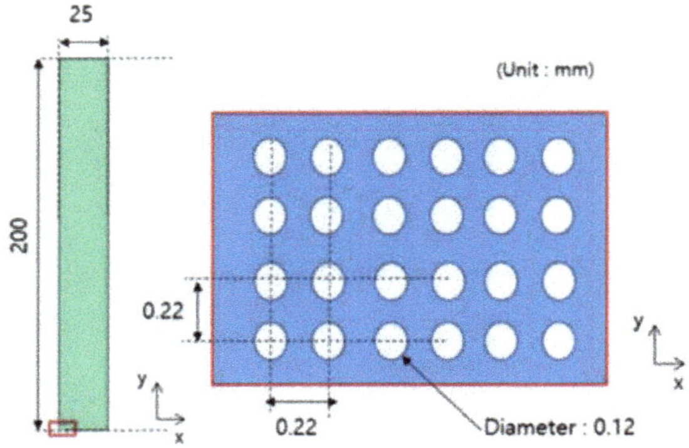

Figure 5. Dimension and distribution of circular pores (Reprinted from Ref. [13]).

Pattern processing was performed by applying an etching technique for pore processing, and a thickness of 30 μm was selected through discussion with the manufacturer for convenience and quality of production.

In addition to the circular pores, verification of the elliptical pores was also performed. The size and distribution of elliptical pores were determined through discussion with the manufacturer. The diameter was set to 120 μm × 240 μm, and the spacing between pores was determined to be 220 μm in the X direction and 440 μm in the Y direction. The size of a specimen was determined to be the same as that of a circular pore specimen, 25 mm × 200 mm × 30 μm, and the shape is shown in Figure 6. However the real dimensions were different from the design data; thus, we checked real dimensions such as the hole diameter. The pattern processing shape of a processed specimen was checked using an optical scope. Its size and shape were measured by applying the optical microscope (i-megascope System2 Mega pixels, Sometech, Seoul, Korea). The shape of an optical microscope is shown in Figure 7. Absolutely, we applied the real dimensions for the simulation. The information is in Table 1. Each unit cell for simulation is defined in Figure 8. The base material of each specimen was SUS304, and its mechanical properties are shown in Table 2.

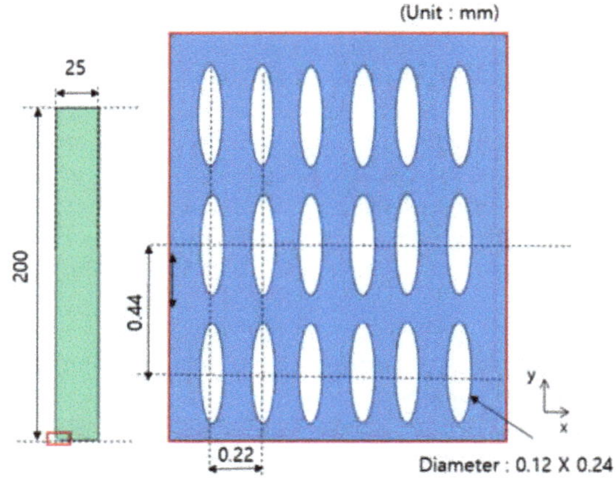

Figure 6. Dimension and distribution of elliptical pores (Reprinted from Ref. [14]).

(**a**)

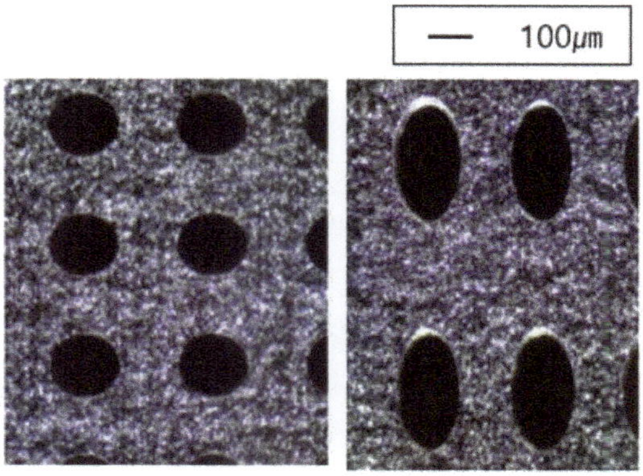

(**b**)

Figure 7. (**a**) Shape of optical microscope. (**b**) Image with optical microscope.

Table 1. Comparison of design proposal and actual results (unit: μm).

			Design	Real
Circular pore	Length of unit cell		220.0	221.25
	Hole diameter		120.0	116.25
Elliptical pore	Length of unit cell		440.0	433.13
	Hole diameter	Short	120.0	120.00
		Long	240.0	233.13

(**a**) (**b**)

Figure 8. The shape of unit cell. (**a**) Circular pore. (**b**) Elliptical pore.

Table 2. Mechanical properties of SUS304 [13,14].

Contents	Value	Unit
Density	8000	kg/m^3
Modulus of elasticity	193.0	GPa
Poisson's ratio	0.29	-

3.4. Simulation Procedure

The unit model described in Section 3.2 was modeled, and the grid was constructed using a hexagonal grid (C3D8R). A displacement of 0.1% relative to the length was applied as a tensile load and a shear load. The equivalent stress was calculated according to the process in Section 3.2, and the elastic modulus matrix was calculated using the equivalent strain. The procedure of this method is shown in Figure 9.

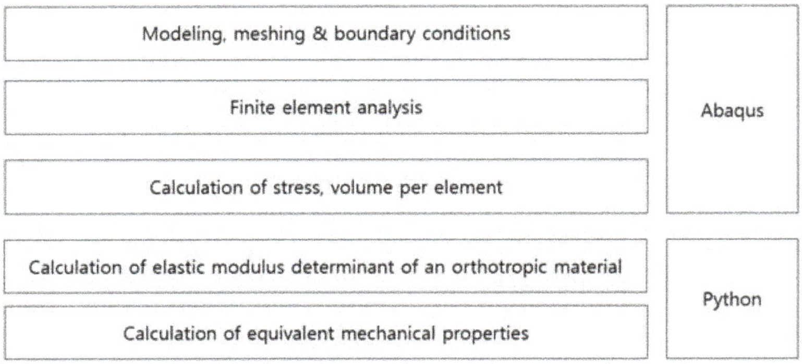

Figure 9. The procedure of calculation of equivalent mechanical properties.

When simulating the circular pores, the length of cubic was 221.25 μm and the diameter of hole was 116.25 mm, which was based on real specimen. The size of mesh was 5 μm, and the shape of mesh was hexagonal grid with reduced integration (C3D8R), which is shown in Figure 10. Moreover, the number was mesh is over 76,000.

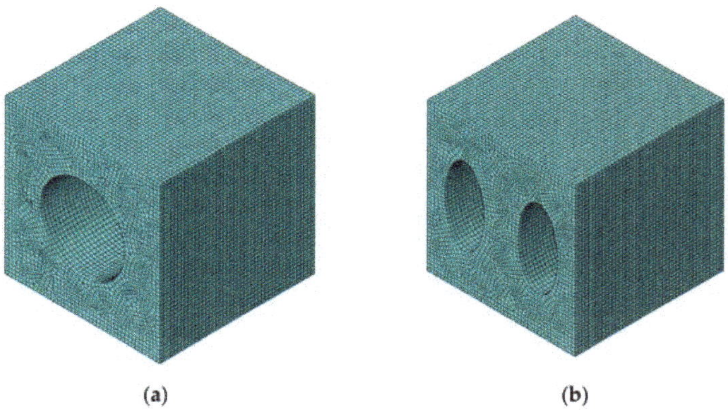

Figure 10. The shape of unit cell after meshing. (**a**) Circular pores. (**b**) Elliptical pores.

In case of elliptical pores, the length of cubic was 433.13 μm and the diameters of holes were 120.0 μm and 233.13 μm. The mesh style was as same as in the case of circular pores, but the size was 10μm and the number of mesh was over 77,000. It is shown in Figure 10.

Tensile strain and shear strain were loaded with boundary condition, and 0.1% of strain was applied. The tensile strain, shear strain by each direction, was applied in each step. Step 1 was for tensile strain of X direction, step 2 was for tensile strain of Y direction, step 3 was for tensile strain of Z direction, step 4 was for shear strain of XY direction, step 5 was for shear strain of XZ direction, and step 6 was for shear strain of YZ direction. Examples are shown in Figure 11. Each step was independent from other steps, and each did not affect other steps. The results of FEA are shown in Figure 12.

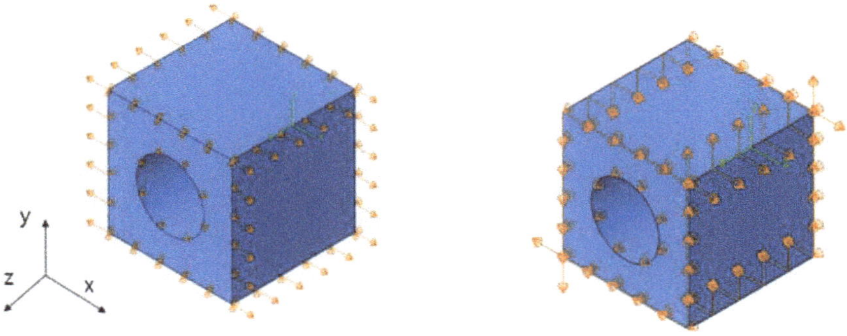

(**a**) (**b**)

Figure 11. The boundary condition for strain. (**a**) X-direction tensile strain. (**b**) XY-direction shear strain.

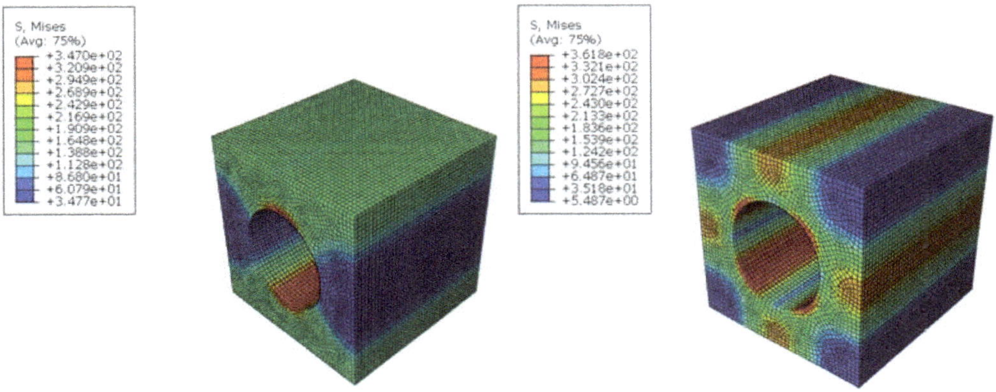

(**a**) (**b**)

Figure 12. The distribution of stress with 0.1% strain condition. (**a**) X-direction tensile strain. (**b**) XY-direction shear scheme 11. S22, S33, S12, S13 and S23 (S11: tensile stress of X-direction, S22: tensile stress of Y-direction, S33: tensile stress of Z-direction, S12: shear stress of XY-direction, S13: shear stress of XZ-direction, S23: shear stress of YZ-direction) of each element were calculated and printed out. The calculation of weighted average concluded the equivalent stress of each direction at each step.

The elastic modulus determinant of an orthotropic material (Equation (7)) can be solved with those stress data at each step. At this stage, the calculation for matrix such as inverse matrix and matrix product should be performed, Python with numpy library can solve this problem. With python, not only elastic modulus matrix but also the equivalent Young's modulus and Poission's ration were calculated

3.5. Simulation Results

Based on the calculation according to Section 3.2, the circular pores are shown in Table 3 and the elliptical pores are shown in Table 4.

Table 3. Equivalent properties with simulation for circular pores.

Contents	Value	Unit
Elastic modulus (E_x)	112.3	GPa
Elastic modulus (E_y)	112.3	GPa
Elastic modulus (E_z)	147.9	GPa
Poisson's ratio (ν_{yx})	0.230	m/m
Poisson's ratio (ν_{xy})	0.230	m/m
Poisson's ratio (ν_{zx})	0.290	m/m
Poisson's ratio (ν_{zy})	0.290	m/m

Table 4. Equivalent properties with simulation for elliptical pores.

Contents	Value	Unit
Elastic modulus (E_x)	106.2	GPa
Elastic modulus (E_y)	127.6	GPa
Elastic modulus (E_z)	150.5	GPa
Poisson's ratio (ν_{yx})	0.235	m/m
Poisson's ratio (ν_{xy})	0.196	m/m
Poisson's ratio (ν_{zx})	0.290	m/m
Poisson's ratio (ν_{zy})	0.290	m/m

4. Verification and Results through Experiments

4.1. Experimental Equipment and Methods

For the verification of this study, a test specimen was prepared and the elastic modulus and Poisson's ratio were measured using a universal testing machine (UTM) (Instron, Norwood, MA, USA). As a universal testing machine, the Instron 5969 was used, and a two-axis strain gauge was used to measure the elastic modulus and Poisson's ratio (Figure 13). Experiments and measurements were performed in compliance with KS M ISO 527-4, and the static material properties were measured at a tensile speed of 1 mm/min. For higher reliability, the experiment was performed six times and the average value was used.

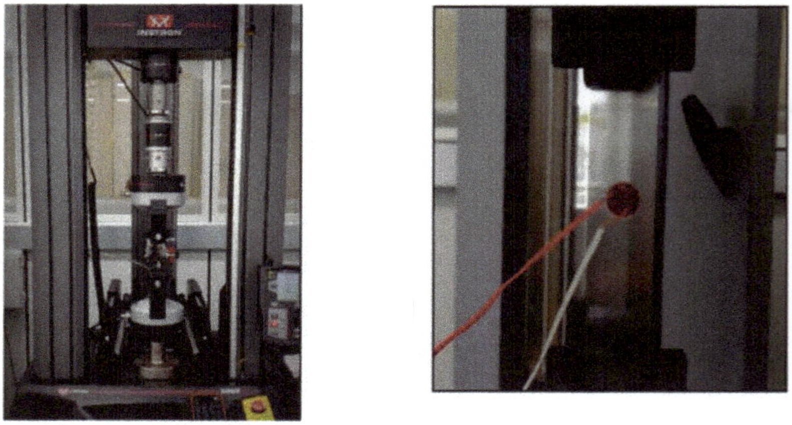

(a) (b)

Figure 13. (**a**) UTM. (**b**) Two-axis strain gauge.

4.2. Shape of Test Specimen

The test specimen was produced in a size of 25 mm × 200 mm × 0.03 mm, and an etching technique was used to prepare the specimen. STS304 was used as a material, and the experiment was performed with three types of specimens. Type I is a test specimen that simulates a circular pore, and Type II and Type III are test specimens that simulate elliptical pores. In each case, a specimen was prepared separately to measure the equivalent elastic modulus and the equivalent Poisson's ratio. When measuring a tensile stress using UTM, it was measured in the Y direction. The shape of a specimen is summarized in Figure 14 [14].

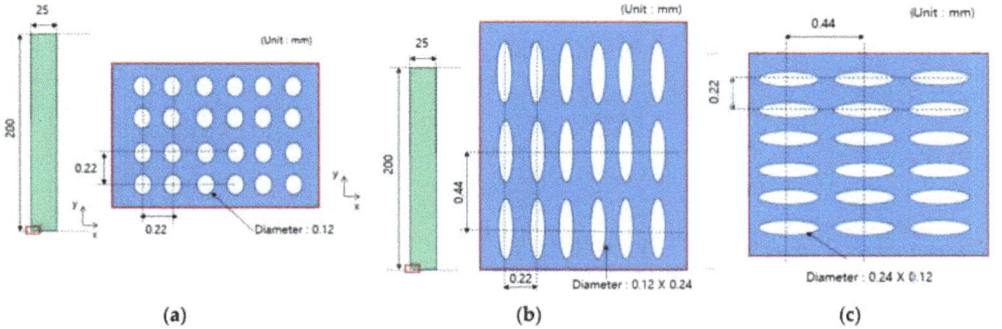

Figure 14. Types of test specimens. (**a**) Specimen with circular pores. (**b**) Specimen with elliptical pores—long axis in the tensile direction. (**c**) Specimen with elliptical pores—short axis in the tensile direction (Reprinted from Ref. [14]).

4.3. Measurement Results

The results of six experiments with three types of specimens are shown in Tables 5–7. It was found that the range and standard deviation of the results of six experiments were not significant.

Table 5. Equivalent properties of Specimen Type I [13].

Test No.	Modulus of Elasticity (GPa)	Poisson's Ratio (mm/mm)
1	116.0	0.226
2	117.0	0.228
3	117.0	0.243
4	118.0	0.248
5	118.0	0.242
6	117.0	0.231
Average	117.17	0.2363
Standard Deviation	0.687	0.0083

Table 6. Equivalent properties of Specimen Type II [14].

Test No.	Modulus of Elasticity (GPa)	Poisson's Ratio (mm/mm)
1	124.0	0.211
2	124.0	0.233
3	124.0	0.212
4	127.0	0.223
5	123.0	0.219
6	122.0	0.226
Average	124.0	0.221
Standard Deviation	1.53	0.0077

Table 7. Equivalent properties of Specimen Type III [14].

Test No.	Modulus of Elasticity (GPa)	Poisson's Ratio (mm/mm)
1	110.0	0.196
2	109.0	0.196
3	111.0	0.188
4	110.0	0.206
5	109.0	0.194
6	109.0	0.199
Average	109.7	0.197
Standard Deviation	0.75	0.0054

4.4. Comparison and Verification

Tables 8–10 compare the equivalent properties estimated from this study with those obtained through actual measurements.

Table 8. Difference between simulation and measurement for Type I.

	Modulus of Elasticity (GPa)	Poisson's Ratio (mm/mm)
Simulation	112.3	0.232
Measurement	117.2	0.236
Difference (%)	4.18 (%)	1.69 (%)

Table 9. Difference between simulation and measurement for Type II.

	Modulus of Elasticity (E_y, GPa)	Poisson's Ratio (ν_{yx}, m/m)
Simulation	127.6	0.236
Measurement	124.0	0.221
Difference (%)	2.82 (%)	6.36 (%)

Table 10. Difference between simulation and measurement for Type III.

	Modulus of Elasticity (E_x, GPa)	Poisson's Ratio (ν_{xy}, m/m)
Simulation	106.2	0.196
Measurement	109.7	0.197
Difference (%)	3.19 (%)	0.51 (%)

In addition, the stress–strain curves for Type I, Type II, and Type III are shown in Figures 15 and 16. As the standard deviation of each case is not large, the difference among tests and average is very small. Moreover, the difference between average and simulation is not big; the curves are similar.

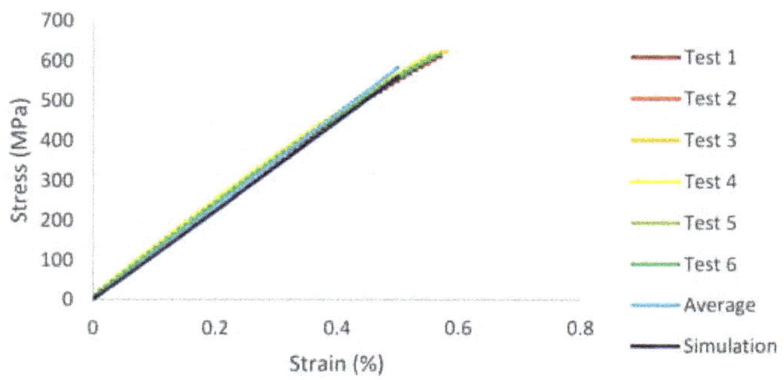

Figure 15. Stress–strain curve for circular pores (Type I).

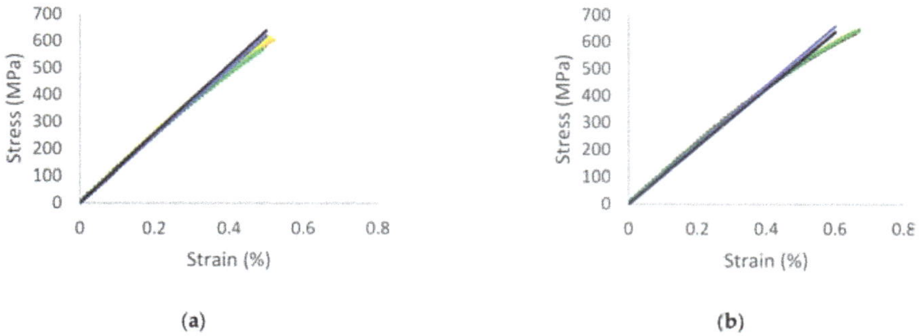

Figure 16. Stress–strain curve for elliptical pores. (**a**) Type II. (**b**) Type III.

5. Discussion

The purpose of this study is to estimate the elastic modulus and Poisson's ratio of a material with irregular pores. When the size and arrangement of pores are irregular, the material properties should be analyzed using a completely anisotropic material. Therefore, more in-depth research is required. In this study, a research method that is feasible in the industrial field was proposed by estimating material properties more easily through working from the assumption that pores have the characteristics of 2D orthogonal anisotropy. To avoid the modeling and analysis of all pores, calculation for equivalent properties was performed using the representative volume element method. A Python program was developed to calculate equivalent properties based on the assumption that the pores are orthogonal anisotropic materials. To verify the results produced by the program, a specimen was produced by simulating circular pores and elliptical pores with horizontal/vertical asymmetry, and the simulation and measurement results were compared after measuring the specimen using UTM. We checked and verified three cases with our method. As differences between simulation and measurement are under 6.4%, our method is reliable and accurate. This research can be applied to many industries such as medicine and dentistry.

In the case of a specimen simulating pores in an elliptical shape, it was found that its production was difficult due to the shape. Thus, the fabrication was not perfect; as shown in Figure 17, the dimensions of top side and bottom side of ellipse were not same. As the shape of upper and lower sides was different, it might affect the results.

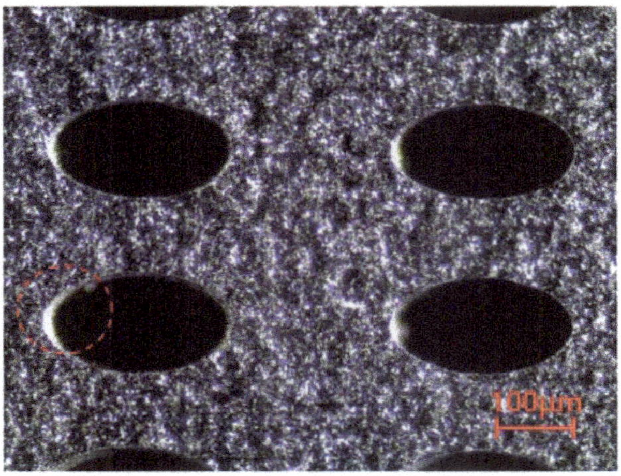

Figure 17. The shape of imperfect fabrication.

Because the shape and arrangement of pores are not uniform, the assumption of an orthotropic material has limitations when it comes to simulating the actual behavior. However, in the case of anisotropic elasticity, it is only possible to estimate the elastic modulus matrix through measurement, and it is difficult to say that the estimated elastic modulus matrix perfectly simulates a porous material. Although it does rely on some assumptions, the results of this study will be of great help in industrial design if they are applied to the finite element method using porous materials in the industrial field. Especially, this method can be useful for analysis of materials in medicine and dentistry, such as bacterial biofilm [21], bones, and porous tantalum. Research considering porosity is very important, as porosity of certain materials can influence the retention and formation of bacterial biofilm

6. Conclusions

In this study, the equivalent properties were calculated and verified by assuming that the material with pores was an orthotropic material while applying the representative volume element method.

1. In the case of a finite element method using porous or composite material, it is inefficient to perform the analysis using material modeling. The equivalent properties of a material were estimated by applying the representative volume element method.
2. Working from the assumption that the pores are horizontally/vertically asymmetrical, an elastic modulus matrix of an orthogonal anisotropic material was constructed. The equivalent elastic modulus and equivalent Poisson's ratio of a representative volumetric element were calculated using the equivalent strain and equivalent stress.
3. Based on the element volume and element stress values derived from the finite element method program, the representative stress value and elastic modulus matrix were calculated using Python. In addition, the equivalent material properties were derived using the calculated elastic modulus matrix.
4. A thin-plate specimen made of STS304 was etched in a specific pattern to simulate pores. The elastic modulus and Poisson's ratio were measured using UTM and verified through comparison with simulation results.
5. This research can be applied to many industries such as medicine and dentistry, which treat porous materials such as bacterial biofilm, bones, teeth, and porous tantalum. As porosity of certain materials can influence the retention and formation of bacterial biofilm, this research is very powerful for analysis for materials with porosity.

Author Contributions: Conceptualization, C.P. and J.K.; methodology, C.P.; software, C.P. and Y.K.; validation, Y.K. and S.K.; formal analysis, S.K.; investigation, C.P.; resources, C.P.; data curation, Y.K.; writing—original draft preparation, C.P.; writing—review and editing, J.K.; visualization, C.P.; supervision, J.K.; project administration, J.K.; funding acquisition, J.K. All authors have read and agreed to the published version of the manuscript.

Funding: This work was supported by Technology Innovation Program (Project name: Equipment technologies for 50 cm2 all solid state battery cell, project Number: 20012349) funded by the Ministry of Trade, Industry and Energy (MOTIE, Korea).

Institutional Review Board Statement: Not applicable.

Informed Consent Statement: Not applicable.

Data Availability Statement: The data presented in this study are available on request from the corresponding author.

Acknowledgments: We gratefully thank Munsung Kim, Dongyun Yeon, and Chanhoe Kang of Dassault System Korea for their assistance.

Conflicts of Interest: The authors declare no conflict of interest.

References

1. Katsoulidis, A.P.; Antypov, D.; Whitehead, G.F.S.; Carrington, E.J.; Adams, D.J.; Berry, N.G.; Darling, G.R.; Dyer, M.S.; Rosseinsky, M.J. Chemical control of structure and guest uptake by a conformationally mobile porous material. *Nature* **2019**, *565*, 213–217. [CrossRef] [PubMed]
2. Slater, A.G.; Cooper, A.I. Porous materials function-led design of new porous materials. *Science* **2015**, *348*, 8075. [CrossRef] [PubMed]
3. Nomura, T.; Okinaka, N.; Akiyama, T. Impregnation of porous material with phase change material for thermal energy storage. *Mater. Chem. Phys.* **2009**, *115*, 846–850. [CrossRef]
4. Trunh, L.; Lee, D. Welding of Thin Tab and Battery Case for Lithium-ion Battery Cylindrical Cell Using Nanosecond Pulsed Fiber Laser. *J. Weld. Jt.* **2020**, *28*, 389–394. [CrossRef]
5. Mour, M.; Das, D.; Winkler, T.; Hoenig, E.; Mielke, G.; Morlock, M.M.; Schilling, A.F. Advances in Porous Biomaterials for Dental and Orthopaedic Applications. *Materials* **2010**, *3*, 2947–2974. [CrossRef]
6. Liu, Y.; Bao, C.; Wismeijer, D.; Wu, G. The physicochemical/biological properties of porous tantalum and the potential surface modification techniques to improve its clinical application in dental implantology. *Mater. Sci. Eng. C* **2015**, *49*, 323–329. [CrossRef] [PubMed]
7. Wang, Z.; Chang, K.; Muzaffer, S. Fatigue Analysis of the Effects of Incomplete Penetration Defects on Fatigue Crack Initiation Points in Butt-Welded Members. *J. Weld. Jt.* **2020**, *38*, 543–550.
8. Wu, C.; Kim, J. Review on Mitigation of Welding-Induced Distortion Based on FEM Analysis. *J. Weld. Jt.* **2020**, *38*, 56–66. [CrossRef]
9. Babu, K.P.; Mohite, P.M.; Upadhyah, C.S. Development of an RVE and its stiffness predictions based on mathematical homogenization theory for short fibre composites. *Int. J. Solids Struct.* **2018**, *130–131*, 80–114. [CrossRef]
10. Breuer, K.; Stommel, M. RVE modelling of short fiber reinforced thermoplastics with discrete fiber orientation and fiber length distribution. *SN Appl. Sci.* **2020**, *2*, 91. [CrossRef]
11. Bargmann, S.; Klusemann, B.; Markmann, J.; Schnabel, J.E.; Schneider, K.; Soyarslan, C.; Wilmers, J. Generation of 3D representative volume elements for heterogeneous materials: A review. *Prog. Mater. Sci.* **2018**, *96*, 322–384. [CrossRef]
12. Pan, Y.; Iorga, L.; Pelegri, A.A. Generation of 3D representative volume elements for heterogeneous materials: A review. *Compsites Sci. Technol.* **2008**, *68*, 2792–2798. [CrossRef]
13. Pyo, C. A Study for Estimation of Equivalent Mechanical Properties of Materials with Porosity Part I. Isotropic Elasticity in Plane Stress. *Korean Soc. Mech. Technol.* **2021**, *23*, 347–353.
14. Pyo, C. A Study for Estimation of Equivalent Mechanical Properties of Materials with Porosity Part II. Orthotropic Elasticity in Plane Stress. *Korean Soc. Mech. Technol.* **2021**, *21*, 412–419.
15. Schmitz, A.; Horst, P. A finite element unit-cell method for homogenized mechanical properties of heterogeneous plates. *Compos. Part A Appl. Sci. Manuf.* **2014**, *61*, 23–32. [CrossRef]
16. Zheng, Q.; Fan, H. Equivalent continuum method of plane-stress dominated plate-lattice materials. *Thin-Walled Struct.* **2021**, *164*, 107865. [CrossRef]
17. Daniel, I.M.; Ishai, O. *Engineering Mechanics of Composite Materials*; Oxford University Press: New York, NY, USA, 2006; pp. 63–92.
18. Pyo, C.; Kim, J.; Kim, J. Estimation of Heat Source Model's Parameters for GMAW with Non-linear Global Optimization—Part I: Application of Multi-island Genetic Algorithm. *Metals* **2020**, *10*, 885. [CrossRef]
19. Deng, D.; Murakawa, H. Numerical simulation of temperature field and residual stress in multi-pass welds in stainless steel pipe and comparison with experimental measurements. *Comput. Mater. Sci.* **2006**, *37*, 269–277. [CrossRef]
20. Sixun, T.; Haipeng, G.; Hao, L.; Lei, Q. The FEM Model Driven Method of the Static Performance Calculation for Gas Joumal Bearing by Python. In Proceedings of the 2018 IEEE International Conference on Mechatronics and Automation, Changchun, China, 5–8 August 2018; pp. 18–22.
21. Stafslien, S.; Bahr, J.; Feser, J.; Weisz, J.; Chisholm, B.; Ready, T.; Boudjou, P. Combinatorial Materials Research Applied to the Development of New Surface Coatings I: A Multiwell Plate Screening Method for the High-Throughput Assessment of Bacterial Biofilm Retention on Surfaces. *J. Comb. Chem.* **2006**, *8*, 156–162. [CrossRef] [PubMed]

materials

MDPI

Article

Mathematical Modeling and Experimental Substantiation of the Gas Release Process in the Production of Non-Autoclaved Aerated Concrete

Evgenii M. Shcherban' [1], Sergey A. Stel'makh [1], Alexey Beskopylny [2,*], Levon R. Mailyan [3], Besarion Meskhi [4], Anatoly Shuyskiy [5], Nikita Beskopylny [6] and Natal'ya Dotsenko [5]

1 Department of Engineering Geology, Bases and Foundations, Don State Technical University, 344003 Rostov-on-Don, Russia; au-geen@mail.ru (E.M.S.); sergej.stelmax@mail.ru (S.A.S.)
2 Department of Transport Systems, Don State Technical University, 344003 Rostov-on-Don, Russia
3 Department of Roads, Don State Technical University, 344003 Rostov-on-Don, Russia; lrm@aaanet.ru
4 Department of Life Safety and Environmental Protection, Don State Technical University, 344003 Rostov-on-Don, Russia; reception@donstu.ru
5 Department of Technological Engineering and Expertise in the Construction Industry, Don State Technical University, 344003 Rostov-on-Don, Russia; a2293613@mail.ru (A.S.); natalya_1998_dotsenko@mail.ru (N.D.)
6 Department Hardware and Software Engineering, Don State Technical University, 344003 Rostov-on-Don, Russia; beskna@yandex.ru
* Correspondence: besk-an@yandex.ru; Tel.: +7-863-273-8454

Citation: Shcherban', E.M.; Stel'makh, S.A.; Beskopylny, A.; Mailyan, L.R.; Meskhi, B.; Shuyskiy, A.; Beskopylny, N.; Dotsenko, N. Mathematical Modeling and Experimental Substantiation of the Gas Release Process in the Production of Non-Autoclaved Aerated Concrete. *Materials* **2022**, *15*, 2642. https://doi.org/10.3390/ma15072642

Academic Editor: Petrica Vizureanu

Received: 18 February 2022
Accepted: 31 March 2022
Published: 3 April 2022

Abstract: The widespread use of aerated concrete in construction has led to the emergence of many types and compositions. However, additional research should fill theoretical gaps in the phenomenon of gas release during the formation of the structure of aerated concrete. Based on theoretical analysis and experimental studies, the article proposes a mathematical model of the swelling process based on the physicochemical laws of convection and molecular diffusion of hydrogen from a mixture and the conditions of swelling, precipitation, and stabilization of the mixture. An improved method for the manufacture of aerated concrete is proposed, which consists of introducing cement pre-hydrated for 20–30 min into the composition of the aerated concrete mixture and providing improved gas-holding capacity and increased swelling of the mixture, reducing the average density of aerated concrete up to 29% and improving heat-shielding properties up to 31%. At the same time, the small dynamics of a decrease in the strength properties of aerated concrete were observed, which is confirmed by an increased structural quality factor (CSQ) of up to 13%. As a result, aerated concrete has been obtained that meets the requirements of environmental friendliness and has improved mechanical and physical characteristics. Economic efficiency is to reduce the cost of production of aerated concrete and construction in general by about 15%.

Keywords: aerated concrete; gassing; swelling; blowing agent; average density; thermal conductivity

1. Introduction

1.1. Relevance and Article Tasks

The study's relevance is due to the current scientific lack of research on the fundamental nature and substantiation of mathematical and physical models to form promising cellular concrete. This material fully meets the world's requirements for energy efficiency and energy saving. It is a suitable material for improving its porous structure with the help of formulation and technological factors to obtain the most efficient building structures with a minimum cross-section and minimum thermal conductivity. In terms of cost characteristics, non-autoclaved cellular concretes significantly outperform autoclaved competitors due to the significantly reduced cost of production and manufacture of such concretes and less environmental impact. From the point of view of technological and recipe factors, there are quite a lot of studies of non-autoclaved and autoclaved aerated concrete [1–35]. However, from the point of view of fundamental science, some gaps should be filled with additional research in the phenomenon of outgassing during the formation of the structure of non-autoclaved aerated concrete.

Studies [2–4] are devoted to the possibility of obtaining aerated concrete products using iron and graphite tailings. The influence of the content and size of iron tailings on the mechanical properties of aerated concrete products has been studied, the products of hydrothermal synthesis of aerated concrete have been considered, and new types of autoclaved aerated concrete have been developed. The production of autoclaved aerated concrete (AAC) using graphite tailings such as silicon is described in [4]. In addition to cement dosage, water to solids ratio, and foaming agent content, the calcium–silicon ratio also plays an essential role in producing graphite waste autoclaved aerated concrete (GT-AAC).

It is known that the recycling of aerated concrete is widely used as raw material in the production of aerated concrete products and structures [5–10]. Thus, it was found in [5] that when replacing sand with recycled aerated concrete, an increase in compressive strength is 16% higher than that of conventional aerated concrete and 29–156% higher than the values obtained using industrial waste.

Studies of the influence of various fibers and microsilica on lightweight concrete's mechanical properties (compressive and bending strength) are also of great importance [11–16]. For example, in [11], the change in the value of thermal conductivity, compressive strength, and bending of autoclaved aerated concrete was studied when polypropylene, carbon, basalt, and glass fibers were added to autoclaved aerated concrete. It was found that the thermal conductivity of AAC with the addition of fibers varies linearly with the thermal conductivity of the fiber and that AAC reinforced with basalt fiber gives the highest thermal conductivity. In addition, the influence of the type and size of the fiber in aerated concrete production, on the values of compressive strength, bending, and thermal conductivity was studied [14]. Aerated concrete samples were made with polypropylene, carbon, basalt, and glass fibers. The study [14] found that adding various reinforcing fibers to aerated concrete resulted in increased flexural and compressive strength, and carbon fiber reinforced aerated concrete gave better flexural and compressive strength than other fiber types.

Complex additives are also used in aerated concrete technology. In [15], the authors studied the effect of a complex additive consisting of basalt fibers and SiO_2 microdust on the strength properties of samples of autoclaved aerated concrete. Based on the results of the studies, it was found that the complex additive affects the process of hydration of cellular concrete. Namely, it leads to calcium hydro silicates of the tobermorite group (C–S–H). The compressive strength of aerated concrete samples containing a complex additive increased by 52%, and the tensile strength in bending by 62% compared with the strength characteristics of samples of the control composition without the additive.

In other studies [17–23], the authors investigated the thermal conductivity of cellular concrete and various factors affecting it. For example, in [18], the authors developed a model for predicting the thermal conductivity of non-autoclaved cellular concrete using the linearization approach. This model has shown good reliability for predicting the thermal conductivity of cellular non-autoclaved concrete at 28 days of age. It has also been found that the main factors affecting thermal conductivity are the water–cement ratio, curing age, and temperature.

Investigation of dependences and modeling parameters of the behavior of AAC products in terms of wind load [24], material moisture [25,26], strength obtained using non-destructive testing methods [26], replacing aluminum powder on natural zeolite [27], as well as studies on porosity parameters [28–30], made a significant contribution to the development of this study and helped to formulate its purpose and scientific novelty.

Several works [31–35] are devoted to the macrostructure formation stage, the analysis of which shows that the determining influence on the formation of the porous structure of aerated concrete is exerted by the kinetics of gas release by the gasifier and the change in the rheological properties of the interpore material. Furthermore, studies of the swelling process to improve the quality of the macrostructure and improve the physical and mechanical properties of aerated concrete are currently aimed at an in-depth study and determination of the optimal combination of these kinetic processes.

1.2. Plan, Main Aim, and Hypothesis of the Study

Diffusion phenomena that occur during swelling, as a result of which part of the hydrogen is removed from the cellular concrete mixture, are the consequence of defect formation and lead to an increase in the average density of cellular concrete and a decrease in the quality of products. Therefore, the creation of conditions under which the diffusion phenomena of the intumescent mixture are manifested to a lesser extent makes it possible to achieve relatively better characteristics of the material structure.

The research plan is presented in a structural–logical block diagram (Figure 1).

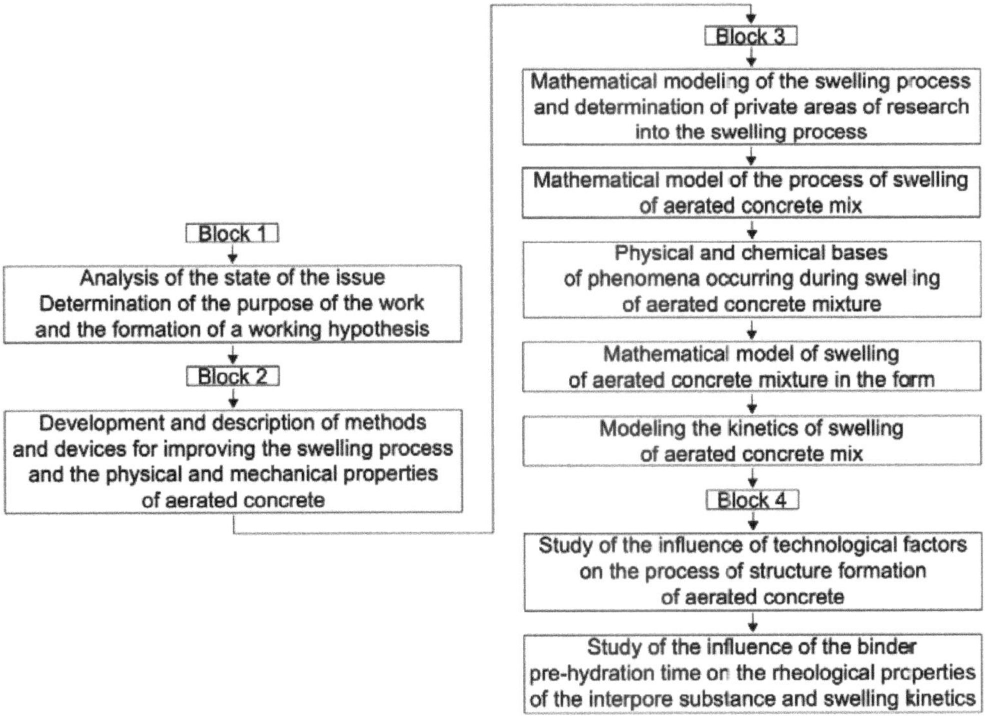

Figure 1. Structural and logical block diagram of the study plan.

This work aims to optimize the technological process for the manufacture of aerated concrete products using injection technology at the molding stage, ensuring a decrease in the average density and an increase in the uniformity of material properties. This goal is achieved by deepening the theoretical understanding of the processes of formation of a cellular structure and by optimizing the kinetics of processes occurring during swelling and improving on this basis the technology for manufacturing aerated concrete products, improving the means of controlling the properties of materials.

The working hypothesis of the work is formulated as follows: the intumescent aerated concrete mixture meets the optimal conditions for the manufacture of aerated concrete with desired properties, which ensures the coincidence in time of the start of coagulation of new cement formations and the start of intensive gas release by the aluminum blowing agent.

Based on the mathematical model, it is necessary to calculate an environmentally friendly, economic non-autoclaved aerated concrete that meets the requirements of the UN global ESG agenda for environmental protection, has an improved structure and improved energy-efficient and energy-saving properties, and better mechanical characteristics.

2. Materials and Methods

2.1. Materials, Test Equipment, and Measuring Instruments

This section lists the characteristics of the raw materials used in the experimental studies. The study of the processes of formation of a cellular structure, the determination of the optimal compositions of cellular concrete, and the physical and mechanical properties of the obtained materials were carried out on cement–sand mixtures.

When carrying out the experiments, Portland cement of the CEM I 42.5N brand produced by JSC Novoroscement (Novorossiysk, Russia) was used. Table 1 presents Portland cement's physical and mechanical characteristics and its chemical composition.

Table 1. Physico-mechanical characteristics and chemical composition of Portland cement CEM I 42.5N.

Property	**Value**
Physical and mechanical	
Compressive strength at the age of 28 days, MPa	44.7
Setting time, min	
- start	155
- end	220
The fineness of grinding, passage through a sieve N 008, %	96.7
Specific surface, m^2/kg	331
Normal density of cement paste, %	23.5
Chemical	
Loss on ignition, wt%	0.70
Silicon oxide content (SiO_2), %	20.89
Content of aluminum oxide (Al_2O_3), %	4.72
Iron oxide content, (Fe_2O_3), %	4.32
Content of calcium oxide (CaO), %	63.27
Mass fraction of magnesium oxide (MgO), %	2.45
Mass fraction of sulfuric anhydride (SO_3), %	2.81
Mass fraction of alkali oxides in terms of Na_2O, %	0.60
Content of free calcium oxide (CaO_{fr}), %	0
Mass fraction of chloride ion (Cl^-), %	0.038
Insoluble residue, %	0.20

Quartz sand produced by Arkhipovsky Quarry OJSC (village Arkhipovskoye, Belorechensky District, Krasnodar Territory, Russia) was used as a fine aggregate, the physical characteristics of which are presented in Table 2.

Table 2. Physical characteristics of fine aggregate.

<table>
<tr><th colspan="7">Grain Composition</th><th rowspan="4">Pass Through a Sieve Mesh 0.16, wt%</th><th rowspan="4">Size Modulus</th><th rowspan="4">Content of Dust and Clay Particles, %</th><th rowspan="4">True Density, kg/m³</th><th rowspan="4">Bulk Density, kg/m³</th></tr>
<tr><th colspan="7">Sizes of Sieve Openings, mm</th></tr>
<tr><th colspan="7">Private and Total Residues on Sieves, %</th></tr>
<tr><th>10</th><th>5</th><th>2.5</th><th>1.25</th><th>0.63</th><th>0.315</th><th>0.16</th></tr>
<tr><td rowspan="2">0</td><td rowspan="2">0</td><td>0.17</td><td>1.39</td><td>8.86</td><td>45.80</td><td>41.03</td><td>2.49</td><td rowspan="2">1.66</td><td rowspan="2">1.1</td><td rowspan="2">2650</td><td rowspan="2">1438</td></tr>
<tr><td>0.17</td><td>1.56</td><td>10.42</td><td>56.21</td><td>97.25</td><td>99.74</td></tr>
</table>

Aluminum powder PAP-1 produced by OOO SKIF (St. Petersburg, Russia) was used as a blowing agent. The physical characteristics and chemical composition of aluminum powder are shown in Table 3.

Table 3. Physical properties and chemical composition of aluminum powder.

Covering Capacity on Water, cm^2/g	Residue on Sieve 0.08, %	Chemical Composition, %								Buoyancy, %
		Active Aluminum	Impurities							
			Fe	Si	Cu	Mn	Moisture	Fat		
7000	1.0	-	0.4	0.3	0.05	0.01	0.2	3.5		80

The compressive strength was determined on 100 mm × 100 mm × 100 mm cube samples following the requirements of GOST 10,180 "Concretes. Methods for strength determination using reference specimens" [36]. The average density of aerated concrete was also determined on cube samples measuring 100 mm × 100 mm × 100 mm by the requirements of GOST 12,730.1 "Concretes. Methods of determination of density" [37].

The thermal conductivity of aerated concrete was determined under the requirements of GOST 7076 [38] on samples of dimensions 100 mm × 100 mm × 20 mm (Figure 2a).

The study used:

- Technological equipment—installation for the production of foam concrete and aerated concrete GBS-250 manufactured by METEM (Perm, Russia); ball planetary mill "Activator-4M" (LLC "Plant of Chemical Engineering", Novosibirsk, Russia);
- Testing equipment—hydraulic press MIP-25 (LLC NPP INTERPRIBOR, Chelyabinsk, Russia);
- Measuring instruments—metal measuring ruler 500 mm (JSC "Stavropol Tool Plant", Stavropol, Russia); laboratory scales HT-5000 (NPP Gosmetr, St. Petersburg, Russia); caliper ShTs-I-250-0.05 (LLC NPP Chelyabinsk Tool Plant, Chelyabinsk, Russia); thermal conductivity meter ITP-MG4 (OOO SKB Stroypribor, Chelyabinsk, Russia) (Figure 2b); viscometer ZM-1001 (Priborelectro LLC, Moscow, Russia); plastometer K-2ZhV (OOO NPP Tochpribor, Rostov-on-Don, Russia) [39–43].

(a) (b)

Figure 2. Specimens (**a**) and device (**b**) for testing aerated concrete for thermal conductivity.

2.2. Devices and Methods for Researching Aerated Concrete Mix

To determine the parameters of the aerated concrete mixture and due to the lack of necessary methods, there was a need for special instruments and methods for studying gas release, swelling, and other characteristics described in this section of the work. The device PGV-2A (DSTU, Rostov-on-Don, Russia) operates in automatic control mode. The device parameters are registered by a multipoint automatic potentiometer of the KSP-4 type (OOO RKPO, Moscow, Russia). The volume of the reaction vessel is 2–3 L, the measured volume of the evolved gas is up to 3 L.

The device PGV-2A (Figure 3a) consists of a reaction vessel (1) with a scale (2) designed to determine the level of filling and expansion of a substance (3). In addition, there is a thermocouple (4) for measuring the temperature of a substance. A lifting table (5) presses the reaction vessel against the upper base (6). To seal the internal volume, the device is equipped with a rubber gasket (7). A variable measuring container (8) is attached to the upper base of the device, made in the form of a thin corrugated pipe (for example, made of polyethylene). The upper part of the displaced measuring container (8) is closed with a lid, on which a contact pressure sensor (9), a valve (10), and a thermocouple (11) are placed to measure the temperature of the vapor–gas medium in the device above the surface of the substance. To measure the volume of the variable measuring capacity (8), a displacement device (12), a contact pressure sensor (9), a displacement control unit (ACU), and a reversible motor (13) are used, which make it possible to maintain the required pressure in the device when the substance releases gas and changes the temperature of the vapor–gas medium. The volume of the variable capacity (8) is measured by a sensor (14). The control of the internal pressure in the device is carried out by the U-shaped pressure gauge (15) (Figure 3b). To study the kinetics of gas evolution of dispersed gas-forming agents in liquid media at different temperatures, the device is equipped with a special reaction vessel (Figure 3c) containing a contact thermometer (16), a stirrer (17), an electric heating element (18), and a device for introducing a gas-forming agent suspension (19).

On the front control panel of the PGV-2A device, there are controls for internal pressure control, stirring rate control (ACU), and temperature control (TCU) of the reaction medium, as well as the control of the movement of the variable measuring capacitance (MCU).

Using the PGV-2A device, the parameters of the swelling kinetics are determined as follows. First, the test substance is placed in a reaction container installed in the device. Then a thermocouple is introduced into the substance, and the device is sealed. After that, for time points 1, 2, 3 ... i ... p with a time interval of no more than 15 s, the process, and device parameters are recorded: the temperature of the substance $t_{B(i)}$, the temperature of the vapor–gas medium above the substance $t_{2(i)}$, the volume of the variable measuring capacitance $V_{p(i)}$, the volume of concrete mix in the container $V_{B(i)}$.

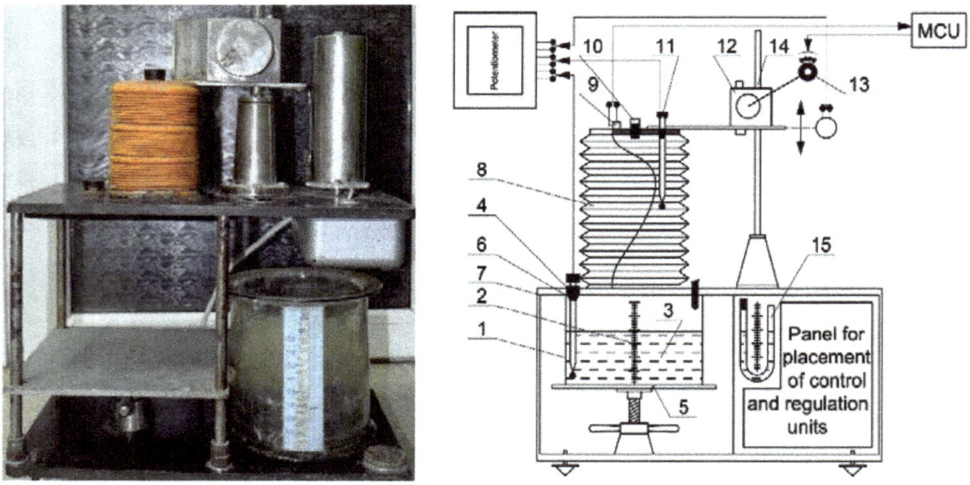

(a) (b)

Figure 3. *Cont.*

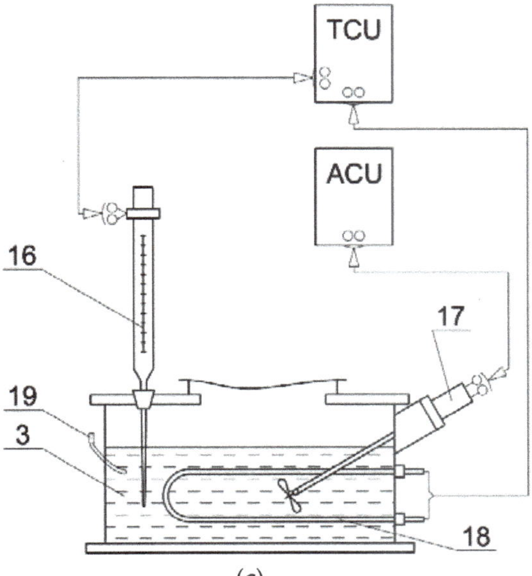

(c)

Figure 3. Device for recording the parameters of gas release and swelling of aerated concrete mixtures PGV-2A: (**a**) photo; (**b**) schematic diagram of PGV-2A; (**c**) connection diagram of the heating element and the stirrer of the reaction container.

After the completion of the process, the calculation of the parameters of the swelling kinetics of the substance—the swelling coefficient $K_{B(i)}$, the gas diffusion coefficient $D(i)$, the gas evolution of $V_{(i)}^{H_2}$, the temperature $t_{B(i)}$, for time points (i) is carried out according to the following formulas:

$$K_{B(i)} = \frac{V_{(i)}^{H_2}}{V_{(1)}^{H_2}} \tag{1}$$

$$V_{B(i)}^{H_2} = \frac{293.15\,P_0\,\beta_{B(i)}\left(V_{B(i)} - V_{B(1)}\right)}{11,325\left(273.15 + t_{B(i)}\right)},\ \text{cm}^3 \tag{2}$$

$$V_{D(i)}^{H_2} = 293.15\,P_0\,\beta_{2(i)}\,\frac{V_{0(i)} - V_{B(i)}}{273.15 + t_{2(i)}} - \frac{V_{0(1)} - V_{B(i)}}{273.15 + t_{2(i)}},\ \text{cm}^3 \tag{3}$$

$$V_i^{H_2} = V_{B(i)}^{H_2} + V_{D(i)}^{H_2},\ \text{cm}^3 \tag{4}$$

$$D_i = \frac{V_{D(i)}^{H_2}}{V_i^{H_2}} \tag{5}$$

where, $V_{0(i)}$ is the total internal volume of the device (the volume of the reaction and variable measuring capacity), cm^3; $V_{B(i)}^{H_2}$ is the volume of hydrogen in the cellular concrete mixture, cm^3; $V_{D(i)}^{H_2}$ is the volume of hydrogen diffused from the cellular concrete mixture, cm^3; $V_i^{H_2}$ is the total volume of hydrogen formed during swelling, cm^3; P_0 is atmospheric pressure, Pa; $\beta_{2(i)} = 1 - \alpha_0\, g_t\left(273 + t_{2(i)}\right)$ is the relative content of gases in the vapor–gas medium above the mixture; $\beta_{B(i)}$ the same, in a mixture; α_0 = 4.555 cm^3/(g × deg); g_t is the absolute humidity of the gaseous medium (at 100% relative humidity), g/cm^3 [33–35].

3. Modeling the Kinetics of Gas Release during the Manufacture of Aerated Concrete

The study considers the fundamental aspects that occur at the level of interaction of various media during gas release and the formation of the aerated concrete structure. The mathematical models of all these processes have been developed. It is proposed to consider aerated concrete mixtures as a system of finite elements, for which one gas bubble is taken. Furthermore, using the apparatus of mathematical physics, the equations of molecular diffusion, and modeling the process of gas release in the emerging aerated concrete, two main hypotheses can be tested:

- Revealing the fundamental-applied connection between the processes of gas release and structure formation of aerated concrete with the performance characteristics of the hardened composite;
- Verification of technological and operational efficiency when using partially hydrated cement in the production of non-autoclaved cellular concrete with improved structural and operational characteristics.

Several processes should first be modeled from the point of view of mathematical physics and then wholly or partially reproduced in the laboratory in a natural experiment. To determine the parameters of the kinetics of gas evolution, it is necessary to stabilize the conditions of the experiment. For this purpose, milk of lime with a density of 1.05–1.2 g/cm^3 is used as the reaction medium, the temperature of the reaction medium and the stirring rate are maintained constant at a given level [33–35].

To determine the parameters of the kinetics of gas evolution, lime milk is poured into the reaction vessel of the PGV-2A device, which is heated to a predetermined temperature. The device is sealed, and then the required amount of the gas-forming suspension is added to the reaction medium, the temperature of the vapor–gas medium $t_2(i)$ is recorded, and the internal volume of the device V_p is changed. After the end of the gas evolution reaction, the values of the volume of released hydrogen ($V_{(i)}^{H_2}$) are calculated, reduced to the conditions of t = 20 °C, P_{atm}= 101,325 Pa, according to the formula:

$$V_{(i)}^{H_2} = \frac{293.15\, P_0\, \beta_{2(i)}}{101,325}\left(\frac{V_i - V_p}{273.15 + t_{2(i)}} - \frac{V_i - V_p}{273.15 - t_{2(i)}}\right) \quad (6)$$

The gas evolution kinetics curve usually has a sigmoid form and can be approximated by a function of the form

$$\frac{V^{H_2}}{V_{\max}^{H_2}} = 1 - \exp\{-[m(\tau - \tau_{ind})^n]\} \quad (7)$$

where, V^{H_2} is an experimental volume of hydrogen formed in the process of swelling; $V_{\max}^{H_2}$ is the maximum calculated volume of hydrogen generated during the swelling process; τ_{ind} is the time of the induction period; m, n—indicators that determine the curvature of the curve of the kinetics of gas evolution.

The content of actual aluminum in the blowing agent is determined by the formula

$$\alpha = \frac{V_{\max}^{H_2}}{1330\, P_{Al}} \quad (8)$$

where P_{Al} is the mass of aluminum in the blower, g; 1330 is the volume of hydrogen, cm^3/g, released during the complete reaction of active aluminum and at t = 20 °C and P_{atm}= 101,325 Pa.

3.1. General Characteristics of the Process of Swelling of Aerated Concrete Mix

The swelling of a cellular concrete mixture is a complex hydrodynamic and physicochemical process that develops over time. It consists of the nucleation and growth of cells due to the reaction of gas evolution, deformation of the mixture at a variable rate under the action of the pressure arising in the cell, vaporization, and condensation of water vapor. At the same time, the temperature, rheological, and diffusion characteristics of the interpore wall and, in general, the entire mixture change. The swelling process is complicated by the boundaries of the expansion in the form and some other conditions.

The process of swelling of aerated concrete mixtures can be conditionally divided into the following periods of the state of aerated concrete mixture—induction, swelling, precipitation, and stabilization of the structure. Each of these periods is characterized by the predominance of individual phenomena that occur during swelling.

The induction period is characterized by slow swelling of the array and an increase in molecular diffusion of hydrogen through the surface of the array to a certain local maximum, the onset of which conditionally indicates the end of this period. The phenomena taking place during this period can be explained as follows. The heterogeneous reaction of the interaction of the liquid (alkaline) phase with aluminum particles having an oxide layer and covered with a stearin film has an induction period that depends on the rate of development of the reactive surface of aluminum particles. The hydrogen released as a result of the reaction diffuses from the surface of a part of aluminum into the liquid phase, saturating it. Due to the high concentration of hydrogen, gas bubbles form at the surface of the particles. At the same time, hydrogen migrates into the depth of the liquid phase and into entrained air bubbles, which are always present in real mixtures. Hydrogen migration also proceeds to the surface of the mixture and into the surrounding atmosphere. A significant hydrogen concentration gradient can explain the predominance of molecular diffusion through the mixture surface on the mixture surface. Supersaturation of the liquid phase with hydrogen is the result of a significant partial pressure of hydrogen in bubbles of a small radius, as well as an insufficient surface of the existing bubbles for the diffusion of molecular hydrogen into them. Achieving the maximum rate of molecular diffusion of hydrogen from the mixture is a conditional transitional moment between the induction period and the intense swelling of the mixture.

The swelling period is characterized by an intensive increase in the volume of the aerated concrete mixture up to a certain maximum value. The phenomena that take place during this period can be explained by the high rate of the gas formation reaction, the increase in the radius of gas bubbles, and their surface area. The limiting factor for swelling is the interpore wall's structural and rheological kinetic characteristics. A decrease in molecular diffusion is a consequence of a decrease in supersaturation of the liquid phase with hydrogen and saturation of the mixture with gas bubbles. During the period of maximum intensity of swelling, sound emission of the mixture is observed, which indicates the beginning of the formation of defects in the interpore walls in the form of cracks and holes. This moment corresponds to the minimum diffusion of gas through the mixture. The subsequent intensive removal of gas from the mixture is a consequence of the convective diffusion of vapor and hydrogen through defects. The damping of swelling can be explained by convective diffusion of the gas, which reduces the pressure in the bubbles of the mixture. When the gas pressure in the bubbles is balanced and the pressure required to deform the interpore wall, the mixture stops swelling.

The period of precipitation is characterized by a decrease in the volume of the mixture and intense convective diffusion of the vapor–gas mixture from it (gas breakthrough). These phenomena proceed at a variable speed. The sedimentation of the mixture occurs due to the deformation of interpore partitions under the forces of the hydrostatic pressure of the overlying layers. The settling decreases as the internal gas pressure in the bubbles, the resistance stress of the interpore walls to compression, and the hydrostatic pressure of the overlying layers of the mixture are balanced.

The stabilization period is characterized by the absence of a decrease in the volume of the mixture and the high diffusion of gas from it. The phenomena of this period are similar to the phenomena of the previous one. Full stabilization of the mixture is achieved under the condition that the stress of resistance of the interpore walls to collapse exceeds the pressure of the overlying layers. The period of precipitation may be absent if the condition of stabilization of the mixture is met immediately after the period of swelling. The swelling process can be considered completed after the end of gas diffusion through the surface of the mixture.

The considered phenomenological concepts of swelling are the basis of the model of this process [33–35].

3.2. Physical and Chemical Bases of Phenomena Occurring during Swelling of Aerated Concrete Mixture

Further, based on [33–35], the physicochemical foundations of the phenomena occurring during the swelling of an aerated concrete mixture were considered.

3.2.1. Main Characteristics of Cells (Bubbles)

The leading indicators characterizing the dispersed phase are the gas content, disperse composition, and chemical composition of the gas phase. Gas bubbles in cellular concrete are not monodisperse. The average size of the bubbles and their polydispersity largely depend on the specific conditions of production, the rheological and surface properties of the interpore substance, and many other factors. Cells of different sizes in aerated concrete have different values of excess surface energy and therefore have different solubilities or vapor pressures. As a result, in polydisperse systems, particles are redistributed in size in the direction of their enlargement, and the thermodynamic limit of this redistribution is the division of the disperse system into two layers. During swelling, cells in aerated concrete can be enlarged in two ways: coalescence and molecular transfer of gas from small cells to large ones. The simultaneous action of these two mechanisms is also possible.

The stability of gas emulsions with respect to coalescence processes is determined by the interaction energy of gas bubbles and the strength of the two-sided film upon their contact. Therefore, it depends on several conditions: the properties of the dispersion medium and the dispersed phase, the nature of the emulsifier used, the presence of impurities, temperature. The phenomenon of coalescence in the aerated concrete mixture, apparently, prevails at the initial stage of swelling, when the size of the bubbles and the rheological properties of the interpore wall (due to the absence of structural bonds between the particles of the substance) are still small.

In gas-filled disperse systems containing bubbles of various sizes, the size of the bubbles is redistributed due to diffusion mass transfer. The mechanism of this phenomenon is based on the fact that the vapor pressure of small bubbles is greater than that of large ones. The thermodynamic regularities of formation, growth, and distribution of bubbles show that bubbles are more stable, the larger they are. The gas in small bubbles dissolves since the gas concentration in the liquid phase of the material of the interpore wall is lower than the equilibrium one. In contrast, with respect to the largest bubbles, the same liquid phase of the interpore substance is supersaturated, and the gas dissolved in the liquid phase is released into large bubbles. In this case, the largest bubbles grow due to the dissolution of small ones and partially due to the liquid phase of the interpore substance. As a result, the gas concentration in the liquid phase decreases, and the critical size of the bubbles, which are in equilibrium at a given gas saturation of the liquid phase, continuously increases. If gas formation occurs during a chemical reaction in the interpore substance, then the degree of gas supersaturation in the liquid phase reaches large values, significantly exceeding the equilibrium concentrations for small and large bubbles. Therefore, it can be assumed that bubbles of different sizes grow at different rates in an intumescent aerated concrete mixture, and growth is more intense for larger ones.

The gas porosity is:

$$\Pi_g = \left(1 + \frac{\delta_{av}}{R}\right)^{-3} \tag{9}$$

where δ_{av} is the average thickness of the interpore wall per cell; R is the average cell radius.

The dependence of the gas porosity of aerated concrete on the average pore radius R and the minimum thickness of the interpore partition $\delta_{\min}$ per one pore, and the coordination number N_k, can be represented by the following formula:

$$\Pi_g = \left(1 + \frac{\delta_{\min}}{R}\right)^{-3} \frac{(N_k - 2)}{N_k(N_k - 1)} \quad (10)$$

Based on Equations (11) and (12), we obtain the dependence of the minimum thickness of the interpore wall $\delta_{\min}$ on the average thickness of the interpore wall ε_{av}, the average radius R of the cell, and the coordination number N_k

$$\delta_{\min} = R\left[\left(\frac{(N_k - 2)^2}{N_k(N_k - 1)}\right)^{\frac{1}{3}} \left(1 - \frac{\delta_{av}}{R}\right)^{\frac{1}{3}} - 1\right] \quad (11)$$

If the batch volume V and the number of cells in the aerated concrete mix N are known, then with an average cell radius R, the average value of the thickness of the interpore wall δ_{av} per cell

$$\delta_{av} = \sqrt[3]{R^3 + \frac{3V_0}{4\pi N}} - R;\ \text{or}\ \frac{\delta_{av}}{R} = \sqrt[3]{1 + \frac{3V_0}{4\pi N R^3}} - 1 \quad (12)$$

at $R \to 0$

$$\delta_{av} = \sqrt[3]{\frac{3V_0}{4\pi N}} \quad (13)$$

The minimum thickness of the interpore partition during swelling of the aerated concrete mixture has a certain critical value, upon reaching which, it is possible to form holes in the interpore partition. This critical thickness depends on the dispersion of the interpore substance and is equal to 3–5 diameters of the coarse component. If we characterize the dispersion of the interpore substance by the specific surface, then the average particle diameter D_p of the interpore substance is equal to

$$D_p = \frac{6}{\gamma\, S_{sp}} \quad (14)$$

where γ is the interpore particle density.

However, in natural materials, the thickness of the partitions is much greater than the critical minimum thickness of the interpore partition $\delta^{cr}_{\min} = 3D_p$, since the packing density of cement and sand particles is insufficient due to their angularity, roughness, and the presence of mixing water in the composition of the mixture. In this regard, it can be assumed that, under the condition $\delta_{\min}$, $\delta^{cr}_{\min}$, the number of defects in interpore walls is equal to

$$N = N_0 \exp\left[-\left(\frac{\delta_{\min}}{\delta^{cr}_{\min}} - 1\right)\right] \quad (15)$$

where N_0 is the total number of cells in the mixture (layer).

Considering that as a result of cement hydration (until the stabilization of rheological properties τ_{cl}), a finely dispersed phase of cement neoplasms appears in the composition of the aerated concrete mixture, which affects the deformative properties of the interpore partition, it can be assumed that this circumstance contributes to a decrease in the critical minimum thickness of the interpore partition:

$$\delta_{\min} = \delta^{cr1}_{\min} \exp\left(-3\frac{\tau}{\tau_{cl}}\right) + \delta^{cr2}_{\min} \quad (16)$$

where $\delta^{cr1}_{\min}$ is the initial critical minimum thickness of the interpore wall; $\delta^{cr2}_{\min}$ is the final critical minimum thickness of the interpore wall.

3.2.2. Interaction of a Single Bubble and Interpore Medium

The growth of vapor–gas bubbles in the aerated concrete mixture occurs due to their expansion under the action of the gas released into the bubbles dissolved in the liquid phase or filling them with steam due to mass transfer and heat exchange processes of the bubble and the liquid phase surrounding it.

Both the processes of bubble expansion and filling it with gas or vapor are determined by the level of pressure or resolution in the liquid phase.

The liquid film surrounding the gas bubbles, due to the gas dissolved in the liquid phase, is supersaturated with gas to a concentration determined by the level of gas pressure in the bubble and the solubility coefficient [44], i.e., on the surface of the bubble

$$C_0 = k_H(P_r T_r)\, P_r \tag{17}$$

where C_0 is the equilibrium concentration; $k_H(P_r T_r)$ is the Henry solubility coefficient, which depends on temperature and, in general, on the pressure level; P_r is the partial pressure of the gas in the bubble.

The solubility of gases is greatly influenced by temperature: usually, increasing it lowers the solubility. It is known that the solubility of gases in liquids obeys the Clausius–Clapeyron equation. The solubility of gases in aqueous solutions of electrolytes decreases compared to pure water.

If the liquid is saturated to a dissolved gas concentration equal to C_∞, and the concentration C_0 on the bubble surface is lower than C_∞, then due to the difference in concentrations C_∞–C_0, a mass flow of the gas dissolved in the liquid into the bubble arises.

At a negative difference C_∞–C_0, the gas in the bubble is dissolved in the liquid phase, and the bubble is compressed.

In the general case, the rate of gas evolution is determined by the rate of gas diffusion in the liquid phase, that is, the rate of transfer to the bubble boundary, the rate of direct release (desorption) of the volume of the gas bubble.

The difference in gas concentrations in the liquid phase leads to the molecular transfer of matter. According to Fick's first law, the diffusion process is described by the dependence [44]:

$$\frac{dm}{S\,dt} = D(T, p)\left(\frac{\partial C}{\partial r}\right)_{r=R} \tag{18}$$

where, $\frac{dm}{S\,dt}$ is the limiting gas mass flow through a surface unit; $D(T, p)$ is the molecular diffusion coefficient; $\left(\frac{\partial C}{\partial r}\right)_{r=R}$ is the derivative of the concentration along the normal to the bubble surface.

The rate of substance transfer at a given temperature and pressure is proportional only to the concentration gradient near the interface.

The value of the concentration derivative in Equation (21) at a fixed interface is determined from the solution of the molecular diffusion equation

$$\frac{\partial C}{\partial r} = D\left(\frac{\partial C}{\partial r} + \frac{2}{r}\frac{\partial C}{\partial r}\right) \tag{19}$$

$$\begin{gathered} \text{at } \tau = 0\ r = R, \text{ at } \tau = \infty\ r \to \infty, C = C_\infty \\ \text{at } \tau > 0\ r = R, C = C_0, \text{ at } \tau = \infty, r \to \infty, C = C_\infty \end{gathered} \tag{20}$$

The solution of Equation (19) under the initial and boundary Conditions (20) makes it possible to determine the concentration gradient at the bubble boundary

$$\left(\frac{\partial C}{\partial r}\right)_{r=R} = (C_\infty - C_0)\left(\frac{1}{R} + \frac{1}{\sqrt{\pi D \tau}}\right) \tag{21}$$

Dependences (18) and (21) the mass flow of gas through a single area, which is determined by the process of molecular diffusion in a liquid, can be found

$$\frac{\partial m}{\partial \tau} = DS(C_\infty - C_0)\left(\frac{1}{R} + \frac{1}{\sqrt{\pi D \tau}}\right) \tag{22}$$

With a time-varying gas concentration in liquid $C_\infty(\tau)$ and $C_0(\tau)$, Equation (25) takes the form

$$\frac{\partial m}{\partial \tau} = DS(T)\,[C_\infty(\tau) - C_0(\tau)]\left(\frac{1}{R} + \frac{1}{\sqrt{\pi D \tau}}\right) \tag{23}$$

In the case of a vapor–gas bubble, the rate of its growth or dissolution depends, in addition to diffusion, on heat transfer, which is described by the control of heat transfer during liquid evaporation or vapor condensation. The equilibrium vapor pressure (vapor pressure) over a concave liquid surface (with radius R) is lower than for a flat interface. It can be determined from the Kelvin–Helmholtz relation, which for the case ρ_l, ρ_v has the form

$$P_v = P_v^1 \exp\left(-\frac{2\sigma}{R}\frac{1}{\rho_l BT}\right) \tag{24}$$

where P_n^1 is the vapor pressure over a flat surface.

For capillaries with a radius greater than 10^{-5} cm, the saturation vapor pressure above the meniscus in the capillary is equal to the saturation vapor pressure over a flat surface with an accuracy of 1%. For the entire set of bubbles in the aerated concrete mixture, the amount of gas released into the bubble is equal to

$$\frac{\partial m}{\partial \tau} = D4\pi NR^2\left[C_\infty(\tau) - C_0(\tau)\frac{1}{R} - \frac{1}{\sqrt{\pi D \tau}}\right] \tag{25}$$

3.2.3. Conditions for the Deformation of the Gas Mixture

Swelling of the mass of aerated concrete mixture occurs as a result of deformation of the interpore wall under the action of gas pressure in the cells, and the sediment is the result of deformation of the interpore material under the action of hydrostatic pressure from the overlying layers of the aerated concrete mixture.

According to the theory of the energy of shape change of a plastic body, the specific energy of shape change for any point remains constant in the plastic state of matter. Based on this, when deforming an array of aerated concrete, you can write the following expression for the pressure of the vapor–gas medium in the cells:

$$P_0 = P_\sigma + P_h + P_{mn} + P_v \tag{26}$$

where P_0 is atmospheric pressure; P_σ—pressure from the action of surface energy forces at the boundary of two phases; P_h—hydrostatic pressure from the overlying layers of aerated concrete mixture; P_{mn} is the pressure required for plastic deformation of the interpore wall; P_v is the pressure caused by the limited swelling of the mixture in the mold and the constraint of cell movement.

The condition of swelling and settling of an aerated concrete mixture and any elastic-plastic system can be justified by the position of the problem of motion stability. The effect of the loss of stability of the interpore partition with an increase in the internal pressure in the cell is the appearance of swelling deformations that occur when a certain critical value of the internal pressure in the cell is exceeded, that is:

$$P_c - P_0 - P_\sigma - P_h \geq P_{mn}^B \tag{27}$$

When the mixture settles, there is a loss of stability of the interpore wall, which occurs under the action of the hydrostatic pressure of the overlying layers when the condition

$$P_c - P_0 - P_\sigma - P_h < P^0_{mn} \quad (28)$$

Cellular concrete mixtures and their interpore partitions under applied stresses not exceeding the limiting shear stress τ_0 behave like a solid body, and when these stresses are exceeded, they irreversibly flow with a strain rate dv/dy due to the plastic viscosity of the mixture and can be described by the Shvedov–Bingham

$$\tau = \tau_0 + \eta \frac{dv}{dy} \quad (29)$$

In an utterly plastic state of a polydisperse medium with spherical inclusions, the pressure causing such a state is expressed by the dependence

$$P_{mn} = \pm 4\tau \ln \frac{r + \delta}{r} \quad (30)$$

The formula determines the critical value of swelling pressure:

$$P^S_{mn} = 4\tau_0 \ln \frac{r + \delta}{r} \quad (31)$$

and the critical value of sediment pressure:

$$P^0_{mn} = -4\tau_0 \ln \frac{r + \delta}{r} \quad (32)$$

The pressure in the bubble from the action of surface energy forces at the bubble–interstitial wall boundary is determined by the formula:

$$P_\sigma = \frac{2\sigma}{R} \quad (33)$$

where σ is the surface tension of the liquid at the phase boundary.

The hydrostatic pressure from the overlying layers is:

$$P_h = \gamma_{mn}\, g \left[1 - \left(1 + \frac{\delta}{R}\right)^{-3}\right] (H - h) \quad (34)$$

where γ_{mn} is the average density of the interpore wall; σ is the surface tension of the liquid at the phase boundary; H is the height of the aerated concrete mixture; h is the height from the bottom of the mold to the layer under study; $g = 9.81\ \mathrm{m/s^2}$.

3.3. Modeling the Kinetics of Swelling of Aerated Concrete Mix

The presented mathematical model [33–35,45–48] was used to simulate the process of swelling of an aerated concrete mixture under conditions that correspond to the hypotheses of this study, to confirm the assumptions made about increasing the homogeneity of material properties, due to the swelling process during the start of coagulation of the colloidal fraction of the liquid phase of the mixture (Figure 4).

For this purpose, the conditions of the process of swelling of the aerated concrete mixture were modeled, under which the time-varying parameters were the outgassing of the gasifier G(j), the limiting shear stress of the interpore substance τ_0(j), and the temperature of the mixture T_0 was constant and equal to 40 °C.

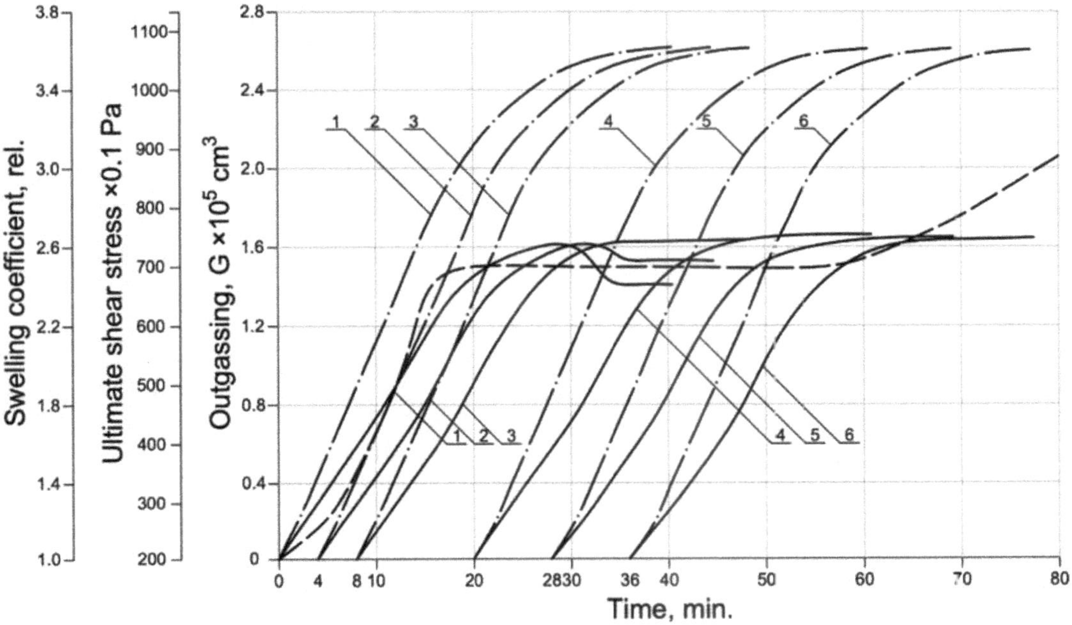

Figure 4. Kinetics of swelling of an aerated concrete mixture (simulation): 1–6 are experiment numbers.

The initial data for the calculation according to the model were also:

1. Mixing volume $V_0 = 10^5$ cm^3.
2. Sectional area of the array $S_0 = 2.5 \times 10^3$ cm^2.
3. Volume of gas saturation of the mixture (nuclei) $\varphi = 0.27\%$.
4. Atmospheric pressure P_0 = 1,013,250 dynes/cm^2.
5. Molecular diffusion coefficient D = 4.8×10^{-5} cm^2/s.
6. Henry coefficient $k_H = 1.1618 \times 10^{-10}$ (cm^3/cm^3) × (cm^2/dyne).
7. Partial pressure of water vapor P_n = 76,196.4 dynes/cm^2.
8. Surface tension of water σ = 70 erg/cm^2.
9. Average density of interpore substance γ_{mn} = 1.762 g/cm^2.
10. W/T—mixture ratio = 0.5.
11. Density of water ρ_l = 1 g/cm^3.
12. The average density of the solid phase ρ_T = 2.874 g/cm^3.
13. Coordination number of cells N_k = 12.
14. Free fall acceleration g = 980 cm/s^2.
15. The thickness of the diffuse film $\delta = 10^{-4}$ cm.
16. Number of layers i = 10.
17. The number of analyzed time intervals j = 40.
18. Time interval $\Delta\tau$ = 60 s.
19. The initial concentration of hydrogen in the liquid phase is $C_\infty(0) = 0$.
20. The average initial radius of steam–air bubbles (nuclei) of the mixture is R(i,0) = 10^{-2}.
21. Coefficient M = 10^{-9} s;
22. The minimum critical thickness of the interpore wall at the beginning of the process per cell $\delta_{\min}^{kp1} = 5 \times 10^{-3}$ cm.
23. The minimum critical thickness of the interpore wall 20 min after mixing the mixture with water $\delta_{\min}^{kp2} = 5 \times 10^{-3}$ cm.

Swelling modeling was carried out for six experiments, differing in the moment of the beginning of the gas evolution of the blowing agent in the mixture.

An analysis of the simulation presented in Figure 4 shows that, under these experimental conditions, a shift in the onset of gas evolution by 4 and 8 min (experiments 1, 2, and 3), relative to the moment of combining the mixture with water, leads to a change in the swelling coefficient of the mixture by more than 8%. Furthermore, an increase in the start time of gas evolution (experiments 4, 5, and 6) leads to stabilizing the swelling parameters. Furthermore, in all experiments in the process of swelling, the heterogeneity of the parameters of individual layers of the aerated concrete mixture is observed, which is determined by the magnitude of the hydrostatic pressure from the overlying layers of the mixture, and the degree of convective diffusion of hydrogen, leading to sedimentation and compaction of the concrete.

The observed inhomogeneity of the properties of the expanded aerated concrete mixture depends on the correspondence between the kinetics of microstructure formation of the interpore substance and the gas release of the blowing agent, and decreases with a shift at the beginning of the kinetics of the blowing out of the blowing agent, in the region of stabilization of the rheological properties of the interpore wall.

The simulation of the swelling of aerated concrete mixture made it possible to identify two practical directions for implementing the provisions of the hypothesis: the first is the use of a blowing agent with a long induction period, the second is the use of pre-hydrated cement as a binder.

4. Results and Discussion

4.1. Study of the Influence of Technological Factors on the Structure Formation of Aerated Concrete, Analysis of the Kinetics of the Properties of the Interpore Substance of the Aerated Concrete Mix

The interpore wall of the aerated concrete mix consists of a multi-component substance. The change in the rheological characteristics is determined mainly by the kinetics of the structure formation of the binder.

To study the swelling processes of aerated concrete mixtures, of particular interest is the initial period of structure formation of the binder, and the period of intense hydration, coinciding in time with the stage of formation of the cellular structure.

To assess the kinetics of the rheological properties of the material of the interpore wall of the aerated concrete mixture, the change in its ultimate shear stress (τ_0), plastic viscosity (η), and plastic strength (P_m) in time equal to the swelling period (0–80 min) was studied.

The test procedure was adopted as follows. Raw materials (Portland cement and ground sand S_{sp} = 3000 cm^2/g) were mixed with water (W/T = 0.5), for 3 min. The mass (at t = 35 °C) was placed in a viscometer and, upon reaching a given point in time, the values τ_0 and η were determined. The experiment was repeated 3 times. The results of the experiment are shown in Figure 5.

An analysis of the obtained curves of the kinetic rheological properties of the mixture shows that in the first minutes after mixing, there is a sharp increase in the values of the ultimate shear stress, observed up to the 20th minute, followed by a slight decline and stabilization. The rate of change of η reaches its maximum value after 10 min, then it decreases by the 20th minute and stabilizes after 70 min (to the value of 40×10^{-2} Pa×s) (Figure 5b). The curve of change in the plastic strength of the interpore material has the only similarity—a rapid increase up to 10 min. Then stabilization and after 40 min a rapid increase. This character of the τ_0 and P_m curves is explained by the intensive processes of hydration of clinker minerals in the initial period of structure formation, the appearance of hydrate neoformations such as ettringite and $Ca(OH)_2$ on their surface, and the appearance of coagulation bonds between the particles of the hardening system (Figure 5a,c), which does not contradict the data [49,50].

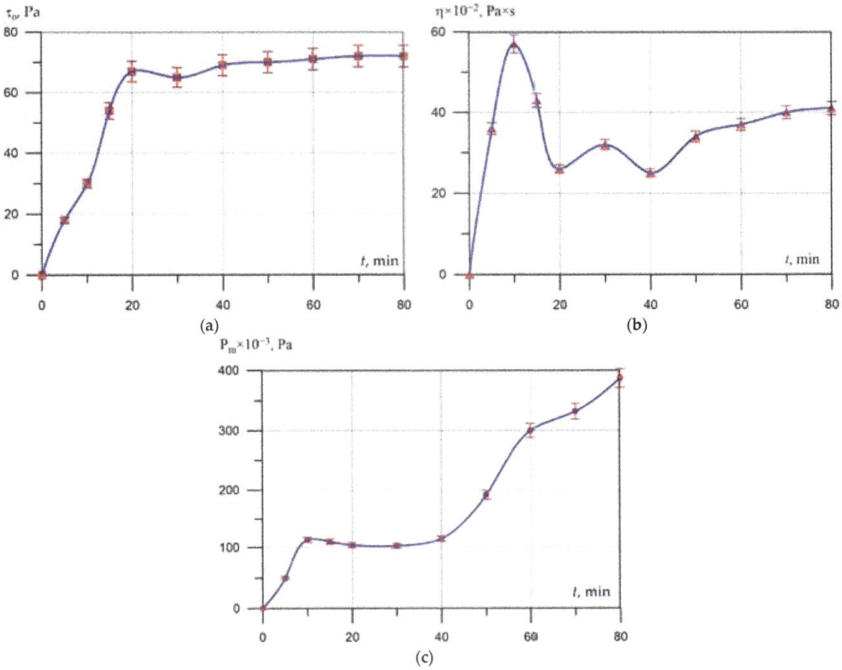

Figure 5. Kinetics of structure formation of the interpore substance of aerated concrete mix (spline approximation): (**a**) ultimate shear stress; (**b**) plastic viscosity; (**c**) plastic strength.

The fact that the values of τ_0 and Pm stabilize in the period of 20–40 min is explained by the saturation of the liquid phase of the $Ca(OH)_2$ interpore substance, fine-grained cement neoplasms, and the appearance of a binder film of hydrated neoplasms on the surface of the grains, which prevents the development of structure formation of the cement system. The instability of the parameters of the hydration kinetics of cement in the initial period, and the impossibility of operational control of hydration, leads to a change in the kinetics of swelling of the aerated concrete and the heterogeneity of the final properties. It is possible to reduce the influence of the instability of the initial period of structure formation on the process of swelling of the aerated concrete mixture, if the gas release of the blowing agent begins during the period of stabilization of the properties of the cement system. This can be achieved by introducing pre-hydrated cement into the aerated concrete mixture or by holding the mixture until the blowing agent is introduced.

4.2. Study of the Influence of the Time of Preliminary Hydration of the Binder on the Rheological Properties of the Interpore Substance, the Swelling Kinetics, and the Properties of Aerated Concrete

The alignment characterizes the traditional casting technology in a time of three kinetic phenomena—intensive hydration of the binder, changes in the rheological properties of the interpore material, and the gas evolution reaction of the aluminum blowing agent, which have a decisive influence on the swelling kinetics and physical and mechanical properties of aerated concrete. The study on the phenomenological model, carried out in the previous section, revealed that if the outgassing occurs during the period of intensive hydration of the binder (up to 30 min from the start of mixing), then the swelling kinetics is unstable, and a slight change in the outgassing parameters, rheological, and other properties lead to significant changes in the swelling process.

To achieve increased physical and mechanical properties, an optimal combination of the kinetics of hydration phenomena, microstructure formation of the interpore wall, and gas release is necessary. To create favorable conditions for swelling, cement previously mixed with water and aged for a given time, was introduced into the composition of the aerated concrete mixture as its component. At the same time, the process of cement structure formation reached the stage of coagulation of hydrate neoplasms, and the rheological properties stabilized.

Preliminary studies of the influence of the water-cement ratio W/C (0.3–0.6) on the regularities of the kinetics of the rheological properties of the cement paste showed that the time for the start of stabilization of the limiting shear stress with an increase in W/T increases from 17 to 24 min. For the convenience of work in further experiments, W/C = 0.5 was taken.

As a result of tests (Figure 6b), it was found that with an increase in the pre-hydration time of the binder, the value of plastic viscosity stabilizes over time and the value of the initial peak of plastic viscosity observed in the control composition decreases significantly. This is explained by the fact that in the cement system with preliminary hydration of the cement, by the time of testing, phenomena have already occurred that lead to the appearance of a peak on the plastic viscosity curve. That is, it is a reflection of the properties of cement. The decrease in the magnitude of the ultimate shear stress (Figure 6a) in pre-hydrated cement systems is explained by the mechanical destruction of the initial contacts between cement particles covered with shells of hydrated neoplasms. The magnitude of the reduction in the ultimate shear stress during the curing of the cement system is determined by the degree of completion of the formation of the primary shell by the testing time. As for the change in plastic strength (Figure 6c), in the interval from 0 to 40 min, its values for both compositions vary from 20 to 150 Pa. However, starting from 50 min, a sharp increase in plastic strength is observed in the non-porous mixture of the control composition.

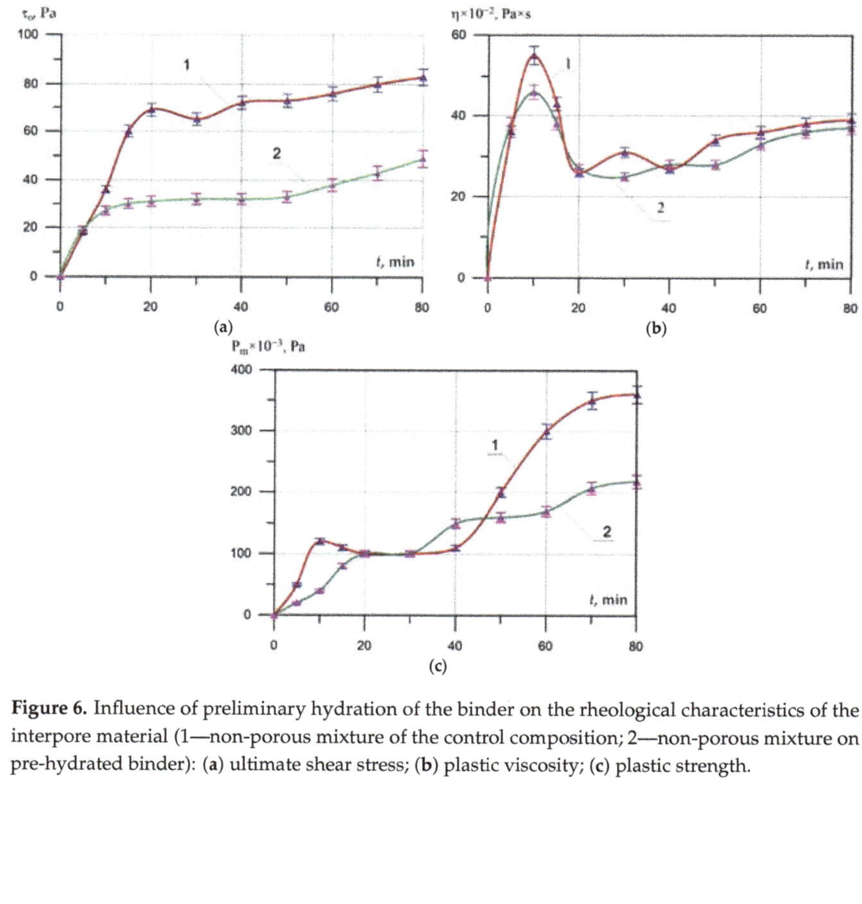

Figure 6. Influence of preliminary hydration of the binder on the rheological characteristics of the interpore material (1—non-porous mixture of the control composition; 2—non-porous mixture on pre-hydrated binder): (**a**) ultimate shear stress; (**b**) plastic viscosity; (**c**) plastic strength.

The results of swelling of an aerated concrete mixture (Figure 7a,b) of the same composition, which do not differ in the time of preliminary hydration of the cement, show that with an increase in the holding time to 30 min, an intensive increase in the swelling coefficient is observed, in the range from 30 to 60 min, the value of the swelling coefficient stabilizes. The maximum value of the diffusion coefficient was fixed at 15 min, and starting from the 20th minute, it decreases and stabilizes from the 30th minute, which is the result of an increase in the gas-retaining capacity of the aerated concrete mixture and a change in the rheological properties of the interpore wall material.

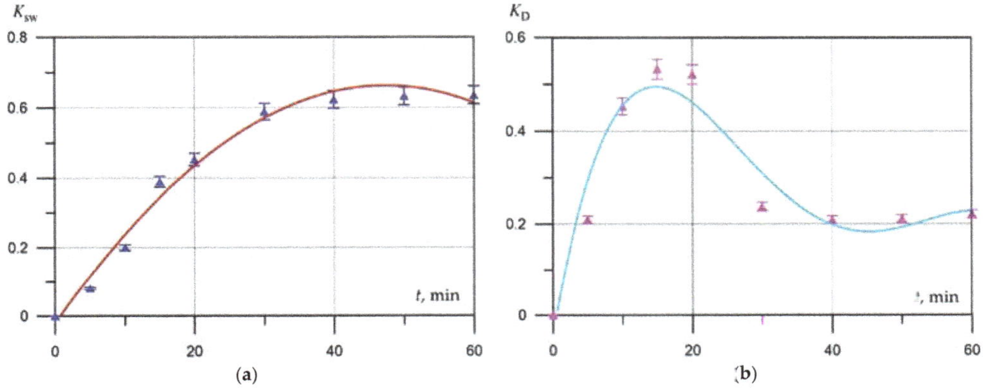

Figure 7. Kinetics of (**a**) swelling of mixture; (**b**) gas diffusion from a mixture, with pre-hydrated cement.

To assess the effectiveness of the pre-hydration time of the binder on the mechanical and physical characteristics of aerated concrete, the coefficient of constructive quality (CSQ) [51,52] was calculated:

$$CSQ = \frac{R_b}{\rho_{av}}\text{, MPa/kg} \times \text{m}^3 \quad (35)$$

where, R_b is compressive strength, MPa; ρ_{av} is average density, kg × m^3.

The values of the average density of aerated concrete and its compressive strength (Table 4) show that the maximum reduction in average density by 29% (from 695 to 497 kg/m^3) and thermal conductivity by 31%, as well as an increase in CSQ, is achieved with a pre-hydration time of 20–30 min. At the same time, according to the CSQ indicator, it is noticeable that the dynamics of strength reduction in this pre-hydration time interval is the smallest CSQ is 7–13% higher compared to other time intervals.

Table 4. Physical and mechanical properties of aerated concrete and their change depending on the time of preliminary hydration of the binder.

Num	Pre-Hydration Time, Min	Average Density ρ_{av}, kg/m^3	Decrease in Average Density $\Delta_{\rho av}$, %	Compressive Strength R_b, MPa	Reduction in Compressive Strength Δ_{Rb}, %	Coefficient of Thermal Conductivity λ, W/m·K	Reduction of the Coefficient of Thermal Conductivity Δ_λ, %	CSQ, MPa/kg × m^3 × 10^{-3}
1C	-	695	0	4.98	0	0.154	0	7.2
2C	10	630	9.4	4.50	9.6	0.145	5.8	7.1
3C	15	564	18.8	4.09	17.9	0.130	15.6	7.3
4C	20	497	28.5	3.84	22.9	0.115	25.3	7.7
5C	30	508	26.9	3.87	22.3	0.106	31.1	7.6
6C	40	548	21.2	3.95	20.7	0.113	26.6	7.2
7C	50	569	18.1	4,01	19.5	0.124	19.5	7.0
8C	60	613	11.8	4.16	16.5	0.151	1.9	6.8

A further increase in the pre-exposure time is impractical since it worsens the quality of the structure. Thus, the determination of the adequate time of preliminary hydration of cement at a given material strength can be determined from the corresponding values of its average density and thermal conductivity.

In order to assess the contribution of the study to the base of world science, it is necessary to determine its place in comparison with the results obtained by other authors. The study has a theoretical and applied character and consists of two main aspects. The first is the filling of research gaps, expressed in the study of the very nature of the process of outgassing and modeling of these processes to obtain more accurate theoretical knowledge and develop existing ideas about the processes occurring during the formation of the microstructure of aerated concrete. The second aspect is the filling of technological deficiencies, expressed in the determination of quantitative and qualitative aspects and values of the recipe and technological parameters that affect the quality of the final material—non-autoclaved aerated concrete of a new level, and obtaining effective building structures, buildings, and structures from them. In comparison with the researchers who previously carried out work to determine the nature of the formation of the aerated concrete structure, in particular with the works [34,35], more accurate data on the process of formation of the aerated concrete structure were established, the nature of the porosity that occurs when using modern building materials and initial components was clarified in comparison with the work carried out about 40 years ago and taking into account all the achievements of modern science. From the point of view of technology, our results have led to an increase in the qualitative and quantitative characteristics of the resulting aerated concrete, expressed in a decrease in average density by 29% and thermal conductivity by 31%, and at the same time in the smallest loss of strength, confirmed by the highest coefficient of structural quality (the difference was up to 13%), compared with the works [34,35].

5. Conclusions

Based on theoretical analysis, experimental studies, and the mathematical model of the swelling process, which take into account the convective and molecular diffusion of hydrogen from the mixture and the conditions of swelling, precipitation, and stabilization of the mixture, the increase in the uniformity of the level of swelling of the aerated concrete mixture was developed.

An improved method for the manufacture of aerated concrete is proposed, which comprises preliminarily hydrated for 20–30 min cement into the composition of the aerated concrete mixture. The method provides an improvement in the gas-holding capacity and an increase in the swelling of the mixture, a decrease in the average density of aerated concrete up to 29%, and an improvement in heat-shielding properties up to 31%. At the same time, the smallest dynamics of the decrease in the strength properties of aerated concrete were observed, which is confirmed by a large CSQ of up to 13%.

Aerated concrete has been obtained that meets the requirements of environmental friendliness and economic efficiency, and has improved mechanical and physical characteristics. The effectiveness of the use of production waste in the form of partially hydrated cement in the compositions of such aerated concrete has been proven, thereby reducing the cost of production of aerated concrete and construction in general, amounting to about 15%.

Thus, the environmental effect of our proposals is to reduce the burden on the environment through the use of waste-free technology. Such technology is the addition of partially hydrated cement, i.e., a waste from aerated concrete production, in a new production. At the same time, the environmental effect develops into an economic one, which is expressed in a reduction in the cost of raw materials, which is especially important for manufacturers.

Author Contributions: Conceptualization, S.A.S., E.M.S., A.B. and A.S.; methodology, A.S., S.A.S., E.M.S. and N.D.; software, S.A.S., E.M.S., A.B. and N.B.; validation, A.S., S.A.S., E.M.S. and A.B.; formal analysis, A.S., S.A.S. and E.M.S.; investigation, A.S., L.R.M., S.A.S., E.M.S., A.B. and B.M.; resources, B.M.; data curation, S.A.S., E.M.S. and N.D.; writing—original draft preparation, S.A.S., E.M.S. and A.B.; writing—review and editing, S.A.S., E.M.S. and A.B.; visualization, S.A.S., E.M.S., A.B. and N.B.; supervision, L.R.M. and B.M.; project administration, L.R.M. and B.M.; funding acquisition, A.B. and B.M. All authors have read and agreed to the published version of the manuscript.

Funding: This research received no external funding.

Institutional Review Board Statement: Not applicable.

Informed Consent Statement: Not applicable.

Data Availability Statement: The study did not report any data.

Acknowledgments: The authors would like to acknowledge the administration of Don State Technical University for their resources and financial support.

Conflicts of Interest: The authors declare no conflict of interest.

References

1. Sharafutdinov, E.; Shon, C.-S.; Zhang, D.; Chung, C.-W.; Kim, J.; Bagitova, S. Frost Resistance Number to Assess Freeze and Thaw Resistance of Non-Autoclaved Aerated Concretes Containing Ground Granulated Blast-Furnace Slag and Micro-Silica. *Materials* **2019**, *12*, 4151. [CrossRef] [PubMed]
2. Wang, C.-L.; Ni, W.; Zhang, S.-Q.; Wang, S.; Gai, G.-S.; Wang, W.-K. Preparation and properties of autoclaved aerated concrete using coal gangue and iron ore tailings. *Constr. Build. Mater.* **2016**, *104*, 109–115. [CrossRef]
3. Cai, L.; Ma, B.; Li, X.; Lv, Y.; Liu, Z.; Jian, S. Mechanical and hydration characteristics of autoclaved aerated concrete (AAC) containing iron-tailings: Effect of content and fineness. *Constr. Build. Mater.* **2016**, *128*, 361–372. [CrossRef]
4. Peng, Y.; Liu, Y.; Zhan, B.; Xu, G. Preparation of autoclaved aerated concrete by using graphite tailings as an alternative silica source. *Constr. Build. Mater.* **2021**, *267*, 121792. [CrossRef]
5. Rafiza, A.R.; Fazlizan, A.; Thongtha, A.; Asim, N.; Noorashikin, M.S. The Physical and Mechanical Properties of Autoclaved Aerated Concrete (AAC) with Recycled AAC as a Partial Replacement for Sand. *Buildings* **2022**, *12*, 60. [CrossRef]
6. Rahman, R.A.; Fazlizan, A.; Asim, N.; Thongtha, A. A Review on the Utilization of Waste Material for Autoclaved Aerated Concrete Production. *J. Renew. Mater.* **2021**, *9*, 61–72. [CrossRef]
7. Lam, N.N. Recycling of AAC waste in the manufacture of autoclaved aerated concrete in Vietnam. *Int. J. Geomate* **2021**, *20*, 128–134. [CrossRef]
8. Abraham, H.B.; Alengaram, U.J.; Alnahhal, A.M.; Haddadian, A.; Karthick, S.; Deboucha, W. Performance evaluation of cellular lightweight concrete using palm oil industrial waste as cement and fine aggregate replacement materials. *Mater. Today Proc.* **2021**, *10*, 301. [CrossRef]
9. El-Didamony, H.; Amer, A.A.; Mohammed, M.S.; El-Hakim, M.A. Fabrication and properties of autoclaved aerated concrete containing agriculture and industrial solid wastes. *J. Build. Eng.* **2019**, *22*, 528–538. [CrossRef]
10. Rafiza, A.R.; Chan, H.Y.; Thongtha, A.; Jettipattaranat, W.; Lim, K.L. An Innovative Autoclaved Aerated Concrete (AAC) with Recycled AAC Powder for Low Carbon Construction. *IOP Conf. Ser. Earth Environ. Sci.* **2019**, *268*, 012050. [CrossRef]
11. Pehlivanlı, Z.O.; Uzun, İ.; Yücel, Z.P.; Demir, İ. The effect of different fiber reinforcement on the thermal and mechanical properties of autoclaved aerated concrete. *Constr. Build. Mater.* **2016**, *112*, 325–330. [CrossRef]
12. Tanyildizi, H. Effect of temperature, carbon fibers, and silica fume on the mechanical properties of lightweight concretes. *New Carbon Mater.* **2008**, *23*, 339–344. [CrossRef]
13. Laukaitis, A.; Kerienė, J.; Mikulskis, D.; Sinica, M.; Sezemanas, G. Influence of fibrous additives on properties of aerated autoclaved concrete forming mixtures and strength characteristics of products. *Constr. Build. Mater.* **2009**, *23*, 3034–3042. [CrossRef]
14. Pehlivanlı, Z.O.; Uzun, İ.; Demir, İ. Mechanical and microstructural features of autoclaved aerated concrete reinforced with autoclaved polypropylene, carbon, basalt and glass fiber. *Constr. Build. Mater.* **2015**, *96*, 428–433. [CrossRef]
15. Sinica, M.; Sezeman, G.A.; Mikulskis, D.; Kligys, M.; Česnauskas, V. Impact of complex additive consisting of continuous basalt fibres and SiO_2 microdust on strength and heat resistance properties of autoclaved aerated concrete. *Constr. Build Mater.* **2014**, *50*, 718–726. [CrossRef]
16. Fan, J.; Cao, D.; Jing, Z.; Zhang, Y.; Jing, Y. Synthesis and microstructure analysis of autoclaved aerated concrete with carbide slag addition. *J. Wuhan Univ. Technol. Mat. Sci. Ed.* **2014**, *29*, 1005–1010. [CrossRef]
17. Chen, G.; Li, F.; Geng, J.; Jing, P.; Si, Z. Identification, generation of autoclaved aerated concrete pore structure and simulation of its influence on thermal conductivity. *Constr. Build. Mater.* **2021**, *294*, 123572. [CrossRef]
18. Ulykbanov, A.; Sharafutdinov, E.; Chung, C.-W.; Zhang, D.; Shon, C.-S. Performance-based model to predict thermal conductivity of non-autoclaved aerated concrete through linearization approach. *Constr. Build. Mater.* **2019**, *196*, 555–563. [CrossRef]

19. Tasdemir, C.; Sengul, O.; Tasdemir, M.A. A comparative study on the thermal conductivities and mechanical properties of lightweight concretes. *Energy Build.* **2017**, *151*, 469–475. [CrossRef]
20. Sun, Y.; Gao, P.; Geng, F.; Li, H.; Zhang, L.; Liu, H. Thermal conductivity and mechanical properties of porous concrete materials. *Mater. Lett.* **2017**, *209*, 349–352. [CrossRef]
21. Jin, H.-Q.; Yao, X.-L.; Fan, L.-W.; Xu, X.; Yu, Z.-T. Experimental determination and fractal modeling of the effective thermal conductivity of autoclaved aerated concrete: Effects of moisture content. *Int. J. Heat Mass Transf.* **2016**, *92*, 589–602. [CrossRef]
22. Ferretti, D.; Michelini, E. The Effect of Density on the Delicate Balance between Structural Requirements and Environmental Issues for AAC Blocks: An Experimental Investigation. *Sustainability* **2021**, *13*, 13186. [CrossRef]
23. Shon, C.-S.; Mukangali, I.; Zhang, D.; Ulykbanov, A.; Kim, J. Evaluation of Non-Autoclaved Aerated Concrete for Energy Behaviors of a Residential House in Nur-Sultan, Kazakhstan. *Buildings* **2021**, *11*, 610. [CrossRef]
24. Lu, J.; Chen, J.; Zhu, K.; Xu, H.; Zhang, W.; Deng, Q. Experimental and Numerical Studies on the Behaviors of Autoclaved Aerated Concrete Panels with Insulation Boards Subjected to Wind Loading. *Materials* **2021**, *14*, 7651. [CrossRef]
25. Garbalińska, H.; Bochenek, M.; Stasiak, M. Experimental and Modeling Investigations on the Water Sorption Behaviors of Autoclaved Aerated Concrete. *Materials* **2021**, *14*, 6235. [CrossRef] [PubMed]
26. Jasiński, R.; Drobiec, Ł.; Mazur, W. Validation of Selected Non-Destructive Methods for Determining the Compressive Strength of Masonry Units Made of Autoclaved Aerated Concrete. *Materials* **2019**, *12*, 389. [CrossRef] [PubMed]
27. Karakurt, C.; Kurama, H.; Topçu, İ.B. Utilization of natural zeolite in aerated concrete production. *Cem. Concr. Compos.* **2010**, *32*, 1–8. [CrossRef]
28. Fang, X.; Wang, C.; Li, H.; Wang, X.; Zhang, S.; Luo, X.; Jia, H. Influence of mesoscopic pore characteristics on the splitting-tensile strength of cellular concrete through deep-learning based image segmentation. *Constr. Build. Mater.* **2022**, *315*, 125335. [CrossRef]
29. Liu, X.; Qian, X.; Pu, S.; Sheng, K.; Sun, D.; Hong, B. Methods for testing the quality of lightweight cellular concrete during pouring. *Constr. Build. Mater.* **2022**, *315*, 125755. [CrossRef]
30. Thakur, A.; Kumar, S. Mechanical properties and development of light weight concrete by using autoclaved aerated concrete (AAC) with aluminum powder. *Mater. Today Proc.* 2022, *in press*. [CrossRef]
31. Kim, T.; Kang, C.; Seo, K. Development and Characteristics of Aerated Alkali-Activated Slag Cement Mixed with Zinc Powder. *Materials* **2021**, *14*, 6293. [CrossRef] [PubMed]
32. Kornienko, P.V. Theoretical foundations for the formation of the optimal structure of cellular concrete. *Sci. Technol. Kaz.* **2010**, *2*, 81–89. Available online: https://cyberleninka.ru/article/n/teoreticheskie-osnovy-obrazovaniya-optimalnoy-struktury-yacheistogo-betona (accessed on 7 February 2022).
33. Kunnos, G.Y.; Lapsa, V.K.; Lindenberg, B.Y.; Solodovnik, A.B.; Shteinert, A.R. *Elements of Technological Mechanics of Cellular Concrete*; Zinatne: Riga, Latvia, 1976; p. 96. Available online: https://pt.pb1lib.org/book/2964197/2b6764 (accessed on 7 February 2022).
34. Shuisky, A.; Stelmakh, S.; Shcherban, E.; Torlina, E. Recipe-technological aspects of improving the properties of nonautoclaved aerate concrete. *Matec Web Conf.* **2017**, *129*, 05011. [CrossRef]
35. Pylaev, A.J.; Minas, A.I.; Savin, E.S. Development of a mathematical model for determining the optimal conditions for swelling of aerated concrete. In Proceedings of the Abstracts of the 2nd All-Union Symposium: Rheology of Concrete Mixtures and its Technological Problems, Riga, Latvia, 5–7 October 1976; pp. 172–173.
36. *GOST 10180*; Concretes. Methods for Strength Determination Using Reference Specimens. Available online: http://docs.cntd.ru/document/1200100908 (accessed on 7 February 2022).
37. *GOST 12730.1*; Concretes. Methods of Determination of Density. Available online: https://docs.cntd.ru/document/1200177299 (accessed on 7 February 2022).
38. *GOST 7076*; Building Materials and Products. Method of Determination of Steady-State Thermal Conductivity and Thermal Resistance. Available online: https://docs.cntd.ru/document/1200005006 (accessed on 7 February 2022).
39. Stel'makh, S.A.; Shcherban', E.M.; Shuiskii, A.I.; Prokopov, A.Y.; Madatyan, S.M.; Parinov, I.A.; Cherpakov, A.V. Effects of the Geometric Parameters of Mixer on the Mixing Process of Foam Concrete Mixture and Its Energy Efficiency. *Appl. Sci.* **2020**, *10*, 8055. [CrossRef]
40. Stel'makh, S.A.; Shcherban', E.M.; Beskopylny, A.N.; Mailyan, L.R.; Meskhi, B.; Butko, D.; Smolyanichenko, A.S. Influence of Composition and Technological Factors on Variatropic Efficiency and Constructive Quality Coefficients of Lightweight Vibro-Centrifuged Concrete with Alkalized Mixing Water. *Appl. Sci.* **2021**, *11*, 9293. [CrossRef]
41. Shcherban', E.M.; Stel'makh, S.A.; Beskopylny, A.; Mailyan, L.R.; Meskhi, B.; Shuyskiy, A. Improvement of Strength and Strain Characteristics of Lightweight Fiber Concrete by Electromagnetic Activation in a Vortex Layer Apparatus. *Appl. Sci.* **2022**, *12*, 104. [CrossRef]
42. Stel'makh, S.A.; Shcherban, E.M.; Zholobov, A.L. Electroactivation of Foam Concrete for Buildings and Structures with Improved Constructive and Energy Efficient Characteristics. *IOP Conf. Ser. Mater. Sci. Eng.* **2018**, *463*, 022034. [CrossRef]
43. Beskopylny, A.N.; Stel'makh, S.A.; Shcherban, E.M.; Mailyan, L.R.; Meskhi, B.; El'shaeva, D.; Varavka, V. Developing Environmentally Sustainable and Cost-Effective Geopolymer Concrete with Improved Characteristics. *Sustainability* **2021**, *13*, 13607. [CrossRef]
44. Brunauer, S.; Deming, L.S.; Deming, W.E.; Teller, E. On a Theory of the van der Waals Adsorption of Gases. *J. Am. Chem. Soc.* **1940**, *62*, 1723–1732. [CrossRef]

45. Butt, Y.M.; Topilskii, G.V.; Vasina, T.P.; Bukatina, T.A. Study of colloidal solutions formed during the hydration of Portland cement in a paste. In *Hardening of Cement: Abstracts of Reports and Messages of the All-Union Conference*; NIIPromstroy: Ufa, Russia, 1974; pp. 98–100.
46. Terentiev, A.E.; Kunnos, G.Y. Structural and strength aspects of the study of highly filled porous bodies (on the example of aerated concrete). In *Theory, Production and Application of Artificial Building Conglomerates: Abstracts of Reports of the All-Union Scientific and Technical Conference*; Vladimir Book Publishing House: Vladimir, Russia, 1982; pp. 405–407.
47. Terentiev, A.E.; Kunnos, G.Y. On the spatial distribution of pores in aerated concrete. In *Proceedings of the Technological Mechanics of Concrete: A Collection of Scientific Papers of the RPI*; Riga Polytechnic Institute: Riga, Latvia, 1982; pp. 143–162.
48. Topilskii, G.V.; Vasina, T.P.; Bukatina, T.A. Study of the composition of the liquid phase during cement hydration. In Proceedings of the Sixth International Congress on the Chemistry of Cement, Moscow, Russia, 23–27 September 1974; Stroyizdat: Moscow, Russia, 1974; Volume 2, pp. 88–91.
49. Smirnova, O.; Kazanskaya, L.; Koplík, J.; Tan, H.; Gu, X. Concrete Based on Clinker-Free Cement: Selecting the Functional Unit for Environmental Assessment. *Sustainability* **2021**, *13*, 135. [CrossRef]
50. Smirnova, O.M. Low-Clinker cements with low water demand. *J. Mater. Civ. Eng.* **2020**, *32*, 06020008. [CrossRef]
51. Stel'makh, S.A.; Shcherban', E.M.; Beskopylny, A.; Mailyan, L.R.; Meskhi, B.; Dotsenko, N. Enchainment of the Coefficient of Structural Quality of Elements in Compression and Bending by Combined Reinforcement of Concrete with Polymer Composite Bars and Dispersed Fiber. *Polymers* **2021**, *13*, 4347. [CrossRef] [PubMed]
52. Mailyan, L.R.; Beskopylny, A.N.; Meskhi, B.; Shilov, A.V.; Stel'makh, S.A.; Shcherban, E.M.; Smolyanichenko, A.S.; El'shaeva, D. Improving the Structural Characteristics of Heavy Concrete by Combined Disperse Reinforcement. *Appl. Sci.* **2021**, *11*, 6031. [CrossRef]

materials

Article

EIS Characterization of Ti Alloys in Relation to Alloying Additions of Ta

Pedro P. Socorro-Perdomo [1], Néstor R. Florido-Suárez [1], Julia C. Mirza-Rosca [1,*] and Mircea Vicentiu Saceleanu [2]

[1] Mechanical Engineering Department, Las Palmas de Gran Canaria University, 35017 Las Palmas de Gran Canaria, Spain; pedro.socorro@ulpgc.es (P.P.S.-P.); nestor.florido@ulpgc.es (N.R.F.-S.)
[2] Neurosurgery Department, Faculty of Medicine, "Lucian Blaga" University, 550024 Sibiu, Romania; vicentiu.saceleanu@gmail.com
* Correspondence: julia.mirza@ulpgc.es

Abstract: The increased popularity of Ti and its alloys as important biomaterials is driven by their low modulus, greater biocompatibility, and better corrosion resistance in comparison to traditional biomaterials, such as stainless steel and Co–Cr alloys. Ti alloys are successfully used in severe stress situations, such as Ti–6Al–4V, but this alloy is related to long-term health problems and, in response, different Ti alloys composed of non-toxic and non-allergic elements such as Nb, Zr, Mo, and Ta have been developed for biomedical applications. In this context, binary alloys of titanium and tantalum have been developed and are predicted to be potential products for medical purposes. More than this, today, novel biocompatible alloys such as high entropy alloys with Ti and Ta are considered for biomedical applications and therefore it is necessary to clarify the influence of tantalum on the behavior of the alloy. In this study, various Ti–xTa alloys (with x = 5, 15, 25, and 30) were characterized using different techniques. High-resolution maps of the materials' surfaces were generated by scanning tunneling microscopy (STM), and atom distribution maps were obtained by energy dispersive X-ray spectroscopy (EDS). A thorough output of chemical composition, and hence the crystallographic structure of the alloys, was identified by X-ray diffraction (XRD). Additionally, the electrochemical behavior of these Ti–Ta alloys was investigated by EIS in simulated body fluid at different potentials. The passive layer resistance increases with the potential due to the formation of the passive layer of TiO_2 and Ta_2O_5 and then decreases due to the dissolution processes through the passive film. Within the Ti–xTa alloys, Ti–25Ta demonstrates excellent passive layer and corrosion resistance properties, so it seems to be a promising product for metallic medical devices.

Keywords: Ti–Ta alloys; corrosion; electrochemical impedance spectroscopy

Citation: Socorro-Perdomo, P.P.; Florido-Suárez, N.R.; Mirza-Rosca, J.C.; Saceleanu, M.V. EIS Characterization of Ti Alloys in Relation to Alloying Additions of Ta. *Materials* **2022**, *15*, 476. https://doi.org/10.3390/ma15020476

Academic Editors: Jun Liang and Shengqiang Ma

Received: 21 October 2021
Accepted: 5 January 2022
Published: 8 January 2022

1. Introduction

Titanium achieves its excellent corrosion protection because of the high stability of the passive layer that develops on its surface [1], which can be re-formed at body temperature and in physiological fluid if it is damaged. The increased popularity of Ti and its alloys as important biomaterials is driven by their low modulus [2–4], greater biocompatibility, and better corrosion resistance in comparison to traditional biomaterials, such as stainless steel and Co–Cr alloys [5].

These desirable qualities were the motivating force for the early insertion of Ti as an implantable material. Ti has low mechanical strength [6,7], and when aluminum and vanadium are incorporated in low amounts, the resistance of the alloy is greatly improved over that of titanium and the alloy could be successfully used in severe stress situations, such as Ti–6Al–4V, which has been predominantly employed.

However, Ti–6Al–4V has significant toxicity; harmful tissue reactions are caused by vanadium and the release of both V and Al ions are related to long-term health disorders such as peripheral neuropathy and Alzheimer's and Parkinson's diseases [8,9].

In response to these health problems, different Ti alloys composed of non-toxic and non-allergic elements such as Nb, Zr, Mo, Ta, etc., have been developed for biomedical applications [4,10–12]. However, the non-toxicity of alloying elements is only the first of the three criteria for metallic materials to be used for medical applications. The second important criterion for biomaterials is their resistance to corrosion, which also dictates the tissue compatibility and eventual osseointegration of the implant—critical aspects to consider for implant alloy design. The last important criterion is the Young's modulus of the implant in comparation with that of the bone. A mismatch between these modules can lead to a reallocation of loads surrounding the implant, causing implant loosening [13].

In this context, binary alloys of titanium and tantalum have been developed and analyzed [14–23] and are predicted to be potential products for medical purposes; as tantalum is a non-toxic element [24], they have better compatibility with bone tissue compared with cp–Ti and Ti–6Al–4V alloys [21], and Ti–Ta alloys exhibit reduced modulus of elasticity and increased relative strength (at equivalent stiffness) compared with commercially pure titanium (cp-Ti) [25]. More than this, today novel biocompatible alloys such as high entropy alloys with Ti and Ta are considered for biomedical applications and, therefore, it is necessary to clarify the influence of tantalum in the behavior of the alloy.

Electrochemical impedance spectroscopy (EIS) is applied to characterize the behavior of different metals and alloys in various environments and to provide new information that previously could not be obtained with classical dc techniques [26,27]. Although a significant amount of research has been performed using EIS to characterize the biomaterials, little research has been conducted on EIS measurements of Ti–Ta alloys [17,28]. It is observed that it is essential for all systems to consider suitable impedance models that can be used to fit the experimental results and to provide the relevant data that characterize the corrosion process.

In this study, various Ti–xTa alloys (with x = 5, 15, 25, and 30) were characterized using different techniques. High-resolution maps of the materials´ surface were generated using scanning tunneling microscopy (STM), and information about atom distribution maps was obtained using energy dispersive X-ray spectroscopy (EDS). A thorough output of chemical composition and hence the crystallographic structure of the alloys were identified by X-ray diffraction (XRD). Additionally, the electrochemical behavior of these Ti–xTa alloys was investigated in simulated body fluid (SBF) at different potentials.

2. Materials and Methods

2.1. Material and Sample Preparation

The studied titanium tantalum alloys were Ti–5Ta, Ti–15Ta, Ti–25Ta, and Ti–30Ta from R&D CS (Research & Development Consulting and Services), Bucharest, Romania. The Ti–Ta ingots (diameter = 20 mm, length = 30 mm) were produced by levitation fusion in a high-frequency induction furnace operating with a cold copper crucible followed by a homogenization heat treatment (heating rate 5 °C/min, homogenized at 1000 °C for 8 h followed by natural cooling) in order to eliminate the segregation. The chemical composition of the alloys was determined by the supplier; for Ti and Ta content, the XRF technique was applied, while for the impurities (e.g., O, N, and C), the inert gas fusion technique was employed. The detailed chemical composition of the alloys is presented in Table 1.

Table 1. Chemical composition of Ti–xTa alloys.

Alloy	Ti (wt. %)	Ta (wt. %)	O (wt. ppm)	N (wt. ppm)	C (wt. ppm)
Ti–5Ta	94.20 ± 0.06	4.92 ± 0.05	180 ± 14	80 ± 7	110 ± 8
Ti–15Ta	84.41 ± 0.06	14.82 ± 0.07	162 ± 11	75 ± 5	101 ± 6
Ti–25Ta	74.52 ± 0.05	24.89 ± 0.09	158 ± 12	82 ± 6	103 ± 5
Ti–30Ta	69.61 ± 0.07	29.67 ± 0.05	172 ± 11	78 ± 5	111 ± 4

The elastic modulus E and tensile strength σ_t of the obtained alloys are Ti–5Ta (E = 142 GPa, σ_t = 381 MPa), Ti–15Ta (E = 101 GPa, σt = 402 MPa), Ti–25Ta (E = 65 GPa, σt = 464 MPa), and Ti–30Ta (E = 94 GPa, σ_t = 445 MPa). The experimental methods followed the ASTM E3-11 (2017) standard for metallo-graphic sample preparation [29]. The ingots were cut with minimal deformation using the Buehler IsoMet 4000 Precision Saw, (Chicago, IL, USA) and one-micron positioning allows for precise sectioning. Then, the specimens were mounted with acrylic (compression hot mounting) in order to protect edges during the polishing process. The next operation involves the grinding up to 2500 grit with SiC paper and then polishing with 0.1 μm alpha-alumina until a mirror finish is obtained in a Struers TegraPol-11 (Copenhague, Denmark) polishing machine. The samples were ultrasonically cleaned using deionized water and rinsed thoroughly with distilled water and ethanol.

2.2. Microstructural Characterization

EDS measurements were carried out with an environmental scanning electron microscope model FEI XL30 ESEM with an LaB6 cathode attached to an energy dispersive X-ray electron sample analyzer, model EDAX Sapphire.

The structures of the Ti–Ta alloys were investigated by high-resolution scanning tunneling microscopy (STM). All determinations were performed in air with a Hitachi TM3030 microscope that had been transversely calibrated by imaging atomically accurate oriented pyrolytic graphite. The tips were obtained by chopping a 0.20 mm Pt0.8Ir0.2 wire. The data were acquired in constant current operation with specific tunneling currents of 0.13–0.3 nA and a specimen polarization of 0.4–1.0 V. No tip-induced shifts were noted.

X-ray diffraction (XRD) determinations were made using an Empyrean diffractometer (Malvern-Panalytical). The device worked with a Cu Kα anode (1.5406 Å) in the range of 2θ = 0–64° with a step size of 0.04° at a power of 45 kV and 40 mA in Bragg–Brentano geometry. The samples have been rotated while collecting data in order to achieve better data capture. The obtained patterns were simulated in order to determine the presence of the crystalline phase, the lattice parameter, and the diameter of the grain with the assistance of Malvern-Panalytical's HighScore Plus software.

2.3. Electrochemical Measurements

Electrochemical measurements were carried out with a PAR 263 A potentiostat coupled with a PAR 5210 (AMETEK, Berwyn, PA, USA) lock-in amplifier. A standard three-electrode electrochemical cell with a Pt grid as a counter electrode and a saturated calomel electrode (SCE) as a reference electrode was used. The mounted, grinded, and polished samples of Ti–xTa alloys were employed as working electrode for the electrochemical measurements. All the measurements were performed in simulated body fluid (SBF) prepared in our laboratory with the pH = 7.8 measured with a multiparameters analyzer CONSORT 831C and the composition is presented in Table 2.

The open circuit potential was recorded for 24 h with the samples immersed in SBF solution. The potentiodynamic polarization curves were obtained with a scanning rate of 0.166 mV/s in the potential range −800 to +2000 mV vs. SCE. The polarization resistance R_p was calculated from traces of the polarization curve at ±10 mV versus open circuit potential.

Table 2. Composition of simulated body fluid (SBF).

Compound	Composition [g/L]
NaCl	6.80
KCl	0.40
$CaCl_2$	0.20
$MgSO_4 \cdot 7H_2O$	0.20
$NaH_2PO_4 \cdot H_2O$	0.14
$NaHCO_3$	2.20
Glucose	1.00

The effect of the potential on the passive film of Ti–Ta alloys operating in a simulated physiological environment was evaluated by electrochemical impedance spectroscopy (EIS). The AC potential amplitude was set at 10 mV, and single sine wave recordings were performed at frequencies in the range of 10^{-1} and 10^{5} Hz for all specimens. To characterize the oxide layer, impedance spectra were registered in the range of −400 to 2000 mV with a step of 100 mV by continuously polarizing the electrodes and letting the system equilibrate for 600 s at every potential. For the numerical fit of the measured impedance data, the software program ZSimWin was used.

All the electrochemical tests were normally repeated three or four times to ensure that they presented reasonable reproducibility.

3. Results and Discussions

The EDS analysis has been performed on micro-zones of the same square area, and the elemental distribution maps (see Figure 1) and chemical composition for the four alloys (see Table 3) are presented. It can be observed that there are no impurities in the metallic mass and titanium and tantalum were the only identified elements.

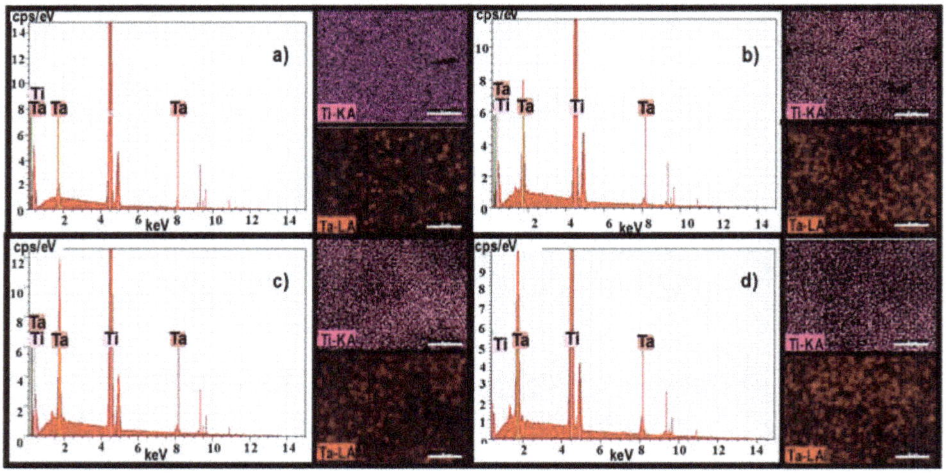

Figure 1. Atoms distribution of Ti and Ta in the studied alloys: (**a**) Ti5Ta, (**b**) Ti15Ta, (**c**) Ti25Ta, and (**d**) Ti30Ta.

In Figure 2, we can observe the surface topography acquired by scanning tunneling microscopy (STM). A three-dimensional analysis of the surface topography reveals nanoscale deviations within an overall homogeneous nanoscale architecture found on the surface of the Ti–Ta alloy.

Table 3. EDS global analysis on micro-areas for Ti–xTa alloys.

Alloy	Ti			Ta		
	at.%	wt.%	Error, %	at.%	wt.%	Error, %
Ti–5Ta	99.01	94.95	1.22	0.88	4.38	3.67
Ti–15Ta	95.62	83.95	1.95	4.38	14.87	2.24
Ti–25Ta	91.76	74.63	2.03	8.23	24.83	2.15
Ti–30Ta	90.06	69.74	1.81	9.95	29.72	1.98

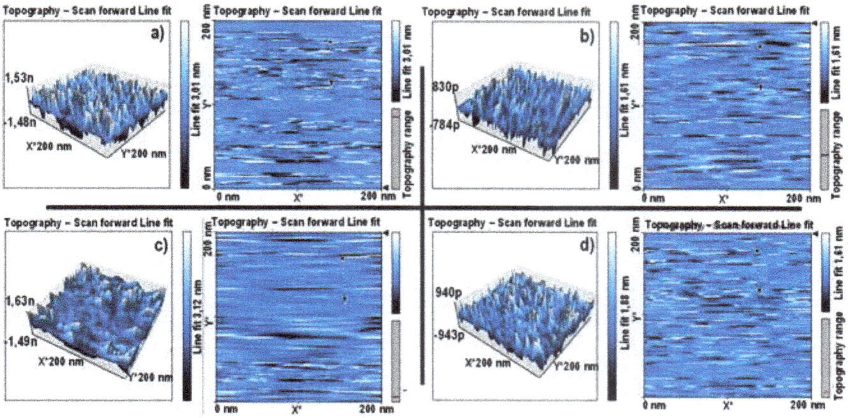

Figure 2. Representative scanning tunneling microscopy (STM) images of surface topographies for (**a**) Ti5Ta, (**b**) Ti15Ta, (**c**) Ti25Ta, and (**d**) Ti30Ta.

Nanoscale deviations, suggestive of the topography present in natural tissue, have been repetitively demonstrated to enhance the protein-specific adsorption, thus further driving cellular activity and cell–cell relationships [30]. Thus, these features enable a very promising interface to enhance cellular interaction because of increased nanoscale roughness of the material interface.

Pure titanium has a closed hexagonal structure (HCP), i.e., an α-phase at room temperature. At temperatures above 883 °C, there is a body-centered cubic (BCC) structure, i.e., a β-phase. The β-phase is stable at temperatures below 883 °C with the addition of β-stabilizers, and its stability depends on the amount of β-alloying elements. The quantity of β-stabilizer required to retain the pure β-phase at room temperature depends on the molybdenum equivalence, a rule derived from the analysis of binary titanium alloys. Molybdenum equivalence is given by [31]:

$$Mo_{eq} = 1.0\ Mo + 0.67\ V + 0.44\ W + 0.28\ Nb + 0.22\ Ta + 1.6\ Cr + \ldots\ \ldots\ -1.0\ Al$$

In general, a molybdenum equivalency of approximately 10 is necessary to stabilize the β-phase while quenching [14], and the critical value to reach a complete β-phase is ap-proximately 25. As we can see for the studied Ti–Ta alloys, the Moeq is below the mini-mum value needed for a fully stable β phase, and the microstructure is made up of α″ grains within β grains. Figure 3 presents the XRD patterns taken from the Ti–xTa alloys.

Because the Ta element is a β phase former in the Ti-based alloys and Moeq is low, the microstructures of all studied samples are clearly a mixture of α″ phase (orthorhombic structure) and β phase. However, the intensity of α″ phase decreased as the concentration of Ta increased, and this is attributed to the variation of the volume fraction of α″ phase in the matrix.

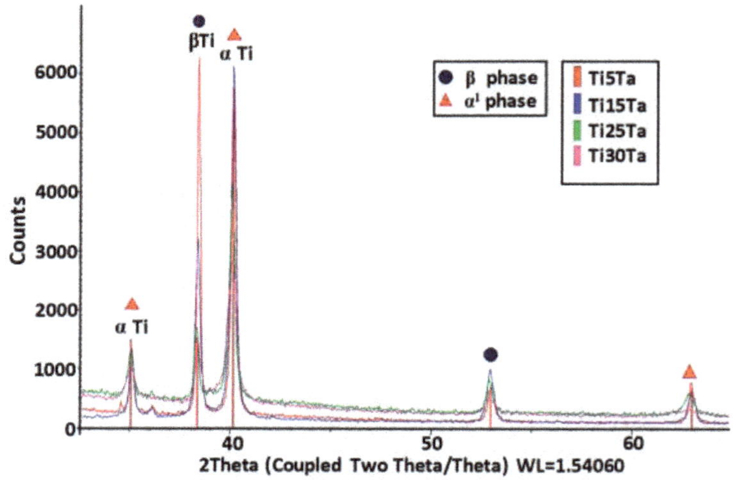

Figure 3. XRD pattern for TixTa alloys.

3.1. Electrochemical Impedance Spectroscopy

Figure 4 shows the open circuit potential curves for all four Ti–xTa alloys immersed for 24 h in simulated body fluid. It can be observed that following the immersion, an abrupt displacement of the potential has taken place towards positive values during a pe-riod of 2–6 h. Afterwards, the open circuit potential continued to increase slowly, suggesting the growth of a passive layer on the metallic surface. The linear polarization curves, in semilogarithmic coordinates for the tested Ti–xTa alloys in simulated body fluid, are displayed in Figure 5, and Table 4 shows the instantaneous corrosion parameters in this physiological environment. All samples are characterized by high values of the polarization resistance R_p (10^5 $\Omega \cdot cm^2$).

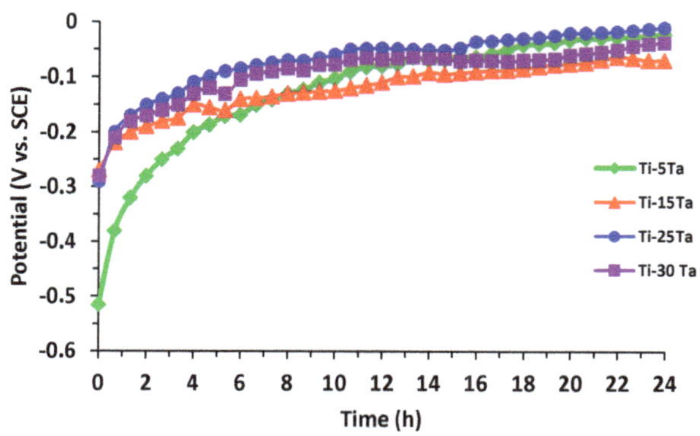

Figure 4. Open circuit potential curves for Ti–xTa alloys during 24 h immersion in SBF.

The electric potential difference between the reference electrode and metal interface is a relevant factor directly related to the surface conditions. EIS tests have been carried out at various potentials in three areas: cathodic–anodic transition, passive transition, and quasi-transpassive transition. The impedance results will be used to compare the effect of the potential on the properties of the passive layer.

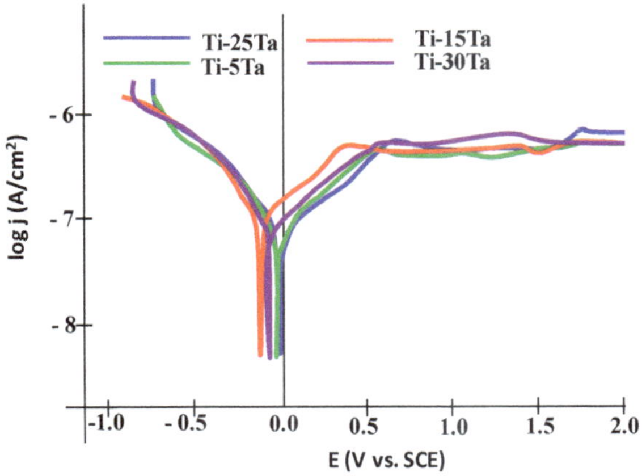

Figure 5. Polarization curves for Ti–xTa alloys.

Table 4. Corrosion parameters from polarization curves (mean ± SD).

Alloy	E_{corr} (mV vs. SCE)	i_{corr} ($\mu A/cm^2$)	i_{pass} ($\mu A/cm^2$)	R_p ($k\Omega \cdot cm^2$)
Ti–5Ta	-22 ± 4	0.67 ± 0.06	0.61 ± 0.11	522 ± 16
Ti–15Ta	-70 ± 3	0.52 ± 0.12	0.58 ± 0.05	501 ± 23
Ti–25Ta	-10 ± 3	0.43 ± 0.04	0.51 ± 0.16	598 ± 28
Ti–30Ta	-38 ± 4	0.58 ± 0.21	0.57 ± 0.18	495 ± 12

3.1.1. Plots Interpretation

The Nyquist plots correspond to the impedance of Ti–5Ta, Ti–15Ta, Ti–25Ta, and Ti–30Ta at different potential values. Even the EIS data were recorded within the -0.4 V $\leq$ E $\leq$ 2.0 V potential range with a step of 100 mV; not all the obtained curves are shown in Figures 6–9.

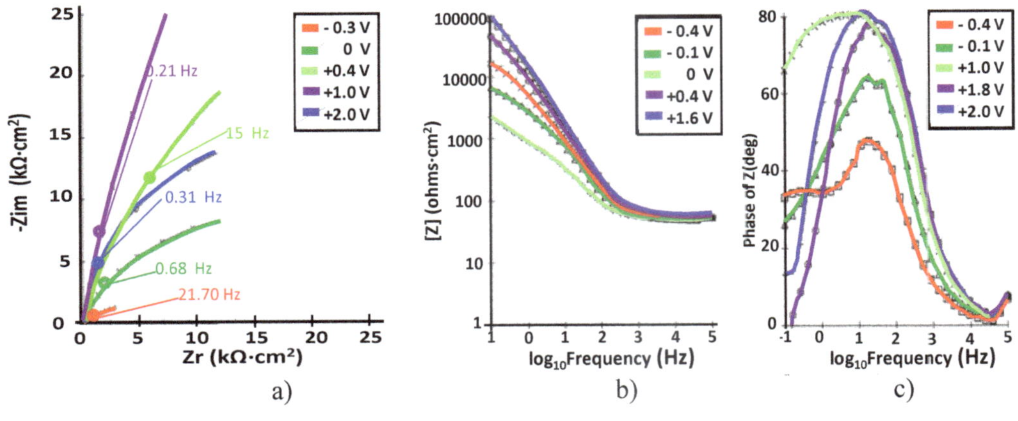

Figure 6. (**a**) Nyquist; (**b**) Bode –IZI and (**c**) Bode-phase spectra at different potentials for Ti–5Ta in simulated body fluid at pH = 7.8.

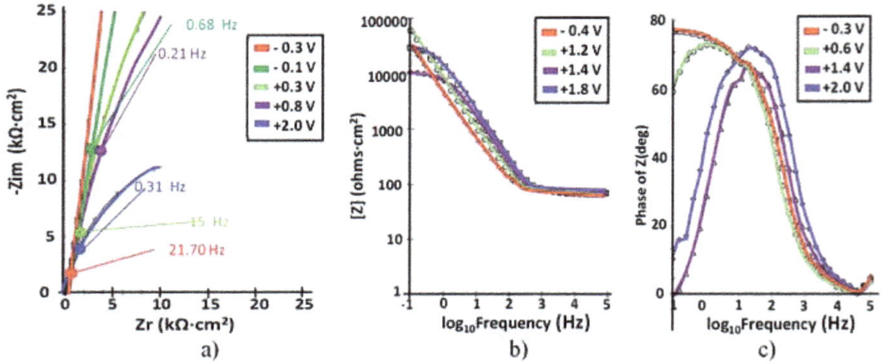

Figure 7. (**a**) Nyquist; (**b**) Bode –IZI and (**c**) Bode-phase spectra at different potentials for Ti–15Ta in simulated body fluid at pH = 7.8.

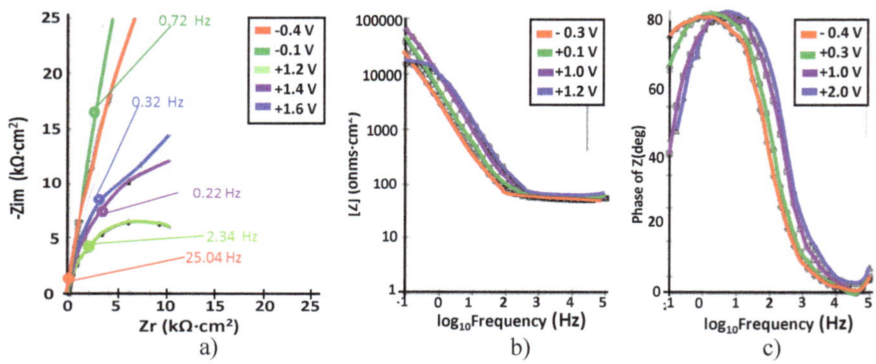

Figure 8. (**a**) Nyquist; (**b**) Bode –IZI and (**c**) Bode-phase spectra at different potentials for Ti–25Ta in simulated body fluid at pH = 7.8.

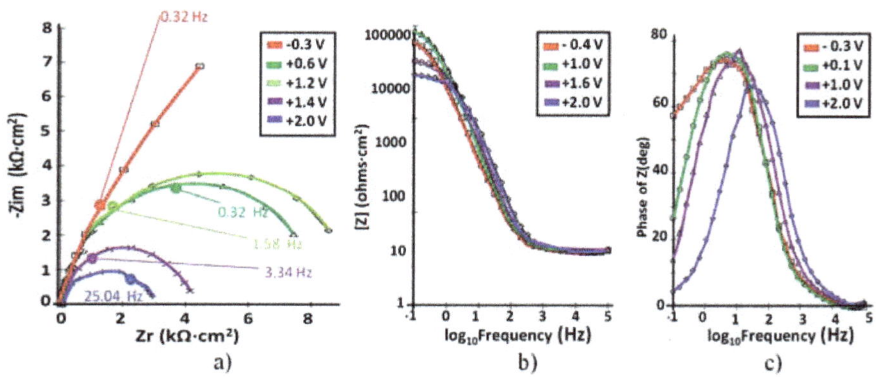

Figure 9. (**a**) Nyquist; (**b**) Bode –IZI and (**c**) Bode-phase spectra at different potentials for Ti–30Ta in simulated body fluid at pH = 7.8.

3.1.2. Ti–5Ta

The corresponding EIS plots are presented in Figure 6.

At all potentials, a near capacitive response was detected, characterized in Nyquist plots (see Figure 6a) by incomplete semicircles. In the higher frequency band (1–100 kHz), the Bode plot (Figure 6b) shows constant values (horizontal line) of log |Z| versus log(f) with a phase angle approaching 0°. From 0 to 1 V, in the wide range of low and medium frequencies, the spectra show a linear slope of approximately –1 in log |Z| as the frequency decreases, while the phase angle values are close to 80°. This is the typical response of a compact passive film capacitor. The acquired values for passive film resistance are high until 1.8 V, after which they decrease with the potential. In Figure 6c, it can be observed that the phase angle observed for Ti–5Ta was encountered in the range of about −65° to −80°, suggesting a highly stable film on Ti–5Ta [19]. At a potential value higher than 0 V, a unique peak is noted in the phase angle graphs, which indicates the engagement of one relaxation time.

3.1.3. Ti–15Ta

The corresponding EIS plots are presented in Figure 7.

For this alloy, the shape of the impedance data is similar to that of Ti–5Ta but the response of the compact passive oxide can be observed until 1.2 V, with no modifications of the value of the electrolytic solution resistance (see Figure 7b). Bode-|Z| spectra displayed a linear slope of about −1 and high impedance values (order of 10^5 Ω·cm^2) in the low and middle frequency ranges, which represent the characteristic response of a capacitive behavior of the passive film [32]; after a potential value of 1.2 V, the impedance values slightly decreased over time as result of the low dissolution of the passive layer [10].

3.1.4. Ti–25Ta

The corresponding EIS plots are presented in Figure 8.

For this alloy, we can also observe a strong change in the shapes of the impedance data, but at lower potential than for Ti–5Ta and Ti–15Ta. This change takes place between 0.8 and 1.0 V. The Nyquist plots illustrate, at all applied potentials, one depressed semicircle, suggesting non-ideal capacitive behavior (high corrosion resistance) for Ti–25Ta. When the semicircles deformed and their diameters decreased over time, it shows some dissolution processes through the passive film. During the potential scan, the electrolytic resistance is constant (60 Ω·cm^2), and the passive layer capacitance slowly decreases with the potential. With the Ti–25Ta alloy, the best results in EIS spectra were obtained, completing the lowest elastic modulus and the highest ratio of strength to modulus among Ti–Ta alloys [33]. It can be observed that Bode-|Z| impedance plots showed linear portions with the slope amounting to −1 (from −0.96 to −0.99) over a large frequency range (from 100 mHz to 100 kHz). At all analyzed potential values, a single peak is observed in phase angle plots, which indicates the involvement of one relaxation time.

3.1.5. Ti–30Ta

The corresponding EIS plots are presented in Figure 9.

It can be observed that the Nyquist plots revealed the same capacitive loops but with the lowest impedance values compared with the other alloys. It can be seen that the Bode-|Z| impedance plots exhibited linear parts with a slope around −1 only at intermediate frequencies. It was found that the maximum phase angle observed for Ti–30Ta was the lowest among all the values obtained for the Ti–Ta alloys. It can be observed that the electrolytic resistance was constant during the experiments (27 Ω·cm^2).

3.2. Equivalent Circuits

The obtained impedance data were matched by employing a nonlinear regression approach using the Randles equivalent circuit (see Figure 10a) [34–36]. This widely used electrical equivalent circuit involves an ohmic solution resistance (R_s) in series with the parallel arrangement of a CPE (Q_l) and a resistor (R_l). R_l is a numerical indicator that provides a value for the corrosion resistance of the passive layer that acts as a damper to the flux of electrons (resistance) across the metallic material/electrolyte interface and gives a measure of the level of protection provided by the passive layer that forms at the inter-face [10,37]. The employment of a constant phase element (Q_l) in place of an "ideal" capacitor was required due to the inhomogeneous nature of the passive layer formed on the alloy surface and because of the differences between the flat surface of an ideal capacitor. From the obtained data, it was found that the value of R_s does not change significantly during the test due to the fact that the controlling factors that affect its magnitude, as the stability of the exhibited surface and the number of ions in the simulated body fluid solution do not change.

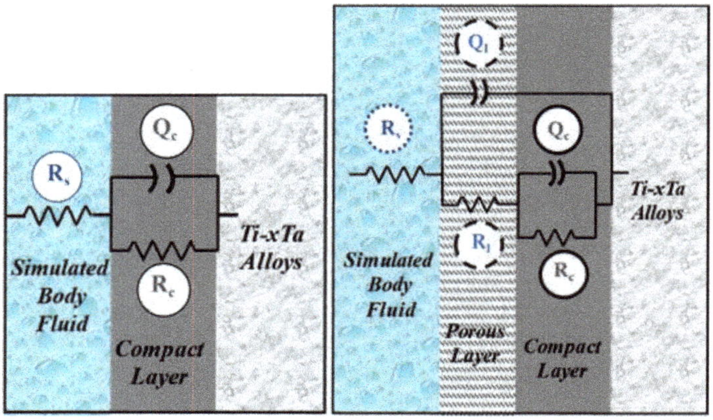

(a) (b)

Figure 10. Equivalent circuits used for fitting the experimental data with: (**a**) one time-constant; (**b**) two time-constants.

In the corrosion process, a passive layer is developed on the metallic surface and the charge transfer reaction that occurs can be disregarded and the resulting measured impedance could be assigned to the impedance of the passive film. The components of the equivalent circuit are:

R_s—ohmic resistance of the physiological solution

R_l—resistance of the passive layer

C_l—capacitance of the passive layer

The analysis of the impedance plots was carried out by fitting these data with ZSimpWin software. The performance of the fit to the equivalent circuit was assessed in the first place by the chi-square value and in the second place by comparing experimental data with simulated data.

In place of capacitance, a constant phase element (CPE), representing the shift from the real capacitive performance, was employed. The impedance of a CPE is characterized by [27]:

$$Z_{CPE} = \frac{1}{Y^0 (j\omega)^n} \quad (1)$$

where:

- Z is the impedance of the constant phase element CPE
- j is the imaginary number ($j^2 = -1$)
- ω is the angular frequency ($rad \cdot s^{-1}$)
- $n\pi/2$ is the constant phase angle of the constant phase element (rad)
- Y^0 is the constant of the constant phase element [$S(s \cdot rad^{-1})^n$]

According to all Bode plots and our previous work [10], only one time constant can be clearly distinguished; however, at low frequencies, no clear definition of the second time constant can be observed so only the high–medium frequencies' data have been analyzed in this work and only the simple circuit was used. Different studies [38,39] have proven that data at low potentials have indicated the growth of a single layer of oxide on titanium characterized by one-time constant equivalent circuit; the same circuit was reported by others who studied the corrosion of tantalum either in its pure form or as a coating [40–43].

When the impedance spectra were fitted to the equivalent circuits presented in Figure 10, a CPE element was used because, generally, a distributed relaxation feature is presented for TiO_2 films.

In order to obtain the total impedance of the equivalent circuit, we determine the admittance of the parallel arrangement (R_lQ_l):

$$\frac{1}{Z_l} = \frac{1}{Z_{R_l}} + \frac{1}{Z_{Q_l}} \tag{2}$$

Because the roughness factor "n" is > 0.85, near to 1, the resulting Y^0 value of Q_l is assumed to be C_l in the following discussion:

$$\frac{1}{Z_l} = \frac{1}{R_l} + j\, w\, C_l \tag{3}$$

The diagnostic considerations for the selection of equivalent circuits for modeling the impedance data can be resumed by visual observations of the shifts in the experimental Bode diagrams with the change of potential and concentration of the alloying metal. It appears that, in the passive zone potentials (mostly strictly capacitive impedance), the Bode graphs provide a good fit if the total impedance is modeled following the circuit in Figure 10a. The obtained results of C_l and R_l are presented in Figures 11 and 12, respectively.

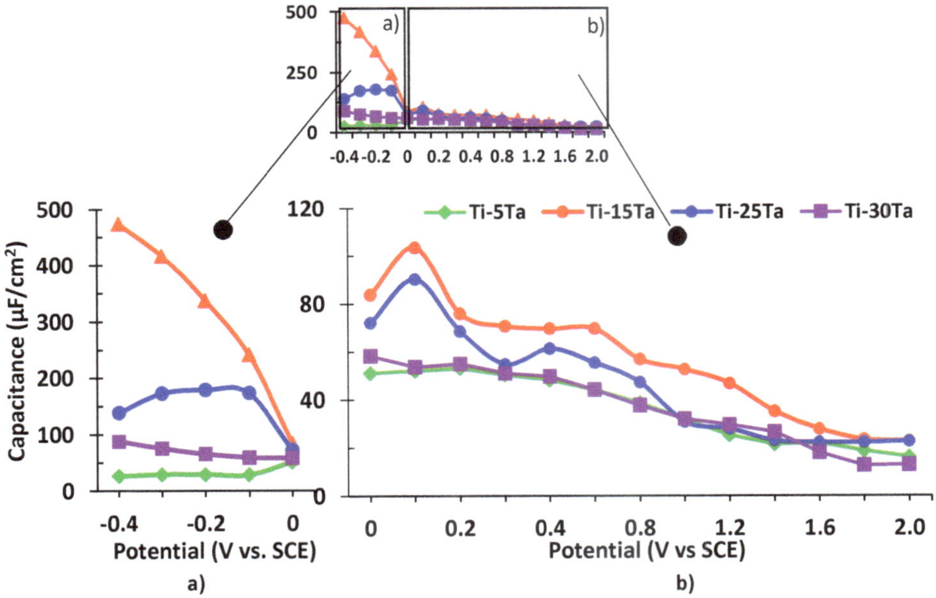

Figure 11. Capacitance of the passive films for Ti–Ta alloys in simulated body fluid: (**a**) at negative potentials; (**b**) at positive potentials.

Figure 11 points to the fast decreasing of the passive film capacitance. A high initial C_l is compatible with a high initial active surface and a high roughness factor. The evolution of C_l with the potential is also compatible with the decreasing of the active surface due to the increasing of the passive layer thickness, with all the alloys eventually reaching the same order of film thickness.

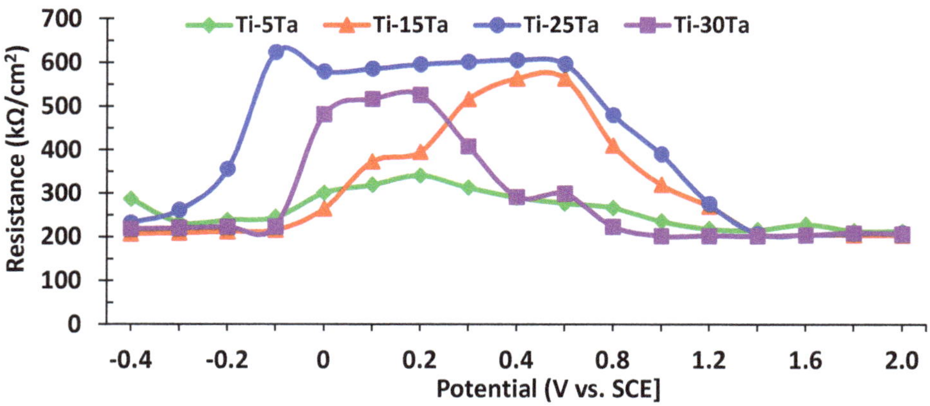

Figure 12. Resistance of the passive films for Ti–Ta alloys in Ringer's solution.

In Figure 12, the resistance of the passive film (k$\Omega \cdot cm^{-2}$) as a function of potential is presented. It can be observed that the value obtained for Ti–25Ta is higher than that of the other alloys, which depicts its highest corrosion resistance in simulated body fluid. This is due to the chemical composition of this alloy, which is closed to the compositional boundary between the α' and α'' phase and has the lowest elastic modulus and the highest ratio of strength to modulus among Ti–Ta alloys [9].

The shape of the potentiodynamic polarization curves indicated that all the Ti–xTa alloys are passivating immediately at immersion in simulated body fluid, a behavior that can be termed stable passivity. In these conditions, after 24 h of immersion, the passive film is thicker and the EIS data can be fitted with good results (chi^2 of 10^{-4} order) by the two-time equivalent electrical circuit presented in Figure 10b. In Table 5, the main parameters of the proposed circuit for all the studied alloys are presented. Polarization resistance R_p is represented by the sum of the resistance of the porous passive film and the compact barrier layer (R_l + R_c) [44]. The value of the "n" parameter corresponds to the extent of dispersion and is attributed to the surface inhomogeneity. The values of n_c are almost 1, and Q_c is similar with a pure capacitor. The values of n_p are lower than those of n_c, and this may be due to the higher roughness and diffusion through the outer layer. The simulated data are in good agreement with the experimental data, and chi-square values of 10^{-4} were obtained.

Table 5. Circuit parameters calculated from the fitting of the EIS spectra at corrosion potential to the equivalent circuit from Figure 10b.

Alloy	R_s ($\Omega \cdot cm^2$)	$Y^0{}_p \cdot 10^5$ ($S \cdot s^n/cm^2$) $0 < n < 1$	n_o	R_l ($\Omega \cdot cm^2$)	$Y^0{}_c \cdot 10^5$ ($S \cdot s^n/cm^2$) $0 < n < 1$	n_c	R_c ($k\Omega \cdot cm^2$)	$\chi^2 \cdot 10^4$
Ti–5Ta	34.2	1.72	0.59	500.6	2.71	0.89	153.1	7.24
Ti–15Ta	59.0	9.90	0.78	887.5	17.6	0.91	322.4	4.37
Ti–25Ta	11.5	1.74	0.72	14.5	4.31	0.94	577.3	4.89
Ti–30Ta	27.1	6.13	0.79	28.1	1.10	0.90	485.9	1.88

The corrosion resistance of an alloy could be characterized by the polarization resistance at the corrosion potential. This parameter can be calculated in two ways: (1) taking the slope of the E vs. I relationship in the range of ± 10 mV vs. open circuit potential; (2) from the fitting results of EIS spectra with all the implicated resistances [45]. Comparing the values obtained by the two techniques, it is observed that, for low concentrations of tantalum (Ti–5Ta and Ti–15Ta), the values of corrosion resistance obtained by EIS are lower than those of the DC experiments. This is probably due to the fact that with the shift of the potential towards more positive values vs Ecorr, titanium forms a more compact oxide layer than it does at the corrosion potential. At higher Ta concentrations (Ti–25Ta and Ti–30Ta), Ta_2O_5 plays a more important role in the compactness of the passive film and the corrosion resistance values from EIS and polarization curves are approximately the same. It can be observed that the highest corrosion resistance of Ti–xTa alloys was obtained for Ti–25Ta.

Titanium is not a noble metal, and if this metal generally exerts a high electrode potential, it is because a passivating film of oxide is formed on its surface; the stability of this layer protects the metal from further deterioration.

During the anodic polarization, a film of TiO_2 is formed:

$$Ti + 2H_2O \rightarrow TiO_2 + 4H^+ - 4e^- \quad (4)$$

For Ti, the Porbaix diagram of Ti–H_2O indicates that the substantial decrease in the passive film resistance at potentials close to 1 V can be attributed to the development of TiO_3–H_2O on the surface of the alloys.

The Pourbaix Ta–H_2O diagram [46] shows that the high strength of tantalum is attributed to the development of a protective film of tantalum pentoxide (Ta_2O_5). Tantalum, during the anodic polarization, tends to cover itself with this layer of Ta_2O_5, which has good qualities (compactness, continuity, etc.). The reaction that takes place is:

$$2Ta + 5H_2O \rightarrow Ta_2O_5 + 10H^+ + 10e^- \quad (5)$$

Tantalum pentoxide can exist in the hydrated state; it has a variable percentage of water of crystallization and is attributed the formula $Ta_2O_5{\cdot}H_2O$ or $HTaO_3$.

4. Conclusions

1. The Nyquist plots for all the Ti–Ta alloys show the same incomplete semicircles with large diameters increasing with the potential (up until a critical value for each alloy) due to the improvement of the protective properties of the passive film formed on the surface of the alloy.
2. For all Ti–Ta alloys, the Bode phase plots exhibited one phase angle—typical for a capacitive barrier passive layer formed on the surface of an alloy.
3. Impedance spectra are fitted with a one-time constant equivalent circuit, common for a compact oxide layer, for all Ti–Ta alloys in extra-cellular fluids. After a long immersion period in simulated body fluid, the passive film is thicker and develops a bi-layer structure: an outer porous layer and an inner compact layer. Tantalum addition increases the stability of the passive film due to the development of Ta_2O_5.
4. Of the Ti–Ta alloys, Ti–25Ta demonstrates excellent passive layer and corrosion resistance properties, and thus it seems to be a promising product for metallic medical devices.

Author Contributions: conceptualization, P.P.S.-P. and J.C.M.-R.; methodology, J.C.M.-R.; software, N.R.F.-S.; validation, M.V.S., N.R.F.-S., and J.C.M.-R.; formal analysis, N.R.F.-S.; investigation, P.P.S.-P.; resources, M.V.S.; data curation, P.P.S.-P.; writing—original draft preparation, J.C.M.-R.; writing—review and editing, P.P.S.-P., M.V.S., and N.R.F.-S.; visualization, N.R.F.-S.; supervision, J.C.M.-R.; project administration, N.R.F.-S.; funding acquisition, M.V.S. All authors have read and agreed to the published version of the manuscript.

Funding: This research was funded by Asociatia Neurochirurgie Sibiu and Gran Canaria Cabildo, project number CABINFR2019-07.

Acknowledgments: We are grateful for the company's support and generosity of R&D CS (Research & Development Consulting and Services) from Bucharest, Romania, in the absence of which this study could not have been carried out.

Conflicts of Interest: The authors declare no conflict of interest.

References

1. Rack, H.J.; Qazi, J.I. Titanium alloys for biomedical applications. *Mater. Sci. Eng. C* **2006**, *26*, 1269–1277. [CrossRef]
2. Lee, T.; Lee, S.; Kim, I.S.; Moon, Y.H.; Kim, H.S.; Park, C.H. Breaking the limit of Young's modulus in low-cost Ti–Nb–Zr alloy for biomedical implant applications. *J. Alloys Compd.* **2020**, *828*, 154401. [CrossRef]
3. Zhou, Y.L.; Niinomi, M.; Akahori, T. Effects of Ta content on Young's modulus and tensile properties of binary Ti-Ta alloys for biomedical applications. *Mater. Sci. Eng. A* **2004**, *371*, 283–290. [CrossRef]
4. Verestiuc, L.; Spataru, M.-C.; Baltatu, M.S.; Butnaru, M.; Solcan, C.; Sandu, A.V.; Voiculescu, I.; Geanta, V.; Vizureanu, P. New Ti–Mo–Si materials for bone prosthesis applications. *J. Mech. Behav. Biomed. Mater.* **2021**, *113*, 104198. [CrossRef] [PubMed]
5. Zhang, L.C.; Chen, L.Y. A Review on Biomedical Titanium Alloys: Recent Progress and Prospect. *Adv. Eng. Mater.* **2019**, *21*, 1–29. [CrossRef]
6. Souto, B.M.; Burstein, G.T. A preliminary investigation into the microscopic depassivation of passive titanium implant materials in vitro. *J. Mater. Sci. Mater. Med.* **1996**, *7*, 337–343. [CrossRef]
7. Kokubo, T.; Kim, H.-M.; Kawashita, M.; Nakamura, T. REVIEW Bioactive metals: Preparation and properties. *J. Mater. Sci. Mater. Med.* **2004**, *15*, 99–107. [CrossRef]
8. Khadija, G.; Saleem, A.; Akhtar, Z.; Naqvi, Z.; Gull, M.; Masood, M.; Mukhtar, S.; Batool, M.; Saleem, N.; Rasheed, T.; et al. Short term exposure to titanium, aluminum and vanadium (Ti 6Al 4V) alloy powder drastically affects behavior and antioxidant metabolites in vital organs of male albino mice. *Toxicol. Rep.* **2018**, *5*, 765–770. [CrossRef]
9. Kim, T.I.; Han, J.H.; Lee, I.S.; Lee, K.H.; Shin, M.C.; Choi, B.B. New titanium alloys for biomaterials: A study of mechanical and corrosion properties and cytotoxicity. *Biomed. Mater. Eng.* **1997**, *7*, 253–263. [CrossRef]
10. Perdomo-Socorro, P.P.; Florido-Suárez, N.R.; Verdú-Vázquez, A.; Mirza-Rosca, J.C. Comparative EIS study of titanium-based materials in high corrosive environments. *Int. J. Surf. Sci. Eng.* **2021**, *15*, 152–164. [CrossRef]
11. Florido-Suarez, N.R.; Verdu-Vazquez, A.; Socorro-Perdomo, P.P.; Mirza-Rosca, J.C. Past Advances and Future Perspective of Ti-Ta Alloys. *Glob. J. Eng. Sci.* **2021**, *7*, 20–22. [CrossRef]
12. Geanta, V.; Voiculescu, I.; Vizureanu, P.; Victor Sandu, A. High Entropy Alloys for Medical Applications. In *Engineering Steels and High Entropy-Alloys*; IntechOpen: London, UK, 2020; pp. 4–12. [CrossRef]
13. Manivasagam, G.; Dhinasekaran, D.; Rajamanickam, A. Biomedical Implants: Corrosion and its Prevention—A Review. *Recent Patents Corros. Sci.* **2010**, *2*, 40–54. [CrossRef]
14. Song, T.; Tang, H.P.; Li, Y.; Qian, M. Liquid metal dealloying of titanium-tantalum (Ti-Ta) alloy to fabricate ultrafine Ta ligament structures: A comparative study in molten copper (Cu) and Cu-based alloys. *Corros. Sci.* **2020**, *169*. [CrossRef]
15. Huang, S.; Sing, S.L.; de Looze, G.; Wilson, R.; Yeong, W.Y. Laser powder bed fusion of titanium-tantalum alloys: Compositions and designs for biomedical applications. *J. Mech. Behav. Biomed. Mater.* **2020**, *108*, 103775. [CrossRef]
16. Wu, C.; Peng, P.; Chou, H.; Ou, K. Microstructural, mechanical and biological characterizations of the promising titanium-tantalum alloy for biomedical applications. *J. Alloys Compd.* **2018**, *735*, 2604–2610. [CrossRef]
17. Mendis, S.; Xu, W.; Tang, H.P.; Jones, L.A.; Liang, D.; Thompson, R.; Choong, P.; Brandt, M. Characteristics of oxide films on Ti-(10–75) Ta alloys and their corrosion performance in an aerated Hank' s balanced salt solution. *Appl. Surface Sci.* **2020**, *506*, 145013. [CrossRef]
18. Chen, G.; Yin, J.; Zhao, S.; Tang, H.; Qu, X. Microstructure and tensile properties of a Ti-28Ta alloy studied by transmission electron microscopy and digital image correlation. *Int. J. Refract. Met. Hard Mater.* **2019**, *81*, 71–77. [CrossRef]
19. Zhao, S.; Yin, J.; Shen, L.; Ge, Y.; Chen, G. Ti-60Ta Powders Produced by PREP and Their Properties. *Xiyou Jinshu Cailiao Yu Gongcheng/Rare Met. Mater. Eng.* **2017**, *46*, 1679–1683.
20. Capellato, P.; Sachs, D.; Vasconcelos, L.V.B.; Melo, M.M.; Silva, G.; Ranieri, M.G.A.; Zavaglia, C.A.d.C.; Nakazato, R.Z.; Alves Claro, A.P.R. Optimization of anodization parameters in Ti-30Ta alloy. *Metals (Basel)* **2020**, *10*, 1–11. [CrossRef]
21. Soro, N.; Attar, H.; Brodie, E.; Veidt, M.; Molotnikov, A.; Dargusch, M.S. Evaluation of the mechanical compatibility of additively manufactured porous Ti–25Ta alloy for load-bearing implant applications. *J. Mech. Behav. Biomed. Mater.* **2019**, *97*, 149–158. [CrossRef]
22. Traxel, K.D.; Bandyopadhyay, A. Modeling and experimental validation of additively manufactured tantalum-titanium bimetallic interfaces. *Mater. Des.* **2021**, *207*, 109793. [CrossRef]
23. Zhao, D.; Liang, H.; Han, C.; Li, J.; Liu, J.; Zhou, K.; Yang, C.; Wei, Q. 3D printing of a titanium-tantalum Gyroid scaffold with superb elastic admissible strain, bioactivity and in-situ bone regeneration capability. *Addit. Manuf.* **2021**, *47*, 102223. [CrossRef]
24. Niinomi, M. Mechanical biocompatibilities of titanium alloys for biomedical applications. *J. Mech. Behav. Biomed. Mater.* **2008**, *1*, 30–42. [CrossRef]
25. Zhou, Y.L.; Niinomi, M.; Akahori, T. Changes in mechanical properties of Ti alloys in relation to alloying additions of Ta and Hf. *Mater. Sci. Eng. A* **2008**, *483–484*, 153–156. [CrossRef]
26. Al-Muhanna, K.; Habib, K. Corrosion behavior of different alloys exposed to continuous flowing seawater by electrochemical impedance spectroscopy (EIS). *Desalination* **2010**, *250*, 404–407. [CrossRef]
27. Escobar Claros, C.A.; Contri Campanelli, L.; Moreira Jorge, A.; Leprêtre, J.C.; Bolfarini, C.; Roche, V. Corrosion behaviour of biomedical β-titanium alloys with the surface-modified by chemical etching and electrochemical methods. *Corros. Sci.* **2021**, *188*. [CrossRef]

28. Mareci, D.; Chelariu, R.; Gordin, D.M.; Ungureanu, G.; Gloriant, T. Comparative corrosion study of Ti-Ta alloys for dental applications. *Acta Biomater.* **2009**, *5*, 3625–3639. [CrossRef]
29. ASTM International. ASTM E3-11(2017), Standard Guide for Preparation of Metallographic Specimens. ASTM International: West Conshohocken, PA, USA, 2017.
30. Smith, B.S.; Capellato, P.; Kelley, S.; Gonzalez-Juarrero, M.; Popat, K.C. Reduced in vitro immune response on titania nanotube arrays compared to titanium surface. *Biomater. Sci.* **2013**, *1*, 322–332. [CrossRef]
31. Welsch, G.; Boyer, R.; Collings, E.W. (Eds.) *Materials Properties Handbook: Titanium Alloys*, 4th ed.; ASM International: Geauga County, OH, USA, 1993.
32. Vaswani-Reboso, J.; Florido-Suarez, N.; Socorro-Perdomo, P.; Mirza-Rosca, J. Study of Biocompatibility, Mechanical Properties and Microstructural Analysis af Ag-Pd Alloy. *Microsc. Microanal.* **2021**, *27*, 550–552. [CrossRef]
33. Zhou, Y.L.; Niinomi, M. Ti-25Ta alloy with the best mechanical compatibility in Ti-Ta alloys for biomedical applications. *Mater. Sci. Eng. C* **2009**. [CrossRef]
34. Garcia-Falcon, C.M.; Gil-Lopez, T.; Verdu-Vazquez, A.; Mirza-Rosca, J.C. Electrochemical characterization of some cobalt base alloys in Ringer solution. *Mater. Chem. Phys.* **2021**, *260*, 124164. [CrossRef]
35. Garcia-Falcon, C.M.; Gil-Lopez, T.; Verdu-Vazquez, A.; Mirza-Rosca, J.C. Corrosion behavior in Ringer solution of several commercially used metal alloys. *Anti-Corrosion Methods Mater.* **2021**, *68*, 324–330. [CrossRef]
36. Li, X.; Wang, L.; Fan, L.; Zhong, M.; Cheng, L.; Cui, Z. Understanding the effect of fluoride on corrosion behavior of pure titanium in different acids. *Corros. Sci.* **2021**, *192*, 109812. [CrossRef]
37. López Ríos, M.; Socorro Perdomo, P.P.; Voiculescu, I.; Geanta, V.; Crăciun, V.; Boerasu, I.; Mirza Rosca, J.C. Effects of nickel content on the microstructure, microhardness and corrosion behavior of high-entropy AlCoCrFeNix alloys. *Sci. Rep.* **2020**, *10*, 1–11. [CrossRef]
38. Yu, X.Y.; Zhu, W.Q.; Chen, W.; Chen, W.Q.; Zhang, S.M.; Qiu, J. Osteoclast-mediated biocorrosion of pure titanium in an inflammatory microenvironment. *Mater. Sci. Eng. C* **2021**, *119*, 111610. [CrossRef] [PubMed]
39. Bojinov, M.; Betova, I.; Karastoyanov, V. Modeling barrier film growth and dissolution on titanium based on EIS, XPS and photocurrent data. *Electrochim. Acta* **2020**, *344*, 136137. [CrossRef]
40. Huo, W.T.; Zhao, L.Z.; Yu, S.; Yu, Z.T.; Zhang, P.X.; Zhang, Y.S. Significantly enhanced osteoblast response to nano-grained pure tantalum. *Sci. Rep.* **2017**, *7*, 40868. [CrossRef] [PubMed]
41. Fattah-alhosseini, A.; Elmkhah, H.; Ansari, G.; Attarzadeh, F.; Imantalab, O. A comparison of electrochemical behavior of coated nanostructured Ta on Ti substrate with pure uncoated Ta in Ringer's physiological solution. *J. Alloys Compd.* **2018**, *739*, 918–925. [CrossRef]
42. Hee, A.C.; Martin, P.J.; Bendavid, A.; Jamali, S.S.; Zhao, Y. Tribo-corrosion performance of filtered-arc-deposited tantalum coatings on Ti-13Nb-13Zr alloy for bio-implants applications. *Wear* **2018**, *400*, 31–42. [CrossRef]
43. Fattah-alhosseini, A.; Pourmahmoud, M. Passive and Semiconducting Properties Assessment of Commercially Pure Tantalum in Hank's Physiological Solution. *J. Mater. Eng. Perform.* **2018**, *27*, 116–123. [CrossRef]
44. Moreto, J.A.; Marino, C.E.B.; Bose Filho, W.W.; Rocha, L.A.; Fernandes, J.C.S. SVET, SKP and EIS study of the corrosion behaviour of high strength Al and Al-Li alloys used in aircraft fabrication. *Corros. Sci.* **2014**, *84*, 30–41. [CrossRef]
45. Sowa, M.; Simka, W. Electrochemical impedance and polarization corrosion studies of tantalum surface modified by DC Plasma electrolytic oxidation. *Materials (Basel)* **2018**, *11*, 545. [CrossRef]
46. Pourbaix, M.; Zhang, H.; Pourbaix, A. Presentation of an Atlas of chemical and electrochemical equilibria in the presence of a gaseous phase. *Mater. Sci. Forum* **1997**, *251–254*, 143–148. [CrossRef]

materials

MDPI

Article

New Titanium Alloys, Promising Materials for Medical Devices

Madalina Simona Baltatu [1], Petrica Vizureanu [1,*], Andrei Victor Sandu [1,2], Nestor Florido-Suarez [3], Mircea Vicentiu Saceleanu [4] and Julia Claudia Mirza-Rosca [3,*]

[1] Department of Technologies and Equipments for Materials Processing, Faculty of Materials Science and Engineering, Gheorghe Asachi Technical University of Iaşi, Blvd. Mangeron, No. 51, 700050 Iasi, Romania; cercel.msimona@yahoo.com (M.S.B.); sav@tuiasi.ro (A.V.S.)

[2] Romanian Inventors Forum, Iasi. Sf. P. Movila 3, 700089 Iasi, Romania

[3] Mechanical Engineering Department, Las Palmas de Gran Canaria University, 35017 Tafira, Spain; nestor.florido@ulpgc.es

[4] Neurosurgery Department, Faculty of Medicine, Lucian Blaga University, 550024 Sibiu, Romania; vicentiu.saceleanu@gmail.com

* Correspondence: peviz@tuiasi.ro (P.V.); julia.mirza@ulpgc.es (J.C.M.-R.)

Abstract: Titanium alloys are used in medical devices due to their mechanical properties, but also for their corrosion resistance. The natural passivation of titanium-based biomaterials, on the surface of which a dense and coherent film of nanometric thickness is formed, composed mainly of TiO_2, determines an apparent bioactivity of them. In this paper, the method of obtaining new Ti20MoxSi alloys (x = 0.0, 0.5, 0.75, and 1.0) is presented, their microstructure is analyzed, and their electrochemical responses in Ringer´s solution were systematically investigated by linear polarization, cyclic potential dynamic polarization, and electrochemical impedance spectroscopy (EIS). The alloys corrosion resistance is high, and no evidence of localized breakdown of the passive layer was observed. There is no regularity determined by the composition of the alloys, in terms of corrosion resistance, but it seems that the most resistant is Ti20Mo1.0Si.

Keywords: Ti–Mo–Si alloys; biomedical alloys; corrosion resistance

Citation: Baltatu, M.S.; Vizureanu, P.; Sandu, A.V.; Florido-Suarez, N.; Saceleanu, M.V.; Mirza-Rosca, J.C. New Titanium Alloys, Promising Materials for Medical Devices. *Materials* **2021**, *14*, 5934. https://doi.org/10.3390/ma14205934

Academic Editor: Yulin Hao

Received: 23 September 2021
Accepted: 5 October 2021
Published: 9 October 2021

1. Introduction

Globally, implantology aims to use materials with specific biological and biomechanical characteristics, as high as possible [1], and titanium alloys have advantages both in terms of reducing risks to patients during and after medical interventions, and in their efficacy and biocompatibility with human tissue [2].

The concern of researchers and physicians in recent years, for the improvement of biomaterials, comes from the reason that biomaterials have to be tolerable by the organism [3] for a significant period and, therefore, have to satisfy the performance requisites in accordance with the medical purposes in which they are destined to be used [4,5].

Implant materials are produced with a high degree of complexity [6], with biological, mechanical and technological characteristics that are specific to the field of application, which must comply with strict quality standards, so as not to affect the health of patients [7,8]. Biomaterials must have special characteristics, including excellent fatigue strength, tensile strength, high corrosion and wear resistance, low modulus of elasticity, good hardness, low density, and high biocompatibility [9].

Titanium and titanium alloys, due to their properties, tend to replace the classic materials, such as stainless steels, Co–Cr alloys, etc., in most orthopedic applications [10,11]. Titanium and its alloys have prevailed because they have the optimal characteristics required for implant materials, namely, very good corrosion resistance, biocompatibility, excellent mechanical and fatigue resistance properties, toughness, low modulus of elasticity, satisfactory strength at wear, and affordable price [12].

Titanium is a non-toxic element, including in high amounts; several investigations have shown the effect of the ingesting of up to 0.8 mg of Ti by humans per day, thus demonstrating that Ti was excreted with no absorption/assimilation [13]. Many studies have shown that titanium implants do not show rejection due to their biocompatibility, good interaction with the host bone, and high resistance to corrosion [10]. An important factor in the mechanical and structural properties depends on the alloying elements added to the titanium and the appropriate design of the alloys (type of element and concentration), which can lead to them being an attractive option for medical applications.

Taking all these into account, for the design of these new alloys, we have chosen two elements, Mo and Si, due to the following considerations:

Molybdenum is an element that is widely used as an alloying element of titanium [14,15]. Some research in the field [16–18] has shown that a percentage of about 20% Mo can reduce the modulus of elasticity, as it is a β-stabilizing element, leading to proper mechanical properties, similar to human bone [5]. It is also a biocompatible element, with a minor level of toxicity, with respect to Cr, Ni, and Co. There is a study [19] of two titanium alloys, Ti10Mo and Ti20Mo, highlighting the formation of equiaxial beta grains of different sizes, highlighting that an alloy with a percentage of 10% Mo has grains that are larger than an alloy with a concentration of 20% Mo.

Silicon is a biocompatible element, a β-stabilizing element for titanium alloys, that reduces ductility, improves resistance to creep, and increases corrosion resistance and turnability. Jiang et al. [20] studied the corrosion resistance of Ti–Si alloys in acid solutions. Research has shown that it exhibits a high resistance to corrosion, due to the passive films formed on Ti–Si alloys, composed of TiO_2/SiO_2 oxides [21,22].

Even Ti–Mo alloys have received extensive attention, including the FDA certification (Food and Drug Administration–USA); however, there are not corrosion studies on the newly obtained alloys from the Ti–Mo–Si system. In this paper, the method of obtaining these new Ti20MoxSi alloys (x = 0.0, 0.5, 0.75, and 1.0) is presented, their microstructure is analyzed, and their electrochemical responses in Ringer´s solution were systematically investigated by linear polarization, cyclic potential dynamic polarization, and electrochemical impedance spectroscopy (EIS).

2. Materials and Methods

2.1. Obtaining Ti–Mo–Si Alloys

Obtaining the Ti–Mo–Si base alloys was performed by vacuum arc remelting (VAR), in MRF ABJ 900 equipment (Materials Research Furnaces, Suncook NH, USA). The raw materials used were high-purity elements such as Ti (99% purity), Mo (99% purity) and Si (99% purity) supplied by Alfa Aesar by Thermo Fisher Scientific.

The solidification structure of the ingots resulting in VAR technology depends on the solidification rate and the temperature gradient at the liquid/solid interface. In order to avoid obtaining the primary dendritic structure, it is necessary to maintain a constant temperature gradient during the entire remelting process. The direction of growth of the cellular structure, according to the direction of the thermal gradient and the maintenance of the direction of heat dissipation in the recrystallization crucible, maintains a constant crystallization front.

There are several conditions that must be met to avoid the formation of defects in ingots, they are as follows:

- Avoidance of uniform and incomplete melting, especially of the areas enriched in alloying elements from the inter-dendritic space of the electrode subjected to re-melting;
- The spring is not strong enough to be able to dissolve and melt the alloy in the melting electrode uniformly, so recrystallization of the melt does not take place before the melting is homogenized;
- The solidification front after remelting should be uniform and at a constant speed.

The design and production of metallic materials frequently used as biomaterials requires a detailed study, both in terms of design and the technological parameters of the development equipment.

The design of the Ti–Mo–Si alloys was based on information about the physical properties of the alloying elements (melting temperature and density), as well as the equilibrium diagrams for the Ti–Mo–Si alloys [23–25].

The technical parameters regarding the vacuum arc remelting furnace used are the following: melting power—min. 55 kVA; melting current—min. 650 A, 60% DS, three-phase voltage; maximum temperature—3700 °C; continuous monitoring of the vacuum level with display of its value; maximum vacuum level obtained with preliminary and diffusion vacuum pumps: 10^{-6} mbar; inert gas supply system—argon; the oven chamber is made of 306 L stainless steel, and the double walls are cooled with water; water-cooled copper base plate measuring 230 mm (diameter) × 13 mm (thickness); non-consumable electrode made of tungsten with 6.5 mm diameter.

The loading of the raw material in the installation was performed in ascending order of their specific density, as follows: tantalum, molybdenum and silicone, followed by the sealing closure of the chamber of the processing furnace. During the melting operations, a vacuum atmosphere of 4.5×10^{-3} mbar was created, followed by purging the chamber with inert gas (Ar), a cycle repeated three times to remove the air from the working chamber of the installation. The development of the process of elaboration of Ti–Mo–Si alloys was controlled through the observation window of the furnace, properly illuminated. The elaboration of Ti–Mo–Si alloys made with the help of the vacuum spring melting installation took place by melting the elements, followed by melting the alloy six times, an operation necessary in order to refine and homogenize the alloys. The melting of the elements occurred in a uniform manner, resulting in alloys with a defined and uniform chemical composition.

The amount of metallic load for experimental alloys is shown in Table 1 and its weight is approximately 30 g per alloy. The obtained Ti–Mo–Si ingots were then cut into spherical samples with a diameter of 10 mm and a thickness of 2 mm for specific tests.

Table 1. Calculation of the metallic load for the experimental alloys.

	Alloy	Element [g]			Batch Weight [g]	Ingot Weight [g]	Efficiency [%]
		Ti	Mo	Si			
S1	Ti20Mo	24.01	6.01	-	30.02	29.97	99.83
S2	Ti20Mo0.5Si	23.88	6.03	0.14	30.05	30.00	99.83
S3	Ti20Mo0.75Si	23.70	6.12	0.22	30.04	30.01	99.90
S4	Ti20Mo1.0Si	23.70	6.05	0.31	30.06	30.03	99.90

2.2. Microstructure and Hardness

To examine the microstructure of the Ti–Mo–Si alloys by optical microscopy, their surface was etched with Kroll reagent containing HF, HNO3 and water in a 2:1:17 volumetric ratio. Surface observations were performed with an OLYMPUS PME 3-ADL microscope. Scanning electron microscope (SEM) observations were performed using an Apreo field emission scanning electron microscope SLoVac (Thermo Fisher Scientific, Co., Waltham, MA, USA) coupled with a Bruker EDAX detector. To ensure the best imaging and analysis conditions in high vacuum, the microscope was set to operate at a voltage of 20 kV and a beam current of 1.6 nA, with a 10.0 mm working distance.

X-ray diffraction experiments were carried out with the use of a Panalytical X'Pert Pro MPD empirical diffractometer (Malvern Panalytical Ltd., Malvern, UK). As in previous studies [26], the instrument worked with a Cu Kα anode at a power of 45 kV and 40 mA in Bragg–Brentano geometry. The specimens were rotated while the data were obtained to guarantee better information capture. The purchased patterns were generated to extract the presence of the crystalline phase, lattice configuration and grain shape with the help of Malvern Panalytical HighScore Plus software.

Samples from Ti–Mo–Si system were analyzed using Wilson Wolpert 751 N (Instron Deutschland GmbH), in order to evaluate the Vickers hardness, providing an average value from eight measured points.

2.3. Electrochemical Tests

All measurements were carried out in Grifols Ringer's solution from Laboratorios Grifols, Barcelona, Spain, with the following components in mmol/l: Na^+ 129.9; K^+ 5.4; Ca^{2+} 1.8; Cl^- 111.7; and $C_3H_5O_3{}^-$ 27.2. This is a physiological solution that was modified, in which part of the Na^+ is substituted by Ca^{2+} and K^+ and part of the Cl^- by $C_3H_5O_3{}^-$. The lactate ions are converted into bicarbonate ions, which allows the pH of the solution to be regulated.

The solution was naturally aerated, tests were performed at 25 °C and the surface of alloys was prepared suitably.

Electrochemical techniques were applied in order to characterize the metallic surfaces, and corrosion studies were performed with the following two types of potentiostats: VoltaLab PGP 201 potentiostat and PGZ 301 dynamic potentiostat.

The VoltaLab PGP 201 potentiostat (VolataLab 21) (Radiometer Analytical SAS - France), a compact potentiostat/galvanostat, is an ideal device for electrochemical corrosion studies. This can be operated in the following two ways: with VoltaMaster 4 data collection and processing software or it can be controlled manually.

The VoltaLab PGP 201 Potentiostat was employed for the following:

- To obtain anodic polarization curves in conditions of low electrode potential sweep rates, curves that were then employed to measure the polarization resistance, the corrosion current and the corrosion rate.
- To obtain cyclic polarization curves with potential sweep rates of 10 mV/s, from which the kind of corrosion and the corrosion rate when the alloy´s potential is far from the corrosion potential were evaluated.

For both the potentiodynamic measurements and the electrochemical impedance spectroscopy determinations, a three-electrode cell was used.

The electrodes used (working electrode) were built in the form of discs (10 mm in diameter and 2 mm thick), being mounted in the cell by means of a screw and a Teflon sealing gasket. Through this mount, the surface of the electrode exposed to the corrosion medium was flat and circular, without edges or tips. The surface exposed to corrosion media was in all cases equal to 0.283 cm^2. The samples were polished on SiC paper up to 2000 grit and then degreasing and washing with distilled water were performed. A standard 3-electrode electrochemical cell was used with Pt as auxiliary electrode and a SCE as reference electrode.

The working conditions used in the measurements were as follows:

- Linear anodic polarization: potential range (−200) ÷ (+500) mV vs. SCE, potential scanning speed—dE/dt = 0.5 mV/s;
- Cyclic polarization: potential range (−500) ÷ (+2000) mV vs. SCE, potential scanning speed = 10 mV/s;
- EIS measurements: working potential = open circuit potential, frequency range = 100 kHz ÷ 100 mHz, AC potential amplitude of 10 mV and single sine wave measurements were conducted.

All potentials were measured against the saturated calomel electrode.

Open circuit potential, E_{OCP} (I = 0), was measured, and Tafel slopes (b_a and b_c), polarization resistance (R_p), corrosion current density (J_{corr}) were calculated using the facilities offered by VoltaMaster 4 software. For smoothing the curve, 9 points were taken into account. The calculation area was ±120 mV from E_{corr} and the linear segment was 30 mV on each of the two branches of the polarization curve (cathodic line and anodic line).

In all studied cases, when interpreting the linear polarization data, a calculation area of 120 mV and a linearity segment of 30 mV were considered.

Electrochemical impedance spectroscopy (EIS) data were used, which were acquired with VoltaMaster 4 software and processed with ZSimpWin software, in which the electrochemical spectra are interpreted by the least squares method developed by Boukamp. For this purpose, VoltaMaster data were converted to ZSimpWin processable data with EIS file converter software (Radiometer Analytical S.A.).

3. Results

3.1. Microstructure and Hardness

The elemental compositions obtained by energy-dispersive X-ray spectroscopy analysis (EDX) for the Ti–Mo–Si alloys present no impurities in the metallic mass, and Ti, Mo, and Si were the main elements identified. Table 2 present the average results of ten EDX points of the Ti–Mo–Si alloys surfaces.

Table 2. Average elemental compositions.

Sample		Ti [wt.%]	Mo [wt.%]	Si [wt.%]
S1	Average	80.05	19.95	-
	Stdev	±0.2	±0.1	-
S2	Average	80.15	19.44	0.41
	Stdev	±0.1	±0.1	±0.01
S3	Average	79.74	19.55	0.71
	Stdev	±0.3	±0.2	±0.02
S4	Average	78.98	20.01	1.01
	Stdev	±0.5	±0.3	±0.01

The microstructure of the Ti–Mo–Si alloys was highlighted by optical microscopy, see Figure 1. Ti20Mo has a structure with a β crystal grain, and Ti20Mo0.5Si, Ti20Mo0.75Si, and Ti20Mo1.0Si show a dendritic structure.

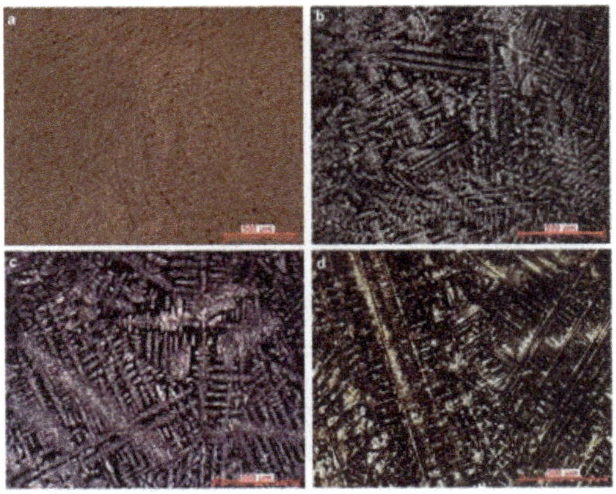

Figure 1. Optical microstructure of the alloys at 100X. (**a**) Ti20Mo, (**b**) Ti20Mo0.5Si, (**c**) Ti20Mo0.75Si, (**d**) Ti20Mo1.0Si.

In addition to the structure of the Ti–Mo–Si alloys, the X-ray diffraction (XRD) was also highlighted, see Figure 2. The effect of the addition of Si can be clearly observed in both Figures 1 and 2.

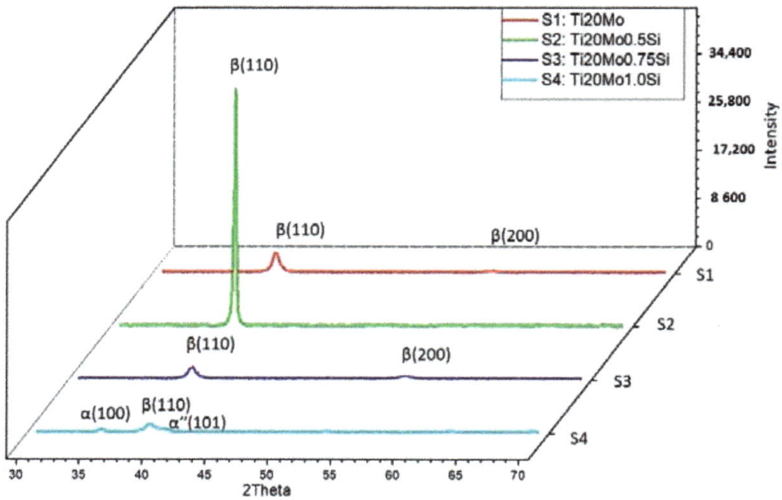

Figure 2. XRD patterns Ti–Mo–Si alloys: S1—Ti20Mo, S2—Ti20Mo0.5Si, S3—Ti20Mo0.75Si, S4—Ti20Mo1.0Si.

All the alloys were mainly composed of the parent β phase, with different precipitates of minor secondary phases, such as the α″ martensite phase and α phase.

Using the Mo–Si and Mo–Ti binary diagrams, the following three stable compounds were determined in these alloys: Ti4.00Si8.00 (reference code: 96-100-9013), Ti4.00 (reference code: 96-901-1601), and Ti2.00 (reference code: 96-901-2925). These compounds that were formed in alloys were identified by XRD analysis. The main characteristics for Ti4.00Si8.00 are as follows: crystal system: orthorhombic, a (Å): 4.4280, b (Å): 4.7790, c (Å): 9.0780, alpha (°): 90.00, beta (°): 90.00, gamma (°): 90.00, calculated density (g/cm^3): 3.60, and volume of cell (106 pm3): 192.10. The main characteristics for Ti4.00 are as follows: crystal system: hexagonal, a(Å): 2.9700, b (Å): 2.9700, c (Å): 4.7200, alpha (°): 90.00, beta (°): 90.00, gamma (°): 120.0000, calculated density (g/cm^3): 8.82, and volume of cell (106 pm^3): 36.06. The main characteristics for Ti2.00 are as follows: crystal system: cubic, a (Å): 3.2820, b (Å): 3.2820, c (Å): 3.2820, alpha (°): 90.00, beta (°): 90.00, gamma (°): 90.00, calculated density (g/cm^3): 4.50, and volume of cell (106 pm^3): 35.35.

The β Ti2.00 phase, which has a body-centered cubic structure, is obtained by alloying titanium with transition elements (Mo, Si), while the α″ (HC) phase, Ti4.00Si8.00, is a martensite, crystallized in the orthorhombic system, and is generated by increasing the amount of stabilizing elements in the β phase, which decomposes during the cooling process.

The hardness measurement results from the Ti–Mo–Si samples are as follows: Ti20Mo: 409HV, Ti20Mo0.5Si: 348 HV, Ti20Mo0.75Si: 170 HV, and Ti20Mo1.0Si: 175HV. Accordingly, with the obtained results, the addition of Si to Ti–Mo alloys significantly decreases the values of hardness. Compared to other biomaterials, Ti20Mo0.75Si alloy is close to the 316 L stainless steel alloy (155HV) value.

3.2. Linear Potentiodynamic Polarization

The corrosion potential, E_{corr}, is a measure of the corrosion tendency of a metal or alloy immersed in an electrolytic medium. High negative values of this parameter indicate a high probability that the metal will corrode in that environment. This value was measured directly versus a saturated calomel electrode (Hg/Hg_2Cl_2/KCl sat.), after 15 min of thermostating the sample in the corrosion medium. The measured parameter represents the open circuit potential (OCP), and its value should be equal to the corrosion potential (E_{corr}) only after the metal/electrolyte balance is reached. The time after which this equilibrium is reached depends on the temperature and the processes that take place at the metal–solution interface (chemical reaction, adsorption of some species from the solution, passivation or dissolution). Several preliminary tests have indicated that this time is of the order of several hours.

For this reason, the corrosion potential was evaluated indirectly from the linear polarization curves, using the Evans diagram, which represents the logarithm of the current density as a function of the electrode potential in a range of ±50 mV. In these coordinates, the intersection of the linear portions of the anodic and cathodic branches of the polarization curve gives the value of the corrosion potential on the axis, E_{corr} (Evans).

The polarization resistance method was used to evaluate the corrosion rate. This method serves to determine the corrosion current, at the corrosion potential of the metal or alloy, from the linear polarization curve obtained for a relatively small overvoltage. The corrosion current determined by this method therefore represents the current that appears at the metal/corrosive medium interface when the metal is immersed in the solution, and represents the instantaneous corrosion current. From a practical point of view, the density of the instantaneous corrosion current is important, ($J_{corr} = I_{corr}/S$), which is directly correlated with the corrosion rate (v_p), according to Equations (1)–(4).

$$R_p = \left(\frac{dE}{dj}\right)_{(E_{corr})} = \frac{b_a \cdot b_c}{2.303 \cdot J_{corr} \cdot (b_a + b_c)} \left[ohm.cm^2\right] \quad (1)$$

$$J_{corr} = \frac{b_a + b_c}{2.303(b_a + b_c) \cdot R_p} \left[\mu A/cm^2\right] \quad (2)$$

$$v_p = 3.27 \cdot \left(\frac{A}{z}\right) \cdot \frac{J_{corr}}{\rho} [\mu m/year] \quad (3)$$

where the following applies:

$$b_a = \left(\frac{\partial E}{\partial logj}\right)_a \; and \; b_c = \left(\frac{\partial E}{\partial logj}\right)_c [mV/dec] \quad (4)$$

The linear polarization curves, in semilogarithmic coordinates (Evans diagram), for the tested specimens in Ringer's solution are displayed in Figure 3, and Table 3 shows the instantaneous corrosion parameters in this physiological environment.

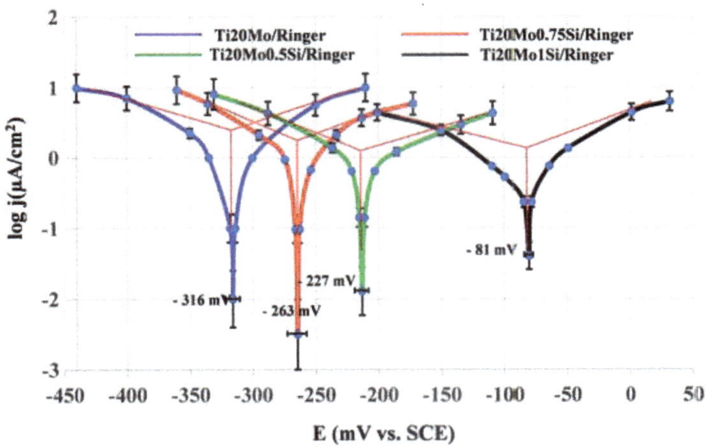

Figure 3. Linear polarization curves in semi-logarithmic coordinates for Ti–Mo–Si alloys in Ringer's solution.

Table 3. Instantaneous corrosion parameters for Ti–Mo–Si alloys in Ringer's solution.

Alloys	E_{corr} [mV vs. SCE]	R_p [kohm.cm^2]	J_{corr} [μA/cm^2]	V_{corr} [μm/year]	b_a [mV/dec]	b_c [mV/dec]
Ti20Mo	−315	13.84	2.71	26.25	192	−205
Ti20Mo0.5Si	−227	17.71	2.09	20.19	310	−142
Ti20Mo0.75Si	−263	16.50	2.68	25.89	289	−219
Ti20Mo1.0Si	−81	21.55	1.61	15.55	168	−224

The presence of silicon produces a decrease in the corrosion rate.

3.3. Cyclic Potentiodynamic Polarization

One of the methods for further characterization of the corrosion process is cyclic potentiodynamic polarization, and additional information on the processes taking place in the system can be obtained.

In the present study, the cyclic polarization curves (cyclic voltammograms) were recorded in the potential range (−500 ... +2000 ... −500) mV vs. SCE, with a potential sweep rate of 10 mV/s. All the measurements were made in naturally aerated solutions. It started from a sufficiently high negative potential, in order to reduce all ionic or molecular species that contaminate the surface of the alloy. The speed of variation in the working electrode potential was relatively high (10 mV/s), in order to obtain current intensities that were high enough to cover possible accidental fluctuations in the system, but low enough to detect all the processes that could occur in the solution or on the surface of the electrode.

Based on these curves, the following characteristic parameters of the passivation process were evaluated: the corrosion potential (E_{corr}), the repassivation potential (E_{rp}), and, in some cases, the potential for transpassivation (E_{tp}). In addition, in order to be able to make a comparison between the process intensities at the various alloy/corrosion medium pairs, the current densities corresponding to the maximum applied overpotential, respectively, 2V (J_{2V}), were evaluated. For the calculations, cyclic voltammograms were used both in linear coordinates, j = f (E), and in semi-logarithmic coordinates, E = f (log j). From the linear diagram, J_{2V} and E_{tp} were evaluated, while, from the semi-logarithmic diagram, E_{corr} and E_{rp} were evaluated.

At potentials higher than the E_{tp} potential on the oxide-coated surface, transpassivation phenomena occur, which, in this case, consist of the release of oxygen.

An additional use of cyclic voltammograms was made, in order to evaluate the corrosion rate (passivation) based on the anodic polarization curve obtained at a high potential scanning speed (10 mV/s). For this purpose, the Tafel method (polarization resistance method), applied for the points on the anodic branch of the polarization curve located in the vicinity of the corrosion potential (±120 mV compared to E_{corr}), was used. The following were evaluated: polarization resistance (R_p), corrosion current density (J_{corr}), and corrosion rate (v_{corr}).

The values of the characteristic potentials and the corrosion parameters evaluated in the case of using a sweep rate of 10 mV/s, obtained for the alloys in the corrosion media, are presented in Table 4, and the cyclic voltammograms in semi-logarithmic coordinates are presented in Figure 4.

Table 4. Corrosion parameters obtained from cyclic polarization curves in Ringer's solution.

Alloys	E_{corr} [mV vs. SCE]	E_{rp} [mV]	E_{tp} [mV]	J_{2V} [mA/cm²]	Tafel Parameters on v_s = 10 mV/s		
					R_p [KΩ.cm²]	J_{corr} [μA/cm²]	V_{corr} [μm/Year]
Ti20Mo	−124	+112	1704	1.28	6.14	5.583	53.96
Ti20Mo0.5Si	−235	+42	1070	0.46	9.41	3.788	36.59
Ti20Mo0.75Si	−183	+52	1500	0.59	10.68	3.424	33.09
Ti20Mo1.0Si	−77	+100	-	0.23	22.06	1.624	15.69

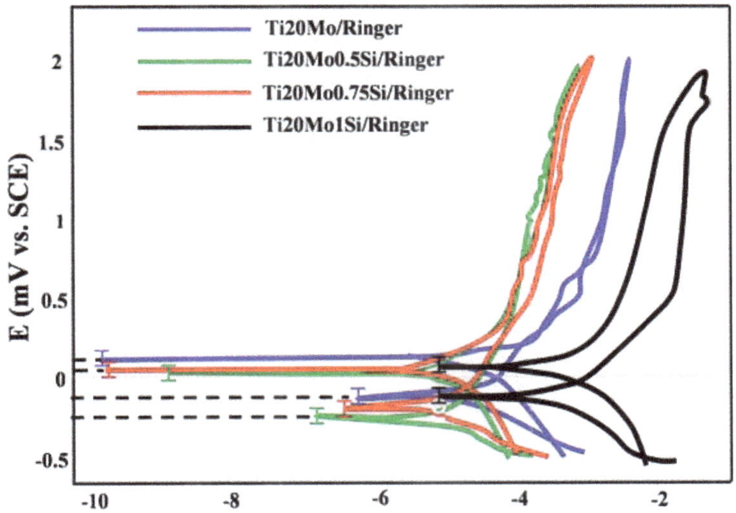

Figure 4. Corrosion parameters obtained from cyclic polarization curves in Ringer's solution.

For most samples, the anodic curve (branch of the cyclic voltammogram obtained by scanning the potential from negative values to positive values, −0.5 V → 2 V) is located above the cathode curve (return branch, +2 V → −0.5 V), and this indicates a passivation of the alloy. As a result of this passivation, the repassivation potentials are more positive than the corrosion potentials, but no correlation can be established between E_{corr} and E_{rp}, and no rule of dependence of the two potentials on the composition of the alloy can be established.

3.4. Electrochemical Impedance Spectroscopy

Using the ZSimWin program, the experimental data obtained with the PGZ 301 potentiostat were converted using the "EIS file converter" software (EISFC150).

The EIS data were represented in the Nyquist diagram (imaginary impedance component as a function of the real impedance component: $-Z_{im}$ = f (Z_{real})) and in the Bode diagram. An example of the Nyquist and Bode diagrams corresponding to the experimental data series obtained for the alloys studied in Ringer's solution is shown in Figure 5. The experimental data can be fitted with different equivalent circuits that are associated with the physical model of the surface.

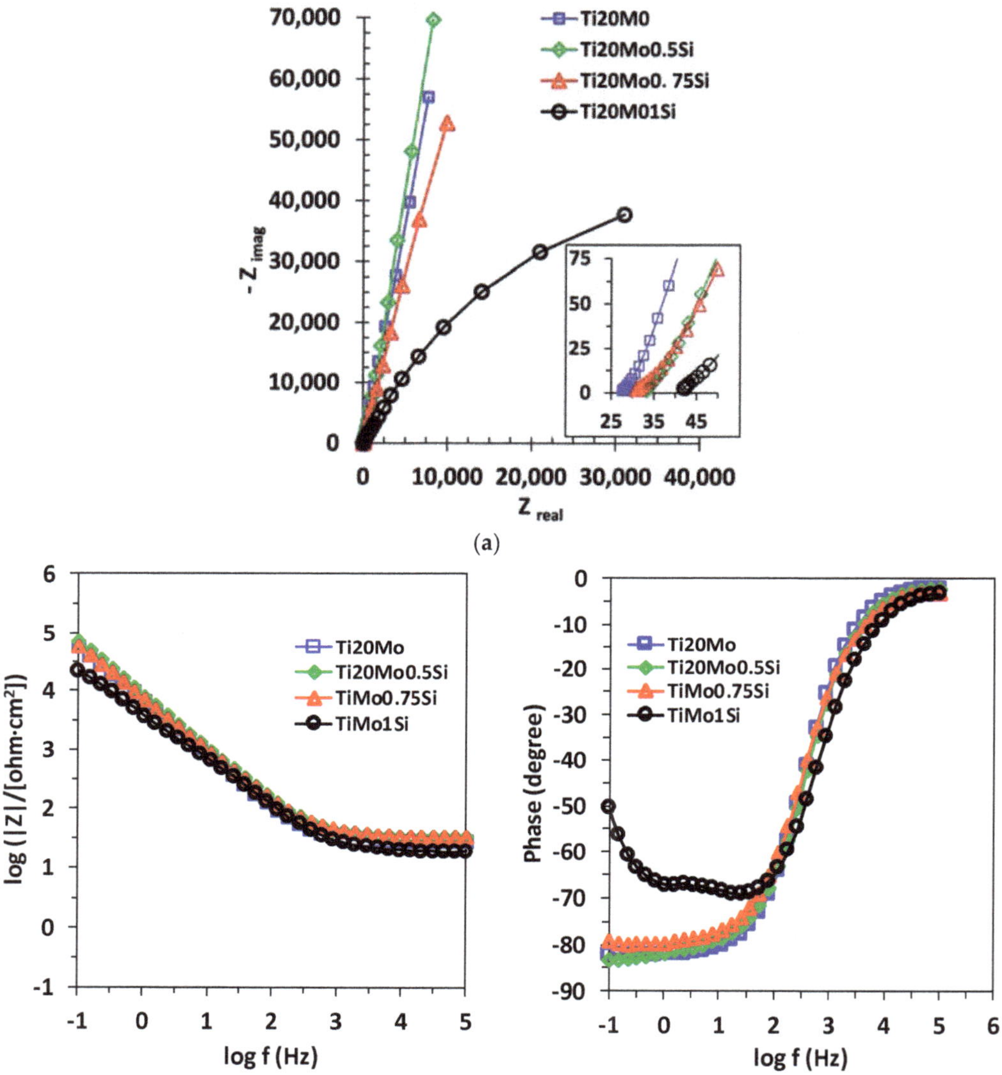

(a)

(b) (c)

Figure 5. (**a**) Nyquist,(**b**) Bode—IZI and (**c**) Bode—phase diagrams for Ti20Mo, Ti20Mo0.5Si, Ti20Mo0.75Si and Ti20Mo1.0Si alloys in Ringer's solution.

The obtained spectra were interpreted by modeling the data with the equivalent circuits shown in Figure 6. The shape of the Bode spectra for the Ti20Mo1.0Si alloy is clearly different in comparison with the other alloys, suggesting a two-layer structure of the solution–alloy interface, as illustrated in Figure 6b.

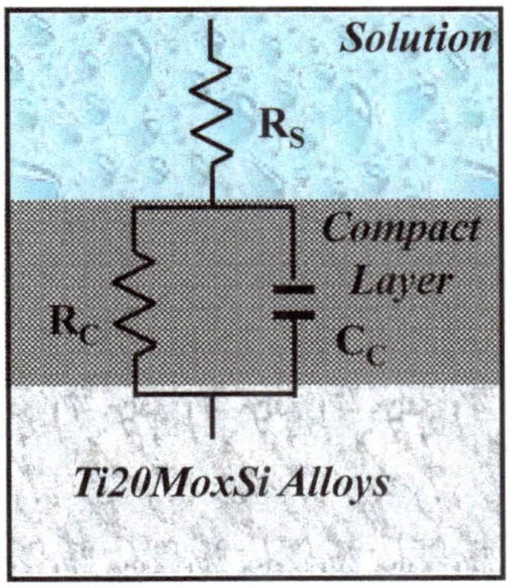

(a)

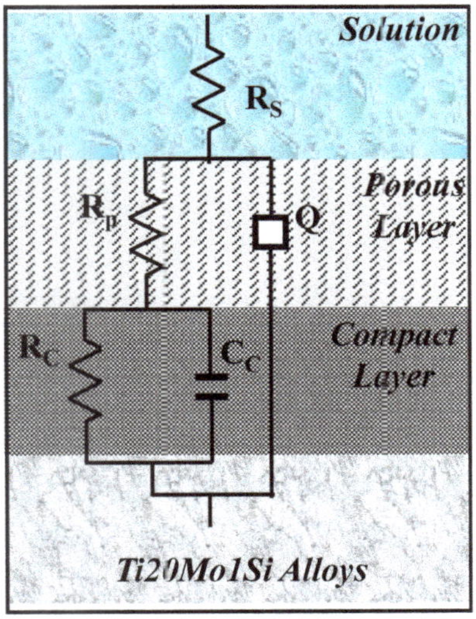

(b)

Figure 6. The equivalent circuit and the solution–alloy interface for the following: (**a**) Ti20Mo, Ti20Mo0.5Si, Ti20Mo0.75 Si; (**b**) Ti20Mo1.0Si.

The fitting results were consistent with the experimental data, and the chi-squared values were of the order 10^{-3}, see Table 5, which demonstrated that good simulation quality was achieved. As shown in the table, comparing Ti20Mo with Ti20Mo0.5Si, it can be observed that with the rise in Si content, the value of R_C increased, while the values of C decreased, suggesting that the corrosion resistance of the Ti20Mo alloy is improved due to the addition of Si.

Table 5. The values of the circuit for Ti20Mo, Ti20Mo0.5Si, Ti20Mo0.75 Si and Ti20Mo1.0Si alloys in Ringer solution.

Alloys	Circuit	R_s [Ω.cm²]	$10^5 \times C_C$ [Ss^n/cm²]	R_C [Ω.cm²]	$10^5 \times Y_0$ [F/cm²]	n_1	R_P [Ω.cm²]	$10^3 \times \chi^2$
Ti20Mo	R(CR)	30.47	1.85	$2.53 \cdot 10^5$	-	-	-	38.4
Ti20Mo0.5Si	R(CR)	37.93	1.45	$3.40 \cdot 10^5$	-	-	-	53.3
Ti20Mo0.75Si	R(CR)	39.22	1.61	$3.89 \cdot 10^5$	-	-	-	71.0
Ti20Mo1.0Si	R(Q(R(CR)))	18.88	1.07	$0.39 \cdot 10^5$	4.75	0.817	$1.81 \cdot 10^4$	0.63

Equivalent circuits that are associated with this physical model have also been proposed for anodized aluminum [27] and for passive films on titanium alloys [28–30].

In this equivalent circuit, R_S represents the resistance of the solution between the sample and the reference electrode, R_P—the resistance of the porous layer, and R_C—the resistance of the compact layer. In this case, in order to widen the scope of application of the model, instead of the ideal capacitance, a constant phase element (CPE) was introduced, the impedance of which is defined by the following relation:

$$Z_{CPE} = \frac{1}{Y_0(j\omega)^n} \tag{5}$$

where Q is a particular combination of circuit elements, which cannot be separated in the electrochemical system, mainly related to surface inhomogeneity and solid-phase mass transport. The introduction of such an element is recommended when, in the Nyquist diagram, the semicircle is open and incomplete, ω is the angular frequency, and j is the imaginary number ($j^2 = -1$). The exponent n is related to the slope of the linear portion in the log frequency – IZI graph of the Bode diagram and can take values between –1 and +1. If the exponent n = 0, CPE is an ideal resistor, while if n = 1, then CPE is an ideal capacitor. If the exponent n has the particular value n = 1/2 at high frequencies, the circuit element is called the diffusive Warburg impedance. Even if a phase constant element was used in this study, it was assimilated with a non-ideal capacitor (C) when $0.8 < n < 1$, and with a Warburg impedance when $0.5 < n < 0.65$.

4. Discussion

4.1. Linear Potentiodynamic Polarization

With the exception of two alloys, the passivation rates are of the same order of magnitude, with the lowest passivation rate being presented by the Ti20Mo1.0Si alloy.

In terms of quantity, the corrosion potentials of the Ti–Mo–Si samples studied are negative and relatively small, between 80 and 300 mV in Ringer's solution. The differences between the corrosion potentials of these samples (below 250 mV) are not significant; such differences may occur as a result of the difference in surface processing, contamination of the surface with organic substances, different aeration of the corrosion medium, temperature variations, etc.

The corrosion rates (actually passivation rates) are of the same order of magnitude, relatively small (15–30 μm/year), for all the alloys. It seems that the most stable alloy is Ti20Mo1.0Si.

The potentiodynamic analysis indicated that the addition of silicon minimized the anodic and cathodic reactions, thus reducing the corrosion current densities up to almost 50% (see Table 3). Also, the addition of silicon affected the displacement of the corrosion potential values to more positive values, with respect to the absence of Si, suggesting that the addition of silicon had a greater effect on the anodic reaction than on the cathodic reaction, by forming SiO_2 in the passive film. The Si–O bond is a strong covalent bond, and is helpful to achieve the loss mid-gap states and passivate the surface of the alloy.

In Ringer´s solution, the corrosion rate values also follow the same trend as the corrosion current density, with the values decreasing due to silicon addition. It is very important to report that the polarization resistance (Rp) values rise with the increase in Si concentration, recording a max. value of 21.55 $k\Omega \cdot cm^2$.

4.2. Cyclic Potentiodynamic Polarization

The study of the behavior of alloys at high potentials, by recording cyclic voltammograms at large potential domains, gives information on the behavior of alloys when they are immersed for a longer period of time.

The maximum current densities (corresponding to E = 2V) are very small, below 1 mA/cm^2, and the vast majority are below 0.5 mA/cm^2. Accordingly, the corrosion rates are equally low at 15–55 μm/year in Ringer's solution.

There is no regularity determined by the composition of the alloys, in terms of corrosion resistance, but it seems that the most resistant is Ti20Mo1.0Si.

In Ringer´s solution (see Table 4), we can observe that without Si, the alloy has a bigger corrosion current density than TiMoxSi alloys, and this can be due to the fact that the SiO_2 serves as a barrier against corrosion.

4.3. Electrochemical Impedance Spectroscopy

The shape of the alloys curves for the electrochemical impedance spectra indicates that all the alloys studied have similar behavior in Ringer's solution, with small differences depending on the composition of the alloy. The Nyquist curves, obtained at E_{corr}, indicate a capacitive behavior, characterized by incomplete semicircles, while the Bode representations show one or two constants for the relaxation time.

The obtained spectra, interpreted by modeling the data with an equivalent circuit with two time constants, suggest a two-layer structure of the passive film.

The fitting results were consistent with the experimental data, and the chi-squared values were of the order 10^{-3} (see Table 5), which demonstrated that good simulation quality was achieved. As the tables show, the addition of Si affects both the characteristics of the external porous passive layer (process dominated by ion diffusion) and the internal compact layer (process dominated by charge transfer).

There are no appreciable variations in the values of ohmic resistance of the solution, which means that no ions were released in the solution during the experiment. The values of R_C are higher than the R_P values, reflecting that the outer porous film exhibits lower resistance than the inner barrier layer.

It can be observed that the fitted values of C_C decrease with the addition of Si, indicating that the thickness and stability of the protective passive film increased due to SiO_2 formation.

5. Conclusions

Four new alloys from the Ti–Mo–Si system were obtained by vacuum arc remelting, and the corrosion behavior of these alloys was studied.

The analysis of corrosion performance was carried out by electrochemical methods (linear potentiodynamic polarization, cyclic potentiodynamic polarization, and electrochemical impedance spectroscopy) in artificial (simulated) physiological environments.

The polarization resistances show very low corrosion rates. The process taking place on the surface of all the alloys is the oxidation of titanium, with the development of an oxide layer on the surface (TiO_2), and Si additions improve the corrosion resistance of the alloys.

Cyclic potentiodynamic polarization revealed that there is no regularity determined by the composition of the alloys, in terms of corrosion resistance, and the most resistant alloy was Ti20Mo1.0Si.

The obtained EIS spectra show a compact passive layer for Ti20Mo, Ti20Mo0.5Si, and Ti20Mo0.75Si, and a two-layer structure of the solution/alloy interface (one porous and another compact passive film for Ti20Mo1.0Si).

It can be concluded that the silicon addition minimized the anodic and cathodic reactions, thus reducing the corrosion current densities up to almost 50% (see Table 3). Also, the addition of silicon affected the displacement of the corrosion potential values to more positive values, with respect to the absence of Si, suggesting that the addition of silicon had a greater effect on the anodic reaction than on the cathodic reaction, by forming SiO_2 in the passive film, which acted as a barrier against corrosion.

All the alloys have shown chemical stability and high corrosion resistance, and, from a chemical point of view, can be recommended to be used for future medical applications.

Author Contributions: M.S.B.: writing—original draft preparation, investigation, management; P.V.: conceptualization, validation, financing; A.V.S.: investigation, writing—review and editing; N.F.-S. and M.V.S.: methodology, investigation, data curation; J.C.M.-R.: methodology, visualization, investigation, data curation. All authors have read and agreed to the published version of the manuscript.

Funding: This research was funded by a publications grant of the Gheorghe Asachi Technical University of Iasi (TUIASI), project number GI/P6/2021 and by Gran Canaria Cabildo, project number CABINFR2019-07.

Conflicts of Interest: The authors declare no conflict of interest.

References

1. Verestiuc, L.; Spataru, M.C.; Baltatu, M.S.; Butnaru, M.; Solcan, C.; Sandu, A.V.; Voiculescu, I.; Geanta, V.; Vizureanu, P. New Ti–Mo–Si materials for bone prosthesis applications. *J. Mech. Behav. Biomed. Mater.* **2021**, *113*, 104198. [CrossRef]
2. Geetha, M.; Singh, A.K.; Asokamani, R.; Gogia, A.K. Ti based biomaterials, the ultimate choice for orthopaedic implants—A review. *Mater. Sci.* **2009**, *54*, 397–425. [CrossRef]
3. Niinomi, M.; Akiyama, S.; Ikeda, M.; Hagiwara, M.; Maruyama, K. Ti-2007 Science and Technology. *Jpn. Inst. Met.* **2007**, 1417–1424. Available online: https://cdn.ymaws.com/titanium.org/resource/resmgr/ZZ-WCTP2007-VOL2/2007_Vol_2_Pres_131.pdf (accessed on 23 June 2021).
4. Bombac, D.M.; Brojan, M.; Fajfar, P.; Kosel, F.; Turk, R. Review of materials in medical applications. *RMZ-Mater. Geoenviron.* **2007**, *54*, 471–499.
5. Chen, Q.; Thouas, G.A. Metallic implant biomaterials. *Mater. Sci. Eng. R* **2015**, *87*, 1–57. [CrossRef]
6. Wang, L.; Zhang, L.C. *Development and Application of Biomedical Titanium Alloys*; Bentham Science Publishers: Sharjah, United Arab Emirates, 2018. [CrossRef]
7. Niinomi, M. Mechanical properties of biomedical titanium alloys. *Mater. Sci. Eng. A* **1998**, *243*, 231–236. [CrossRef]
8. Breme, H.J.; Biehl, V.; Helsen, J.A. Metals and implants. In *Metals as Biomaterials*; John Wiley and Sons Ltd.: Hoboken, NJ, USA, 1998; ISBN 0471 969354.
9. Elias, C.N.; Lima, J.H.C.; Valiev, R.; Meyers, M.A. Biomedical applications of titanium and its alloys. *JOM* **2008**, *60*, 46–49. [CrossRef]
10. Sidambe, A.T. Biocompatibility of Advanced Manufactured Titanium Implants—A Review. *Materials* **2014**, *7*, 8168–8188. [CrossRef]
11. Smyrnova, K.V.; Pogrebnjak, A.D.; Kassenova, L.G. Structural Features and Properties of Biocompatible Ti-Based Alloys with β-Stabilizing Elements. In *Advances in Thin Films, Nanostructured Materials, and Coatings*; Lecture Notes in Mechanical Engineering; Pogrebnjak, A., Novosad, V., Eds.; Springer: Singapore, 2019. [CrossRef]
12. Baltatu, M.S.; Tugui, C.A.; Perju, M.C.; Benchea, M.; Spataru, M.C.; Sandu, A.V.; Vizureanu, P. Biocompatible Titanium Alloys used in Medical Applications. *Rev. Chim.* **2019**, *70*, 1302–1306. [CrossRef]
13. Niinomi, M. Titanium Alloys. In *Encyclopedia of Biomedical Engineering*; Elsevier: Amsterdam, The Netherlands, 2019; pp. 213–224. ISBN 9780128051443.
14. Eisenbarth, E.; Velten, D.; Müller, M.; Thullb, R.; Breme, J. Biocompatibility of β-stabilizing elements of titanium alloys. *Biomaterials* **2004**, *25*, 5705–5713. [CrossRef]
15. Beresnev, V.M.; Klimenko, S.A.; Sobol', O.V.; Litovchenko, S.V.; Pogrebnjak, A.D.; Srebnyuk, P.A.; Kolesnikov, D.A.; Meilekhov, A.A.; Postel'nik, A.A.; Nemchenko, U.S. Influence of the high-temperature annealing on the structure and mechanical properties of vacuum–arc coatings from Mo/(Ti + 6 wt % Si)N. *J. Superhard Mater.* **2017**, *39*, 172–177. [CrossRef]
16. Mareci, D.; Fernandez-Perez, B.M.; Trinca, L.C.; Fotea, L.; Souto, R.M. Electrochemical Investigation of the Corrosion Resistance of Ti20Mo Alloys in Simulated Physiological Solution with Added Proteins for Biomaterial Application. *Int. J. Electrochem. Sci.* **2016**, *11*, 6922–6932. [CrossRef]
17. Mareci, D.; Chelariu, R.; Dan, I.; Gordin, D.M.; Gloriant, T. Corrosion behaviour of β-Ti20Mo alloy in artificial saliva. *J. Mater. Sci. Mater. Med.* **2010**, *21*, 2907–2913. [CrossRef] [PubMed]
18. Mareci, D.; Chelariu, R.; Gordin, D.M.; Romas, M.; Sutiman, D.; Gloriant, T. Effect of Mo content on electrochemical behaviour of TiMo alloys for dental applications. *Mater. Corros. Werkst. Korros.* **2010**, *61*, 829–837. [CrossRef]
19. Zhou, Y.L.; Luo, D.M. Microstructures and mechanical properties of Ti–Mo alloys cold-rolled and heat treated. *Mater. Charact.* **2011**, *62*, 931–937. [CrossRef]
20. Jiang, Z.; Dai, X.; Middleton, H. Effect of silicon on corrosion resistance of Ti–Si alloys. *Mater. Sci. Eng. B* **2011**, *176*, 79–86. [CrossRef]
21. Minciună, M.G.; Vizureanu, P.; Geantă, V.; Voiculescu, I.; Sandu, A.V.; Achiței, D.C.; Vitalariu, A.M. Effect of Si on the mechanical properties of biomedical CoCrMo alloys. *Rev. Chim.* **2015**, *66*, 891–894.
22. Weng, W.; Biesiekierski, A.; Li, Y.; Wen, C. Effects of selected metallic and interstitial elements on the microstructure and mechanical properties of beta titanium alloys for orthopedic applications. *Materialia* **2019**, *6*, 100323. [CrossRef]
23. Baltatu, M.S.; Vizureanu, P.; Sandu, A.V.; Munteanu, C.; Istrate, B. Microstructural Analysis and Tribological Behavior of Ti-Based Alloys with a Ceramic Layer Using the Thermal Spray Method. *Coatings* **2020**, *10*, 1216. [CrossRef]
24. Zhang, L.C.; Chen, L.Y. A Review on Biomedical Titanium Alloys: Recent Progress and Prospect. *Adv. Eng. Mater.* **2019**, *21*, 1801215. [CrossRef]
25. Fatoba, O.S.; Jen, T.C.; Akinlabi, E.T. Experimental study on microstructural evolution, mechanical property, and corrosion behaviour of laser additive manufactured (LAM) titanium alloy grade 5. *Int. J. Adv. Manuf. Technol.* **2021**, *114*, 655–669. [CrossRef]
26. Ríos, M.L.; Perdomo, P.P.S.; Voiculescu, I.; Geanta, V.; Crăciun, V.; Boerasu, I.; Rosca, J.C.M. Effects of nickel content on the microstructure, microhardness and corrosion behavior of high-entropy AlCoCrFeNix alloys. *Sci. Rep.* **2020**, *10*, 21119. [CrossRef] [PubMed]
27. Hitzig, J.; Jüttner, K.; Lorenz, W.J.; Paatsch, W. AC-Impedance Measurement on Corroded Porous Aluminum Oxide Films. *J. Electrochem. Soc.* **1986**, *133*, 887–892. [CrossRef]

28. Perdomo, P.P.S.; Suarez, N.R.F.; Verdu-Vazquez, A.; Rosca, J.C.M. Comparative EIS study of titanium-based materials in high corrosive environments. *Int. J. Surf. Sci. Eng.* **2021**, *15*, 152–164. [CrossRef]
29. Bălțatu, M.S.; Vizureanu, P.; Mareci, D.; Burtan, L.C.; Chiruță, C.; Trincă, L.C. Effect of Ta on the electrochemical behavior of new TiMoZrTa alloys in artificial physiological solution simulating in vitro inflammatory conditions. *Mater. Corros. Werkst. Korros.* **2016**, *67*, 1314–1320. [CrossRef]
30. Oliveir, N.T.C.; Guastaldi, A.C. Electrochemical stability and corrosion resistance of Ti–Mo alloys for biomedical applications. *Acta Biomater.* **2009**, *5*, 399–405. [CrossRef]

materials

MDPI

Article

Potential of Rapid Tooling in Rapid Heat Cycle Molding: A Review

Nurul Hidayah Mohamad Huzaim [1], Shayfull Zamree Abd Rahim [1,2,*], Luqman Musa [2,3], Abdellah El-hadj Abdellah [4], Mohd Mustafa Al Bakri Abdullah [2,3], Allan Rennie [5], Rozyanti Rahman [2,3], Sebastian Garus [6], Katarzyna Błoch [7], Andrei Victor Sandu [8,9], Petrica Vizureanu [8,10,*] and Marcin Nabiałek [7]

1 Faculty of Mechanical Engineering and Technology, Universiti Malaysia Perlis, Arau 02600, Malaysia; hidayahhuzaim@gmail.com
2 Center of Excellence Geopolymer and Green Technology (CEGeoGTech), Universiti Malaysia Perlis, Kangar 01000, Malaysia; luqman@unimap.edu.my (L.M.); mustafa_albakri@unimap.edu.my (M.M.A.B.A.); rozyanty@unimap.edu.my (R.R.)
3 Faculty of Chemical Engineering and Technology, Universiti Malaysia Perlis, Kangar 01000, Malaysia
4 Laboratory of Mechanics, Physics and Mathematical Modelling (LMP2M), University of Medea, Medea 26000, Algeria; lmp2m_cum@yahoo.fr
5 Lancaster Product Development Unit, Engineering Department, Lancaster University, Lancaster LA1 4YW, UK; a.rennie@lancaster.ac.uk
6 Faculty of Mechanical Engineering and Computer Science, Częstochowa Univerity of Technology, 42-201 Częstochowa, Poland; sebastian.garus@pcz.pl
7 Department of Physics, Częstochowa University of Technology, 42-201 Częstochowa, Poland; katarzyna.bloch@pcz.pl (K.B.); nmarcell@wp.pl (M.N.)
8 Faculty of Materials Science and Engineering, Gheorghe Asachi Technical University of Iasi, 41 D. Mangeron Street, 700050 Iasi, Romania; sav@tuiasi.ro
9 Romanian Inventors Forum, Str. Sf. P. Movila 3, 700089 Iasi, Romania
10 Technical Sciences Academy of Romania, Dacia Blvd 26, 030167 Bucharest, Romania
* Correspondence: shayfull@unimap.edu.my (S.Z.A.R.); peviz@tuiasi.ro (P.V.)

Citation: Huzaim, N.H.M.; Rahim, S.Z.A.; Musa, L.; Abdellah, A.E.-h.; Abdullah, M.M.A.B.; Rennie, A.; Rahman, R.; Garus, S.; Błoch, K.; Sandu, A.V.; et al. Potential of Rapid Tooling in Rapid Heat Cycle Molding: A Review. *Materials* **2022**, *15*, 3725. https://doi.org/10.3390/ma15103725

Academic Editor: Nicolas Sbirrazzuoli

Received: 9 April 2022
Accepted: 12 May 2022
Published: 23 May 2022

Abstract: Rapid tooling (RT) and additive manufacturing (AM) are currently being used in several parts of industry, particularly in the development of new products. The demand for timely deliveries of low-cost products in a variety of geometrical patterns is continuing to increase year by year. Increased demand for low-cost materials and tooling, including RT, is driving the demand for plastic and rubber products, along with engineering and product manufacturers. The development of AM and RT technologies has led to significant improvements in the technologies, especially in testing performance for newly developed products prior to the fabrication of hard tooling and low-volume production. On the other hand, the rapid heating cycle molding (RHCM) injection method can be implemented to overcome product surface defects generated by conventional injection molding (CIM), since the surface gloss of the parts is significantly improved, and surface marks such as flow marks and weld marks are eliminated. The most important RHCM technique is rapid heating and cooling of the cavity surface, which somewhat improves part quality while also maximizing production efficiencies. RT is not just about making molds quickly; it also improves molding productivity. Therefore, as RT can also be used to produce products with low-volume production, there is a good potential to explore RHCM in RT. This paper reviews the implementation of RHCM in the molding industry, which has been well established and undergone improvement on the basis of different heating technologies. Lastly, this review also introduces future research opportunities regarding the potential of RT in the RHCM technique.

Keywords: rapid tooling; rapid heat cycle molding; additive manufacturing; injection molding process

1. Introduction

Time to market is a critical aspect of product development strategies [1–3]. In general, speed is balanced with other factors such as functionality, innovation, or quality [4,5].

Today, with the emergence of various new technologies, global competition in product development is developing rapidly. In addition, various industries are constantly in search for the newest competitive technologies in order to meet rising demand, reduce costs, and be able to produce products in small batches, while ensuring high quality and the capability to reach sustainable development goals [6–8]. The rapid growth of RT technology in today's global market requires rapid product development to replace conventional methods in order to improve manufacturing processes, particularly for tooling fabrication [7–9]. RT offers a faster manufacturing speed that reduces cost and saves project time, which is critical for successfully completing testing and moving into full production, as illustrated in Figure 1 [10].

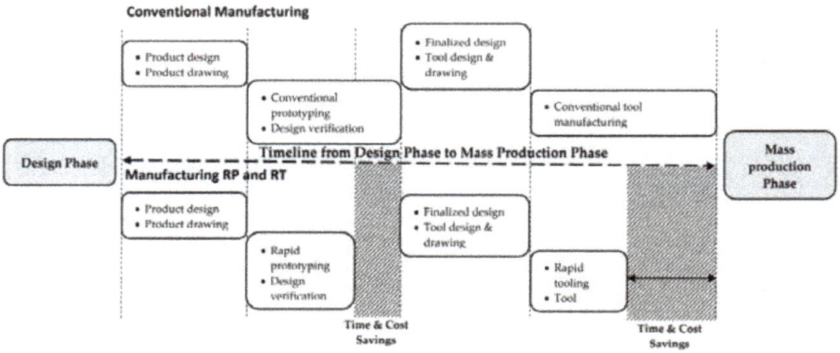

Figure 1. Summary of RT integration in manufacturing.

Regardless of the size of the industry, there will be times when immediate tooling is required to address certain issues. Moreover, an intermediate tooling system is required while producing a small quantity of functional prototypes to test the performance of the product designed. Such a small volume of product is generally used as a market sample, evaluation requirement, and production process design [11,12]. Although 3D printing technologies are available to produce the prototype, the properties of the parts produced are far from the properties of the parts produced using actual manufacturing processes for mass production. Therefore, RT is definitely required in order to evaluate the actual performance of the product designed, especially related to the material properties.

For any scientific research, functional tools or prototypes need to be launched prior to mass production. These are not launched in a mass volume for the consumers but in a small amount for the researchers [11,13]. RT is extremely beneficial for this situation, as it allows the launching of the products in a short time. Furthermore, since the manufacturing cost is low, users can get the products at a low price compared to the prices of those mass-produced using production tooling. This is why several new start-up or even big companies prefer using this tool to expand their revenue and achieve a competitive edge over fellow companies in the market.

Normally, mold-making or prototyping industries use mild steel or aluminum as material for the mold inserts in RT [14–16]. However, the use of these types of materials requires the same machining processes used to manufacture the production tooling such as computer numerical control (CNC), electrical discharge machining (EDM), and wire EDM machining, which are costly and time-consuming [6,17]. Recently, additive manufacturing (AM) was used to fabricate mold inserts for RT [9,18–21]. Typically, RT uses a model or prototype fabricated by AM technology as a pattern to form mold inserts, or mold inserts are directly manufactured through the AM method for a minimum number of prototypes [8,22]. In industry, there are variety of RT technologies available, such as the combined process of RT and AM to lessen the time for making the RT.

According to Huang et al. [23], RT differs from conventional tooling because the time required to manufacture tools is much quicker and it can be categorized as indirect or direct processes. Direct tooling techniques use AM processes to directly manufacture mold inserts without the need for a master pattern in the process [6,24–26]. The processes involved in direct tooling are selective laser sintering (SLS), direct metal laser sintering (DMLS), fused deposition modeling (FDM), stereolithography (SLA), jetted photopolymer (PolyJet), and 3D printing (3D-P) [6,8,27,28]. On the other hand, indirect processes use AM parts as the master pattern to make a mold for castings or plastic molding processes. These processes include cast metal tooling, silicone molding, spray metal tooling, and 3D KelTool tooling [6,8,9,18,29].

Generally, mold inserts for RT produced using AM technologies such as SLS, FDM, SLA, PolyJet, and 3D-P exhibit low thermal and mechanical properties; therefore, they are prone to failure throughout the molding process [30–32]. On the other hand, the DMLS process is not suitable for a simple plastic part design as it is time-consuming compared to machining processes. Moreover, for complex parts, DMLS has a few limitations for certain feature designs. In addition, past research on RT mold inserts discovered that the stress implemented in the mold inserts throughout the injection molding cycle has a significant impact on mold life [6,17,33–35].

On the other hand, a new technology called rapid heat cycle molding (RHCM) has been proven to eliminate weld line and molding defects in CIM, as well as the high glossy surface, by heating the core and/or cavity insert beyond the glass transition temperature of the plastic resin [36–38]. Thus, by raising the mold temperature to the glass transition temperature, weld lines can be considerably eliminated [36,39–42].

Table 1 shows various types of RCHM technologies investigated by researchers to improve weld line strength, tensile strength, and surface quality using various types of RHCM technology such as electromagnetic induction heating, induction heating, hot oil, a combination of hot oil and induction heating, steam heating (vapor chamber), and electric heating in hard/production tooling.

Table 1. RCHM Technologies.

No.	RHCM Technology	Researchers
1	Electromagnetic induction heating	Chen et al., 2006 [39]
2	Induction heating	Huang and Tai, 2009 [43]; Huang et al., 2010 [44]; Nian et al., 2014 [45]
3	Hot oil	Huang and Tai, 2009 [43]
4	Combination of hot oil and induction heating	Huang and Tai, 2009 [43]
5	Steam heating (vapor chamber)	Tsai, 2011 [46]; Wang, 2014 [47]
6	Electric heating	Li, 2016 [48]; Xie, 2017 [49]; Liu, 2020 [50]

Tosello et al. [51] investigated the impact of injection molding parameters on weld line formation and strength reduction. According to the studies, the most important factor to reduce weld lines is increasing the mold temperature. Wang et al. [52] analyzed the heat transfer during heating and cooling phases of RHCM process. The results of production showed that the RHCM process completely eliminates the weld lines on the surface of the parts, and the surface gloss of the parts is also greatly improved. The influence of the RHCM process on part surface quality has been studied [40,42,50,53,54]. The results showed that the high cavity surface temperature near to the polymer's glass transition temperature can assist in reducing surface roughness, significantly enhancing surface gloss, and eliminating the weld line. Studies have shown that, by using the RHCM process to eliminate weld lines, the surface roughness of molded parts can be significantly reduced. As a result, processes such as spraying and coating can be eliminated, reducing energy, material, and production costs while also protecting the environment [55–58]. Thus, it is being considered as a possible manufacturing technology in the plastic injection molding industry.

Therefore, this paper intends to provide a brief review on the role of RT technology in manufacturing methods of tooling toward its potential in the RHCM technique. The review of the literature included previous studies that were conducted on the application of RHCM in the injection molding process, which has been well established and continues to improve on the basis of different heating technologies. In addition, most of the studies proved that RT can be used to produce low-volume products. However, despite the potential advantages of RT in industry, there is a lack of studies and verification of the weld line and surface quality of plastic molded parts. Furthermore, the potential of RHCM in RT has not been explored, despite the use of RT in CIM having been investigated and research having been published. RT, on the other hand, is a great choice for low-volume production.

2. Injection Molding Process

The injection molding process is divided into several levels, i.e., operating cycles, including plasticization, filling, packing, and cooling [59–62]. The cooling stage is designated as the most significant of all stages due to its critical factors in developing productivity and molding quality, especially for complex molded parts [63–65]. Otherwise, defects such as sink marks, shrinkages, residual stresses, and warpage can occur if not addressed [66–70].

Initially, the plastic pellets are heated by screw shearing and a feed pipe before being plasticized to melt. The melt is then injected into the runner system to fill the mold cavities. After the cooling process, the molded part is ejected from the mold [38,71,72]. The quality of final molded parts is uncontrollable because it is influenced by many factors in the injection molding process. Some important factors which may have an impact on the finishing are the process parameters, mold design, and properties of plastic materials [52,73,74]. One of the factors that determines the quality of injection molded parts is the weld line [50,75,76]. The weld line refers to the process in which the liquid material is divided into two or more flows in the cavity during the injection process and merged together after a period of time [51,77–79].

There are five main phases in the injection molding process which are melting, filling, packing, cooling, and demolding [80–82]. It is desirable to keep the filling and packing phases at a high temperature to promote polymer melt flow and enhance reproducibility of the molded product [80,83,84]. A lower mold temperature is preferred during the subsequent cooling phase to allow for rapid cooling of the polymer melt, which occupies the majority of the molding cycle time [47,80,85,86]. As a result, to improve part quality and reduce molding cycle time, a dynamic mold temperature control system for rapid heating and cooling of the mold is required [80,83,87]. As shown in Figure 2, after the gate is solidified, the mold is cooled down rapidly by water to solidify and demold the polymer part [88]. However, in the CIM, the mold temperature control technique uses a concurrent cooling method, where a constant-temperature coolant circulates in a cooling channel to cool the mold and polymer melt [80,89,90]. Even though RHCM gives a longer cycle time, it improves the strength, eliminates the weld line, and improves the surface quality (high glossy surface) [38,91,92].

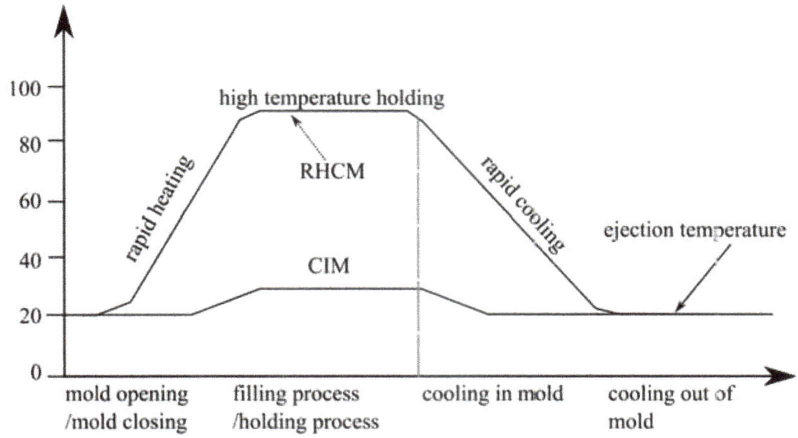

Figure 2. The cycle in the RHCM process [88].

3. Rapid Tooling

Today, with different kinds of new technologies, global product development competition is becoming increasingly fierce. In addition, industries are constantly on the lookout for the latest technology that can compete to meet future demands, reduce costs, and produce small batches of high-quality products while also being able to meet sustainability goals [93]. In today's market globalization, rapid tooling (RT) technologies have been steadily evolving with the demand for rapid product development in order to substitute conventional methods for better manufacturing processes, particularly for tool manufacturing [8,94,95].

RT is the AM technology that refers to the manufacturing methods of tooling [94,96,97]. RT utilizes the AM model as a pattern to construct a mold, or the mold for a small volume of prototypes is made directly from the AM process [8,27,98–100]. There are a variety of RT technologies accessible in the industry, such as the combination process of RT and AM, to lessen the time for making tools. Chua et al. [98] investigated several of RT and AM technologies using direct processes for tool production. Nevertheless, the majority of RT technologies implement indirect processes and models or master patterns produced by the AM process. According to Pouzada [101], there is an outline for the industry to take into consideration when choosing alternative production methods including RT, which includes a shorter delivery time, higher quality, shorter product development phase, and adaptation to global technology. This is a markedly new technique that has the potential to have a dramatic impact on the engineering practice, particularly during the product development stage, mostly in the casting and plastic molding industries [95,102]. As supported by Boparai et al. [27] and Mendible et al. [6], RT is also among the rapid prototyping (RP) variants, which have been proven to be cost-effective and time-efficient for the development of RT.

According to Pontes et al. [103], RT differs from conventional tooling in that it takes much less time to fabricate tooling and can be classified into indirect or direct processes. Direct tooling techniques use AM processes to directly manufacture the molds without requiring a master pattern. The processes involved in direct tooling are SLS, DMLS, FDM, SLA, PolyJet, and 3D-P [8,96]. On the other hand, indirect processes utilize AM as master patterns to make a mold for castings. These processes include metal casting, resin casting, silicone rubber molding, and 3D KelTool process [6,97,100]. Jurkovic et al. [104] introduced the emergence of products and the evolution phase of RT technology that began in the early 1990s. Product design takes a long time. The standards start from the construction of the workshop, involving applications of computer-aided construction and geometric models, the rapid development of computer-oriented products (modeling and simulation) and virtual product development, rapid prototyping, rapid tooling, and the product development digitization process.

As the quantity of RT technologies with numerous techniques has increased, there has been a trend to categorize them, as can be seen in Figure 3. Indirect tools, direct tools, and casting patterns can be distributed into the implemented groups [98]. After all, the division of these groups may change depending on AM's capability to speed up the production process. Some of the processes have been simplified with new technical capabilities, from indirect tooling to direct tooling, such as the use of mold inserts in an epoxy mold system rather than a conventional process [8]. In order to produce a good plastic injection mold, the selection of the material in mold making should be highly considered and understood.

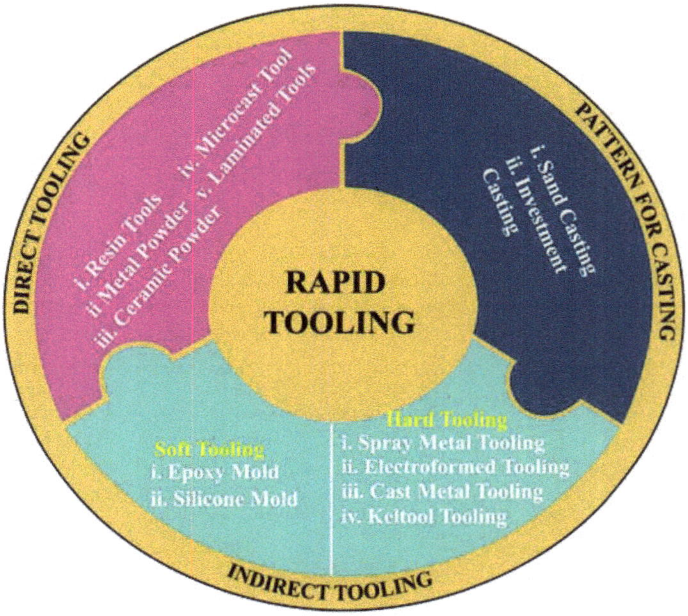

Figure 3. Classification of RT.

4. Materials for Injection Mold

Several criteria need to be considered when selecting materials for the injection mold, including cycle time, production volume, cost, maintenance, materials to be molded, part requirements, manufacturability, and life cycle costs [69,105,106]. Hardness, capability to achieve the requisite surface finish, corrosion resistance, wear resistance, thermal stability, and thermal conductivity are all important properties of a material to consider when designing an injection mold [69,107].

Metals are the most frequently used material in the industry due to high machinability, great variety of composition and properties, heat treatment possibility, and high thermal conductivity [69]. Steel and aluminum are often used, contingent on the project's requirements [69]. To ensure chemical resistance, strength, and hardness, steels are often made of chromium and nickel alloys, which make it a more cost-effective option [69]. On the other hand, aluminum alloy is a better thermal conductor than steel, reducing cycle time by up to 30% [69,107]. However, aluminum alloy molds only last around 2000 cycles, whereas steel molds last at least 50,000 cycles, making them sturdier than aluminum molds [69,107].

The mechanical properties of mold products do not change proportionally to the thermal conductivities or temperature of the mold material. The thermal conductivities of mold materials differ significantly, but changes in the mechanical properties of injected products from mold materials are barely noticeable. Steel mold material had four times the thermal conductivity of aluminum mold material in one study. When molding materials with different thermal conductivities were used, mechanical properties varied by 10–20% [108].

Usually, the mold is produced by a CNC machine with subtractive manufacturing. Harder materials including steel alloys necessitate specialized tools and further milling work, increasing the cost of the injection mold, as well as the cost of plastic parts to be produced [69].

4.1. Mold Base Material

A standard mold base is made up of four parts: the mold plate, the guide pin bushing, the return pin, and the screw. The mold plate is classified into two types according to its application: main plates (A plate and B plate) and structural plates (top clamping plate, bottom clamping plate, support plate, ejector plate, space block and runner stripper plate, etc.). Mold bases are made up of the A plate, B plate, and various structural plates assembled in a specific order [109].

The most common steel types used as mold base material are pre-hardened mold and holder steel, through hardening mold steel, and corrosion-resistant mold steel [110–112]. The pre-hardened mold and holder steels are mainly utilized for large molds, molds that do not require high wear resistance, extrusion dies, and high-strength holder plates [110–112]. These steels are usually supplied in a hardened and tempered state in the 270–400 HB range and do not require heat treatment [111]. Pre-hardened mold steel is usually utilized for large molds and medium production-run molds [111].

On the other hand, through hardened steels are mainly utilized for: long-term production, as well as specific molding materials with resistance to abrasion, high closing pressure or injection pressure, and suitability for high-pressure processes such as compression molding [111]. This type of steel is supplied in a soft annealed state. It normally undergoes rough machining, stress relieving, semifinished machining, hardening and tempering to the requisite hardness, finishing/grinding, and polishing or photoetching [111].

The choice of mold material may have an impact on the mold's performance [111]. Different people define performance differently in terms of mold life, plastic part quality, and productivity. Wear, surface defects, deformation, and corrosion wear may occur as a result of reinforced plastic or long service life, and surface defects may occur due to polishing or EDM defects in the mold manufacturing process [111], while a plastic part's quality is determined by its function, as well as its appearance. For highly polished molds, the choice of steel is significant. Steel should be clean and extremely low in impurities [111]. Due to the uneven temperature of the mold, tolerance may occur, which of course largely depends on the size and position of the cooling channel, as well as the mold material chosen. Mold materials with high thermal conductivity, such as aluminums or copper alloys, can be used in some cases. Sometimes, it is even possible to achieve productivity by selecting mold materials. The most obvious situation is to choose a material with high thermal conductivity [111].

The mold material requirements depend on the number of injections, the plastic material used, the mold size, and the desired surface finish; many different materials can be used. The following basic mold material properties must be considered: strength and hardness, toughness, wear resistance, cleanliness, corrosion resistance, and thermal conductivity [111].

A summary of the material properties commonly used as mold base materials is listed in Table 2. It can be seen that steel and aluminum are frequently used as mold base materials as they have high thermal conductivity and hardness, meeting the requirements of mold base materials in most studies [108,113,114].

Table 2. Physical and mechanical properties of mold base materials.

Type of Material	Density (g/cm^3)	Specific Heat (J/g·°C)	Thermal Conductivity (W/m·K)	Hardness (HB)	Authors
Aluminum (2000) series	2.78	0.869	139	135	Ozcelik et al. [108]
AISI (1020) steel	7.87	0.486	51.9	170	Ozcelik et al. [108]
Aluminum (QC-10)	2.85	N/A	159.12	150–170	Raus et al. [113]
Copper Alloy (C18000)	8.81	N/A	225	94	Raus et al. [113]
Steel (PC-20)	7.87	N/A	34.59	264–331	Raus et al. [113]
Carbon steel AISI 1050	7.85	0.486	49.8	196	Tang et al. [114]

4.2. Mold Insert Material

There are different types of mold inserts materials used in the fabrication of molds for the injection molding process. The fully hardened steel used for cavity and core inserts is typically held in a pre-hardened steel holder block (such as RoyAlloy or Ramax HH) [111]. Hardened molds or cavity inserts, for example, in the 48–60 Rockwell C range, can improve wear resistance, deformation and indentation resistance, and polishability [111]. When using filled or reinforced plastic materials, better wear resistance is particularly important. The ability to resist deformation and indentation in the cavity, gate area, and parting line contributes to maintaining the quality of the part. When a high surface finish on the molded part is required, better polishability is significant [111].

Aluminum is also a preferred metal for the fabrication of industrial mold inserts, although its strength and wear resistance [115] lower the benefits of aluminum. A good alternative for quick prototyping mold inserts is aluminum 6061-T6. It can be used very quickly [115] and subjected to diamond turning [116]. Its low specific density makes the molding machine easy to manufacture, assemble, and install, consequently reducing the production costs [115]. However, its strength is lower than that of copper-coated beryllium or nickel steels [115].

Steel alloys are often used for the production of molds because of their resilience, reliable mold operation, and long service lives, whereas their high strength impedes machinability and increases production costs [115]. Polishing of steel molds is often necessary [117,118]. Moreover, the rapid tool wear requiring the use of a special vibration-assisted cutting systems to reduce the cutting strength and wear in the tool [119] prevents steel alloys from being turned using a single point-diamond tool. Coating a nickel layer on steels can improve processability, but it also increases the cost and time of manufacturing [115]. The materials used for mold inserts in CIM, RHCM, and RT are discussed in further detail in the next section.

4.2.1. Mold Inserts for Conventional Injection Molding

Mold making is an essential supplementary industry, and its related products correspond to more than 70% of consumer product components. The high demand for shorter design and manufacturing production lines, good dimensionality and product quality, and rapid changes in design has created bottlenecks in the mold industry [114]. It is a difficult process that necessitates the use of a skilled and experienced mold maker [114].

Mold maintenance requirements are reduced by ensuring that the core and cavity surfaces maintain their original finishing for long operation cycles. Classic stainless steel is the best choice for unacceptable production of rust and high hygiene requirements, such as in the medical industry, optical industry, and other industries requiring high-quality transparent parts [120]. Table 3 contains a summary of the physical and mechanical properties of mold insert materials commonly used in CIM. According to the summary in Table 3, steel is widely used as mold insert material in CIM due to its dimensional accuracy, surface finish, and productive capability required for plastic products [121].

Table 3. Physical and mechanical properties of mold insert materials for CIM.

Type of Material	Density (g/cm^3)	Specific Heat (J/g·°C)	Thermal Conductivity (W/m·K)	Authors
P20 mold steel	7.8	460	29	Nasir et al. [122]
AISI P20	7.85	460	34	Xiao and Huang [123]
P20 tool steel	7.86	N/A	41.5	Sateesh [124]
Tool steel SKD-61	7.76	460	25	Chiang and Chang [121]
NAK80 (pre-hardened steel)	7.72	N/A	41.3	Chung [125]
(45–55 HRC, STAVAX) (pre-hardened steel)	7.8	460	16	Okubo et al. [126]

4.2.2. Mold Inserts for Rapid Heat Cycle Molding

Mold materials must be specifically chosen due to the rapid heating and cooling characteristics of RHCM. Compared with CIM mold materials, there are at least three additional factors to consider when selecting materials for electric heating molds. First, as the temperature of a low-heat-capacity material changes easily, a mold material with this characteristic is preferable for electrically heated RHCM. Secondly, because the mold requires frequent heating and cooling, fatigue cracks caused by cyclic thermal stress are more likely to appear on the surface of the cavity compared to other molding processes. Therefore, mold materials with a low thermal expansion coefficient and high thermal fatigue strength are needed.

Furthermore, due to the high mold temperature, several corrosive gases such as hydrogen chloride (HCl), hydrogen fluoride (HF), and sulfur dioxide (SO_2) generated by the melt decomposition will corrode the surface of the cavity. Thus, the cavity/core material must be highly corrosion-resistant. High-thermal-conductivity materials, such as copper, copper alloys, and aluminum alloys, are often used as cooling plates for electric heating molds to enhance the cooling efficiency [127].

Steel is often chosen as the mold material. Steel has a lower heat transfer than aluminum plates (due to its higher density and lower conductivity), although steel is more commonly used in large series (often up to 106 cycles) because aluminum has poorer mechanical properties under injection pressure (not only strength but also stiffness) [128].

Table 4 summarizes the physical and mechanical properties of the mold insert used in the RHCM. A higher thermal conductivity of the cavity/core material results in higher heating and cooling efficiencies and a shorter molding cycle time [87].

Table 4. Physical and mechanical properties of mold insert materials for RHCM.

Type of Material	Density (g/cm^3)	Specific Heat (J/g·°C)	Thermal Conductivity (W/m·K)	Authors
Hot work tool steel (Vidar Superior)	7.78	460	30	Li et al. [55]
AISI P20	7.85	460	34	Wang et al. [87]

Table 4. *Cont.*

Type of Material	Density (g/cm^3)	Specific Heat (J/g·°C)	Thermal Conductivity (W/m·K)	Authors
AMPCO 940	8.71	380	208	Wang at al. [87]
CENA 1	7.78	495	22.9	Wang et al. [47]

4.2.3. Mold Inserts for Rapid Tooling

Rapid tooling is among the rapid prototyping applications in the manufacturing industry. It allows building molds for small batch production products quickly and at low cost. It can be divided into direct and indirect tooling, as well as hard and soft tooling. Direct tooling is attained directly through a rapid prototyping process which uses soft materials (such as stereolithography materials) [3,9,15,129–131]. Several other tools are made of hard materials, such as powder metal [9,132], while, in the indirect tooling method, a casting pattern is produced through a rapid prototyping process and then used to build the necessary tool. Aluminum-filled epoxy resin [103,133] is a prevalent soft material because of its ease of use in fabricating mold inserts. Silicon rubber [134] is mostly used to make indirect tools.

The tool life is the most important consideration for injection molds formed using the rapid tooling approach. Since rapid prototyping technologies have matured, tools directly produced by rapid prototyping machines can accurately and precisely depict all the specific details and features of the model. Nevertheless, certain soft rapid prototyping materials have lower thermal conductivity and are typically unable to withstand high injection pressure and melt temperature [134]. Therefore, the lifespan of the tool is shorter. Even though additional processes such as metal laser sintering can be used to coat soft materials with a layer of metal [135] to increase its hardness, they increase the difficulty of the manufacturing process. In contrast, epoxy resin is a commonly used material in indirect tooling method because it can be easily molded with the casting pattern. The addition of metal powder can significantly increase its hardness and thermal conductivity, thereby further extending its tool life. Nevertheless, it also makes the epoxy resin mold cavity brittle. Thus, tools created through the indirect tooling method do not last long [9]. Some of the studies focusing on rapid tooling are presented in Table 5.

Tomori et al. [136] investigated the impact of mold performance and part quality on the composite tooling board material formulation and processing parameters. A sample preparedness flow chart is displayed in Figure 4. The boards were made of three ingredients: RP4037 (fluid), RP4037 hardener, and silicon carbide (SiC) filler (powder). Three levels of tooling board formulations were chosen for the six molds: 28.5, 34.7 and 39.9 wt.% of the SiC filler, along with two cutting speed levels (1.00 and 1.66 m/s). The variable parameter of this study was cutting speed, whereas the surface roughness of molded parts was the response variable. Since no apparent mold damage was found, the physical structure of the mold was unaffected by SiC concentration and cutting speed. This finding concluded that SiC concentration in the mold greatly impacts the surface roughness of the molded parts. In addition, the flexural strength increased (58.75 to 66.49 MPa) as SiC filler concentration increased, following a pattern similar to the mold material's thermal conductivity. However, this investigation did not discuss the consequence of filler concentration on the weld line of molded parts.

Table 5. Research on mold inserts in RAPID TOOLING.

	Researcher	Epoxy Resin/Hardener	Particles/Fillers Used	Weight Percentage of Filler (wt.%)	Particle Size	Mechanical Test											
						Arithmetic Mean Roughness (Ra) (µm)	Flexural Strength (MPa)	Hardness Test (R_H)	Thermal Conductivity (W/m·K)	Fatigue Test	Tensile Strength (MPa)	Compressive Strength (MPa)	Vickers Hardness, (kgF/mm^2)	Shore D Hardness Test	Density (g/cm^3)	Thermal Diffusivity (mm^2/s)	Surface Roughness
1.	Tomori et al. (2004) [136]	• RP4037 (resin) • RP4037 (hardener)	• SiC	• 28.5 • 34.7 • 39.9	N/A	1.03 to 1.35	58.75 to 66.49	N/A	N/A	N/A	N/A	N/A	N/A	N/A	N/A	N/A	N/A
2.	Senthilkumar et al. (2012) [137]	• Araldite LY 556 (resin)	• Al	• 40 • 45 • 50 • 55 • 60	45–150 µm	N/A	N/A	69 to 89	3.97 to 5.39	15,786 to 734	N/A	N/A	N/A	N/A	N/A	N/A	N/A
3.	Srivastava and Verma (2015) [34]	• PL-411 (resin) • PH-861 (hardener)	• Cu • Al	• 1 • 5 • 8 • 10	N/A	N/A	N/A	N/A	N/A	N/A	<85 (pure epoxy)	Cu = 65 at 10 wt.%	Cu = 22.4 at 8 wt.%.	N/A	N/A	N/A	N/A
4.	Fernandes et al. (2016) [35]	• RenCast 436 (resin with Al filler) • Ren HY 150 (hardener)	• Al	• 21.4	N/A	N/A	N/A	N/A	N/A	N/A	• Steel AISI P20 inserts = 20.0 ± 4.5 • Epoxy resin/Al inserts = 22.0 ± 5.0	N/A	N/A	• Steel AISI P20 inserts = 66 ± 3.2 • Epoxy resin/Al inserts = 61 ± 1.6	N/A	N/A	N/A
5	Khushairi et al. (2017) [138]	• RenCast CW 47 (resin with Al filler) • Ren HY 33 (hardener)	• Brass • Cu	• 10 • 20 • 30	N/A	N/A	N/A	N/A	• Brass: 10% = 1.18, 20% = 1.21, 30% = 1.37 • Cu: 10% = 1.66, 20% = 1.73, 30% = 1.87	N/A	N/A	• Brass: 10% = 95.61, 20% = 93.23, 30% = 92.69 • Cu: 10% = 80.83, 20% = 81.51, 30% = 73.17	N/A	N/A	• Brass: 10% = 1.85, 20% = 2.01, 30% = 2.22 • Cu: 10% = 1.83, 20% = 1.96, 30% = 2.08	• Brass: 10% = 0.644, 20% = 0.657, 30% = 0.740 Cu: 10% = 0.837, 20% = 0.923, 30% = 1.112	N/A
6	Kuo and Lin (2019) [139]	• TE-375 (Al filled epoxy resin)	N/A	N/A	N/A	N/A	N/A	N/A	N/A	N/A	N/A	N/A	N/A	N/A	N/A	N/A	• Average microgroove depth of Al-filled epoxy resin was 90.5% • Average microgroove width of Al-filled epoxy resin was 98.9%

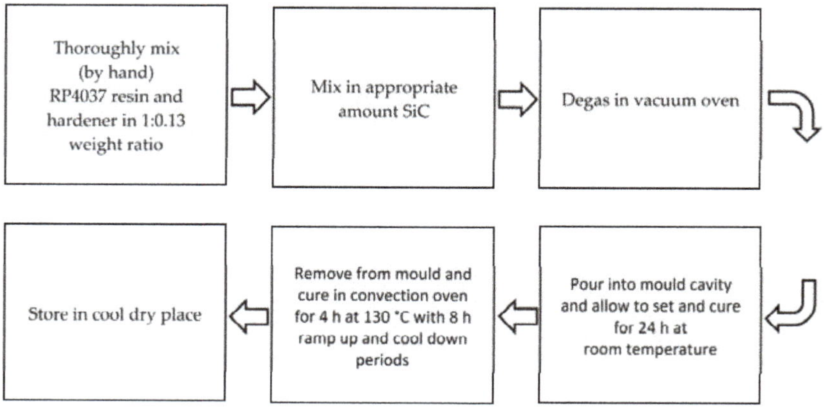

Figure 4. Procedure for tooling board preparation [136].

Senthilkumar et al. [137] studied the mechanical behavior of aluminum (Al) particles added with epoxy resin. A specimen was cast using various percentages of Al filler mixed into the epoxy resin. Optical microscopy results showed that the distribution of Al particles in the epoxy resin matrix was uniform. According to this finding, increasing the proportion of Al particles in epoxy resin matrix resulted in a significant increase in thermal conductivity (3.97 to 5.39 W/m·K) and hardness value (69 to 89 R_{HL}). However, fatigue life of the specimen decreased from 15,786 to 734 cycles as the percentage of Al in the epoxy resin increased. The best proportion to improve mold durability and performance was 45–55 wt.% Al filler particle. This parameter improved in terms of hardness, fatigue life, and thermal conductivity by 72 R_{HL}, 10,011 cycles, and 4.06 W/m·K, respectively. The fatigue life could be reduced by 36.58% for every 5 wt.% increase in Al filler particles, while the hardness value increased by 4.34%. Nonetheless, no further research has been conducted on flexural strength, compressive strength, tensile strength, and surface appearance of the produced molded parts.

Srivastava and Verma [34] carried out a study on the impact of particles on the mechanical properties of reinforced epoxy resin composites containing Copper (Cu) and Al particles. Cu and Al particles were mixed separately in epoxy resin as fillers in different compositions (1, 5, 8, and 10 wt.%). The mechanical test results showed that Al-reinforced epoxy resin has great tensile properties, i.e., 104.5 MPa at 1 wt.%, while Cu filler epoxy resin composites performed the best in the hardness test (22.4 kgF/mm^2 at 8 wt.%), with a compressive strength of 65 MPa at 10 wt.%. In addition, Cu-filled epoxy resin composites had a lower hardness than Al-filled composites, but still performed better than Al. This finding concluded that tensile strength and wear loss decreased steadily as filler content increased, whereas hardness, compressive strength, and friction coefficient increased as Cu and Al filler weight percentages increased. In short, the effect on the weld line on the surface of the molded parts was still lacking in this study.

Fernandes et al. [35] investigated the mechanical and dimensional features of molded PP injection parts for rapid tooling in epoxy resin/Al inserts. The circular geometrical component used for the work had a diameter of 140 mm and five central cavities connected by a 2 mm diameter segment. The runner was 60 mm long, the entrance diameter was 6.5 mm, and the draft angle is 2°. In this study, a new hybrid mold made of epoxy resin and Al was used to inject polypropylene (PP) parts to test the proposed mold. Moreover, similar parts were injected using an AISI P20 (ordinary mold) steel mold, just like the real application. While the tensile strength at yield of the parts injected with epoxy resin/Al inserts (22.0 ± 5.0 MPa) was higher than that of the parts injected with injection steel AISI P20 inserts (20.0 ± 4.5 MPa), this difference was not statistically significant. Other properties (ultimate tensile strength, elongation at break, and modulus of elasticity) in epoxy resin/Al-injected parts were lower than in steel AISI P20-injected parts. Furthermore, the Shore D hardness of parts molded with AISI P20 steel inserts increased by 8.5% compared to epoxy/Al inserts. The geometric deviation of parts injected with AISI P20 steel mold showed less shrinkage than parts injected using the epoxy/Al mold. According to these findings, epoxy/Al molding blocks could be a high-quality substitute for rapid tooling in the production of small series of products. Moreover, the findings with regard to the effect on the weld line of molded parts produced was not discussed in this investigation.

Khushairi et al. [138] analyzed the mechanical and thermal properties of various Al-filled epoxy compositions with the addition of Cu and brass fillers. Different compositions of brass and Cu filler were mixed in Al-filled epoxy (10, 20, and 30 wt.%). At the highest filler composition, the density of brass and Cu was 2.22 g/cm^3 and 2.08 g/cm^3, respectively. Adding 30% Cu fillers to an epoxy matrix resulted in the highest average thermal diffusivity (1.12 mm^2/s) and thermal conductivity (1.87 W/m·K), whereas adding brass had no effect on thermal properties. The compressive strength increased from 76.8 MPa to 93.2 MPa when 20% brass filler was used, and from 76.8 MPa to 80.8 MPa when 10% Cu filler was used. Due to the presence of porosity, further addition of metal fillers reduced the compressive strength. This finding concluded that fillers improve the mechanical, thermal, and density properties of Al-filled epoxy. Nonetheless, an extensive study on the surface appearance, particularly the weld line of the molded parts, is needed to analyze the quality of the molded part.

Kuo and Lin [139] investigated the manufacture of Fresnel lenses using rapid injection molding with liquid silicone rubber. The experiment was developed with RT and LSR parts for the development of a horizontal LSR molding equipment (Allrounder 370S 700–290, ARBURG). Al-filled epoxy resin could be used to produce injection molds for LSR injection molding. The Al-filled epoxy resin mold's average microgroove depth and width had replication rates of 90.5% and 98.9%, respectively. The average microgroove depth and width transcription rates of LSR molded parts were approximately 91.5% and 99.2%, respectively. The microgroove depth and width variations of LSR molded parts could be controlled within ±1 µm. After 200 LSR injector test runs, the average surface roughness of the Al-filled epoxy resins improved by approximately 12.5 nm. However, further tests on the tensile strength, compressive strength, hardness, and density, as well as observations of the weld line, are required to understand the impact of rapid injection molding on the proposed mold in terms of the quality of the molded parts.

According to the above review, several aspects (flexural strength, hardness, thermal conductivity, tensile strength, compressive strength, density, thermal diffusivity, and surface roughness) are important in manufacturing new potential mold inserts material for the injection molding process, as well as in the rapid tooling technique. However, most previous studies did not consider an observation of the surface appearance (weld line) of the molded parts when attempting to find the best proportion of materials in manufacturing mold inserts. Findings from previous research have shown that, by implementing a suitable material, mold inserts can achieve excellent machinability and high compressive strength, combined with sufficient toughness, good resistance to heat and wear, and high thermal conductivity. Nevertheless, the implementation of the rapid tooling technique in rapid heat cycle molding (RHCM) is yet to be investigated in terms of its contribution in the molded parts produced.

5. Rapid Heat Cycle Molding

In conventional injection molding, the mold temperature is significantly lower than the glass transition temperature or melt temperature of the polymer to reduce the molding cycle and enhance output efficiencies. Under such molding conditions, it is inevitable that the polymer melt will solidify prematurely during the filling stage, resulting in various molding defects such as short shots, flow marks, weld marks, jetting marks, swirl marks, and fiber-rich surfaces. In RHCM, rapid mold heating and cooling technology must be implemented to heat the cavity surface in a reasonably wide temperature range. Before performing melt filling in RHCM, the cavity surface is rapidly heated to a much higher temperature than CIM, typically above the glass transition temperature or polymer melt temperature. Therefore, the phenomenon of premature freezing of the polymer melt in the filling stage can be eliminated in the RHCM, thereby eliminating the frozen layer formed in the CIM. As a result, RHCM will effectively resolve the aforementioned inherent CIM molding defects. To shorten the molding cycle, the injection mold and shaped polymer melt cools faster in RHCM than in CIM after the filling stage [37].

Before the filling process in RHCM, the mold is rapidly heated to above the glass transition temperature, and then molten plastic is injected into the cavity. The mold is rapidly cooled after the filling stage to harden the molten plastic. After the cooling stage, the plastic parts are finally ejected [76]. The solid line in Figure 5 shows a demonstrative example of a mold temperature profile, where Tg denotes the material's glass transition temperature. The heating time is indicated by the symbol t_h. The injection, packaging, and cooling times are denoted by the letters t_{inj}, t_p, and t_c, respectively [76].

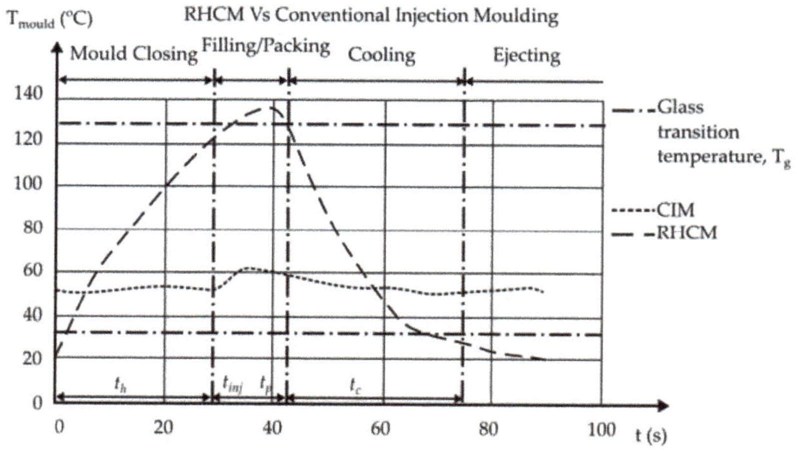

Figure 5. Example of mold temperature profile in RHCM [128].

6. RHCM versus Conventional Injection Molding

The rapid heating cycle molding (RHCM) injection method has been used to overcome product surface defects generated through conventional injection methods, such as weld marks and flow marks, by contrasting the advanced technology with conventional injection methods and products formed [56]. In this section, three comparisons are made in terms of the working process, mold design manufacture, and part quality.

The polymer melt is thoroughly combined during the injection molding process because RHCM uses high temperature, along with a rapid heating and cooling method. The melt flows much better than in the conventional injection method, with lower viscosity. The surface of the mold cavity is made with a high-gloss surface. As a result, the product surfaces are bright and smooth as a mirror, with no weld marks or flow marks.

In general, the conventional injection molding process consists of five stages, namely, the plasticization stage of the polymer, the injection stage, the filling stage, the cooling stage, and the ejection stage. Once the polymer has been plasticized into a molten state, it is injected at a higher pressure and speed into the mold cavity through a nozzle. Following the packing stage, the polymer melt is cooled to a low temperature using water. After that, the mold is opened, and an ejector pin is used to eject the polymer. At this point, the conventional injection molding process has completed one cycle before beginning the next cycle [56].

In contrast to CIM, RHCM injection technologies can be classified into six phases. The stationary mold is heated prior to injecting the polymer melt into the mold cavity.

During this stage, the surface of the mold cavity is heated to the injected polymer's glass transition temperature. The cavity is then filled with the polymer melt. The cavity temperature remains unchanged during the subsequent packing process until the cooling phase starts. Cooling water is utilized as a coolant to preserve the mold and polymer melt inside it to a certain low temperature. The plastic part can now be ejected to complete the RHCM injection cycle. The cavity surface is heated quickly again before the next injection process begins. Then, the next RHCM cycle of injection starts [56].

In RHCM, the initial mold cavity temperature is significantly higher than that in conventional molding, and the temperature is significantly different. However, since the mold temperature rises and falls in a short period of time, the molding cycle is nearly equal to that of a conventional injection molding process [55]. To shorten the molding cycle time, the heating stage for the RHCM mold, which requires heating of both the cavity and the core sides, can begin at the same time as part ejection. If only the cavity side of the mold needs to be heated, the mold can be heated as it opens. During the heating stage, the mold is heated to a preset high temperature, which is usually higher than the polymer's glass transition temperature [87]. Figure 6 shows the RHCM technology principle and the RHCM mold's temperature variation and injection cycles for conventional injection molding. In this figure, a is the heating stage, b is the injection and packing stage, c is the cooling stage, d is the mold opening and part ejection stage, and a', b', and c' are the next stages of the injection cycle, while line 1 represents the heat distortion temperature of polymer, line 2 is the mold temperature of RHCM, line 3 is the ejection temperature of the part, and line 4 is the mold temperature in the conventional injection method.

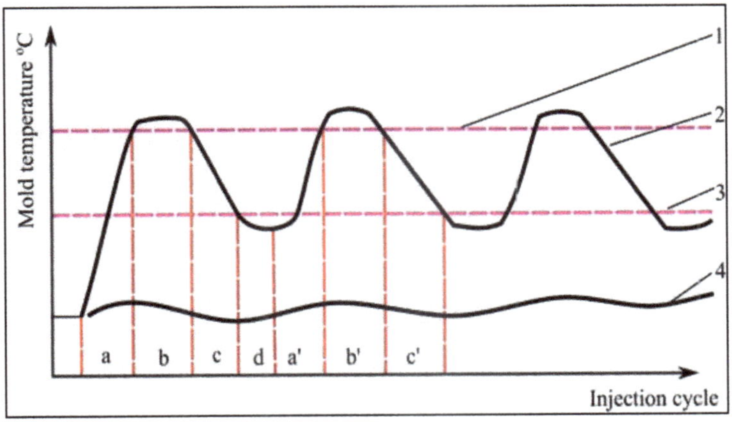

Figure 6. Mold temperature of RHCM technology and conventional injection method [56].

This figure shows the mold temperature changes during RHCM injection cycles, including the polymer's heat distortion temperature and the ejection temperature of part. Although conventional molds are still at the same temperature, the RHCM technology work process is more complex than the conventional injection process. Furthermore, the temperature control of RHCM molds is much more complicated than that in conventional molding, and it is normally controlled by conventional temperature controllers [56]

6.1. Application of RHCM in CIM

Some defects in plastic parts produced by conventional injection molding (CIM) can be solved by RHCM, such as flow marks, silver marks, jetting marks, weld marks, exposed fibers, and short shots [43]. However, RHCM is not a solution for all injection-molded part flaws. Warpage is one of the defects that cannot be solved using RHCM. Li et al. [88] investigated a method for predicting the warpage of crystalline parts molded using the RHCM process. To predict warpage, multilayer models with the same thicknesses as the skin–core structures in the molded parts were developed. The thicknesses of the layers varied with each molding process. The upper skin layer of each injection-molded part was heated on the stationary side. The prediction results were compared to the experimental results, which revealed that the average errors between predicted warpage and average experimental warpage were respectively 7.0%, 3.5%, and 4.4% in CIM, RHCM under a 60 °C heating mold (RHCM60), and RHCM under a 90 °C heating mold (RHCM90). Apart from that, the microstructure and temperature in CIM were symmetrical along the thickness direction and asymmetrical in RHCM. The predicted warpage was influenced by crystallinity, and the warpage predicted with crystallinity was greater than the warpage predicted without crystallinity, especially in RHCM-molded parts.

Furthermore, Li et al. [48] investigated the effect of the weld line on the tensile strength of RHCM and CIM components. Tensile testing results showed that, for specimens with a weld line, the strength of the RHCM specimen was significantly higher than that of the CIM specimen due to the smaller dimensions and higher bonding strength of the weld line. Moreover, the RHCM specimen's weld line was smaller than that of the CIM specimen. Furthermore, the two melt fronts that formed the weld line were welded together better due to the higher temperature and pressure in the RHCM process compared to the CIM process, resulting in a higher bonding strength of the weld line. The tensile strength of specimens without a weld line showed no discernible difference between the two molding processes. In addition, a thin surface layer of material containing the V-notch at the weld line was removed from the RHCM and CIM parts to investigate its effect on tensile properties. When compared to specimens with the original full thickness, thinned out specimens with a weld line showed improved tensile strength, while specimens without a weld line showed decreased tensile strength.

On the basis of the results from the previous researchers [48,88], it can be said that the application of RHCM in CIM has a big impact on the molded part. Previous studies used the same mold insert for both molding processes. The part molded by RHCM showed greater warpage predicted with crystallinity compared to that predicted without crystallinity. Additionally, the strength of the RHCM molded part was higher than that of the CIM molded part with the same shape and size, while the weld line of the RHCM molded part was smaller compared to that of the CIM molded part. Thus, it can be concluded that using RHCM in CIM yields a significant difference in the molded parts produced, demonstrating that RHCM produces a better result.

6.2. RHCM Technologies

Many studies have been performed to enhance the surface quality of plastic products by optimizing the process parameters. Contrary to popular belief, optimized parameters do not eliminate defects, but improve the surface quality of the molded parts. It was recently discovered that raising the mold surface temperature eliminates defects, increases flow length, and improves surface quality [43,87,140–144]. RHCM is an innovative technology that allows for dynamic mold temperature control during the injection molding process. RHCM technology requirements for the temperature of the heat distortion for the injecting polymer should be met before the mold cavity is injected. At the ejected temperature, it must be quickly cooled down. The difference in the mold temperature is significant. Therefore, if the cooling and heating methods are the same as in conventional methods, the production cycle must be extended. The RHCM mold is heated and cooled rapidly to ensure heating and cooling efficiency. Consequently, the structure of the mold differs from a conventional mold, as does the heating method. Companies such as Foreshot Industrial Corporation, Taoyuan City, Taiwan and Letoplast S.R.O., Letovice, Czech Republic have developed and applied RHCM technology, offering the benefits of a perfect product with a glossy surface that does not require painting [145,146]. Letoplast specializes in the production of visual and technical plastic parts made of PC-ABS, ABS, PP, PPE, PC, PA6, and other materials [146]. RHCM technology is used in plastic parts such as network communication equipment computers/communications/consumer electronics, appearance parts, and liquid crystal display televisions (LCD TVs) [145]. This technology allows engineers, technologists, or mold designers to avoid unnecessary premature melt freezing throughout the filling stage, which lowers the melt flow resistance and improves the molten plastic filling capability [38].

Chang and Hwang [147] proposed an infrared heating method for the mold cavity surface. For thermal surface condition assessment, a transient thermal simulation was developed. A change in the mold structure is not necessary with this method. This method has a high construction cost, and it is hard to consistently heat the longitudinal or complex mold cavity. Fischer et al. [148] investigated the effect of a locally different cooling behavior on the injection molding process of semicrystalline thermoplastics. An innovative dynamically tempered mold technology with different temperature zones was investigated, allowing the production of thin-wall components with locally different component properties. The preliminary results showed that, by influencing the inner component properties, significant differences in the optical and mechanical component properties could be achieved. On the other hand, Yao and Kim [142] evaluated the benefits of heating the mold surface through the use of a thin metal coating and thermal insulation. Copper poles were used to control the temperature by heating the thin metal layer. This method was shown to be efficient in terms of energy usage and temperature control. However, the complex coating and tiny parts of the cavity make this process more challenging; in addition, there are safety hazards, such as in the resistor layer's insulation, with the resistor having a limited life time. Table 6 shows some of the studies conducted on RHCM technology.

Chen et al. [39] studied the feasibility, efficiency, and effects of induction heating on the weld line surface appearance for dynamic mold surface temperature control. A simulation and experiment were used to analyze the induction heating with the spiral coil coinciding with the coolant cooling to evaluate its rapid heating/cooling ability and the uniformity of the mold plate surface temperature. Initially, ANSYS 3D thermal analysis was performed on a mold plate with a heating source to verify the analytical capabilities and accuracy. Then, induction heating was performed on the double-gate tensile sample mold to control the surface temperature of the acrylonitrile butadiene styrene (ABS) melt injection mold and determine its influence on the surface marking and weld line strength. From the results, it was indicated that it took 3 s for the center temperature of the plate to rise from 110 °C to 180 °C and 21 s to rise to 110 °C during induction heating and cooling, while, for the second heating/cooling cycle, it took 4 s to reach 200 °C and 21 s to return to 110 °C. The surface temperature of the mold plate could be elevated to around 22.5 °C and cooled to 4.3 °C. The study also proved that the temperature-controlled induction heating of the mold surface also contributed to the elimination of surface markings and to enhancing the strength of the weld line on injection molded ABS tensile bars. However, this study did not yield a positive correlation between the heating time by induction heating and the temperature distribution pattern on mold plates produced by the rapid tooling technique. Therefore, further investigation is required.

Table 6. Summary of research on RHCM technologies.

No.	Researcher	Plastic Material Used	Parameter Settings	Output Response	Technology Used	Material for Mold Inserts	Type of Analysis		Result
							Simulation	Experiment	
1	Chen et al., 2006 [39]	ABS	• Heating stage (110–180 °C and 110–200 °C) • Cooling stage (180–110 °C and 200–110 °C) • Mold temperature (50 °C)	• Surface marks • Weld line strength	Electromagnetic induction heating	AISI 4130 steel	ANSYS	Yes	• Heating times, 3–4 s for mold surface temperature to rise from 110 to 180 °C and 200 °C, along with 21 s for cooling time (return to 110 °C) • Eliminated the surface marks of the weld line and enhanced the strength of the related weld line
2	Huang and Tai, 2009 [43]	PMMA	• Mold temperature (25 °C) • Cooling time (20 s) • Melt temperature (260 °C)	• Mold surface temperature • Replication heights of LGP's microstructures • Residual stress in LGP	N/A	Not specified	No	Yes	• Induction heating the mold surface to 110 °C could increase the replication rate of the microstructure's height by up to 95% • There was no residual stress in the LGP produced by induction heating
			• Mold temperature (80 °C) • Cooling time (20 s) • Melt temperature (260 °C)		Hot oil		No	Yes	
			• Mold temperature (110 °C) • Cooling time (20 s) • Melt temperature (260 °C)		Combination of hot oil and induction heating		No	Yes	
			• Mold temperature (110, 130, and 150 °C) • Cooling time (20 s) • Melt temperature (260 °C)		Induction heating		No	Yes	

Table 6. *Cont.*

No.	Researcher	Plastic Material Used	Parameter Settings	Output Response	Technology Used	Material for Mold Inserts	Type of Analysis		Result
							Simulation	Experiment	
3	Huang et al., 2010 [44]	PMMA	• Injection speed (180–200 mm/s) • Packing pressure (1st stage 50–70 Mpa, 2nd stage 40 Mpa) • Packing time (4–8 s) • Mold temperature (60–80 °C) • Cooling time (30–40 s) • Mold surface temperature (110–150 °C)	• Power rates • Optimum processing parameters • Quality of microfeature heights and angles	Induction heating	Ni	No	Yes	• Optimum process parameters: injection speed (180 mm/s) packing pressure (70 Mpa), packing time (8 s), mold temperature (70 °C), cooling time (30 s), and mold surface temperature (150 °C). • Replication effect on microfeatures was significantly improved by induction heating
4	Tsai, 2011 [46]	ABS	• Two gates, cavity temperature = 75 °C, no vapor chamber • Two gates, cavity = 75 °C temperature, with vapor chamber • Two gates, cavity temperature = 110 °C, with vapor chamber • One gate, cavity temperature = 75 °C, no vapor chamber	• Tensile strength	Steam heating (vapor chamber)	P20 mould steel	No	Yes	• The two gate/vapor chamber system's tensile strength increased by 3.2% when preheating temperatures increased from 75 to 110 °C
5	Wang, 2013 [54]	• ABS/PPMA Fiber-reinforced plastic • PP + 20% glass fiber	• Heating time (10, 20, 30, 40, 50, and 60 s) • Cooling time (20,30,40, 50, and 60 s) • High- and low-temperature holding time (10 s)	• Weld line • Tensile strength	Electric heating (cartridge heater)	AISI H13	ANSYS	Yes	• RHCM process could improve the weld line factor for both materials • RHCM process reduced the tensile strength of the part without weld line

Table 6. *Cont.*

No.	Researcher	Plastic Material Used	Parameter Settings	Output Response	Technology Used	Material for Mold Inserts	Type of Analysis		Result
							Simulation	Experiment	
6	Wang, 2014 [47]	PC	• Mold heating time (18, 24, 25, and 36 s) • Mold cooling time (25, 32, 38, and 46 s)	• Weld line	Steam heating	CENA1	• Moldflow Insight • ANSYS	Yes	• Weld marks on the part surface could be significantly reduced by increasing the cavity surface's temperature just before filling • Surface gloss of product produced by RHCM was more than 90%.
7	Nian et al., 2014 [45]	Not specified	• Mold temperature (between 120 and 150 °C) • Thicknesses of the heated target (10–20 mm) • Pitch of the coil turns (10–20 mm) • Heating distance (5–9 mm) • Position of the induction coil (0–12 mm) • Working frequency (30–40 kHz) • Waiting time (2–6 s)	• Heating rate • Temperature difference	Induction heating	SKD61	COMSOL Multiphysics	Yes	• Heating rate was increased by 19.5%, from 3.3 °C/s to 4 °C/s • Heating uniformity was increased by 62.9%
8	Li, 2016 [48]	iPP	• Heated mold temperature for RHCM (120 °C) • Mold temperature for CIM (25 °C) • Packing pressure (50 Mpa) • Cooling time (30 s)	• Weld line • Tensile strength	Electric heating (electrical heating rods)	Not specified	N/A	Yes	• Weld line decreased tensile strength, but RHCM reduced the weld line's tensile strength reduction effect.

Table 6. *Cont.*

No.	Researcher	Plastic Material Used	Parameter Settings	Output Response	Technology Used	Material for Mold Inserts	Type of Analysis		Result
							Simulation	Experiment	
9	Xie, 2017 [49]	PP	• Silicon insert surface temperature (20, 60, 100, and 140 °C) • Melt temperature (230 °C) • Injection pressure (30 Mpa) • Injection speed (60 mm/s) • Screw back (20 mm) • Sample thickness (0.6 mm)	• Weld line	Electric heating (thin-film resistance heater, graphene coating)	Silicon insert (coated with carbide-bonded graphene)	N/A	Yes	• Width of weld lines: 16.4 μm at 20 °C, 11.24 μm at 60 °C, and 5.6 μm at 100 °C • Weld line disappeared completely at 140 °C
		PS	• Silicon insert surface temperature (20, 40, 80, and 100 °C) • Melt temperature (200 °C) • Injection pressure (30 Mpa) • Injection speed (5 mm/s) • Screw back (15 mm) • Sample thickness (1 mm)	• Residual internal stress • Replication fidelity				Yes	• Residual stress decreased as the surface temperature of silicon insert increased • Coating the silicon insert with carbide-bonded graphene could improve replication fidelity
10	Liu, 2020 [50]	PP with 30% short glass fiber	• Melt temperature (240 °C) • Injection pressure (60 Mpa) • Injection velocity (45%) • Packing time (9 s) • Packing pressure (50 Mpa) • Mold heating temperature (60/90/120 °C) • Cooling time (30 s)	• Microstructure • Tensile properties • Surface quality	Electric heating (electrical heating rods)	Not specified	Autodesk Moldflow	Yes	• Tensile strength of RHCM parts reached peak at 60 °C mold heating temperature • The sample's surface gloss increased as the mold cavity surface temperature increased, but decreased as the mold heating temperature increased above 90 °C

Huang and Tai [43] studied the use of induction heating to rapidly heat the mold surface during the injection molding process, thereby improving the microstructure replication effect of light-guided plates (LGP). First, oil was used to heat the mold to 80 °C, and then induction heating was used to heat the surface of the LGP mold above the glass transition temperature. Induction heating was implemented to achieve the high mold temperature essential for the short filling stage throughout the microstructure replication process of LGP, as well as to rapidly lower the mold temperature in the cooling process, allowing the molded part to cool faster. The heating coil was a single copper wire used for small area heating. The geometry of the induction coil is presented in Figure 7. The molded part of PMMA plastic was a flat LGP of 40 mm × 30 mm, with length and width of 1 mm. Results from these studies showed that the induction heating system only achieved a temperature uniformity of 6.7–12.7 °C. In addition, the combination of an oil heater (80 °C) and an induction heating system (up to 110 °C) offered the best replication performance, significantly raising the replication ratio up to 95%, which was 6% higher than when the mold was heated with an oil heater and 7.8% higher than when the mold was left unheated. Thus, these studies prove that induction heating can improve the LGP microstructure filling efficiency. However, in this study, only the performance of the heating technology when heating the mold surface was discussed, without further details on the strength and surface appearance (weld line) of the molded parts produced. Therefore, a further investigation of the correlation between heating technology and the properties of the molded part produced is required.

Huang et al. [44] used the same experimental setup as Huang and Tai [43] to perform rapid induction heating on the surface of a 2 inch LGP mold, but with a different coil design, using three-layered copper wire, to study the influence of high mold temperature in induction heating, as shown in Figure 8, to enhance the replication rate of LGP microfeatures. The temperature profiles of the Ni heating plate used as the mold insert in 2 inch LGP were investigated in order to identify the optimal power rate setting in the induction heating system; these temperature profiles correlated with the different power rates. Additional testing was carried out to determine the efficiency of the optimal process parameters defined by the Taguchi method. The ANOVA results indicated that the most significant parameters affecting the replication ability of microfeatures were the mold surface temperature and injection speed. As the surface temperature of the mold increased to 150 °C, the replication rate of the average nine-point height of the microfeatures and the average nine-point angle increased to 91.6% and 98.1%, respectively. However, the effects on the strength of the molded parts produced and the temperature distribution surface of the mold insert were not discussed in detail in this study.

Tsai et al. [46] reported the vapor chamber's effect on part tensile strength, which was made using plastic (ABS) material in an injection molding process. This experiment was developed with a copper heat pipe of the flat-plate type. Water was used as the working fluid. Then, to circulate the working fluid, a copper mesh formed using 50 μm wire with 100/200 mesh pore and a solid copper cylinder with a diameter of 2 mm were used to support the vacuum and loading forces. The experiment used the shape of the specimen shown in Figure 9 for tensile testing. The experiment was conducted with two distinct mold designs, one with a single gate and another with two opposite gates. The cavity temperature was used as a parameter in the experiment, while the output was the molded part's tensile strength. As a result of the weld line, the tensile strength of the test part formed by the two gate/no vapor chamber heating system was 11.1% lower than that of the test part formed by the one gate/no vapor chamber heating system (Figure 10). The two gate/vapor chamber system produced a fine weld line, as well as a tensile strength 6.8% higher than that of the two gate/no vapor chamber system. Furthermore, the tensile strength increased by 3.2% when the preheating temperature increased from 75 to 110 °C in the two gate/vapor chamber system. Due to the extended fluid flow, the use of the vapor chamber heating system and the increase in the preheating temperature seemingly improved the tensile strength of parts molded with two opposing gates. However, despite the fact that this study focused on conventional mold steel, additional research on the effect of heating on the mold insert produced by the rapid tooling technique should be considered.

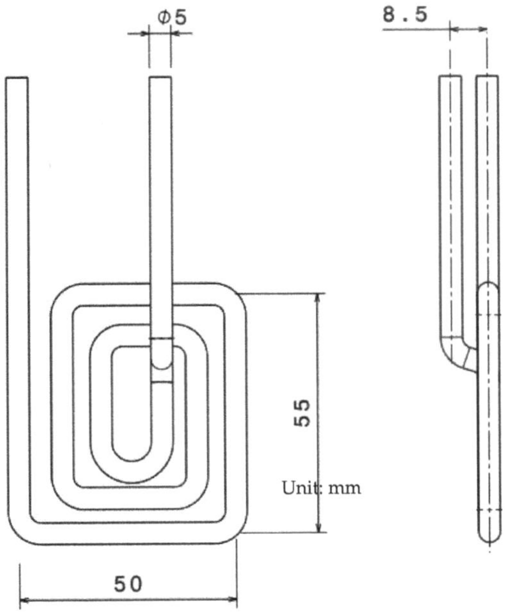

Figure 7. Induction coil's geometry and dimensions.

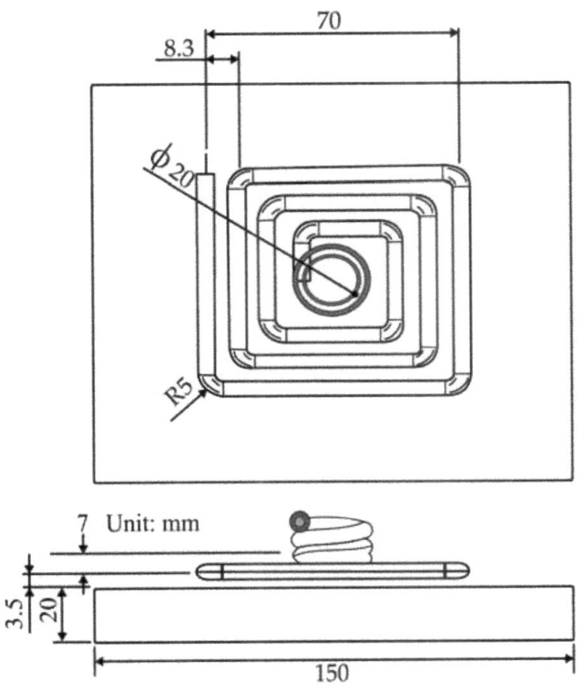

Figure 8. The geometry and dimensions of induction coil.

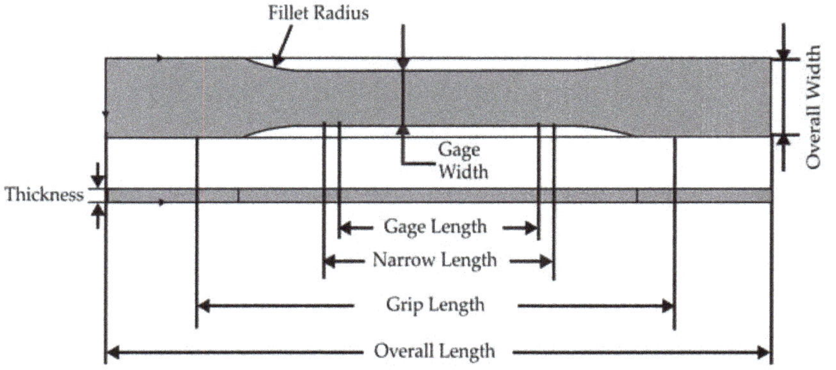

Figure 9. Specimen of tensile testing [149].

The heating and cooling system designs of electric heating molds are critical for RHCM with electric heating because they directly affect the heating/cooling efficiency and temperature uniformity, which have a decisive influence on the molding cycle and part quality [127]. Electric heating in RHCM can be accomplished through the deployment of electrical heating rods and cooling tunnels in the stationary mold, which can be rapidly heated via electric heating prior to filling and cooled via circulating water after filling [50,88]. In another study, Wang et al. [54] investigated the thermal response of the electric heating with the deployment of a cartridge heater in rapid heat cycle molding, as well as its impact on the surface appearance and tensile strength of the molded part. ABS/PPMA and fiber-reinforced PP were the materials used in the molded part experiment. The mold material used for the mold holder was HT 350, while that for the mold plate was AISI 1045, and that for the cavity plate was AISI H13. ANSYS thermal modules were used to investigate the temperature distribution and thermal response of the cavity surface. The thermal cycling process caused the surface temperature of the cavity to rise gradually. The maximum cavity surface temperature in thermal cycling could reach 185 °C after 30 experiment iterations with 60 s of mold heating time and 20 s of mold cooling time. This implied that the power density of the heater must be high enough to shorten the RHCM molding cycle by reducing the mold heating time. As the power density of the heater increased from 15 to 30 W/cm^2, the heating time of the cavity surface to be heated from 30 to 120 °C could be decreased from 58 to 19 s. Experimental results showed that, compared with CIM, the RHCM method reduced the tensile strength of parts without the weld line (Figure 11). In the case of ABS/PMMA plastics, the tensile strength of the weld line part was also significantly lowered due to RHCM. The RHCM method enhanced the weld line factor for ABS/PMMA and 20% glass fiber-reinforced PP. Findings were limited to the effect of RHCM heating technology on the cavity surface and the parts molded using conventional mold inserts. Nonetheless, more research into the effect of RHCM heating technology on mold inserts produced by rapid tooling technique is needed.

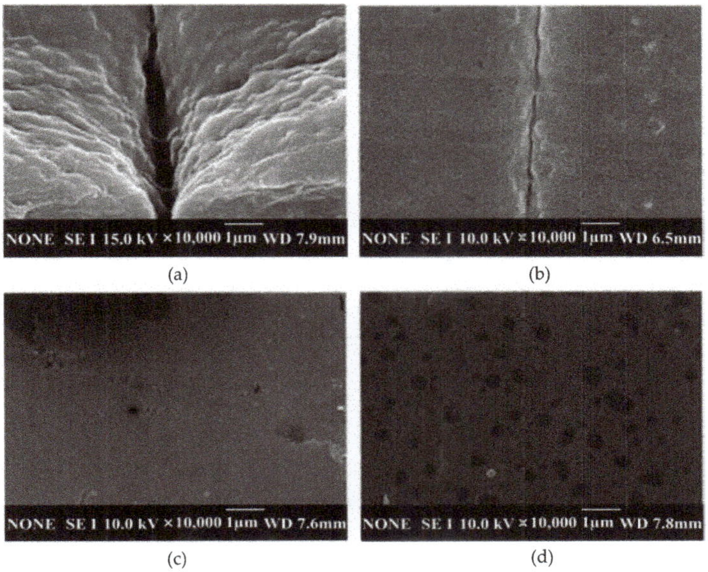

Figure 10. SEM images of one gate and two opposite gates: (**a**) two opposite gates at the temperature of 75 °C at point O without vapor chamber system; (**b**) two opposite gates at the temperature of 75 °C at point O with vapor chamber system; (**c**) two opposite gates at the temperature of 110 °C at point O with vapor chamber system; (**d**) one gate at the temperature of 75 °C at point O without vapor chamber system [46].

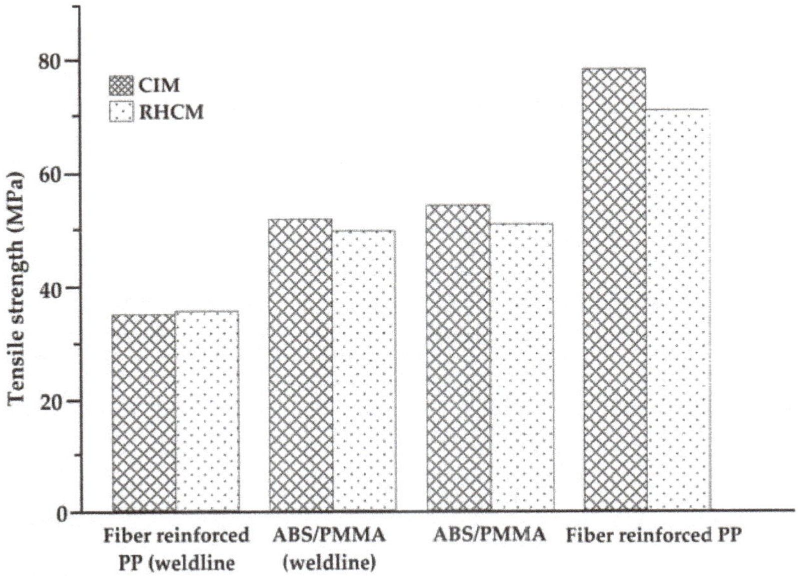

Figure 11. The tensile strength with and without the weld line of the CIM and RHCM parts.

Wang et al. [47] studied the design of heating/cooling channels for an automotive interior part, as well as its evaluation in rapid heat cycle molding. Polycarbonate (PC) material was used for the molded part. The general commercial ANSYS was used to simulate the transient thermal response of the mold cavity throughout the heating process to analyze the efficiencies of the original heating/cooling channel design. The designed heating/cooling systems with baffles could significantly improve the cavity surface thermal response efficiency. Steam was used at a temperature of 180 °C and a pressure of 1.0 Mpa. When experimental and simulation results were compared, the reasonable heat transfer coefficient was determined to be 14,000 W/m^2·°C. Heat exchange occurred between the cavity plate and the environment via natural air convection. The convective heat transfer coefficient was determined to be 25 W/m^2·°C. According to the results, the mold time constant could be lowered from 5 s to 2.5 s, resulting in a 50% decrease in time. Heating efficiency could be improved by 27.3%. The total surface temperature differences could be reduced to 20–30 °C. Moreover, by increasing the temperature of the cavity surface just prior to filling, the weld marks on the surface of the part could be significantly reduced (Figure 12). The weld marks could be completely removed as the cavity surface temperature approached a critical level. The critical cavity surface temperature of the plastic material used was approximately 130 °C. Findings from this study proved that the surface gloss of RHCM products was more than 90%. However, further studies on the implementation of RT in RHCM are required.

Nian et al. [45] studied the key parameters and optimized design of single-layer induction coils for external rapid mold surface heating by controlling the process parameters of induction heating (suitable for applications involving mold plates with different thicknesses and coil positions). Figure 13 depicts the spatial dimensions used in the simulation. An experiment was performed to validate the thickness of the heated workpiece (SKD61) and the induction coil (copper) position influencing the heating rate and temperature uniformity. Furthermore, the Taguchi method and principal component analysis were used to identify the best control factor combination to obtain high heating rates with low temperature deviations. According to the simulation results, the position of the induction coil and the thickness of the workpiece were indeed important design parameters. The optimal parameters indicated heating a 10 mm thick workpiece when designing a single-layer induction coil with a 15 mm turn pitch at a distance of 5 mm from the heating target. Additionally, the coil position should not be offset, and the operating frequency and waiting time must be set to 35 kHz and 6 s, correspondingly. Furthermore, the results of the experiments showed that an optimized design of the induction coil outperformed conventional single-layer coils in terms of temperature uniformity. The heating rate increased from 3.3 °C/s to 4 °C/s (an increase of 19.5%), and the heating uniformity increased by 62.9%. In terms of model validity, all MAPE values indicated high prediction accuracy. However, other important tests such as strength tests, as well as an observation of the appearance of the molded part produced by the induction heated mold surface were not performed.

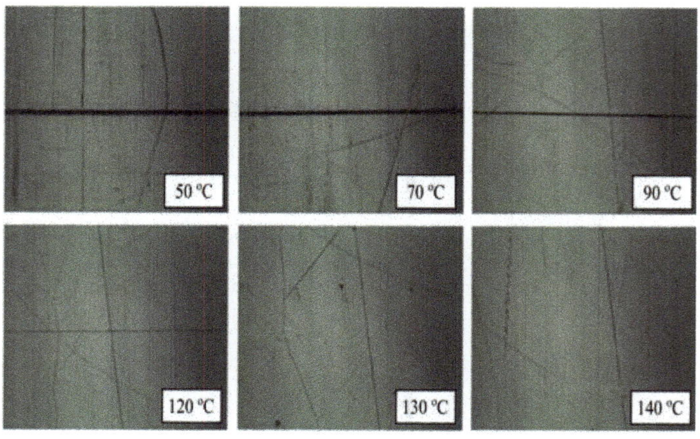

Figure 12. Micrograph of the weld marks or weld lines on the surface of the molded specimen under different cavity surface temperatures just before filling [47].

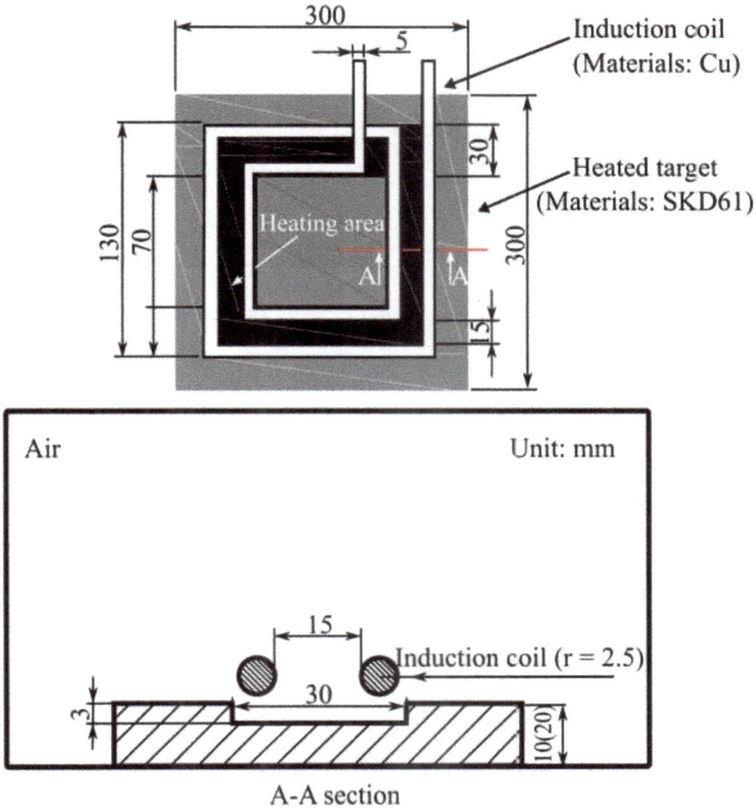

Figure 13. Physical model of induction heating [45].

Li et al. [48] investigated the effects of the weld line and its structure on tensile strength, as well as its process dependence, in CIM and RHCM. The specimens were made of isotactic polypropylene (iPP). Figure 14 shows the dimensions of the dumbbell-shaped specimen; the thickness was 2.5 mm. An RHCM mold with a maximum clamping force of 3800 kN, a maximum melt volume of 1239 cm^3, and a maximum injection pressure of 182 Mpa was used to mold the CIM and RHCM parts. Electrical heating rods were used to quickly heat the mold before filling it; after filling, cooling tunnels cooled the mold with circulating water. The only parameter used for the experiment was the mold temperature, and the output was the tensile strength of the weld line on the molded part. According to the results, the RHCM process reduced the dimensions of the weld line, particularly the width of the weld line adjacent to the heated stationary half. The tensile strength of dumbbell-shaped specimens cut from RHCM parts with and without a weld line is shown in Figure 15. Due to the presence of skin layers or weld lines on both RHCM and CIM specimens, their trends were similar. The RHCM specimen with the weld line had a higher tensile strength than the CIM specimen, but the strength without the weld line was remarkably similar. The weld line reduced the overall tensile strength, while RHCM reduced the decrease in tensile strength. Meanwhile, in the RHCM process, there was no discernible influence on the tensile strength of the specimens not having a weld line. The removal of the surface layer improved the tensile strength of specimens with weld lines but reduced it in specimens without a weld line. Findings from this study prove that RHCM can reduce the effects of weld line on the tensile strength. Nonetheless, further studies on the potential of RT in RHCM are yet to be conducted.

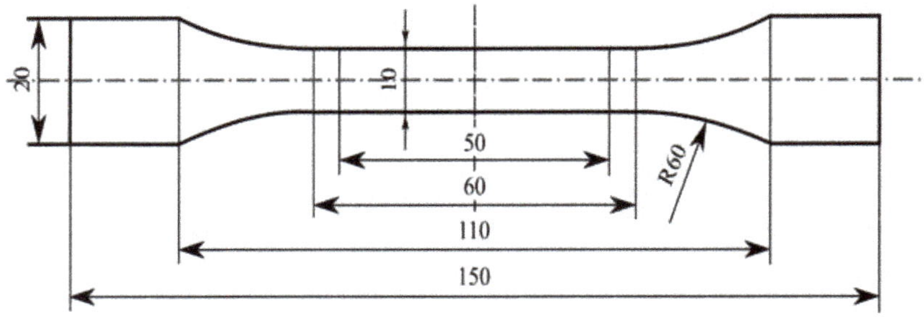

Figure 14. The dimensions of the dumbbell-shaped specimens [48].

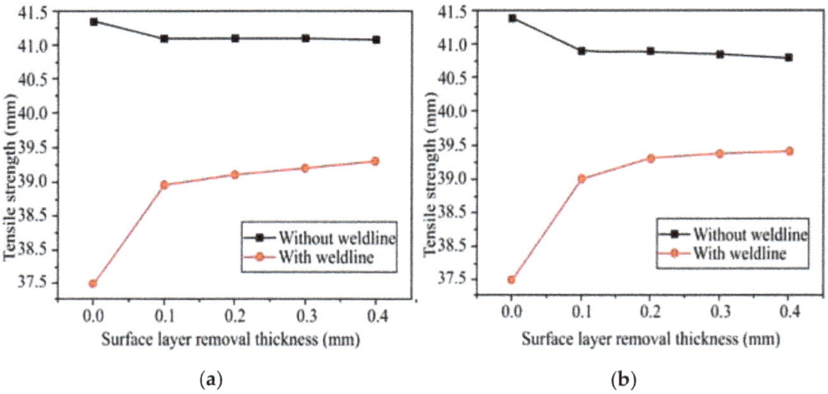

Figure 15. Tensile strength of RHCM parts: (**a**) when the surface layer was removed from the surface adjacent to the heated stationary half, and (**b**) when the surface layer was removed from the surface adjacent to the moving half of the surface [48].

Xie et al. [49] studied the thermal response of the graphene layer, as well as the impact of rapid thermal cooling with carbide bonded-graphene coating on molded product defects such as weld lines, internal stress, and replication fidelity. In this study, the heaters and heating channels implanted in the CIM were detached. They could be rapidly heated to a critical temperature by fine-tuning the voltage applied to the carbide-bonded graphene coating layer, or they could be cooled by a coolant channel. The semicrystalline polymer PP was used for the weld mark experiments, while PS was used in both the residual stress and the replication fidelity experiments. According to the results, with increasing voltage, the rate of heating of the insert surface increased. Once the voltage was set to 240 V, the coating heated up to 145.6 °C in 10 s, indicating that the average and transient heating rates could reach 11.6 °C/s and 16.1 °C/s, respectively. Therefore, the graphene coating could act as a thin-film cavity resistance heater, heating the polymer above Tg in seconds. Additionally, experimental results indicated that improving the surface temperature of the silicon insert could reduce the width of weld lines on molded products. If the temperature was high enough, these lines could also disappear entirely. Figure 16 shows that tensile strength and tensile elongation yields increased with silicon insert surface temperature. The average tensile strength and tensile elongation yields for silicon inserts at room temperature were 27.14 MPa and 128.30% respectively. However, at 140 °C, the average tensile strength and tensile elongation yields increased to 37.39 MPa and 468.77%, correspondingly. Findings from this study concluded that the rapid thermal cooling method could produce sample plates with uniform size thickness (600 μm) and mold products with fewer weld marks, less minimal internal stress, and enhanced replication fidelity. The tensile strength and tensile elongation yield of the products were increased by 37.77% and 265.11%, respectively, with even less energy consumption. Nonetheless, more research on the temperature distribution of the cavity surface of the mold insert is needed.

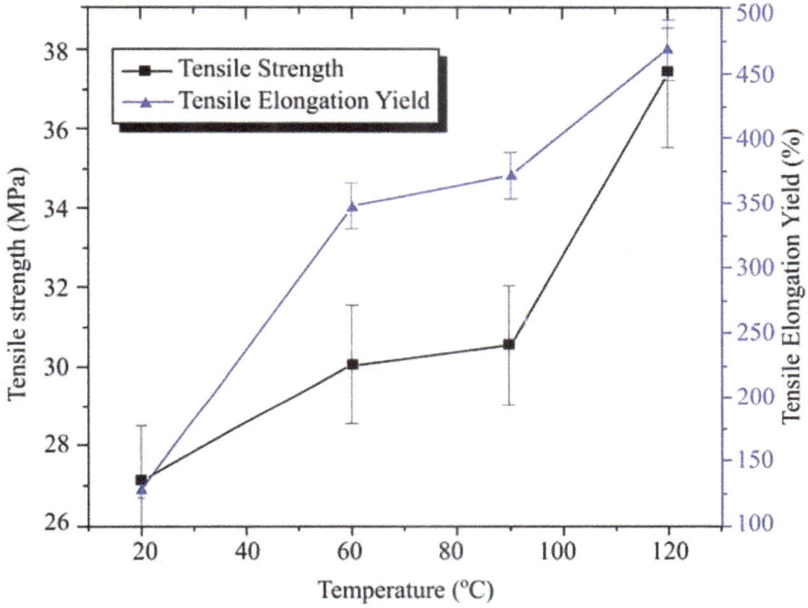

Figure 16. The mechanical properties of samples under different mold surface temperatures [49].

Liu et al. [50] studied the microstructure, tensile properties, and surface quality of glass fiber-reinforced PP parts molded by RHCM, using PP with 30% short glass fibers. The employed electric heating RHCM apparatus consisted of electric heating rods, a K-type thermocouple sensor, and a mold temperature controller MTS-32II system for measuring, regulating, and displaying the mold surface heating temperature. In the sample thickness direction, Autodesk Moldflow was used to evaluate the glass fiber orientation distribution. The simulation's material properties and boundary conditions were based on the CIM and RHCM experiment parameters. RHCM60, RHCM90, and RHCM120 were the sample mold part designations for the cavities heated to 60, 90, and 120 °C. The tensile tests were conducted according to ASTM D638, and the results showed that, as the cavity surface temperature increased prior to filling, the tensile strength increased first, before decreasing. Tensile strength was the highest in the RHCM60 sample, followed by the RHCM90 and CIM samples, while it was lowest in the RHCM120 sample (Figure 17). These variations were due to the microstructure of GFRPP composites being highly dependent on the cavity's surface temperature. As the cavity surface temperature increased, the thickness of the skin layer with medium fiber orientation along the flow direction decreased. RHCM samples had a thicker shear layer than CIM samples, with RHCM60 having the thickest shear layer. Therefore, the overall crystallinity and fiber orientation of RHCM60 in the cross-section were greater than those of the CIM sample. When an external force applied tensile load to the sample, the increase in fiber orientation caused more fibers to bear the load transmitted by the substrate. The increase in matrix crystallinity indicated an increase in the contact area between the crystal and amorphous regions, which aided in load transfer from the matrix to the fiber, resulting in RHCM60 having a higher tensile strength than the CIM sample. With a weak fiber orientation, the core layer thickness decreased initially, before increasing; the RHCM120 and RHCM60 samples had the thickest and thinnest core layers, respectively. It was also reported that the sample's surface gloss tended to increase as the surface temperature of the mold cavity increased, but the changes in surface gloss were noticeably reduced as the mold heating temperature increased above 90 °C. Furthermore, this study took into account the temperature distribution on the mold cavity surface of various mold types of mold inserts in CIM and RHCM.

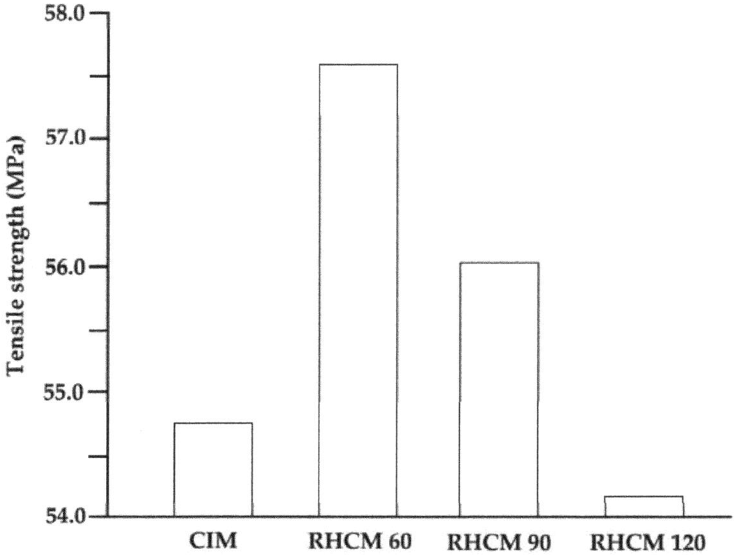

Figure 17. Tensile strength of RHCM and CIM samples.

According to the above review, RHCM has the potential to eliminate weld lines and improve surface gloss. The main distinction between RHCM and CIM is that the mold is heated and rapidly cooled to the ejection temperature after the filling process. The mold is heated above the molten plastic's glass transition temperature to achieve the optimum surface quality on the molded parts. In the reviewed studies, electric heating, induction heating and steam heating were extensively used as heating methods in RHCM in order to heat the mold in the quickest time. According to the research on parts produced by RHCM, it was found that defects caused by weld lines not only affect the surface appearance, but also affect the strength of the molded parts. A lower strength is obtained due to the stress concentration point between the V-notch and the weak bonding area, and the bonding interface strength at the weld line is not as strong as that in the matrix. In addition, different levels of mold surface temperature were introduced into RHCM research to obtain the optimum molded part quality. However, there is a lack of studies on the temperature distribution of the surface of the mold insert, which will give further information on the effect of heating on the parts produced. Nevertheless, most of the research only focused on using conventional materials as mold inserts, and there is still lack of support for and verification of the application of RT in producing mold inserts that are less expensive than conventional material molds in the RHCM method. The next section in this paper focuses on the effects of RCHM on molded parts. As confirmed in previous research, the RHCM method can increase the strength of the molded parts by reducing weld lines.

7. Effect of RHCM on the Molded Parts

Weld lines reduce the strength of injection molded parts; therefore, removing them is important for producing materials with appropriate structural integrity. However, this has not been implemented so far. Consequently, some researchers conducted studies to reduce the four adverse effects of weld lines. The following methods were proposed: optimizing the composition of the materials, improving the design of the molds, and changing the parameters of the injection molding process [39,150]. Furthermore, it is important to detect weld line defects because they may occur in unexpected areas. During the injection molding process, two or many flows meet, while the thickness and temperature may change. Destructive testing can identify these weld line flaws. If no obstacles exist, such as unforeseen circumstances in molding conditions (flux and temperature) or uneven materials during injection molding, for a large number of identical products, weld line defects are expected to occur at the same position [151]. Thermal assistance has become a popular technique for improving the mechanical properties and aesthetic properties of molded products during the injection molding process. RHCM is the most famous thermal assistance method to improve the weld line strength and surface defects (Table 7).

Table 7. Research on improved weld line strength performance and surface quality using RHCM.

No.	Authors	Plastic Material	Heating Technology	Parameters	Response	Results
1	Wang et al. [37]	• PS • PP • ABS • ABS/PMMA • ABS/PMMA/nano-C_aCO_3 • FRPP	• Electric heating	Cavity surface temperature, Tcs	Tensile strength	• The tensile strength for molded specimens without weld lines except for PP was decreased slightly and gradually
2	Zhao et al. [127]	• ABS	• Electric heating	Designed of RHCM mold structures	Weld marks	• The LCD TV panel did not have any surface defects such as flow marks and weld marks
3	Wang et al. [54]	• ABS/PPMA • Fiber-reinforced plastic, PP + 20% glass fiber	• Electric heating	Thermal responses	• Weld lines • Tensile strength	• The RHCM process could greatly increase the surface gloss of the part, especially for the fiber-reinforced plastics. • The RHCM process reduced the tensile strength of the part without weld line

Wang et al. [37] investigated the mechanical and aesthetic properties of molded parts by investigating the effects of cavity surface temperature during the filling phase of the injection molding process in RHCM. PS, ABS, PP, ABS/polymethylmethacrylate (ABS/PMMA), reinforced polypropylene glass fiber (FRPP), and ABS/PMMA/nano-C_aCO_3 were the materials used. The results showed that the tensile strength of PS, ABS, ABS/PMMA, ABS/PMMA/nano-C_aCO_3, and FRPP in the molded samples with or without weld lines gradually decreased, while the tensile strength of PP was restricted in a small range. Random fluctuations increased with the increase in T_{CS}. For the formed samples with weld lines, with the rise in T_{CS}, the tensile strengths of PS and PP gradually increased, while the tensile strengths of ABS, ABS/PMMA, ABS/PMMA/nano-C_aCO_3, and FRPP gradually decreased. Since its T_{CS} is higher than 100 °C, the tensile strength of FRPP with weld lines reduced gradually as T_{CS} increased.

Zhao et al. [127] investigated the efficiency of the electric heating technology applied during the RHCM process to heat the mold surface. Two different mold structures were designed and compared with and without a detached cooling plate for a large LCD TV panel. As a result, the mold's heating efficiency with a separate cooling plate was better than that of the mold without a cooling plate. It was discovered that the panel molded using the CIM process had visible weld marks as a result of the gating system's multiple gates and a low surface gloss. The electric heating RHCM process completely eliminated the weld marks on the surface of the panel and greatly improved the surface gloss. The high cavity surface temperature avoided premature melt freezing during filling and packing and improved polymer molecule chain entanglement for better aesthetics and weld strength.

Wang et al. [54] studied the impact of RHCM process on surface appearance and strength on the plastic parts. The cooling channel was placed perpendicular to the cartridge heater to enhance the structural strength and rigidity of the cavity plate, as shown in Figure 18. The results showed that the weld lines on the surface of RHCM parts vanished entirely compared to CIM. The jetting marks were also completely eliminated on RHCM part surface. Thus, compared to CIM, RHCM could greatly improve surface gloss, especially for fiber-reinforced plastics. Furthermore, compared to CIM, RHCM reduced the part's tensile strength. RHCM decreased the tensile strength of ABS/PMMA plastics by weld lines. RHCM increased the tensile strength of the parts with weld lines in fiber-reinforced PP. The RHCM process could increase the weld line factor of ABS/PMMA and 20% glass fiber-reinforced PP.

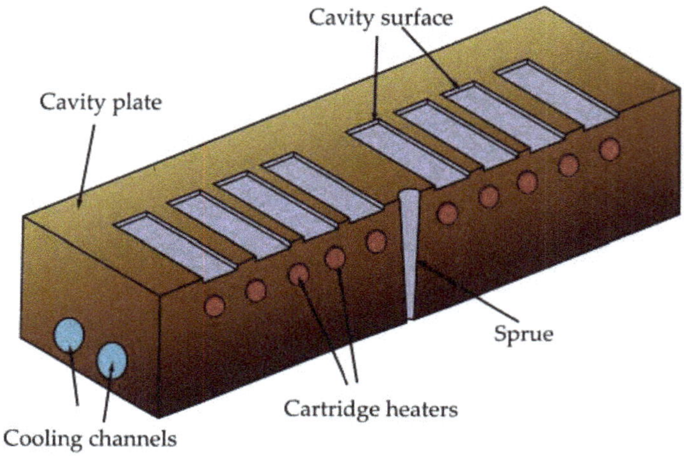

Figure 18. Location in RHCM for heaters and cooling line.

In conclusion, thermal assistance through the use of RHCM is an alternative to increasing the quality of the product by improving its mechanical and aesthetic properties. In order to obtain reliable and optimum results, more research on this technique needs to be conducted. For mold manufacturing and injection processes, complex mold configurations involve an additional cost. However, there are still limited studies available on molds fabricated using RT in RHCM to better understand the mechanical properties and aesthetic properties (surface appearance) of the molded part, as most studies focused on the use of conventional molds in the RHCM process.

8. Summary and Future Works

It can be concluded from the completed review that the RT technique has not been well established and implemented in RHCM technology. RT provides a faster manufacturing speed, which lowers costs and saves project time, both of which are critical for successfully completing testing and moving into full production. The review on various materials in fabricating mold inserts in RT was studied in terms of several aspects (flexural strength, hardness, thermal conductivity, tensile strength, compressive strength, density, thermal diffusivity, and surface roughness). The study found that, by using appropriate materials, the essential properties of mold insert materials were improved, including excellent machinability and high compressive strength, combined with sufficient toughness, good resistance to heat and wear, and high thermal conductivity. Furthermore, the review highlighted the ability of filler particles in epoxy resin to improve the properties of epoxy resin mold materials according to the application requirements (such as compressive strength, density, thermal diffusivity, and thermal conductivity).

Moreover, the effectiveness of RHCM in removing weld lines and improving cosmetic appearance of the product was discussed, and the efficiencies of conventional mold inserts in RHCM were also presented. This review highlighted the efficiency of RHCM with the application of different heating technologies in reducing weld lines and enhancing the strength of the molded parts produced. However, most of the research only focused on using conventional mold material as mold insert cavities, and there remains a lack of support for and verification of the application of RT technique in fabricating mold inserts involving a lower cost than conventional material molds. Considering the gaps in the literature, this review proposes the following future work:

- Previous researchers attempted to integrate new compositions of mold insert materials such as epoxy resin composites containing aluminum (Al) particles as filler to improve the mechanical properties of the injected parts. Thus, various types of metal fillers that can be used as fillers in epoxy resin composites should be studied. Studies should be aimed at looking for a new metal epoxy composites as the mold insert materials in RT with the best compositions of epoxy resin ratio and metal filler ratio.
- The performance of new metal epoxy composite designs in terms of flexural strength, hardness, thermal conductivity, tensile strength, compressive strength, density, thermal diffusivity, and surface roughness must be studied before they can be used as mold inserts in the injection molding process.
- Previous research on RHCM parts produced with conventional mold inserts discovered that the weld line was reduced while the strength was increased. Thus, there is a lot of potential in researching the use of mold inserts made by RT in RHCM technology to yield molded parts with good surface appearance and properties.
- Previous research did not yield a positive correlation between the heating time by induction heating and temperature distribution pattern on mold plates produced by the RT technique. Therefore, further investigation is required.
- Lastly, investigating the temperature distribution on the surface of the mold insert is critical in determining the effect of heating on the molded parts. Thus, further studies into the temperature distribution on the mold insert surface of different materials of mold inserts in RHCM should be conducted.

Author Contributions: Conceptualization, N.H.M.H., S.Z.A.R., M.M.A.B.A., A.E.-h.A. and A.R.; data curation, N.H.M.H., S.Z.A.R., M.M.A.B.A., A.E.-h.A. and A.R.; formal analysis, L.M., R.R., S.G., K.B., M.N., A.V.S. and P.V.; investigation, A.R., L.M., R.R., S.G., K.B. and M.N.; methodology, S.Z.A.R., M.M.A.B.A., A.E.-h.A. and A.R.; project administration, S.Z.A.R., L.M., M.M.A.B.A., R.R., A.V.S. and P.V.; validation, L.M., R.R., S.G., K.B. and M.N.; writing—review and editing, N.H.M.H., S.Z.A.R., M.M.A.B.A., A.E.-h.A., A.R., A.V.S. and P.V. All authors have read and agreed to the published version of the manuscript.

Funding: This study was supported by Center of Excellence Geopolymer and Green Technology (CEGeoGTECH) UniMAP and Faculty of Technology Mechanical Engineering, UniMAP. The authors wish to thank the Ministry of Education, Malaysia, for their financial support of this study through Fundamental Research Grant Scheme (FRGS), FRGS/1/2019/TK03/UNIMAP/02/20. This work was also supported by Gheorghe Asachi Technical University of Iasi TUIASI Internal Grants Program (GI_Publications/2021), financed by the Romanian Government.

Institutional Review Board Statement: Not applicable.

Informed Consent Statement: Not applicable.

Data Availability Statement: Not applicable.

Acknowledgments: We would like to acknowledge the reviewers for the helpful advice and comments provided. The authors wish to thank the Ministry of Education, Malaysia, for their financial support of this study through Fundamental Research Grant Scheme (FRGS), FRGS/1/2019/TK03/UNIMAP/02/20.

Conflicts of Interest: The authors declare no conflict of interest.

References

1. MacDonald, E.; Salas, R.; Espalin, D.; Perez, M.; Aguilera, E.; Muse, D.; Wicker, R.B. 3D Printing for the Rapid Prototyping of Structural Electronics. *IEEE Access* **2014**, *2*, 234–242. [CrossRef]
2. Matzler, K.; Hinterhuber, H.H. How to make product development projects more successful by integrating Kano's model of customer satisfaction into quality function deployment. *Technovation* **1998**, *18*, 25–38. [CrossRef]
3. Rahmati, S.; Dickens, P. Rapid tooling analysis of Stereolithography injection mould tooling. *Int. J. Mach. Tools Manuf.* **2007**, *47*, 740–747. [CrossRef]
4. Rodríguez-Pinto, J.; Carbonell, P.; Rodríguez-Escudero, A.I. Speed or quality? How the order of market entry influences the relationship between market orientation and new product performance. *Int. J. Res. Mark.* **2011**, *28*, 145–154. [CrossRef]
5. Rajapathirana, R.P.J.; Hui, Y. Relationship between innovation capability, innovation type, and firm performance. *J. Innov. Knowl.* **2018**, *3*, 44–55. [CrossRef]
6. Mendible, G.A.; Rulander, J.A.; Johnston, S.P. Comparative study of rapid and conventional tooling for plastics injection molding. *Rapid Prototyp. J.* **2017**, *23*, 344–352. [CrossRef]
7. Costabile, G.; Fera, M.; Fruggiero, F.; Lambiase, A.; Pham, D. Cost models of additive manufacturing: A literature review. *Int. J. Ind. Eng. Comput.* **2016**, *8*, 263–283. [CrossRef]
8. Equbal, A.; Sood, A.K.; Shamim, M. Rapid tooling: A major shift in tooling practice. *Manuf. Ind. Eng.* **2015**, *14*, 1–9. [CrossRef]
9. Rajaguru, J.C.; Duke, M.B.; Au, C. Development of rapid tooling by rapid prototyping technology and electroless nickel plating for low-volume production of plastic parts. *Int. J. Adv. Manuf. Technol.* **2015**, *78*, 31–40. [CrossRef]
10. Andrew, Y.C.; Nee, A. *Handbook of Manufacturing Engineering and Technology*; Springer: London, UK, 2015.
11. Berman, B. 3-D printing: The new industrial revolution. *Bus. Horiz.* **2012**, *55*, 155–162. [CrossRef]
12. Rayna, T.; Striukova, L. From rapid prototyping to home fabrication: How 3D printing is changing business model innovation. *Technol. Forecast. Soc. Chang.* **2016**, *102*, 214–224. [CrossRef]
13. Attaran, M. The rise of 3-D printing: The advantages of additive manufacturing over traditional manufacturing. *Bus. Horiz.* **2017**, *60*, 677–688. [CrossRef]
14. Volpato, N.; Solis, D.M.; Costa, C.A. An analysis of Digital ABS as a rapid tooling material for polymer injection moulding. *Int. J. Mater. Prod. Technol.* **2016**, *52*, 3–16. [CrossRef]
15. Harris, R.A.; Newlyn, H.; Dickens, P.M. Selection of mould design variables in direct stereolithography injection mould tooling. *Proc. Inst. Mech. Eng. Part B J. Eng. Manuf.* **2005**, *216*, 499–505. [CrossRef]
16. Altaf, K.; Qayyum, J.A.; Rani, A.M.A.; Ahmad, F.; Megat-Yusoff, P.S.M.; Baharom, M.; Aziz, A.R.A.; Jahanzaib, M.; German, R.M. Performance Analysis of Enhanced 3D Printed Polymer Molds for Metal Injection Molding Process. *Metals* **2018**, *8*, 433. [CrossRef]
17. Altaf, K.; Rani, A.M.A.; Ahmad, F.; Baharom, M.; Raghavan, V.R. Determining the effects of thermal conductivity on epoxy molds using profiled cooling channels with metal inserts. *J. Mech. Sci. Technol.* **2016**, *30*, 4901–4907. [CrossRef]
18. Thomas, P.A.; Aahlada, P.K.; Kiran, N.S.; Ivvala, J. A Review on Transition in the Manufacturing of Mechanical Components from Conventional Techniques to Rapid Casting Using Rapid Prototyping. *Mater. Today Proc.* **2018**, *5*, 11990–12002. [CrossRef]
19. Tuteski, O.; Kočov, A. Mold Design and Production by Using Additive Manufacturing (AM)–Present Status and Future Perspectives. *Industry* **2018**, *3*, 82–85. Available online: https://www.eos.info (accessed on 7 December 2021).
20. Bagalkot, A.; Pons, D.; Clucas, D.; Symons, D. A methodology for setting the injection moulding process parameters for polymer rapid tooling inserts. *Rapid Prototyp. J.* **2019**, *25*, 1493–1505. [CrossRef]
21. Legesse, F.; Kapil, S.; Vithasth, H.; Karunakaran, K. Additive manufacturing of H13 tooling element with conformal cooling channel using MIG cladding. *Int. J. Rapid Manuf.* **2018**, *7*, 1. [CrossRef]
22. Kuo, C.-C.; Li, M.-R. Development of sheet metal forming dies with excellent mechanical properties using additive manufacturing and rapid tooling technologies. *Int. J. Adv. Manuf. Technol.* **2017**, *90*, 21–25. [CrossRef]
23. Huang, Y.; Leu, M.C.; Mazumder, J.; Donmez, A. Additive Manufacturing: Current State, Future Potential, Gaps and Needs, and Recommendations. *J. Manuf. Sci. Eng.* **2015**, *137*, 014001. [CrossRef]
24. Kampker, A.; Ayvaz, P.; Lukas, G. Direct Polymer Additive Tooling-Economic Analysis of Additive Manufacturing Technologies for Fabrication of Polymer Tools for Injection Molding. *Key Eng. Mater.* **2020**, *843*, 9–18. [CrossRef]
25. Udroiu, R.; Braga, I.C. Polyjet technology applications for rapid tooling. *MATEC Web Conf.* **2017**, *112*, 03011. [CrossRef]
26. Kuo, C.-C.; Qiu, S.-X.; Lee, G.-Y.; Zhou, J.; He, H.-Q. Characterizations of polymer injection molding tools with conformal cooling channels fabricated by direct and indirect rapid tooling technologies. *Int. J. Adv. Manuf. Technol.* **2021**, *117*, 343–360. [CrossRef]
27. Boparai, K.S.; Singh, R. Development of Rapid Tooling Using Fused Deposition Modeling. *Addit. Manuf. Emerg. Mater.* **2018**, *1*, 251–277. [CrossRef]
28. Sathies, T.; Senthil, P.; Anoop, M.S. A review on advancements in applications of fused deposition modelling process. *Rapid Prototyp. J.* **2020**, *26*, 669–687. [CrossRef]
29. Afonso, D.; de Sousa, R.A.; Torcato, R.; Pires, L. Fundamentals of Rapid Tooling. In *Incremental Forming as a Rapid Tooling Process*; Springer Nature: Cham, Switzerland, 2019; pp. 1–22. [CrossRef]
30. Bagalkot, A.; Pons, D.; Symons, D.; Clucas, D. Analysis of Raised Feature Failures on 3D Printed Injection Moulds. *Polymers* **2021**, *13*, 1541. [CrossRef]

31. Krizsma, S.; Kovács, N.; Kovács, J.; Suplicz, A. In-situ monitoring of deformation in rapid prototyped injection molds. *Addit. Manuf.* **2021**, *42*, 102001. [CrossRef]
32. Zink, B.; Kovács, N.K.; Kovács, J.G. Thermal analysis based method development for novel rapid tooling applications. *Int. Commun. Heat Mass Transf.* **2019**, *108*, 104297. [CrossRef]
33. Sarkar, P.; Modak, N.; Sahoo, P. Mechanical and Tribological Characteristics of Aluminium Powder filled Glass Epoxy Composites. *Mater. Today Proc.* **2018**, *5*, 5496–5505. [CrossRef]
34. Srivastava, V.K.; Verma, A. Mechanical Behaviour of Copper and Aluminium Particles Reinforced Epoxy Resin Composites. *Am. J. Mater. Sci.* **2015**, *5*, 84–89. [CrossRef]
35. Fernandes, A.D.C.; Souza, A.F.D.; Howarth, J.L.L. Mechanical and dimensional characterisation of polypropylene injection moulded parts in epoxy resin / aluminium inserts for rapid tooling. *Int. J. Mater. Prod. Technol.* **2016**, *52*, 37–52. [CrossRef]
36. Park, K.; Sohn, D.-H.; Cho, K.-H. Eliminating weldlines of an injection-molded part with the aid of high-frequency induction heating. *J. Mech. Sci. Technol.* **2010**, *24*, 149–152. [CrossRef]
37. Wang, G.; Zhao, G.; Wang, X. Effects of cavity surface temperature on mechanical properties of specimens with and without a weld line in rapid heat cycle molding. *Mater. Des.* **2013**, *46*, 457–472. [CrossRef]
38. Shayfull, Z.; Sharif, S.; Zain, A.M.; Ghazali, M.F.; Saad, R.M. Potential of Conformal Cooling Channels in Rapid Heat Cycle Molding: A Review. *Adv. Polym. Technol.* **2014**, *33*, 1–24. [CrossRef]
39. Chen, S.-C.; Jong, W.-R.; Chang, J.-A. Dynamic mold surface temperature control using induction heating and its effects on the surface appearance of weld line. *J. Appl. Polym. Sci.* **2006**, *101*, 1174–1180. [CrossRef]
40. Zhang, A.; Zhao, G.; Guan, Y. Effects of mold cavity temperature on surface quality and mechanical properties of nanoparticle-filled polymer in rapid heat cycle molding. *J. Appl. Polym. Sci.* **2015**, *132*, 1–9. [CrossRef]
41. Yang, H.; Yilmaz, G.; Han, G.; Eriten, M.; Zhang, Z.; Yu, S.; Shi, M.; Yan, H.; Yang, W.; Xie, P.; et al. A quick response and tribologically durable graphene heater for rapid heat cycle molding and its applications in injection molding. *Appl. Therm. Eng.* **2020**, *167*, 114791. [CrossRef]
42. Su, Q.; Zhang, N.; Gilchrist, M.D. The use of variotherm systems for microinjection molding. *J. Appl. Polym. Sci.* **2016**, *133*, 1–17. [CrossRef]
43. Huang, M.-S.; Tai, N.-S. Experimental Rapid Surface Heating by Induction for Micro-Injection Molding of Light-Guided Plates Ming-Shyan. *J. Appl. Polym. Sci.* **2009**, *113*, 1345–1354. [CrossRef]
44. Huang, M.-S.; Yu, J.-C.; Lin, Y.-Z. Effect of rapid mold surface inducting heating on the replication ability of microinjection molding light-guided plates with V-grooved microfeatures. *J. Appl. Polym. Sci.* **2010**, *118*, 3058–3065. [CrossRef]
45. Nian, S.-C.; Tsai, S.-W.; Huang, M.-S.; Huang, R.-C.; Chen, C.-H. Key parameters and optimal design of a single-layered induction coil for external rapid mold surface heating. *Int. Commun. Heat Mass Transf.* **2014**, *57*, 109–117. [CrossRef]
46. Tsai, Y.-P.; Wang, J.-C.; Hsu, R.-Q. The effect of vapor chamber in an injection molding process on part tensile strength. *Exp. Tech.* **2011**, *35*, 60–64. [CrossRef]
47. Wang, G.-L.; Zhao, G.-Q.; Wang, X.-X. Heating/cooling channels design for an automotive interior part and its evaluation in rapid heat cycle molding. *Mater. Des.* **2014**, *59*, 310–322. [CrossRef]
48. Li, J.; Yang, S.; Turng, L.S.; Xie, Z.; Jiang, S. Comparative study of weldline strength in conventional injection molding and rapid heat cycle molding. *Mater. Plast.* **2016**, *53*, 448–453.
49. Xie, P.; Yang, H.; Zhao, Y.; Yu, W.; Cheng, L.; Yang, W.; Yan, H.; Tan, J. Carbide-bonded graphene coating of mold insert for rapid thermal cycling in injection molding. *Appl. Therm. Eng.* **2017**, *122*, 19–26. [CrossRef]
50. Liu, F.; Li, T.; Xu, F.; Li, J.; Jiang, S. Microstructure, Tensile Property, and Surface Quality of Glass Fiber-Reinforced Polypropylene Parts Molded by Rapid Heat Cycle Molding. *Adv. Polym. Technol.* **2020**, *2020*, 3161068. [CrossRef]
51. Tosello, G.; Gava, A.; Hansen, H.N.; Lucchetta, G.; Marinello, F. Characterization and analysis of weld lines on micro-injection moulded parts using atomic force microscopy (AFM). *Wear* **2009**, *266*, 534–538. [CrossRef]
52. Guilong, W.; Guoqun, Z.; Huiping, L.; Yanjin, G. Analysis of thermal cycling efficiency and optimal design of heating/cooling systems for rapid heat cycle injection molding process. *Mater. Des.* **2010**, *31*, 3426–3441. [CrossRef]
53. Wang, G.; Zhao, G.; Wang, X. Effects of cavity surface temperature on reinforced plastic part surface appearance in rapid heat cycle moulding. *Mater. Des.* **2013**, *44*, 509–520. [CrossRef]
54. Wang, G.; Zhao, G.; Guan, Y. Thermal response of an electric heating rapid heat cycle molding mold and its effect on surface appearance and tensile strength of the molded part. *J. Appl. Polym. Sci.* **2012**, *128*, 1339–1352. [CrossRef]
55. Li, X.-P.; Zhao, G.-Q.; Guan, Y.-J.; Ma, M.-X. Optimal design of heating channels for rapid heating cycle injection mold based on response surface and genetic algorithm. *Mater. Des.* **2009**, *30*, 4317–4323. [CrossRef]
56. Li, X.-P.; Zhao, G.-Q.; Guan, Y.-J. Characteristic of rapid heating cycle moulding and warpage analysis of products. *Plast. Rubber Compos.* **2011**, *40*, 425–432. [CrossRef]
57. Wang, W.; Zhao, G.; Wu, X.; Li, X.; Wang, C. Investigation on phosphorus halogen-free flame-retardancy systems in short glass fiber-reinforced PC/ABS composites under rapid thermal cycle molding process condition. *Polym. Compos.* **2015**, *36*, 1653–1663. [CrossRef]
58. Zhang, A.; Hui, Y.; Hou, J. Effect of Resin Viscosity and Lubricants on Surface and Mechanical Properties of Glass Fiber Reinforced Polymer in Rapid Heat Cycle Molding. *Polym. Korea* **2018**, *42*, 974–981. [CrossRef]

59. Seaman, C.; Desrochers, A.; List, G. Multiobjective optimization of a plastic injection molding process. *IEEE Trans. Control. Syst. Technol.* **1994**, *2*, 157–168. [CrossRef]
60. Chen, W.; Wang, M.; Fu, G.; Chen, C. Optimization of plastic injection molding process via Taguchis parameter design method, BPNN, and DFP. In Proceedings of the 2008 International Conference on Machine Learning and Cybernetics, Kunming, China, 15 July 2008; Volume 6, pp. 3315–3321. [CrossRef]
61. Xu, G.; Yang, Z.-T.; Long, G.-D. Multi-objective optimization of MIMO plastic injection molding process conditions based on particle swarm optimization. *Int. J. Adv. Manuf. Technol.* **2012**, *58*, 521–531. [CrossRef]
62. Yang, J.K.; Xu, Y.J. Warpage Analysis of Injection Molding Based on Mold Flow. *Adv. Mater. Res.* **2012**, *538–541*, 1192–1196. [CrossRef]
63. Park, H.S.; Dang, X.P. Technology for Improving Productivity and Quality of Injection Molding. In *DAAAM International Scientific Book*; EBSCO Publishing, Inc.: Ipswich, MA, USA, 2018.
64. Sachs, E.; Wylonis, E.; Allen, S.; Cima, M.; Guo, H. Production of injection molding tooling with conformal cooling channels using the three dimensional printing process. *Polym. Eng. Sci.* **2000**, *40*, 1232–1247. [CrossRef]
65. Ciofu, C.; Mindru, D.T. Injection and micro injection of polymeric plastics materials: A review. *Int. J. Mod. Manufact. Technol.* **2013**, *1*, 49–68.
66. Agrawal, A.R.; Pandelidis, I.O.; Pecht, M. Injection-molding process control? A review. *Polym. Eng. Sci.* **1987**, *27*, 1345–1357. [CrossRef]
67. Jansen, K.M.B.; van Dijk, D.J.; Husselman, M.H. Effect of processing conditions on shrinkage in injection molding. *Polym. Eng. Sci.* **1998**, *38*, 838–846. [CrossRef]
68. Jansen, K.M.B.; Titomanlio, G. Effect of pressure history on shrinkage and residual stresses? Injection molding with constrained shrinkage. *Polym. Eng. Sci.* **1996**, *36*, 2029–2040. [CrossRef]
69. Dizon, J.R.C.; Valino, A.D.; Souza, L.R.; Espera, A.; Chen, Q.; Advincula, R.C. Three-dimensional-printed molds and materials for injection molding and rapid tooling applications. *MRS Commun.* **2019**, *9*, 1267–1283. [CrossRef]
70. Zhou, H.; Li, D. Numerical Simulation and Experimental Study of Warpage of Injection-Molded Parts. *Polym. Technol. Eng.* **2005**, *44*, 603–617. [CrossRef]
71. Sánchez, R.; Aisa, J.; Martinez, A.; Mercado, D. On the relationship between cooling setup and warpage in injection molding. *Measurement* **2012**, *45*, 1051–1056. [CrossRef]
72. Rahimi, M.; Esfahanian, M.; Moradi, M. Effect of reprocessing on shrinkage and mechanical properties of ABS and investigating the proper blend of virgin and recycled ABS in injection molding. *J. Mater. Process. Technol.* **2014**, *214*, 2359–2365. [CrossRef]
73. Park, K. A Study on Flow Simulation and Deformation Analysis for Injection-Molded Plastic Parts Using Three-Dimensional Solid Elements. *Polym. Technol. Eng.* **2005**, *43*, 1569–1585. [CrossRef]
74. Kurt, M.; Kaynak, Y.; Kamber, O.S.; Mutlu, B.; Bakir, B.; Koklu, U. Influence of molding conditions on the shrinkage and roundness of injection molded parts. *Int. J. Adv. Manuf. Technol.* **2010**, *46*, 571–578. [CrossRef]
75. Khan, R.M.; Acharya, G. Plastic Injection Molding Process and Its Aspects for Quality: A Review. *Eur. J. Adv. Eng. Technol.* **2016**, *3*, 66–70. Available online: www.ejaet.com (accessed on 30 October 2020).
76. Kitayama, S.; Ishizuki, R.; Takano, M.; Kubo, Y.; Aiba, S. Optimization of mold temperature profile and process parameters for weld line reduction and short cycle time in rapid heat cycle molding. *Int. J. Adv. Manuf. Technol.* **2019**, *103*, 1735–1744. [CrossRef]
77. Kagitci, Y.C.; Tarakcioglu, N. The effect of weld line on tensile strength in a polymer composite part. *Int. J. Adv. Manuf. Technol.* **2016**, *85*, 1125–1135. [CrossRef]
78. Kuo, C.-C.; Li, D.-Y.; Lin, Z.-C.; Kang, Z.-F. Characterizations of Polymer Gears Fabricated by Differential Pressure Vacuum Casting and Fused Deposition Modeling. *Polymers* **2021**, *13*, 4126. [CrossRef]
79. Dobránsky, J.; Běhálek, L.; Baron, P. Gate Location and its Impact to Flowing Characteristics of Plastic Moldings. *Key Eng. Mater.* **2016**, *669*, 36–43. [CrossRef]
80. Wang, G.; Zhao, G.; Li, H.; Guan, Y. Research on optimization design of the heating/cooling channels for rapid heat cycle molding based on response surface methodology and constrained particle swarm optimization. *Expert Syst. Appl.* **2011**, *38*, 6705–6719. [CrossRef]
81. Zhou, H.; Zhang, Y.; Wen, J.; Cui, S. Mould cooling simulation for injection moulding using a fast boundary element method approach. *Proc. Inst. Mech. Eng. Part B J. Eng. Manuf.* **2009**, *224*, 653–662. [CrossRef]
82. Calaon, M.; Baruffi, F.; Fantoni, G.; Cirri, I.; Santochi, M.; Hansen, H.N.; Tosello, G. Functional Analysis Validation of Micro and Conventional Injection Molding Machines Performances Based on Process Precision and Accuracy for Micro Manufacturing. *Micromachines* **2020**, *11*, 1115. [CrossRef]
83. Correia, L.; Brito, A.M.; Faria, L.; Félix, M.J.; Santos, G.; Laranjeira, J.; Simoes, R. Dynamic temperature control influence on pressure during injection molding of plastic parts to improve part quality. *Int. J. Qual. Res.* **2020**, *14*, 635–646. [CrossRef]
84. Zhou, J.A. Design of Injection Mold for a Large LCD TV Panel. *Appl. Mech. Mater.* **2013**, *423–426*, 1982–1989. [CrossRef]
85. Hussain, A.R.J.; Alahyari, A.A.; Eastman, S.A.; Thibaud-Erkey, C.; Johnston, S.; Sobkowicz, M. Review of polymers for heat exchanger applications: Factors concerning thermal conductivity. *Appl. Therm. Eng.* **2017**, *113*, 1118–1127. [CrossRef]
86. Ogorodnyk, O.; Martinsen, K.; Ogorodnyk, O.; Martinsen, K. Monitoring and Control for Thermoplastics Injection Molding a Review. *Procedia CIRP* **2018**, *67*, 380–385. [CrossRef]

87. Wang, G.; Zhao, G.; Li, H.; Guan, Y. Research of thermal response simulation and mold structure optimization for rapid heat cycle molding processes, respectively, with steam heating and electric heating. *Mater. Des.* **2010**, *31*, 382–395. [CrossRef]
88. Li, J.; Bei, J.; Liu, W.; Xia, X.; Zhou, B.; Peng, X.; Jiang, S. Warpage Prediction of RHCM Crystalline Parts Based on Multi-Layers. *Polymers* **2021**, *13*, 1814. [CrossRef]
89. De Santis, F.; Pantani, R. Development of a rapid surface temperature variation system and application to micro-injection molding. *J. Mater. Process. Technol.* **2016**, *237*, 1–11. [CrossRef]
90. Wang, G.; Hui, Y.; Zhang, L.; Zhao, G. Research on temperature and pressure responses in the rapid mold heating and cooling method based on annular cooling channels and electric heating. *Int. J. Heat Mass Transf.* **2018**, *116*, 1192–1203. [CrossRef]
91. Mrozek, K.; Muszyński, P.; Poszwa, P. Application of Magnetic Concentrator for Improvement in Rapid Temperature Cycling Technology. *Polymers* **2020**, *13*, 91. [CrossRef]
92. Barahate, M.R.; Shete, M.T. Modeling and Simulation of Rapid Heating and Cooling of Injection Mold by FEM-A Review. *Int. Res. J. Eng. Technol.* **2017**, 2718–2723.
93. Thompson, M.K.; Moroni, G.; Vaneker, T.; Fadel, G.; Campbell, R.I.; Gibson, I.; Bernard, A.; Schulz, J.; Graf, P.; Ahuja, B.; et al. Design for Additive Manufacturing: Trends, opportunities, considerations, and constraints. *CIRP Ann.* **2016**, *65*, 737–760. [CrossRef]
94. Ma, S.; Gibson, I.; Balaji, G.; Hu, Q. Development of epoxy matrix composites for rapid tooling applications. *J. Mater. Process. Technol.* **2007**, *192*, 75–82. [CrossRef]
95. Atzeni, E.; Iuliano, L.; Minetola, P.; Salmi, A. Redesign and cost estimation of rapid manufactured plastic parts. *Rapid Prototyp. J.* **2010**, *16*, 308–317. [CrossRef]
96. Rosochowski, A.; Matuszak, A. Rapid tooling: The state of the art. *J. Mater. Process. Technol.* **2000**, *106*, 191–198. [CrossRef]
97. Ingole, D.S.; Kuthe, A.M.; Thakare, S.B.; Talankar, A.S. Rapid prototyping–a technology transfer approach for development of rapid tooling. *Rapid Prototyp. J.* **2009**, *15*, 280–290. [CrossRef]
98. Chua, C.K.; Hong, K.H.; Ho, S.L. Rapid tooling technology. Part 1. A comparative study. *Int. J. Adv. Manuf. Technol.* **1999**, *15*, 604–608. [CrossRef]
99. Karapatis, N.P.; van Griethuysen, J.-P.S.; Glardon, R. Direct Rapid Tooling. *Rapid Prototyp. J.* **1998**, *4*, 77–89. [CrossRef]
100. Khushairi, M.T.M.; Sharif, S.; Ani, J.S.M. Evaluation of Mechanical Properties of Filled Epoxy Composite for Improving Mould Performance—A Review. *Appl. Mech. Mater.* **2015**, *735*, 13–18. [CrossRef]
101. Pouzada, A.S. Hybrid moulds: A case of integration of alternative materials and rapid prototyping for tooling. *Virtual Phys. Prototyp.* **2009**, *4*, 195–202. [CrossRef]
102. Jain, P.; Kuthe, A. Feasibility Study of Manufacturing Using Rapid Prototyping: FDM Approach. *Procedia Eng.* **2013**, *63*, 4–11. [CrossRef]
103. Pontes, A.J.; Queirós, M.P.; Martinho, P.G.; Bártolo, P.J.; Pouzada, A.S. Experimental assessment of hybrid mould performance. *Int. J. Adv. Manuf. Technol.* **2010**, *50*, 441–448. [CrossRef]
104. Jurković, M.; Mahmić, M.; Jurković, Z. Evolution and Application of Rapid Prototyping Technologies. *J. Technol. Plast.* **2005**, *30*, 1–2.
105. Kerkstra, R. "TOOLING: How to Select the Right Tool Steel for Mold Cavities", Plastic Technology. 2016. Available online: https://www.ptonline.com/articles/tooling-how-to-select-the-right-tool-steel-for-mold-cavities (accessed on 30 October 2020).
106. Atesmesh, "Mold Design and Tooling for Injection Molding", Design World. 2013. Available online: https://www.designworldonline.com/mold-design-tooling-for-injection-molding (accessed on 30 October 2020).
107. Jaycon Systems. "Injection Mold Tooling Materials: Aluminum vs. Steel | by Jaycon Systems | Jaycon Systems | Medium", Jaycon Systems. 2017. Available online: https://medium.com/jaycon-systems/injection-mold-tooling-materials-aluminum-vs-steel-7b5f64ee1112 (accessed on 30 October 2020).
108. Ozcelik, B.; Ozbay, A.; Demirbas, E. Influence of injection parameters and mold materials on mechanical properties of ABS in plastic injection molding. *Int. Commun. Heat Mass Transf.* **2010**, *37*, 1359–1365. [CrossRef]
109. Lau, J. The Types of Mold Base for Plastic Injection Mold. Available online: https://www.injectionmould.org/2019/03/23/the-types-of-mold-base/ (accessed on 1 May 2022).
110. Asnafi, N. Application of Laser-Based Powder Bed Fusion for Direct Metal Tooling. *Metals* **2021**, *11*, 458. [CrossRef]
111. ASSAB, ASSAB Tool Steel for Plastic Moulding. 2019. Available online: https://www.assab.com/app/uploads/sites/133/2019/05/ASSAB_Plastic-Mould_EN.pdf (accessed on 28 October 2021).
112. Britton, P.W. *Considerations for Mold Base Material Selection*; International Mold Steel, Inc.: Hebron, KY, USA, 2013. Available online: https://www.moldmakingtechnology.com/articles/considerations-for-mold-base-material-selection (accessed on 28 October 2021).
113. Raus, A.A.; Wahab, M.S.; Ibrahim, M.; Kamarudin, K.; Ahmed, A.; Sáude, N. A Comparative Study of Mould Base Tool Materials in Plastic Injection Moulding to Improve Cycle Time and Warpage Using Statistical Method. *J. Mech. Eng.* **2017**, *4*, 1–17.
114. Tang, S.; Tan, Y.; Sapuan, M.S.; Sulaiman, S.; Ismail, N.; Samin, R. The use of Taguchi method in the design of plastic injection mould for reducing warpage. *J. Mater. Process. Technol.* **2007**, *182*, 418–426. [CrossRef]
115. Menges, G.; Michaeli, W.; Mohren, P.; Menges, G.; Michaeli, W.; Mohren, P. *How to Make Injection Molds*, 3rd ed.; Hanser Gardner Publications, Inc.: Cincinnati, OH, USA, 2001.
116. Zhong, Z.; Khoo, L.; Han, S. Prediction of surface roughness of turned surfaces using neural networks. *Int. J. Adv. Manuf. Technol.* **2006**, *28*, 688–693. [CrossRef]

117. Zhong, Z.W. Recent Advances in Polishing of Advanced Materials. *Mater. Manuf. Process.* **2008**, *23*, 449–456. [CrossRef]
118. Wan, Y.H.; Wang, G. Polishing Robot: Status and Outlook. *Adv. Mater. Res.* **2009**, *69–70*, 311–315. [CrossRef]
119. Kim, H.-S.; Kim, S.-I.; Lee, K.-I.; Lee, D.-H.; Bang, Y.-B.; Lee, K.-I. Development of a programmable vibration cutting tool for diamond turning of hardened mold steels. *Int. J. Adv. Manuf. Technol.* **2009**, *40*, 26–40. [CrossRef]
120. Stavax, U. Stavax Esr. *Alloy Dig.* **1989**, *38*. [CrossRef]
121. Chiang, K.-T.; Chang, F.-P. Analysis of shrinkage and warpage in an injection-molded part with a thin shell feature using the response surface methodology. *Int. J. Adv. Manuf. Technol.* **2007**, *35*, 468–479. [CrossRef]
122. Nasir, S.; Ismail, K.; Shayfull, Z. Application of RSM to Optimize Moulding Conditions for Minimizing Shrinkage in Thermoplastic Processing. *Key Eng. Mater.* **2016**, *700*, 12–21. [CrossRef]
123. Xiao, C.-L.; Huang, H.-X. Development of a rapid thermal cycling molding with electric heating and water impingement cooling for injection molding applications. *Appl. Therm. Eng.* **2014**, *73*, 712–722. [CrossRef]
124. Sateesh, N.; Reddy, S.D.; Kumar, G.P.; Subbiah, R. Optimization of Injection Moulding Process in Manufacturing the Top Cap of Water Meter. *Mater. Today Proc.* **2019**, *18*, 4556–4565. [CrossRef]
125. Chung, C.-Y. Integrated Optimum Layout of Conformal Cooling Channels and Optimal Injection Molding Process Parameters for Optical Lenses. *Appl. Sci.* **2019**, *9*, 4341. [CrossRef]
126. Okubo, K.; Tanaka, S.; Ito, H. The effects of metal particle size and distributions on dimensional accuracy for micro parts in micro metal injection molding. *Microsyst. Technol.* **2010**, *16*, 2037–2041. [CrossRef]
127. Zhao, G.; Wang, G.; Guan, Y.; Li, H. Research and application of a new rapid heat cycle molding with electric heating and coolant cooling to improve the surface quality of large LCD TV panels. *Polym. Adv. Technol.* **2011**, *22*, 476–487. [CrossRef]
128. Sánchez, R.; Martinez, A.; Mercado, D.; Carbonel, A.; Aisa, J. Rapid heating injection moulding: An experimental surface temperature study. *Polym. Test.* **2021**, *93*, 106928. [CrossRef]
129. Hopkinson, N.; Dickens, P. A comparison between stereolithography and aluminium injection moulding tooling. *Rapid Prototyp. J.* **2000**, *6*, 253–258. [CrossRef]
130. Harris, R.; Newlyn, H.; Hague, R.; Dickens, P. Part shrinkage anomilies from stereolithography injection mould tooling. *Int. J. Mach. Tools Manuf.* **2003**, *43*, 879–887. [CrossRef]
131. Ferreira, J. Manufacturing core-boxes for foundry with rapid tooling technology. *J. Mater. Process. Technol.* **2004**, *155–156*, 1118–1123. [CrossRef]
132. Gibbons, G.J.; Hansell, R.G. Direct tool steel injection mould inserts through the Arcam EBM free-form fabrication process. *Assem. Autom.* **2005**, *25*, 300–305. [CrossRef]
133. Ong, H.; Chua, C.; Cheah, C. Rapid Moulding Using Epoxy Tooling Resin. *Int. J. Adv. Manuf. Technol.* **2002**, *20*, 368–374. [CrossRef]
134. Kovács, J.G.; Bercsey, T. Influence of Mold Properties on the Quality of Injection Molded Parts. *Period. Polytech. Mech. Eng.* **2005**, *49*, 115–122. Available online: https://pp.bme.hu/me/article/view/1346/750 (accessed on 30 October 2020).
135. Rossi, S.; Deflorian, F.; Venturini, F. Improvement of surface finishing and corrosion resistance of prototypes produced by direct metal laser sintering. *J. Mater. Process. Technol.* **2004**, *148*, 301–309. [CrossRef]
136. Tomori, T.; Melkote, S.; Kotnis, M. Injection mold performance of machined ceramic filled epoxy tooling boards. *J. Mater. Process. Technol.* **2004**, *145*, 126–133. [CrossRef]
137. Senthilkumar, N.; Kalaichelvan, K.; Elangovan, K. Mechanical behaviour of aluminum particulate epoxy composite-experimental study and numerical simulation. *Int. J. Mech. Mater. Eng.* **2012**, *7*, 214–221.
138. Khushairi, M.T.M.; Sharif, S.; Jamaludin, K.R.; Mohruni, A.S. Effects of Metal Fillers on Properties of Epoxy for Rapid Tooling Inserts. *Int. J. Adv. Sci. Eng. Inf. Technol.* **2017**, *7*, 1155. [CrossRef]
139. Kuo, C.-C.; Lin, J.-X. Fabrication of the Fresnel lens with liquid silicone rubber using rapid injection mold. *Int. J. Adv. Manuf. Technol.* **2019**, *101*, 615–625. [CrossRef]
140. Chen, S.-C.; Lin, Y.-W.; Chien, R.-D.; Li, H.-M. Variable mold temperature to improve surface quality of microcellular injection molded parts using induction heating technology. *Adv. Polym. Technol.* **2008**, *27*, 224–232. [CrossRef]
141. Wadhwa, R.R.; Kim, B.H. Experimental Results of Low Thermal Inertia Molding. I. Length of Filling. *Polym. Technol. Eng.* **1988**, *27*, 509–518. [CrossRef]
142. Yao, D.; Kim, B. Development of rapid heating and cooling systems for injection molding applications. *Polym. Eng. Sci.* **2002**, *42*, 2471–2481. [CrossRef]
143. Saito, T.; Satoh, I.; Kurosaki, Y. A new concept of active temperature control for an injection molding process using infrared radiation heating. *Polym. Eng. Sci.* **2002**, *42*, 2418–2429. [CrossRef]
144. Yao, D.; Chen, S.-C.; Kim, B.H. Rapid thermal cycling of injection molds: An overview on technical approaches and applications. *Adv. Polym. Technol.* **2008**, *27*, 233–255. [CrossRef]
145. FORESHOT, "RHCM | Plastic Injection Molding Manufacturer". Available online: https://www.foreshot.com.tw/en/product/rhcm_rapid-heat-cycle-molding.html (accessed on 29 April 2022).
146. Letoplast, "Injection Moulding". Available online: https://www.letoplast.cz/en/technologies/injection-moulding/ (accessed on 29 April 2022).
147. Chang, P.-C.; Hwang, S.-J. Simulation of infrared rapid surface heating for injection molding. *Int. J. Heat Mass Transf.* **2006**, *49*, 3846–3854. [CrossRef]

148. Fischer, C.; Jungmeier, A.; Peters, G.; Drummer, D. Influence of a locally variable mold temperature on injection molded thin-wall components. *J. Polym. Eng.* **2018**, *38*, 475–481. [CrossRef]
149. Miller, A.; Brown, C.; Warner, G. Guidance on the Use of Existing ASTM Polymer Testing Standards for ABS Parts Fabricated Using FFF. *Smart Sustain. Manuf. Syst.* **2019**, *3*, 122–138. [CrossRef]
150. Selden, R. Effect of processing on weld line strength in five thermoplastics. *Polym. Eng. Sci.* **1997**, *37*, 205–218. [CrossRef]
151. Oh, G.-H.; Jeong, J.-H.; Park, S.-H.; Kim, H.-S. Terahertz time-domain spectroscopy of weld line defects formed during an injection moulding process. *Compos. Sci. Technol.* **2018**, *157*, 67–77. [CrossRef]

MDPI
St. Alban-Anlage 66
4052 Basel
Switzerland
Tel. +41 61 683 77 34
Fax +41 61 302 89 18
www.mdpi.com

Materials Editorial Office
E-mail: materials@mdpi.com
www.mdpi.com/journal/materials

www.ingramcontent.com/pod-product-compliance
Lightning Source LLC
Chambersburg PA
CBHW061604220926
51850CB00005B/424